GALAXIES IN 3D ACROSS THE UNIVERSE

IAU SYMPOSIUM No. 309

COVER ILLUSTRATION:

The cover illustration was designed to bring together an aesthetic view of the main science and technology topics of the conference "Galaxies in 3D across the Universe" held in the city of Vienna. In this day and age we are able to investigate our 3D Universe and its constituents from different perspectives, through observations and simulations, to portray its content and evolution through different eyes, and to follow leads from selected information given in narrow wavelength ranges or excitation states.

We are visualizing the analysis of light - once split up in ESO's 3D spectrograph VIMOS, here depicted as a flat field observation of the HR blue grism in black and white in the background and overlaid by HI VLA contours of NGC 6046 from the THINGS survey. Vienna's skyline represents the space where communication, discussion and continuing education take place, which are the foundation of any IAU symposium.

There is no *one* way to answer today's science questions, as there is no *one* way to communicate them. An artistic approach to science leads to finding the largest common denominator of art and science: creativity. In the words of Werner Heisenberg, the mathematically simplest answer is usually the most beautiful one (Carl Friedrich von Weizsäcker: The Unity of Nature, 1971). The beauty of creating something new - in fundamental research as well as in art - is an inherent drive to all of us. While for artists "intellectual property is the oil of the 21st century" (Michalis Pichler: Statements on Appropriation, 2009), it is also the intense incentive that leads us scientists to continue research - and to continue sharing steps and results along the way.

"I want people to treasure light the way we treasure gold" said visual artist James Turrell. For us astronomers, light is the passage to knowledge while getting together to communicate our findings is a way of building the future.

Cover art and text: Ulrike Kuchner

IAU SYMPOSIUM PROCEEDINGS SERIES

Chief Editor

THIERRY MONTMERLE, IAU General Secretary
Institut d'Astrophysique de Paris,
98bis, Bd Arago, 75014 Paris, France
montmerle@iap.fr

Editor

PIERO BENVENUTI, IAU Assistant General Secretary
University of Padua, Dept of Physics and Astronomy,
Vicolo dell'Osservatorio, 3, 35122 Padova, Italy
piero.benvenuti@unipd.it

INTERNATIONAL ASTRONOMICAL UNION

UNION ASTRONOMIQUE INTERNATIONALE

GALAXIES IN 3D ACROSS THE UNIVERSE

PROCEEDINGS OF THE 309th SYMPOSIUM OF THE INTERNATIONAL ASTRONOMICAL UNION HELD IN VIENNA, AUSTRIA JULY 7–11, 2014

Edited by

BODO L. ZIEGLER
University Vienna, Department of Astrophysics, Austria

FRANÇOISE COMBES
Observatoire de Paris, LERMA, France

HELMUT DANNERBAUER
University Vienna, Department of Astrophysics, Austria

and

MIGUEL VERDUGO
University Vienna, Department of Astrophysics, Austria

Shaftesbury Road, Cambridge CB2 8EA, United Kingdom

One Liberty Plaza, 20th Floor, New York, NY 10006, USA

477 Williamstown Road, Port Melbourne, VIC 3207, Australia

314–321, 3rd Floor, Plot 3, Splendor Forum, Jasola District Centre, New Delhi – 110025, India

103 Penang Road, #05–06/07, Visioncrest Commercial, Singapore 238467

Cambridge University Press is part of Cambridge University Press & Assessment, a department of the University of Cambridge.

We share the University's mission to contribute to society through the pursuit of education, learning and research at the highest international levels of excellence.

www.cambridge.org
Information on this title: www.cambridge.org/9781107078666

First published 2014

A catalogue record for this publication is available from the British Library

ISBN   978-1-107-07866-6   Hardback

Table of Contents

Feedback and Environment

High-redshift Galaxies

Contents

Preface

The IAU symposium 309 took place in Vienna from 7th to 11th July 2014. According to the title "Galaxies in 3D across the Universe" the common theme of the 109 oral presentations and 103 posters was the aspect of spatially and energetically resolved galaxy properties. This topic was very timely for a world-wide international audience of 218 participants from 28 countries because astrophysics has just entered a new epoch with unbiased investigations of intrinsic properties of galaxies and their evolution. High performance technology in 3D-spectroscopy in the optical/NIR regime and in radio interferometry allows for the first time the efficient mapping of stars, gas, and dust, in galaxies near and far. Detailed measurements of individual objects are complemented by surveys aiming at a full census of galaxies across the local Universe. Reaching out to the limits of the Universe, the evolution of spatially resolved properties is traced along the whole cosmic history. Likewise to these observational campaigns, new computer technology and highly advanced algorithms are exploited for detailed simulations to probe the underlying physical and cosmological connection.

That "3D" is not only a hot topic for cinema enthusiasts but also for scientists is already evident from the fact that the year 2014 has seen three major conferences emphasizing this aspect, with the other two held in Garching bei München and Oxford. To distinct the IAUS309 in Vienna SOC had decided to lay the focus on star formation spanning the whole range from its gaseous source and fuel, via its manifestations, to its global dependence and influence on galactic scales as well as its overall cosmic evolution. Following this observers from both the optical and radio communities as well as theoreticians and simulators lively discussed the interplay between the gas and dust content and star formation in galaxies taking into account spatially resolved properties of gas, dust and stellar populations of galaxies, including their kinematics, distribution and evolution. Another topic covered was analysis and visualization of data cubes as well as new development in instrumentation like KMOS and MUSE as second-generation instruments on ESO's VLT and the submm-/radio-array ALMA.

To meet the overwhelming interest of about 300 astronomers a diversity of contributions was set up from review, invited, contributed, and high-light talks to posters displayed next to the 350-persons lecture hall in the Juridicum building of the University of Vienna in the centre of the city. The program and additional information can still be found on the conference webpage galaxy3d.univie.ac.at. To increase the awareness of posters participants could vote for their favourite poster resulting in three additional talks by the winners of this poster competition.

This symposium was also timely for the hosting Department of Astrophysics since among its recent buzzing activities research fields have come into existence based on 3D-spectroscopy as well as radio interferometry. Such new projects encompass local galaxies (mainly within the CALIFA survey), clusters of galaxies at half the age of the Universe, and very distant galaxies first detected by their submm emission. It was also a milestone on the way to the IAU General Assembly 2018 in Vienna.

The whole week of intense scientific presentations and discussions surrounded by social events focusing on special features of Vienna was engraved by a strong spirit of collegiality and enthusiasm, which the editors wish to be revived whenever this proceedings book is taken off the shelf.

We acknowledge the support of International Astronomical Union, University of Vienna (in particular from Event Management Gerald Schneider, Florian Krug, Margarethe Jurentisch, Benedikt Burkhardt, Mina Jaramaz), Juridicum, Vienna Convention Bureau, Department of Astrophysics, Faculty of Earth Sciences, Geography and Astronomy, Raiffeisenlandesbank Niederösterreich–Wien, Restaurant Kahlenberg, and Caterer Reithofer.

Special thanks go to our SOC and LOC members, without whom this symposium could not have happened!

B.L. Ziegler, F. Combes, H. Dannerbauer, M. Verdugo (eds)
Vienna, October 10th, 2014

THE ORGANIZING COMMITTEE

Scientific

Francoise Combes – Observatoire de Paris, France
Reinhard Genzel – Max-Planck Institut für Extraterrestrische Physik, Germany
Rosa Gonzalez-Delgado – Instituto de Astrofísica de Andalucía, Spain
Roberto Cid-Fernandes – Universidade Federal de Santa Catarina, Brazil
Helmut Dannerbauer – University of Vienna, Department of Astrophysics, Austria
Miguel Verdugo – University of Vienna, Department of Astrophysics, Austria
Joss Bland-Hawthorn – Sydney Institute for Astronomy, Australia
Thorsten Naab – Max-Planck Institut für Astrophysik, Germany
Robert Kennicutt –Institute of Astronomy, University of Cambridge, UK
Jacqueline van Gorkom – Department of Astronomy, Columbia University, USA
Lisa Kewley – The Australian National University, Australia
Bodo Ziegler (chair) –University of Vienna, Department of Astrophysics, Austria

Local

Bodo Ziegler (chair)
Helmut Dannerbauer (co-chair)
Miguel Verdugo (co-chair)
Ulrike Kuchner
Oliver Czoske
Christian Maier
Joao Alves

CONFERENCE PHOTOGRAPH

Conference photograph

LOCAL ORGANIZING COMMITTEE

Oliver Czoske, Miguel Verdugo, Helmut Dannerbauer, Christian Maier, Ulrike Kuchner and Bodo Ziegler.

CONFERENCE, POSTER SESSIONS AND BREAKS

lecture hall.

CONFERENCE, POSTER SESSIONS AND BREAKS

Left: Adam Schaefer, Nicholas Scott and Scott Croom. *Right:* Ana Monreal Ibero and Angel Lopez-Sanchez.

Miguel Verdugo and Veronica Menacho explain Dave Murphy and Jeff Kenney their poster.

Conference photograph

CONFERENCE, POSTER SESSIONS AND BREAKS

Left: Polychronis Papaderos and Thierry Contini. *Right:* Tom Richtler and Bärbel Koribalski.

Left: Poster session. *Right:* Christian Maier and Nadja Lampichler counting the votes for the poster competition.

WELCOME RECEPTION

Thaisa Storchi-Bergmann and Rosa Gonzalez Delgado toasting.

Left: Rhea-Silvia Remus, Marja Seidel and Barry Madore. *Right:* Bodo Ziegler and Raffaella Anna Marino.

RECEPTION

Left: Tucker Jones, Robert Singh and Carlos Eduardo Barbosa *Right:* Laurie Rousseau-Nepton, Alfredo Zenteno, Rebecca Davies and Caroline Foster.

Tobias Goerdt, Michael Hilker, Sylvia Ploeckinger and Helmut Dannerbauer.

RECEPTION

Left: Roberto Cid-Fernandes, Carlos Lopez Sanjuan and Michael Opitsch *Right:* Alfredo Zenteno, Rene Fassbender and Miguel Verdugo.

Marja Seidel, Ulrike Kuchner, Katherine Alatalo and Rhea-Silvia Remus.

EXCURSION TO THE UNIVERSITY OBSERVATORY

Cristina Catalán-Torrecilla , Raffaella Anna Marino and Alexander Fritz.

CONFERENCE DINNER

Etsuko Mieda, Minju Lee, Tomoko Suzuki, and Toshiki Saito.

CONFERENCE DINNER

Left: Wendy Williams, Moses Mogotsi, David Carton, Lucas Ellmeier, Miguel Verdugo and Carmen Toribio *Right:* Ulrike Kuchner and Bodo Ziegler.

Mischa Schirmer, Noelia Jimenez, Kevin Xu and Philip Hopkins.

CONFERENCE DINNER

Left: Francoise Combes thanks Bodo Ziegler for organizing this conference. *Right:* Conference dinner speech by Bodo Ziegler.

View of Vienna from Kahlenberg mountain.

Participants

Alfonso **Aguerri**, Instituto de Astrofísica de Canarias, Spain — jalfonso@iac.es
Katherine **Alatalo**, Caltech, USA — kalatalo@ipac.caltech.edu
Flor **Allaert**, Ghent University, Belgium — flor.allaert@ugent.be
James **Allen**, University of Sydney, Australia — j.allen@physics.usyd.edu.au
Joao **Alves**, University of Vienna, Austria — joao.alves@univie.ac.at
Philippe **Amram**, Laboratoire d'Astrophysique de Marseille, France — philippe.amram@lam.fr
Alfonso **Aragon-Salamanca**, University of Nottingham, United Kingdom — alfonso.aragon@nottingham.ac.uk
Santiago **Arribas**, CSIC, Spain — arribas@cab.inta-csic.es
Junichi **Baba**, Tokyo Institute of Technology, Japan — babajn@elsi.jp
Paramita **Barai**, INAF - Osservatorio Astronomico di Trieste, Italy — pbarai@oats.inaf.it
Carlos Eduardo **Barbosa**, Universidade de Sao Paulo, Brazil — carlos.barbosa@usp.br
Matteo **Barnab**, Dark Cosmology Center, Denmark — mbarnabe@dark-cosmology.dk
Andrew **Battisti**, University of Massachusetts at Amherst, USA — abattist@astro.umass.edu
Matthew **Bayliss**, Harvard University, USA — mbayliss@cfa.harvard.edu
Joss **Bland-Hawthorn**, University of Sydney, Australia — jbh@physics.usyd.edu.au
Frederic **Bournaud**, CEA Saclay, France — frederic.bournaud@cea.fr
Denis **Burgarella**, Laboratoire d'Astrophysique de Marseille, France — denis.burgarella@lam.fr
Leonard **Burtscher**, Max Planck Institute for Extraterrestrial Physics, Germany — burtscher@mpe.mpg.de
Gerold **Busch**, I. Physikalisches Institut der Universität zu Köln, Germany — busch@ph1.uni-koeln.de
Eleanor **Byler**, University of Washington, USA — ebyler@astro.washington.edu
Paula **Calderon**, Universidad de Concepcion, Chile — pcalderon@astro-udec.cl
Anahi **Caldu Primo**, Max Planck Insitute for Astronomy, Germany — caldu@mpia.de
Daniela **Calzetti**, University of Massachusetts, USA — calzetti@astro.umass.edu
Artemi **Camps-Fariña**, Instituto Astrofisico de Canarias, Spain — artemic@iac.es
David **Carton**, Leiden Observatory, Netherlands — carton@strw.leidenuniv.nl
Cristina **Catalán-Torrecilla**, Universidad Complutense de Madrid - UCM, Spain — ccatalan@ucm.es
Renyue **Cen**, Princeton University, USA — cen@astro.princeton.edu
Jieun **Choi**, UC Santa Cruz, USA — jchoi37@ucsc.edu
Ekaterina **Chudakova**, Lomonosov Moscow State University, Russian Fed. — artenik@gmail.com
Roberto **Cid Fernandes**, Universidade Federal de Santa Catarina, Brazil — robertocidfernandes@gmail.com
Lodovico **Coccato**, European Southern Observatory, Germany — lcoccato@gmail.com
Francoise **Combes**, Observatoire de Paris, France — francoise.combes@obspm.fr
Thierry **Contini**, Institut de Recherche en Astrophysique et Planétologie, France — thierry.contini@irap.omp.eu
Stephane **Courteau**, Queen's University, Canada — courteau@astro.queensu.ca
Giovanni **Cresci**, INAF - Osservatorio di Arcetri, Italy — gcresci@arcetri.astro.it
Scott **Croom**, University of Sydney, Australia — scroom@physics.usyd.edu.au
Anna **Curir**, Astrophysical Observatory of Turin, Italy — curir@oato.inaf.it
Michael **Curtis**, Institute of Astronomy, United Kingdom — mc636@ast.cam.ac.uk
Oliver **Czoske**, University of Vienna, Austria — oliver.czoske@univie.ac.at
Mauro **D'Onofrio**, University of Padova, Italy — mauro.donofrio@unipd.it
Helmut **Dannerbauer**, University of Vienna, Austria — helmut.dannerbauer@univie.ac.at
Rebecca **Davies**, Australian National University, Australia — Rebecca.Davies@anu.edu.au
Gert **De Geyter**, Universiteit Gent, Belgium — gert.degeyter@ugent.be
Ilse **De Looze**, Ghent University, Belgium — ilse.delooze@ugent.be
Roberto **Decarli**, Max Planck Insitute for Astronomy, Germany — decarli@mpia.de
Ricardo **Demarco**, Universidad de Concepcion, Chile — rdemarco@astro-udec.cl
Boris **Deshev**, Tartu Observatory, Estonia — boris@to.ee
Miroslava **Dessauges-Zavadsky**, Geneva Observatory, Switzerland — miroslava.dessauges@unige.ch
Catrina **Diener**, ESO Vitacura, Chile — cdiener@eso.org
Michael **Dopita**, RSAA, The Australian National University, Australia — Michael.Dopita@anu.edu.au
Loretta **Dunne**, University of Canterbury, New Zealand — loretta.dunne@canterbury.ac.nz
Arthur **Eigenbrot**, University of Wisconsin Madison, USA — eigenbrot@astro.wisc.edu
Paul **Eigenthaler**, Pontificia Universidad Catolica de Chile, Chile — eigenth@astro.puc.cl
Simon **Ellis**, Australian Astronomical Observatory, Australia — sellis@aao.gov.au
Lucas **Ellmeier**, University of Vienna, Austria — lucas.ellmeier@univie.ac.at
Peter **Erwin**, Max Planck Institute for Extraterrestrial Physics, Germany — erwin@mpe.mpg.de
Maximilian **Fabricius**, Max Planck Institute for Extraterrestrial Physics, Germany — mxhf@mpe.mpg.de
Rene **Fassbender**, Observatory of Rome (INAF-OAR), Italy — rene.fassbender@oa-roma.inaf.it
Rose **Finn**, Siena College, USA — rfinn@siena.edu
David **Fisher**, Swinburne University, Australia — dfisher@swin.edu.au
Lisa **Fogarty**, University of Sydney, Australia — l.fogarty@physics.usyd.edu.au
Natascha M. **Forster Schreiber**, Max Planck Institute for Extraterrestrial Physics, Germany — forster@mpe.mpg.de
Caroline **Foster**, Australian Astronomical Observatory, Australia — cfoster@aao.gov.au
Alexander **Fritz**, INAF-IASF Milano, Italy — afritz@iasf-milano.inaf.it
Isaura **Fuentes-Carrera**, Instituto Politcnico Nacional, Mexico — isaura.fuentescarrera@gmail.com
Reinhard **Genzel**, Max Planck Institute for Extraterrestrial Physics, Germany — genzel@mpe.mpg.de
Wlodzimierz **Godlowski**, Institut of Physics Opole University, Poland — godlowski@uni.opole.pl
Tobias **Goerdt**, University of Vienna, Austria — tobias.goerdt@uam.es
Rosa **Gonzalez Delgado**, Instituto de Astrofisica de Andalucia, Spain — rosa@iaa.es
Kathryn **Grasha**, University of Massachusetts, USA — kgrasha@astro.umass.edu
Loretta **Gregorini**, University of Bologna, Italy — loretta.gregorini@unibo.it
Brent **Groves**, Max Planck Institute for Astronomy, Germany — brent@mpia.de
Elaine **Grubmann**, University of Vienna, Austria — elaine.grubmann@hotmail.com
Madusha **Gunawardhana**, Durham University, United Kingdom — madusha.gunawardhana@durham.ac.uk
Francois **Hammer**, GEPI - Paris Observatory, France — francois.hammer@obspm.fr
Lei **Hao**, Shanghai Astronomical Observatory, China — haol@shao.ac.cn
George **Heald**, ASTRON, Netherlands — heald@astron.nl
Gerhard **Hensler**, University of Vienna, Austria — gerhard.hensler@univie.ac.at
Michael **Hilker**, European Southern Observatory, Germany — mhilker@eso.org
Michaela **Hirschmann**, Institut d.astrophysique de Paris, France — mhirsch@oats.inaf.it
I-Ting **Ho**, Institute for Astronomy, University of Hawaii, USA — itho@ifa.hawaii.edu
Eric **Hooper**, WIYN Observatory & U. Wisconsin-Madison, USA — ehooper@wiyn.org
Philip **Hopkins**, CalTech, USA — phopkins@caltech.edu

Annie **Hughes**, Max Planck Insitute for Astronomy, Germany — hughes@mpia.de
Thomas **Hughes**, Universiteit Gent, Belgium — thomas.hughes@ugent.be
jai-chan **Hwang**, Kyunpook National University, South Korea — jchan@knu.ac.kr
Pavel **Jachym**, Astronomical Institute, Prague, Czech Rep. — jachym@ig.cas.cz
Noelia **Jimenez**, University of St Andrews, United Kingdom — nj22@st-andrews.ac.uk
Evelyn **Johnston**, European Southern Observatory, Chile — ppxej@nottingham.ac.uk
Tucker **Jones**, UC Santa Barbara, USA — tajones@physics.ucsb.edu
Bruno **Jungwiert**, Astronomical Institute ASCR, Czech Rep. — bruno.jungwiert@ig.cas.cz
Maria **Kapala**, Max Planck Institute for Astronomy, Germany — kapala@mpia.de
Carolina **Kehrig**, Instituto de Astrofisica de Andalucia, Spain — kehrig@iaa.es
Jeff **Kenney**, Yale University, USA — jeff.kenney@yale.edu
Lisa **Kewley**, Australian National University, Australia — lisa.kewley@anu.edu.au
Tadayuki **Kodama**, National Astronomical Observatory of Japan, Japan — t.kodama@nao.ac.jp
Baerbel **Koribalski**, CSIRO Australia Telescope National Facility, Australia — Baerbel.Koribalski@csiro.au
Kathryn **Kreckel**, Max Planck Institute for Astronomy, Germany — kreckel@mpia.de
Ulrike **Kuchner**, University of Vienna, Austria — ulrike.kuchner@univie.ac.at
Charles **Lada**, Smithsonian Astrophysical Observatory, USA — clada@cfa.harvard.edu
Nadja **Lampichler**, University of Vienna, Austria — nadja.lampichler@gmx.at
Ronald **Lsker**, Max Planck Insitute for Astronomy, Germany — laesker@mpia.de
Minju **Lee**, The University of Tokyo/NAOJ, Japan — minju.lee@nao.ac.jp
Myung Gyoon **Lee**, Seoul National University, South Korea — mglee@astro.snu.ac.kr
Sarah **Leslie**, RSAA, Australian National University, Australia — sarah.k.leslie@gmail.com
Alexia **Lewis**, University of Washington, USA — arlewis@astro.washington.edu
Timothy **Licquia**, University of Pittsburgh, USA — tcl15@pitt.edu
Jeremy **Lim**, University of Hong Kong, Hong Kong — jjlim@hku.hk
Charles **Liu**, CUNY College of Staten Island/AMNH, USA — cliu@amnh.org
Angel **Lopez-Sanchez**, Australian Astronomical Observatory, Australia — alopez@aao.gov.au
Carlos **Lopez-Sanjuan**, CEFCA, Spain — clsj@cefca.es
Chung-Pei **Ma**, University of California at Berkeley, USA — cpma@berkeley.edu
Juan **Macias-Perez**, LPSC, France — macias@lpsc.in2p3.fr
Barry F. **Madore**, Carnegie Observatories, USA — barry@ociw.org
Christian **Maier**, University of Vienna, Austria — christian.maier@univie.ac.at
Millicent **Maier**, Australian Astronomical Observatory, Australia — mmaier@aao.gov.au
Katarzyna **Malek**, Nagoya University, Japan — malek.kasia@nagoya-u.jp
Andrzej **Marecki**, Torun Centre for Astronomy, Poland — amr@astro.uni.torUnitedpl
Victor **Marian**, University of Vienna, Austria — marian.victor@gmail.com
Raffaella Anna **Marino**, Universidad Complutense de Madrid, UCM, Spain — ramarino@ucm.es
Claire **Max**, University of California Observatories, USA — max@ucolick.org
Daniel **May**, Universidade de So Paulo, Brazil — dmay@usp.br
Nicholas **McConnell**, IfA, University of Hawaii, USA — nmcc@ifa.hawaii.edu
Richard **McDermid**, Macquarie University, Australia — richard.mcdermid@mq.edu.au
Veronica **Menacho**, University of Vienna, Austria — veronica.menacho@univie.ac.at
Claudia **Mendes de Oliveira**, University of Sao Paulo, Brazil — oliveira@astro.iag.usp.br
Etsuko **Mieda**, University of Toronto, Canada — mieda@astro.utoronto.ca
Yosuke **Minowa**, National Astronomical Observatory of Japan, USA — minoways@naoj.org
Moses **Mogotsi**, University of Cape Town, South Africa — moses.mog@gmail.com
Ana **Monreal Ibero**, Observatoire de Paris, France — amivagzci@gmail.com
Leah **Morabito**, Leiden Observatory, Netherlands — morabito@strw.leidenuniv.nl
Lorenzo **Morelli**, University of Padova, Italy — lorenzo.morelli@unipd.it
Jorge **Moreno**, University of Victoria & CITA, Canada — moreno@uvic.ca
Alessia **Moretti**, University of Padova, Italy — alessia.moretti@oapd.inaf.it
Kana **Morokuma-Matsui**, National Astronomical Observatory of Japan, Japan — kana.matsui@nao.ac.jp
David **Murphy**, Pontificia Universidad Catolica de Chile, Chile — dmurphy@astro.puc.cl
Thorsten **Naab**, Max Planck Insitute for Astrophysics, Germany — naab@mpa-garching.mpg.de
Desika **Narayanan**, Haverford College, USA — dnarayan@haverford.edu
Giovanni **Natale**, University of Central Lancashire, United Kingdom — gnatale@uclan.ac.uk
Hyerim **Noh**, Korea Astronomy and Space Science Institute, South Korea — hr@kasi.re.kr
Natalia **Nowak**, Jagiellonian University, Poland — nala@oa.uj.edu.pl
Michael **Opitsch**, Max Planck Institute for Extraterrestrial Physics, Germany — mopitsch@mpe.mpg.de
Shinobu **Ozaki**, National Astronomical Observatory of Japan, Japan — shinobu.ozaki@nao.ac.jp
Polychronis **Papaderos**, Centro de Astrofisica da Universidade do Porto, Portugal — papaderos@astro.up.pt
Rodrigo **Parra**, European Southern Observatory, Chile — parrar@eso.org
Reynier **Peletier**, Kapteyn Astronomical Institute, Groningen, Netherlands — peletier@astro.rug.nl
Debora **Pelliccia**, Laboratoire d'Astrophysique de Marseille, France — debora.pelliccia@lam.fr
Sylvia **Plöckinger**, University of Vienna, Austria — sylvia.ploeckinger@univie.ac.at
Yu **Pui Ling**, The University of Hong Kong, Hong Kong — aliceypl@hku.hk
Johannes **Puschnig**, Department of Astronomy, Sweden — johannes.puschnig@astro.su.se
Miguel **Querejeta**, Max Planck Institute for Astronomy, Germany — querejeta@mpia.de
Rhea-Silvia **Remus**, University Observatory Munich, Germany — rhea@usm.lmu.de
Tiago **Ricci**, IAG - University of Sao Paulo, Brazil — tvricci@iag.usp.br
Tom **Richtler**, Universidad de Concepcion, Chile — tom@astro-udec.cl
Rogemar A. **Riffel**, Universidade Federal de Santa Maria, Brazil — rogemar.riffel@gmail.com
Bruno **Rodriguez Del Pino**, University of Nottingham, United Kingdom — ppxbr@nottingham.ac.uk
Martin **Roth**, Leibniz-University of Vienna Potsdam, Germany — mmroth@aip.de
Huub **Röttgering**, Leiden University, Netherlands — rottgering@strw.leidenuniv.nl
Bernhard **Röttgers**, Max Planck Insitute for Astrophysics, Germany — broett@mpa-garching.mpg.de
Laurie **Rousseau-Nepton**, Laval University, Canada — laurie.r-nepton.1@ulaval.ca
Sambit **Roychowdhury**, Max Planck Institute for Astrophysics, Germany — sambit@mpa-garching.mpg.de
Alexander **Rudy**, University of California, Santa Cruz, USA — arrudy@ucsc.edu
Agnieszka **Rys**, Instituto de Astrofisica de Canarias, Spain — arys@iac.es
Toshiki **Saito**, University of Tokyo/NAOJ, Japan — toshiki.saito@nao.ac.jp
Kazushi **Sakamoto**, ASIAA, Taiwan — ksakamoto@asiaa.sinica.edu.tw
Sebastian F. **Sanchez**, Institto de Astrofisica de Andalucia, Spain — sebastian.f.sanchez@gmail.com
Marc **Sarzi**, University of Hertfordshire, United Kingdom — m.sarzi@herts.ac.uk
Adam **Schaefer**, University of Sydney, School of Physics, Australia — schaefer@physics.usyd.edu.au
Julia **Scharwaechter**, Observatoire de Paris, LERMA, France — julia.scharwaechter@obspm.fr
Eva **Schinnerer**, Max Planck Institute for Astronomy, Germany — schinner@mpia.de
Mischa **Schirmer**, Gemini Observatory, Chile — mschirme@gemini.edu
Nicholas **Scott**, University of Sydney, Australia — nscott@physics.usyd.edu.au

Marja **Seidel**, Instituto de Astrofisica de Canarias, Spain — mseidel@iac.es
Ray **Sharples**, Durham University, United Kingdom — r.m.sharples@durham.ac.uk
Rhythm **Shimakawa**, Subaru telescope, USA — rhythm@naoj.org
Olga **Silchenko**, Lomonosov Moscow State University, Russian Fed. — olga@sai.msu.su
Robert **Singh**, Max Planck Institute for Astronomy, Germany — singh@mpia.de
Rory **Smith**, Universidad de Concepcion, Chile — rsmith@astro-udec.cl
David **Sobral**, Leiden Observatory/CAAUL Lisbon, Portugal — sobral@strw.leidenuniv.nl
Thaisa **Storchi-Bergmann**, Instituto de Fsica, UFRGS, Brazil — thaisa@ufrgs.br
John **Stott**, Durham University, United Kingdom — j.p.stott@durham.ac.uk
Lorrie **Straka**, University of Chicago, USA — straka.lorrie@gmail.com
Tomoko **Suzuki**, The Graduate University for Advanced Studies, Japan — suzuki.tomoko@nao.ac.jp
Mark **Swinbank**, Durham University, United Kingdom — a.m.swinbank@dur.ac.uk
Linda **Tacconi**, Max Planck Institute for Extraterrestrial Physics, Germany — linda@mpe.mpg.de
Matthew **Taylor**, Pontificia Universidad Catlica de Chile, Chile — mtaylor@astro.puc.cl
Rhys **Taylor**, Astronomical Institute of the Czech Academy of Sciences, Czech Rep. — rhysyt@gmail.com
Christina **Thöne**, IAA - CSIC, Spain — cthoene@iaa.es
Carmen **Toribio**, ASTRON, Netherlands — toribio@astron.nl
Sergio **Torres-Flores**, Universidad de La Serena, Chile — storres@dfuls.cl
Christy **Tremonti**, University of Wisconsin-Madison, USA — tremonti@astro.wisc.edu
Virginia **Trimble**, University of California, USA — vtrimble@astro.umd.edu
Junko **Ueda**, National Astronomical Observatory of Japan, Japan — junko.ueda@nao.ac.jp
Glenn **van de Ven**, Max Planck Institute for Astronomy, Germany — glenn@mpia.de
Remco **van den Bosch**, Max Planck Insitute for Astronomy, Germany — bosch@mpia.de
Miguel **Verdugo**, University of Vienna, Austria — miguel.verdugo@univie.ac.at
Anne **Verhamme**, Geneva Observatory, Switzerland — anne.verhamme@unige.ch
Joaquin **Vieira**, University of Illinois at Urbana Champaign, USA — jvieira@illinois.edu
David **Wake**, The Open University, United Kingdom — david.wake@open.ac.uk
Heidi **White**, University of Toronto, Canada — white@astro.utoronto.ca
Anna **Williams**, University of Wisconsin-Madison, USA — williams@astro.wisc.edu
Rebecca **Williams**, University of Cambridge, United Kingdom — rw480@mrao.cam.ac.uk
Wendy **Williams**, Sterrewacht Leiden, Netherlands — wwilliams@strw.leidenuniv.nl
David **Wilman**, Max Planck Institute for Extraterrestrial Physics, Germany — dwilman@mpe.mpg.de
Marsha **Wolf**, University of Wisconsin, USA — mwolf@astro.wisc.edu
Jong-Hak **Woo**, Seoul National University, South Korea — woo@astro.snu.ac.kr
Jianghua **Wu**, Beijing Normal University, China — jhwu@bnu.edu.cn
Eva **Wuyts**, Max Planck Institute for Extraterrestrial Physics, Germany — evawuyts@mpe.mpg.de
Stijn **Wuyts**, Max Planck Institute for Extraterrestrial Physics, Germany — swuyts@mpe.mpg.de
Kevin **Xu**, California Institute of Technology, Austria — cxu@ipac.caltech.edu
Kijeong **Yim**, Kapteyn Astronomical Institute, Netherlands — tinker330@gmail.com
Lisa **Young**, New Mexico Tech, USA — lyoung@physics.nmt.edu
Tiantian **Yuan**, RSAA Australian National University, Australia — tiantian.yuan@anu.edu.au
Michal **Zajacek**, Astronomical Institute of the Czech Academy of Sciences, Czech Rep. — michal_zajacek@yahoo.com
Javier **Zaragoza Cardiel**, Instituto de Astrofsica de Canarias, Spain — jzc@iac.es
Alfredo **Zenteno**, Cerro Tololo Inter-American Observatory, Chile — azenteno@ctio.noao.edu
Bodo **Ziegler**, University of Vienna, Austria — bodo.ziegler@univie.ac.at
Almudena **Zurita**, Universidad de Granada, Spain — azurita@ugr.es

Galaxies in 3D across the Universe
Proceedings IAU Symposium No. 309, 2014
B. L. Ziegler, F. Combes, H. Dannerbauer, M. Verdugo, eds.

© International Astronomical Union 2015
doi:10.1017/S1743921314009211

Interferometry meets the third and fourth dimensions in galaxies

Virginia Trimble

University of California Irvine,
Department of Physics and Astronomy
Irvine, California 92697, USA
email: vtrimble@uci.edu

Abstract. Radio astronomy began with one array (Jansky's) and one paraboloid of revolution (Reber's) as collecting areas and has now reached the point where a large number of facilities are arrays of paraboloids, each of which would have looked enormous to Reber in 1932. In the process, interferometry has contributed to the counting of radio sources, establishing superluminal velocities in AGN jets, mapping of sources from the bipolar cow shape on up to full grey-scale and colored images, determining spectral energy distributions requiring non-thermal emission processes, and much else. The process has not been free of competition and controversy, at least partly because it is just a little difficult to understand how earth-rotation, aperture-synthesis interferometry works. Some very important results, for instance the mapping of HI in the Milky Way to reveal spiral arms, warping, and flaring, actually came from single moderate-sized paraboloids. The entry of China into the radio astronomy community has given large (40-110 meter) paraboloids a new lease on life.

Keywords. Interferometry, Radio Astronomy, Superluminal Velocities, VLA, ALMA

1. Introduction

Critics have claimed that astronomy is not a science, because you cannot repeat an experiment changing one variable at a time. At least part of the answer is, no, but we can find stars that differ from our sun only in mass, or age, or metal content, or rotation rate, or magnetic field and carry out comparisons that achieve the same goal. This works for stars because most of them are nearly spherical. Manifestly that is not true for galaxies of most types. Thus the possibility of determining three-dimensional structures for galaxies and correcting for orientation addresses the question of whether galactic astronomy can claim to be science.

A seemingly-simple example is to extract the real distribution of elliptical shapes from the statistics of eccentricities on the plane of the sky, assuming that a large sample will be randomly oriented relative to us. That assumption is clearly not true in a magnitude-limited sample, because an end-on prolate spheroid will look brighter than a face-on oblate spheroid covering the same area of the sky. Triaxiaity, existence of thick disk galaxies, bars, and dominance of dark matter (whose shape is not necessarily traced by luminous matter any more than its density is) all make this worse.

Three dimensional studies of galaxies and their parts generally address much more detailed questions than "Is galactic astronomy science?" and correspondingly have much better chances of answering the questions asked. The role of interferometry has generally been to enhance angular and occasionally spectral resolution, not without other costs, and has generally meant radio and millimeter interferometry when applied on galactic or Galactic scales, though one early design for Gaia, called GAIA-ROEMER was intended to have interferometric capability.

The oral version of this presentation included images of a very large number of radio telescopes, some interferometers, some monoliths, and some not quite sure. The cited references not specifically mentioned in the text for a particular idea or result are the ones from which most of those images were taken. Krauss (1986) was a particularly rich source. Kellermann & Moran (2001) have provided a truly expert overview of the evolution of radio interferometry from Ryle & Vonberg's (1946) 0.5 km, two-element Michelson-analog device up to the epoch of publication and indeed somewhat beyond, with discussions of larger arrays, space interferometry, and moving further toward short wavelengths in the future. They do not mention that Thomas Gold (Gold & Mitton 2012, p. 102) believed that he was the first in 1955 to suggest using separate, very accurate clocks at widely separated antennas to permit baselines as large as the earth.

2. What is an interferometer?

The primordial version was the 1801 Young two-slit experiment, which demonstrated the wave nature of light, and, as in his case, if you see fringes, it means your interferometer is working. Fizeau thought of applying interferometric concepts in astronomy in 1868, and Stephan at Marseilles in 1874 concluded that stars have angular diameters less than 0.158 arc sec. He masked the middle of a mirror, as did Michelson at Mt. Wilson to measure Jupiter's moons in 1891. Michelson extended the diameter of the Mt. Wilson 100″ and Pease & Anderson used this kluge to find a diameter of 0.047 arcsec for Betelgeuse in 1920. All of these are, in effect, time reversals of the Young experiment. Optical interferometry is undoubtedly a growth industry, but has, so far, been applied largely to stars rather than galaxies (Lindemann 2011).

Fabry-Perot interferometers are the exception. In these, two flat glass (etc.) surfaces very close together bounce light back and forth and so pick out some one particular wavelength with constructive interference. They can be tuned by changing the spacing. Thus the first successful application of interferometry in astronomy came in 1910 when Fabry & Buisson (1914) examined the Orion Nebula with a small telescope at Marseilles (Lequeux 2013). Their goal was to get a sufficiently precise wavelength for an emission feature near 5007Å to permit its identification with some laboratory substance. In this they failed and came to the end of their paper still calling it Nebulium. Emission lines at 3726.1 and 3728.8Å also remained unidentified. The line widths (a real achievement of the F-P method) suggested atomic weights between those of hydrogen and helium, and they suggested identification with two such elements that had been postulated by Rydberg.

It is not true that F-P applications belong exclusively to Marseilles (my thesis advisor was very fond of them), but Marcelin et al. (1983) reported a very interesting scan through the width of the $H\alpha$ feature in the nearly face-on spiral NGC 2903. Different gas blobs show up at different velocities, and we are presumably seeing some combination of inflow, outflow, and galactic fountain processes.

Most of the rest of galactic interferometry has occurred at radio wavelengths, increasingly with the use of aperture synthesis and gradually pushing to shorter wavelengths.

3. What is a dimension?

Figure 1 is a perfectly possible two-dimensional image, but you cannot instruct one in 3-d. It is called a three-pronged blivet, and in one-d it would be merely a line of finite width. The standard one-dimensional joke is, however, the story of the engineer who was called into consultation at a dairy farm and began by saying "Consider a spherical cow." This works for stars (where 1-d is the radius), and T, P, ρ, and composition can be written as functions of r alone. But even a spherical-looking galaxy almost certainly has $\sigma(v)$ - the equivalent of temperature - that is unlikely to be isotropic, so 1-d will not do.

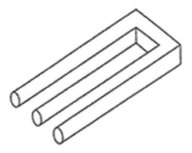

Figure 1. A three-pronged blivet. The radio interferometric equivalent is a map that turns out with negative flux in some places.

Two dimensions could be either r and θ or r and z and provide a plausible next step for study of rotating and/or magnetic stars (where angular momentum is apparently conserved on cylinders during evolution). This still won't do for galaxies, where light clearly does not trace matter, and Schwarzschild's (1982) scheme was therefore doomed to failure. His idea was this: assume a potential; put down a large number of point stars and let them orbit in that potential (in circular, radial, banana, and many other orbit shapes); spread the mass of each star around its orbit, proportionate to the dwell time at each point; now add up the masses and see whether they reproduce the potential. They won't.

The bipolar cow is a bit more promising. Very many astronomical images do look somewhat like what you would expect if a rapidly rotating Maclauren spheroid distorted unstably into unequal lobes and then split. I showed 10 images of this general sort, interferometric ones of both interacting galaxies and star formation regions and a couple of simulations. Fred Adam's drawing of the bipolar cow appears as Fig. 1 of Trimble (2002) and also has the morphology of two blobs with some tenuous connection. Again as in the astronomical case, higher angular resolution reveals more features, whether disks, arms, jets, and gas streams or ears, hooves, and tails.

4. Earliest radio astronomy and the Milky Way

Jansky's (1932) device was a sort of array, and Reber (1940), the world's first dedicated radio astronomer, initially built a paraboloid of revolution, though many of his later devices were arrays and interferometers, often at very low frequency. Both pioneers observed continuous emission and ended up with maps dominated by the Milky Way. If you already knew that there is a substructure, a warp, and outer flaring, you can probably see them in their maps. The next continuum steps, separating out disk and corona and recognizing the North Polar Spur were already dominated by interferometer results from Cambridge and Australia (Pawsey 1964). And some bright spots were arms end-on.

In contrast, the 21 cm maps of neutral hydrogen emission that firmly established our existence as a spiral galaxy came from paraboloids in Dwingeloo and Sydney (Kerr & Westerhout 1964). This was also true for the evidence for disk warping (up to north of sun-center line, down to the south) and flaring at galactic radii larger than the solar circle. Many (not all) other spiral/disk galaxies have similar features, and very many of the best maps have come from the VLA. Another interferometer, BIMA (Berkeley-Illinois-Maryland array, now part of CARMA) provided HI velocity maps confirming a Galactic central bar, whose existence had been found first from star counts.

5. My paraboloid can lick your interferometer (or perhaps conversely)

Radio astronomy grew out of World War II radar technology and was nurtured by people who were not typically astronomers and who had different vocabulary, goals, and mathematical and engineering skills. The world was never quite so firmly bifurcated between single large dishes and arrays as the N-S split in German-speaking Europe between Gute Morgen and Grüss Gott or as impenetrable as the E-W split between the parts of eastern Europe where only Pepsi Cola was available and the land of the Coke. But even the US, which came late to the game, began with a single 1953 dish at Harvard and a pair of 90-foot dishes at Ovens Valley, belonging to Caltech, who had had the excellent sense to import John Bolton from Australia.

Important starting points in the Netherlands were the use of radio-telephone to communicate with distant colonies in the 1930s and Würzburg radar antennas left behind at the end of the war (Strom 2005). In contrast, Australia had already been using Lloyd telescopes (or sea-cliff interferometers) for radar monitoring during the war. The two beams that interfere are the one coming directly to your dish (or Yagis or whatever) from the source and one reflected off ocean water a few hundred feet below the cliff where your telescope sits. A big part of the fun of reading about the beginnings of Australian radio astronomy is the wonderful set of names of their field stations: Badgerys Creek, Dapto, Fleurs, Potts Hill, Dover Heights, Hornsby Valley, and more. (Orchiston & Slee 2005).

Fairly clearly, the first successful use of radio interferometry for astronomical purposes happened in Australia in 1945-46. The target was the sun, and some association of radio emission with active surface regions was quickly established. Joe Pawsey was the senior member of the group (at age 37), but a great deal of the work was done by Ruby Payne Scott, though she never made the transition to galactic and extragalactic radio astronomy (Goss & McGee 2010).

By the early 1950s, both the Australian stations and UK ones (that is, Cambridge and Jodrell bank) had an assortment of facilities engaged in providing accurate positions (for optical identifications), source fluxes (for counting mostly), the first, poorly resolved maps, and spectra (for initial studies of emission mechanisms). Mainstream astronomy was, perhaps, rather slow to accept radio astronomy in general (Jarrell 2005 and many other sources), but interferometry presented some special difficulties. Figures 2 and 3 illustrate these in frivolous and serious ways. Figure 2 embodies a remark by Peter A.G Scheuer (the first theorist attached to the Cambridge group) that "interferometry is like being led blindfolded up a single path to the top of the mountain, and then being asked to describe the entire mountain and its Fourier transform." Figure 3 (adapted from Krauss 1986) shows both algebraically and geometrically the relationship between brightness temperature on the sky (in right ascension and declination for instance) and where your dishes are on the ground, the u-v plane. For a fixed baseline, rotation of the earth will trace out for you some sort of arc in the u-v plane, and the more thorough the uv coverage, the more you will learn about $T_b(\alpha, \delta)$. N antennas will give you $N(N-1)/2$ arcs in the uv plane. And the enormous achievements of the VLA, other similar arrays, and now ALMA arise from careful choice of antenna spacings and the ability to vary those spacing, until the VLA looking at Sgr B2 and ALMA looking almost anywhere in its field of view fill the uv plane like squashed spiders (Wilson 20133, pp 255 & 268).

Early radio maps nearly always showed T_b contour lines, unfamiliar to optical astronomers, and the transition to grey-scale (or even colored) images that has accompanied the filling up of the UV plane has made those maps look very much more like astronomical images at other wavelengths. This is an aspect of the history that does not seem to have been much mentioned.

In contrast, the propensity of interferometers to resolve out diffuse flux and to "confuse" two or more adjacent weak sources into a single, brighter one, have been extensively discussed (Kellermann & Moran 2001). The astronomical context in which confusion in this technical sense caused much anger was the counting of radio sources as a function

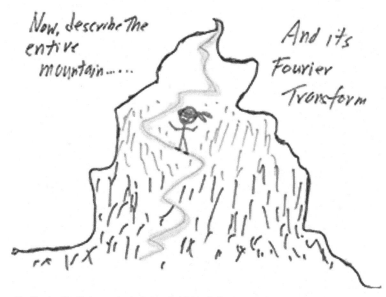

Figure 2. P. A. G. Scheuer's informal 1968 definition of two-element, earth-rotation, aperture-synthesis interferometer. (author's drawing)

of the flux received. Because volume scales as distance cubed and the flux you receive will scale as $(\text{distance})^{-2}$, a homogeneous, isotropic, non-evolving source population will give you a number of source, N, seen down to flux, S, varying as $N(S) \propto S^{-\frac{3}{2}}$. Steeper relationships mean more sources in the past (for an expanding universe) and flatter ones fewer sources in the past. There is no doubt that the early Cambridge reports of $N(S)$ were steeper than reality, and the sorting out took several years, with input from Australia, Jodrell Bank, and elsewhere, and with hard feelings left many places (Gold & Mitton 2012, Longair 2006).

There are, to this day, single-paraboloid people and interferometer people, though I think the latter are winning. The Dwingeloo 25-m is being reconditioned, but the Parkes 64-m (as part of the ATNF), the Nobeyama 45-m, the Lovell 76-m, the Effelsberg 100m, the Greekbank 140-foot, the Algonquin 46-m, and others continue to soldier on. A few years ago, I would have bet that the 100-meter Greenbank Telescope (the Byrd in the hand), completed in 2000 and the largest fully-steerable dish at present, would be the last of its breed. But in the last few years, China has commissioned 40, 50, and 65-meter dishes with 110-meter fully-steerable planned. FAST in Guizhou, their successor to Arecibo (that is a hole in the ground with radio-reflecting surfaces and receivers), will exceed the 1000-foot diameter of the Puerto Rico telescope.

As has happened many times before in radio astronomy, all bets are off when a new player enters the game!

6. One very important VLBI discovery and the havoc it wrought

Very Long Baseline Interferometry generally means the sort where the several collecting areas (from 2 to 13 or more) cannot be connected by cables or even radio links, and the first versions used only two or three antennas. Now the correct explanation for the enormous powers emitted by quasars, radio galaxies, and other active galactic nuclei had been out there since 1964: accretion onto a very massive, compact central object, according to papers by E. E. Salpeter and by Ya. B. Zeldovich and I. D. Novikov. But a VBI result confused the story for several years.

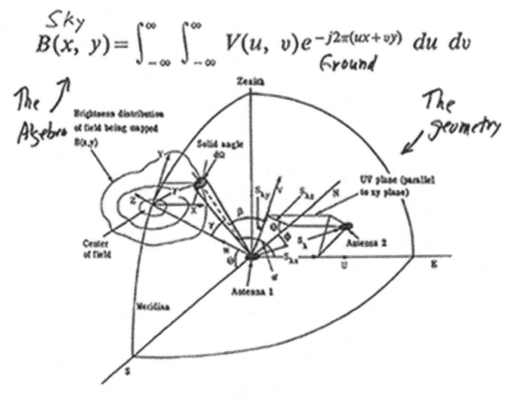

Figure 3.

The quasar 3C279 ($z = 0.538$ and distance debated by hold-out supporters of steady state cosmology and non-cosmoloical redshifts) and radio galaxy 3C 120 ($z = 0.033$ and distance not doubted) both had been displaying rapid, erratic variability. Their sizes were measured by VLBI between Haystack in Massachusetts and Goldstone in California. Both were most simply fit by two components of changing strength and changing separation (Whitney *et al.* 1971, Cohen *et al.* 1971, Shaffer *et al.* 1972), which looked like expansion. If so, then the components were moving, it seemed, at twice or more times the speed of light. Should this have been expected? Yes, because it had been predicted (Rees 1966, 1967) and can best be described as a manifestation of special relativity, not a violation.

But not all were happy. Dent (1972, 1973) continued to monitor the variability (from Haystack) and concluded that he could fit what had happened with a succession of uncorrelated outbursts, happening at a rate of one or two per year, popping up and down at random places in a compact (but not Schwarzschild radius) core. Chains of supernovae or stellar collisions came to mind, and the picture was called the "Christmas tree model;" for the style of (American) holiday lights that flash off and on. Like many foolish things that were originally American, lights of this sort have spread and decorated the restaurant of the conference center where the 2012 Texas Symposium took place in Saõ Paulo, Brazil.

Just when the superluminal/Christmas tree issue was settled is a bit fuzzy, but Kellermann & Moran (2001) suggest that it was images using multiple baselines collected and published by Pearson *et al.* (1981). The implied expansion velocity was about 10c. Since then, many observers have plotted behavior of many superluminal jets where the components break loose from a central core, accelerate, decelerate, bend, and twist. M87 is particularly impressive because it is fairly close (Giovanni *et al.* 2011, Asada *et al.* 2011).

Silence in the hall suggested that many participants did not agree with me that this discovery and confirmation of relativistic jet motion in AGNs was the single most important 3-d contribution of interferometry to the study of galaxies. Perhaps narrowing the field to VLBI rather than all of radio interferometry would have increased agreement. Changes with time of course constitute the 4th dimension in this story.

7. Midcourse corrections and digressions

Although we primarily associate Bernard Lovell with the 250-ft steerable paraboloid at Jodrell Bank, one of his earlier radio telescopes was an array of Yagis, working at a wavelength of 8.2 meters and designed to look for radar echoes from meteors. (Gunn 2005). The first Japanese radio telescope was Tanaka's 1951 open mesh paraboloid, but by the 1970s there was an array of small, solid-surface dishes at Toyokawa (Kaifu 2013). Grote Reber himself, though he constructed the first radio astronomy paraboloid, in due course built a sea cliff interferometer atop Mt. Haleakala, and, after his moves to Tasmania and to very low frequencies (that is, long wavelengths) of necessity switched to extended arrays of dipoles and such. But the much later 23-meter University of Tasmania parabolic dish actually looks a good deal like the primordial Wheaton, IL installation (Kellermann 2005).

The Netherlands, having started with a Würzburg 7.5 meter German radar dish just after the war, to look for predicted 21 cm radiation from neutral hydrogen, built the 25-m Dwingeloo paraboloid for that purpose, but then developed the Westerbork Synthesis Radio Telescope, an array of open-mesh paraboloids (Strom 2005). They operate both to this day. Australia also had a mix of technologies starting very early.

Asada *et al.* (2013) have suggested reconfiguring an ALMA test paraboloid and placing it in Greenland. Operated in tandem with SMA, SMTO, LMT, IRAM, and ALMA itself this would provide 10 new arcs in the uv plane, extending to nearly 9000 km in the north-south (v) direction. Coverage at the longest baselines will once again be sparse, as it was in the early days of VLBI. And there is obviously a price to pay in having to observe sources very close to the horizon, and having large portions of the sky permanently invisible.

8. High angular resolution radio instrumentation in the Decadal Surveys

There have now been six of these surveys, generally named for the chairs of the main panels and titled, or subtitled, "Astronomy for the 1980s" and so forth, There were early calls for more and larger paraboloids (Leo Goldberg said at one point that 500 feet was not too large for studying solar emission) and also for support of moderate sized paraboloids at universities. Some of these things happened (See Trimble 2011 for details); some did not. Table 1 shows the requests for interferometers and, more or less, what became of them. The VLA clearly stands out, for prompt completion the first time around and the later requests for extensions, expansions, and incorporation into a VLBI network. In its various incarnations, it has been, for most of its life, the most productive (of papers and citations) radio telescope in the world. The VLBA has not, perhaps, been as successful as was expected. Interferometry with one or more of the collecting areas in space was requested repeatedly, but the territory eventually handed over to our Japanese and Russian colleagues. The Japanese HALCA lost one of its wavebands; the Russian RadioAstron, after many delays, was launched in July 2011.

On the paraboloid side, the Whitford report asked for two 300′ fully-steerable dishes operable down to 3 cm. Whether the Greenbank 300′ (commissioned in 1965) and the Goldstone 210′ (mostly used for satellite tracking) met this goal is arguable. They also requested a design study for the largest possible steerable paraboloid, which never really happened. Bahcall asked for a 300′ radio telescope in Brazil, after which filled dishes slipped to lower priority.

Table 1. Interferometry in the US Decadal Surveys

Report	Requests and Outcomes
1964 Whitford	Pencil beam array, 100 x 85' dishes, to 3 cm (no)
	OVRO expanded to 6-8 dishes (eventually, as CARMA)
1972 Greenstein	VLA highest priority (construction began 1978)
	large mm array (ARO, ALMA)
	large cm array (VLA eventually pushed there)
1982 Field	VLB Array (1992)
	Space VLBI (to Japan, Russia)
1991 Bahcall	Millimeter array (ALMA, APEX, ASTE)
	Extended VLA (2012)
	Space radio interferometry
2001 McKee-Taylor	Expanded VLA (2012)
	SKA (MeerKAT, ASKAP, not US)
	BIMA merge with OVRO (CARMA 2007)
	Space radio interferometry
	LOFAR (NL dedicated 2010)
2010 Blandford	No explicit requests on high priority list

9. Some (relatively) recent instruments and results

One can easily collect a dozen or more acronyms representing recent, current, or planned radio interferometers of various sorts. The VLA has become the J (for Janksy) VLA, BIMA and OVRO have merged to become CARMA. And there are the SMA, PdB, IRAM, GMRT, ALMA, LOFAR, SKA, MeerKat, ACTA, WSRT, MERLIN, EVN, JVN, JCMT + CSO, and the VLBA Many of these, including CARMA, PdB, SMA, and installations at Nobeyama and Stanford look very much like the VLA — two-dimensional patterns of paraboloids (that would individually have seemed very large to Reber in 1932) on the ground, often on rail lines to permit changing of baselines more or less continuously, while the earth rotates providing lots more coverage of the u-v plane. Some others — LOFAR starting in the Netherlands, the SKA prototype ASKAP, the long wavelength array (LWA) near the VLA, and the MWA do not look like the VLA, and Kellermann & Moran (2001) foresee a future in which digital processing can turn very large numbers of small receiver elements into interferometric phased arrays that can produce maps of large sky regions at many wavelengths all more or less at once. The optical analogy is, I think, the IFU, much discussed at this meeting, but I wouldn't bet money on it.

From 1991 to 2006, I (and a few long-suffering friends) provided yearly reviews of the entire literature of astrophysics, mostly in PASP, and all published as Trimble or Trimble & friend(s). In the first nine years, the following interferometric, 3-d items were picked as highlights:

- Bipolar outflows from galactic nuclei common and a signature of activity
- Orientation of jets and unified models of AGNs
- Many of the jets superluminal, not all (in sense of Sect. 6)
- A (very) few counterrotating gas disks
- A (very) few leading spiral arms
- Galaxies with 1, 2, and 4 arms (the Milky Way still undecided)
- Use of HALCA and RadioAstron for space VLBI

- Gas structure in Milky Way core with two-sided jet and 3-armed spiral
- Issue of correlation or anti correlation of bright spots in jets between radio and optical still undecided (a registration problem, of some importance for understanding shock acceleration of relativistic electrons)
- One shrinking superlumnial source
- Resolution of lensed structures
- Polarization of jets parallel vs perpendicular to jet length
- Magellanic-Stream-like gas in other galaxies
- Double bars
- Maps of non-spherical accretion through cosmic web structures

In the present century, interferometric results have continued to pour forth, though my choice of ones to highlight may well seem perverse. First was a WSRT HI image of NGC 4244, an edge-on disk galaxy that optically looks a good deal like NGC 891. But NGC 891 has lots of HI out of its plane, including what seem to resemble the Milky Way's High Velocity Clouds. NGC 4244 on the other hand has very blobby HI structure in a very thin plane and almost nothing at z more than a couple hundred pc. From a suitable point of view, almost a one-dimensional galaxy.

Then there is Arp 220 imaged with ALMA (Scoville *et al.* 2013). The 349 GHz continuum image seems to have returned to us to a bipolar cow. Ah, but it was really the spectrum they wanted, and, with the frequency selectivity of the array, they were able to pull out a line profile of HCN (4 - 3) that reveals H26 alpha on its low-frequency wing.

Ueda *et al.* (2013 in Kawabe *et al.* 2013, p. 61) have imaged 37 galaxies in CO that they describe a merger remnants, with CARMA, ALMA, SMA, and PdB. No two are alike. There are single central blobs, long worms, extended ellipses, a couple of bipolar cows, some apparent partial spirals, and a few total non-detections. Someone else's IRAM image of 3C84 shows fairly clearly that the radio jets are pushing the CO outward into a couple of crescents. And Espada (2013), who has mapped CO in the inner couple of kpc of Cen A has found a disk with spiral arms. Cen A is, of course, a giant elliptical galaxy, which traditionally, therefore, should not have spiral arms. Perhaps it belongs in with Ueda *et al.*'s merger remnants.

To bring these lists further up to date, please consult the rest of these proceedings and the descriptions of the sessions planned for the IAU General Assembly in Hawaii in August, 2014!

Acknowledgements

I am grateful to the organizers for the invitation to review the topic of history of interferometers and galactic structure and for partial support of my travel expenses. The rest, of course, comes from The Government, in the form of a Schedule A tax deduction. Ms. Alison Lara turned my usual scruffy typescript into a submittable file in her usual expert fashion.

References

Alves, J. *et al.* Eds. 2011, *Computational Star Formation*, Cambridge University Press
Asade, D. *et al.* 2011, in Romero *et al.* 2011, p. 198
Asade, D. *et al.* 2013, in Kawabe *et al.* 2013, p. 243
Cohen, M. H. *et al.* 1971, *ApJ* 170, 207
Dent, W. A. 1972, *Sci* 175, 1105*ApJ* 175, L55
Dent, W. A. 1973, in D. J. Hegyi Ed. Sixth Texas Symposium on Relativistic Astrophysics, Ann. NY Acad. Sci. 224, 65
Espada, E. 2013, in Kawabe *et al.* 2013, p. 69
Fabry, C. & Buisson, H. 1914, *ApJ* 40, 241

Funes, J. G. & Corsini, E. M. Eds. 2008, *Formation and Evolution of Galaxy Disks,* Astronomical Society of the Pacific Publication No. 480

Giovanni, G. *et al.* 2011, in Romero *et al.* 2011, p. 150

Glindemann, A. 2011, *Principals of Stellar Interferometry,* Springer

Gold, T. & Mitton, S. 2012, *Taking the Back off the Watch,* Springer

Goss, W. M. & McGee, R. X. 2010, *Under the Radar,* Springer

Gunn, A. G. 2005, in Orchiston 2005, p. 107

Hoskin, M. Ed. 1997, *Cambridge Illustrated History of Astronomy,* Cambridge University Press

Jansky, K. G. 1932, *Proc. IRE* 20, 1920

Jarrell, R. 2005, in Orchiston 2005, p. 191

Kaifu, N. 2013, in Kawabe *et al.* 2013, p. 243

Kawabe, R. *et al.* Eds. 2013, *New Trends in Radio Astronomy in the ALMA Era: 30th Anniversary of Nobeyama Radio Observatory,* ASP Conf. Proc. No 476

Kellermann, K. 2005, in Orchiston 2005, p. 43

Kellermann, K. & Moran, J. M. 2001, *ARA&A* 39, 457

Kerr, F. J. & Westerhout, G. 1964, in A. Blaauw & M. Schmidt, Eds., *Galactic Structure* Univ. of Chicago Press, p. 167

Kraus, J. D. 1986, *Radio Astronomy, 2nd Edition,* Cygnus-Quasar Books, Powell, Ohio

Lena, P. *et al.* 2012, *Observational Astrophysics, 3rd Ed,* Springer

Lequeux, J. 2013, *Le Verrier - Magnificant and Detestable Astronomer,* Springer

Longair, M. 2006, *The Cosmic Century,* Cambridge Univ. Press

Marcelin, M. *et al.* 1983, *A&A* 128, 140

Melia, F. 2007, *The Galactic Supermassive Black Hole,* Princeton Univ. Press

Munns, D. P. D. 2013, *A Single Sky,* MIT Press, Cambridge Massachusetts

North, J. 2008, *Cosmos,* Univ. of Chicago Press

Orchiston, W. Ed. 2005, *The New Astronomy,* Springer

Orchiston, W. & Slee, B. 2005, in Orchiston 2005, p. 119

Pawsey, J. L. 1964, in A. Blaauw & M. Schnidt Eds. *Galactic Structure,* Univ. of Chicago Press, p. 219

Pearson, T. J. *et al.* 1981, *Nature* 290, 365

Reber, G. 1940, *ApJ* 91, 621

Rees, M. J. 1966, *Nature* 211, 468

Rees, M. J. 1967, *MNRAS* 135, 345

Romero, G. E. *et al.*Eds. 2011, *Jets at All Scales (IAUS 275),* Cambridge Univ. Press

Ryle, M. & Vonberg, D. D. 1946, *Nature* 158, 339

Saha, S. K. 2011, *Aperture Synthesis,* Springer

Schwarzschild, M. 1982, *ApJ* 263, 599

Scoville, N. *et al.* 2013, in Kawabe *et al.* 2013, p. 1

Seigar, M. S. & Treuthardt, P. 2014, *Structure and Dynamics of Disk Galaxies,* ASP Vol 480, San Francisco

Shaffer, D. B. *et al.* 1972, *ApJ* 173, L147

Sjouwerman, L. O. *et al.* Eds. 2014, *The Galactic Center, IAUS 303,* Cambridge Univ. Press

Strom, R. 2005, in Orchiston 2005, p. 93

Sullivan, W. 1984, *The Early Years of Radio Astronomy,* Cambridge Univ. Press

Trimble, V. 2002, in J. F. Alves & M. J. McCaughrean eds. *The Origin of Stars and Planets: The VLT View,* Springer, p. 493

Trimble, V. 2011, in J-P. Lasota, Ed, *Astronomy at the Frontiers of Science,* Springer, p. 13

Ueda, J. *et al.* 2013, in Kawabe *et al.* 2013, p. 61

Waller, W. H. 2013, *The Milky Way,* Princeton Univ. Press

Whitney, A. R. *et al.* 1971, *Science* 173, 225

Wilson, T. L. *et al.* 2009, *Tools of Radio Astronomy, 5th Edition,* Springer

Wilson, T. L. *et al.* 2013, *Tools of Radio Astronomy, 6th Edition,* Springer

Galaxies in 3D across the Universe
Proceedings IAU Symposium No. 309, 2014
B. L. Ziegler, F. Combes, H. Dannerbauer, M. Verdugo, eds.

© International Astronomical Union 2015
doi:10.1017/S1743921314009223

KMOS @ the VLT: Commissioning and Early Science

Ray Sharples[1] and the KMOS Team†

[1]Department of Physics, University of Durham, Durham, UK
email: r.m.sharples@durham.ac.uk

Abstract. KMOS is a multi-object near-infrared integral field spectrograph built by a consortium of UK and German institutes for the ESO Paranal Observatory. KMOS completed commissioning at the ESO VLT in Spring 2013 and was made available to the user community from Oct 2013. We present the unique capabilities of KMOS for 3D galaxy surveys and report on the on-sky performance verification measured during three commissioning runs on the ESO VLT in 2012/13 and some of the early science results.

Keywords. instrumentation, infrared spectroscopy, integral field spectroscopy

1. Introduction

KMOS is a unique cryogenic near-infrared (0.8 to 2.5 μm) multi-object spectrograph that uses deployable integral field units (d-IFUs) to obtain spatially-resolved spectra for multiple targets selected from within an extended field of view. The top-level scientific requirements are: (i) to support spatially-resolved (3-D) spectroscopy; (ii) to allow multiplexed spectroscopic observations; (iii) to allow observations across any of the IZ, YJ, H, and K infrared atmospheric windows in a single exposure. These requirements have been flowed down to the final system design specifications (Sharples *et al.* 2004). The final design employs 24 robotic arms that position fold mirrors at user-specified locations within a 7.2 arcmin diameter field-of-view. Each arm selects a sub-field on the sky of 2.8×2.8 arcseconds. The size of these sub-fields is chosen to match the compact sizes of high redshift galaxies, with a spatial sampling of 0.2 arcsec per pixel to properly sample the excellent infrared seeing at Paranal (median FWHM $\sim$ 0.5 arcsec in the K-band). The sub-fields are optically relayed to 24 advanced image slicer d-IFUs that partition each sub-field into 14 slices with 14 spatial pixels along each slice. Light from the output of the d-IFUs is dispersed by three cryogenic grating spectrometers which generate 14x14 spectra with $\sim$ 1000 Nyquist-sampled spectral resolution elements for all of the 24 independent sub-fields. The spectrometers each employ a single 2kx2k Hawaii-2RG HgCdTe detector and are each fed by 8 d-IFUs. The optical layout for the whole system has a threefold symmetry about the Nasmyth optical axis which has allowed a staged/modular approach to assembly, integration and test. The following sections present the technical overview of the instrument and the as-built performance during commissioning and science verification.

† Ralf Bender, Alex Agudo Berbel, Richard Bennett, Naidu Bezawada, Roberto Castillo, Michele Cirasuolo, Paul Clark, George Davidson, Richard Davies, Roger Davies, Marc Dubbeldam, Alasdair Fairley, Gert Finger, Natascha Frster Schreiber, Reinhard Genzel, Reinhold Haefner, Achim Hess, Ives Jung, Ian Lewis, David Montgomery, John Murray, Bernard Muschielok, Jeff Pirard, Suzanne Ramsey, Phil Rees, Josef Richter, David Robertson, Ian Robson, Stephen Rolt, Roberto Saglia, Ivo Saviane, Joerg Schlichter, Linda Schmidtobreik, Alex Segovia, Alain Smette, Matthias Tecza, Stephen Todd, Michael Wegner, Erich Wiezorrek and others.

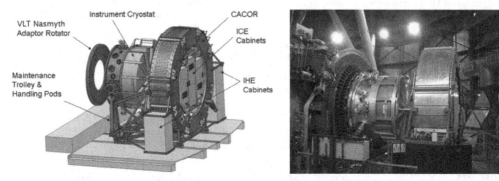

Figure 1. Schematic view of KMOS (left) and an image of KMOS on the Nasmyth platform of VLT UT1 (right).

2. Technical Description

2.1. *Hardware Overview*

From a hardware perspective the instrument neatly partitions into the following key subsystems:

- Pickoff subsystem
- IFU subsystem
- Spectrograph subsystem
- Detector subsystem
- Infrastructure subsystems and electronics

The opto-mechanical parts of the instrument are all contained within a single cryostat mounted on the instrument rotator at the Nasmyth focus of the VLT UT1 telescope. The total rotating mass is 2500kg of which 1100kg is the cold mass. At the temperature and pressure conditions on Paranal, the optical bench was stabilized at a temperature of 115K. Because of the mass-limit on the instrument rotator (3000kg) the electronics cabinets are mounted on a separate instrument rotator (CACOR) sitting on the Nasmyth platform (see Fig. 1). The total mass during operations is 6000kg. KMOS was designed to have three identical modules, each of them comprising eight pickoff arm systems, eight d-IFUs, one spectrograph and one near-IR detector. This ensures that in case of failure of one system, 2/3 of the functionality will still remain. The following sections briefly describe the different sub-systems of KMOS in the order they are encountered along the optical path going from the telescope to the detector; further details can be found in Sharples *et al.* (2006).

2.2. *Pickoff System*

The selection of targets is achieved by means of 24 cryogenic $r - \theta$ robots (pickoff arms), which together patrol a 7.2 arcmin field of view. This is located inside the cryostat below a vacuum-spaced doublet corrector which produces a flat telecentric focal plane. The pickoff arms are arranged into two different layers (top and bottom) of twelve arms each patrolling on either side of the focal plane, so that adjacent arms cannot interfere with each other. Each layer of twelve pickoff arms can patrol 100% of the field. The two motions (radius and angle) of the pickoff arms are both driven open-loop by stepper motors which are modified for cryogenic operation. LVDT encoders on each arm are used to check for successful movement. All 24 arms can be commanded to move simultaneously. A typical field re-configuration takes ~ 4 minutes and is accomplished whilst the telescope is slewing to the next field to avoid any additional dead time.

2.3. *Integral Field Units and Filters*

The deployable integral field units contain a set of fore-optics that collect the output beam from each of the 24 pickoff arms and reimage it with 2:1 anamorphic magnification onto the image slicers. Immediately in front of the fore-optics is a filter wheel, which is used for order-sorting with the selected gratings. The d-IFU sub-system itself has no moving parts and uses gold-coated aluminium mirrors for optimum performance in the near-infrared and at cryogenic temperatures. The anamorphic magnification is required in order that the spatial sampling pixels (spaxels) on the sky are square whilst maintaining Nyquist sampling of the spectral line profile on the detector in the spectral dimension. All of the micro-optics in the IFUs are produced using a combination of diamond-turning and raster fly-cutting techniques (Dubbeldam *et al.* 2012).

2.4. *Spectrographs*

The spectrograph sub-system is comprised of three identical units, which feed three detector sub-systems (Masters *et al.* 2010). Each spectrograph uses a single off-axis toroidal mirror to collimate the incoming light, which is then dispersed via a reflection grating and refocused using a 6-element refractive camera. The five available gratings (IZ, YJ, H, K, H+K) are mounted on a 6-position wheel which allows optimized gratings to be used for the individual bands. A blank position in the filter wheel is used for calibrations.

2.5. *Detectors*

KMOS uses three Teledyne substrate-removed Hawaii 2RG HgCdTe detectors with 2kx2k 18μm pixels (one detector for each spectrograph). The detectors are operated at a temperature of 35K to minimize dark count and persistence (Finger *et al.* 2008). Typical characteristics are QE> 90%, readout noise < 3e$^-$ rms for long integrations with Fowler sampling, and dark current < 0.003e$^-$/sec/pixel). Each detector is adjusted manually for tip-tilt, but is mounted on a remotely operated focus stage which was used to set the optimum focus position during commissioning. Sample-up-the-ramp (non-destructive) readout is used as standard, with Threshold Limited Integration (TLI) to extend the dynamical range for long exposure times. In this mode, if one pixel is illuminated by a bright source and reaches an absolute value above a certain threshold (close to detector saturation), only detector readouts before the threshold is reached are used to compute the slope and the counts written in the FITS image for this pixel are extrapolated to the full exposure time. This is an effective way to remove many cosmic rays on these devices.

2.6. *Infrastructure and Electronics*

The instrument housekeeping electronics (IHE) are mounted directly on the Nasmyth platform in two electronics cabinets (Bezawada *et al.* 2008), whilst the instrument control electronics (ICE) are mounted in three electronics cabinets which co-rotate with the instrument on the CACOR (Hess *et al.* 2010).

2.7. *Software*

In addition to the above hardware components, a customised data reduction pipeline has been provided for KMOS (Davies *et al.* 2013) which allows the observer to evaluate the data quality after each readout using real-time reconstruction of the data cubes and apply sophisticated algorithms for co-adding of data cubes and subtraction of the sky background. With over 4000 spectra per readout, automatic data processing and reduction methods have proved essential to fully exploit the scientific potential of KMOS. KMOS also has an optimised pickoff arm allocation tool, known as KARMA (Wegner *et al.* 2010), which links directly to the ESO observation preparation software (P2PP).

KARMA assigns arms to targets in a prioritised way, whilst ensuring that no invalid arm positions are selected and allows the astronomer to manually reconfigure the list of allocated targets if required.

3. Integration and Commissioning

Following a successful laboratory test campaign (Sharples *et al.* 2012), and a test readiness review by ESO, KMOS was shipped to Paranal in Aug 2012 to begin re-integration and installation at VLT UT1. The main cryostat and associated electronics were transferred via air freight, whilst the larger CACOR consignment was shipped as sea cargo. KMOS was fully re-assembled and re-calibrated in the new instrument integration hall at Paranal over an 8-week period in Sep-Nov 2012 by a dedicated team of technical experts from the consortium partners, working closely with ESO personnel. After an extensive set of verification tests, the instrument was taken up the final stretch of the mountain road at walking pace, before being installed on the Nasmyth platform of UT1 (Antu). Because of the size of the CACOR (about four metres high), this item had to be lifted directly in through the dome aperture using an external crane.

3.1. *First Light*

First light with KMOS at the ESO VLT occurred on 21st Nov 2012 (Sharples *et al.* 2013). Initially this involved pointing the telescope at a relatively bright star and taking short exposures with one arm at a time placed at the centre of the field. All 24 arms revealed a star image close to the centre of the IFU field of view (2.8×2.8 arcsec) much to the relief of the commissioning team. Even this relatively simple observation required a large number of systems to be working together such as the real-time display which showed the positions of the target objects in the pipeline reconstructed data cubes. The real-time display is a key feature of KMOS which allows the telescope pointing to be refined by placing a subset of the pickoff arms onto bright targets which are then centered automatically using a shift and rotation of the telescope field of view (in much the same way that bright reference stars are used to align the slit masks in a multi-slit spectrograph). Once the field is aligned, these acquisition arms can then be redeployed to fainter science targets if desired.

3.2. *Astrometric Performance*

One of the first calibrations to be addressed was the astrometric positioning of the arms. This included both the mapping of the plate scale in the corrected focal plane(s) and also the local corrections (lookup tables) which are used to take out the individual variations in the $r - \theta$ drives for each arm. The latter have to be defined over the whole pie-shaped region accessible to each arm, and were based on an astrometric calibration procedure using densely populated star fields in open or globular star clusters. The absolute positioning accuracy of all the arms over the whole 7.2 diameter field was such that 83% of the targets lay within 1 spaxel (0.2 arcsec) of the centre of the d-IFU. The repeatability was < 0.1 arcsec.

3.3. *KMOS Throughput*

A second key performance parameter is the system throughput. KMOS has a large number of optical surfaces in the optical train (18 refractive and 15 reflective surfaces, excluding the grating and telescope mirrors) in order to deliver the performance capability of a multi-object integral field spectrograph. The throughput was measured on a number of nights using standard stars and is summarized in Fig. 2. The horizontal red (dashed) lines

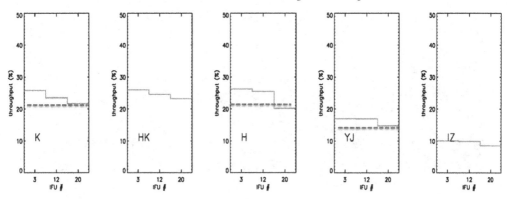

Figure 2. Total system throughput (excluding telescope and detector) in the different KMOS bands for the three different spectrographs (Spectrograph#1 is d-IFUs 1-8, Spectrograph#2 is d-IFUs 9-16, Spectrograph#3 is d-IFUs 17-24)

show the minimum requirements from the instrument technical specification (Sharples *et al.* 2004). Since the IZ-band and HK-bands were added later, on a best-efforts basis (with no impact on the design), there are no specific requirements for these modes.

3.4. *Sky Subtraction*

One of the key performance metrics of any multi-object integral field spectrograph is the ability to do accurate sky-subtraction, particularly in the infrared bands longer than 1 micron where the night sky spectrum is dominated by extremely bright and variable OH emission lines. With KMOS we have experimented with various sky-subtraction strategies including conventional offsets to blank sky with a fixed d-IFU pattern, and also using dedicated sky IFUs to monitor the time-variability. Tests using long exposures of faint emission-line galaxies indicate that residuals can be reduced to < 1% except at the centres of the strongest lines, using a combination of these methods.

4. KMOS Science

Following three on-sky commissioning runs in Nov 2012, Jan 2013 and Mar 2013, KMOS was used for a set of science demonstration programmes proposed by the ESO community in Jun and Sep 2013. It was offered for community open time, and consortium guaranteed time, access from Oct 2013. Since then the instrument has been in heavy demand, leading to UT1 being the most oversubscribed (x6.7) telescope at Paranal in Period 92 (Oct 2013 - Mar 2014).

KMOS has now been used for a wide of scientific programmes ranging from exoplanet transit measurements through to searches for Lyman-α emission in extremely high redshift (z > 10) galaxies. As an example we show in Fig. 3 the velocity fields measured for a sample of z $\sim$ 1 galaxies. Previous observations of such objects have relied on single-IFU observations, making the acquisition of representative complete samples prohibitively expensive in telescope time. Observations with KMOS have already more than doubled the number of such systems with good measurements of star formation rates and internal kinematics. Also in this figure we show a H-band mosaic of Jupiter obtained by stacking 16 dithered KMOS pointings with the arms in a fixed sparse grid pattern. This produces a unique wide-field infrared integral field capability with $\sim$ 75,000 spectra covering a field of 65x43 arcsec at 0.2 arcsec spatial sampling.

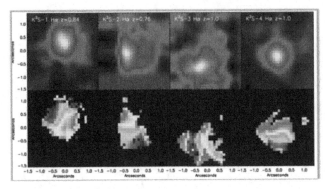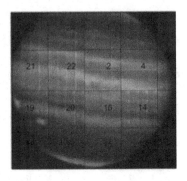

Figure 3. *(left)* Hα emission-line maps (upper) and derived velocity fields (lower) for a sample of faint z ∼ 1 emission-line galaxies in the GOOD-S field. The brightest targets have an observed integrated H-alpha flux of 1.0×10^{-16}ergscm^{-2}s^{-1}. These data were obtained with only 30 mins of on-source exposure time during KMOS commissioning and demonstrate the power of this facility instrument for such surveys. *(right)* Reconstructed spectral image of Jupiter (pseudo colours refer to narrow bands extracted from the spectrum to highlight different molecular bands). This was created using the 24-arm mapping template with non-sidereal tracking and comprises nearly 75,000 spectra. The numbers refer to zones covered by specific d-IFUs during the dithered pointings

5. Conclusions

KMOS is a powerful new multi-object near- infrared integral field spectrograph now available at the ESO VLT. Its performance has been extensively verified through a series of technical and scientific tests at Paranal Observatory.KMOS is already beginning to make its mark in 3D spectroscopic surveys covering a wide variety of different science areas.

Acknowledgements

I acknowledge support from STFC grants PP/E000207/1, ST/J003034/1 and ST/K001175/1.

References

Sharples, R., *et al.*, *Proc. SPIE* 5492, 1179 (2004)
Sharples, R., *et al.*, *Proc. SPIE* 6269, 44 (2006)
Rees, P., *et al.*, *Proc. SPIE* 7735, 166 (2010)
Dubbeldam, M., *et al.*, *Proc. SPIE* 8450, 84501M (2012)
Masters, R., *et al.*, *Proc. SPIE* 7735, 168 (2010)
Finger, G., *et al.*, *Proc. SPIE* 7021, 20 (2008)
Bezawada, N., *et al.*, *Proc. SPIE* 7014, 121 (2008)
Hess, A., *et al.*, *Proc. SPIE* 7735, 93 (2010)
Davies, R., *et al.*, *Astronomy & Astrophysics*, Vol 558, A56 (2013)
Wegner, M., *et al.*, *Proc. SPIE* 7740, 28 (2010)
Sharples, R., *et al.*, *Proc. SPIE* 8446, 84460K (2012)
Sharples, R., *et al.*, *The Messenger* 151, 21 (2013)
Davies R., *MNRAS375, 1009 (2007)*

Galaxies in 3D across the Universe
Proceedings IAU Symposium No. 309, 2014
B. L. Ziegler, F. Combes, H. Dannerbauer, M. Verdugo, eds.

© International Astronomical Union 2015
doi:10.1017/S1743921314009235

Galaxies in 3D across the Universe

Denis Burgarella[1], Toru Yamada[2], Giovanni Fazio[3] and Marcin Sawicki[4]

[1] Aix-Marseille Universit, CNRS, LAM (Laboratoire d'Astrophysique de Marseille) UMR 7326, 13388, Marseille, France
email: denis.burgarella@lam.fr
[2] Astronomical Institute, Tohoku Univ., Japan
email: yamada@astr.tohoku.ac.jp [3] Harvard Smithsonian Center for Astrophysics, Cambridge Massachussets, USA
email: gfazio@cfa.harvard.edu [4] Department of Astronomy and Physics and Institute for Computational Astrophysics, Saint Mary's University, Halifax, Nova Scotia, Canada
email: sawicki@ap.smu.ca and the WISH team

Abstract. WISH is a new space science mission concept whose primary goal is to study the first galaxies in the early universe. The primary science goal of the WISH mission is to push the high-redshift frontier beyond the epoch of reionization by utilizing its unique imaging and spectrocopic capabilities and the dedicated survey strategy. WISH will be a 1.5m telescope equipped with a 1000 arcmin^2 wide-field Near-IR camera to conduct unique ultra-deep and wide-area sky imaging surveys in the wavelength range 1 - 5 μm. A spectroscopic mode (Integral-Field Unit) in the same Near-IR range and with a field of view of 0.5 - 1 arcmin and a spectral resolution R = 1000 is also planned. The difference between WISH and EUCLID in terms of wavelength range explains why the former concentrates on the reionization period while the latter focuses on the universe at $z < 3$. WISH and JWST feature different instantaneous fields of view and are therefore also very complementary.

Keywords. universe: first light - reionization - galaxies: evolution - galaxies: formation - space telescope - near-infrared

1. Where should we statistically search and study the first galaxies?

WISH is designed to detect the first star-forming galaxies in the universe and to follow them up in the early phases of their evolution. One of the first questions that we need to address is which wavelength range would be the best to reach the above objective: the rest-frame far-ultraviolet (FUV) in the rest-frame far-infrared (FIR)? In a recent paper, Burgarella *et al.* (2013) showed that the FUV dust attenuation (A$_{FUV}$) reaches the same level at $z = 3.6$ than at z = 0 (Fig. 1). This suggests that the early universe would be better studied in the rest-frame FUV. At $z > 5$, this FUV range moves into the near-infrared (1 - 5μm), in the WISH spectral range. Of course, it does not mean that dusty galaxies do not exist at high redshifts as shown by the recent discoveries of, e.g., FLS3 (Riechers *et al.*2013) at $z = 6.34$. The above statement should be accepted as a statistical one unless we misunderstand the early star formation history of the universe.

2. What is WISH?

WISH will adopt an optimized strategy to build the photometric surveys and finally reach the planned depth (Fig. 2). This strategy will make the best use of the individual images to detect transient objects like, e.g., type Ia Supernova. A unique optical layout (Fig. 3) has been developed to achieve the diffraction-limited imaging at 1 - 5 μm over the

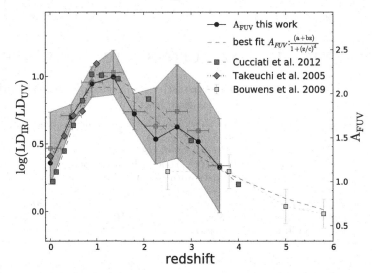

Figure 1. Left axis: The ratio of the FIR-to-FUV luminosity density (IRX) is shown here. The right axis indicates the FUV dust attenuation (A_{FUV}) after converting the FIR-to-FUV ratio using Burgarella *et al.* (2005). The red dotted line with red diamonds is taken from Takeuchi, Buat & Burgarella (2005). The green filled area and green dots are the associated uncertainties estimated through bootstrapping. Black dots denote the values directly computed from the luminosity functions. At $z = 3.6$, A_{FUV} reaches about the same value as at $z = 0$. Takeuchi *et al.* (2005) (red diamonds) used an approach identical to ours while a SED analysis (no FIR data) is performed in Cucciati *et al.* (2012) (blue boxes). Bouwens *et al.* (2009) are estimates based on the UV slope β.

	Depth (3 σ) (AB mag)	Area	Example of the Filters (a plan, to be determined)
Ultra Deep Survey	28	100 deg^2	1.4,1.8, 2.3, 3.0 micron
Multi-Band Survey	28	10 deg^2	1.0,4.0
Ultra Wide Survey	24-25	1000 deg^2	1.4, 1.8, 2.3
Extreme Survey	29-30	0.25 deg^2	1.0, 1.4, 1.8

Figure 2. WISH strategy with the different surveys designed to optimize the photometric detection of $z > 10$ galaxies.

required large area with an angular resolution of about 0.2 arcsec FWHM at 1.5μm. The 1.5m primary mirror will be cooled by passive cooling in the stable thermal environment near the Sun-Earth L2 at about 100K to achieve the zodiacal light limited imaging.

3. An IFU spectrograph for WISH

The spectroscopic mode is studied in France under the responsability of the Laboratoire d'Astrophysique de Marseille, France. The baseline is a 1x1 arcmin^2 Integral-Field Unit (IFU) using 40 (TBC) slicers already studied for the SNAP project. The design shows that the spectrograph could observe in a parallel mode (Fig. 3) which translates into spectroscopic exposure times as long as the telescope lifetime (within uncertainties due to the overheads).

WISH-Spec will allow to acquire very large samples of high-redshift and primordial galaxies by combining spectroscopic detections within total (non contiguous) 1-deg^2 and 10-deg^2 fields of views (Fig. 4). Therefore, the first strong advantage of WISH-Spec wrt JWST is that 10^4 - 10^5 faint emission-line galaxies in 1 deg^2 (Fig. 5 left) and 10^3 - 10^6

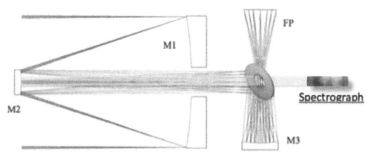

Figure 3. WISH optical design. Light will go to the spectrograph WISH-Spec through a hole in a flat mirror. WISH-Spec could therefore observe in a parallel mode and make the best use of the telescope lifetime to build wide and deep spectroscopic surveys over 10 and 1 deg^2, respectively.

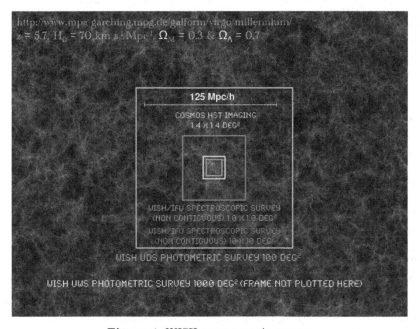

Figure 4. WISH spectroscopic surveys

bright emission-line galaxies in 10 deg^2 in the redshift range $3 < z < 8$ (Fig. 5 right) will be detected.

4. Conclusions

It is important to stress that WISH will be complementary to EUCLID since the former will concentrate on the reionization times at $z > 6$ in the wavelength range 1 - 5 μm while the latter will focus on lower redshifts ($z < 2 - 3$) by construction since EUCLID's wavelength range is limited to $\sim 2\mu$m, and is not optimized to study the first galaxies.

WISH will also be complementary to JWST. The wavelength ranges are the same but the field of view is much larger for WISH both photometrically (900 arcmin^2 for WISH and 9 arcmin^2 for NIRCam) and spectroscopically (even though JWST will feature a MOS over 3×4 arcmin^2, JWST's IFU will only cover 3×3 arcsec^2 while WISH's IFU will be a factor of 10 larger to 0.5 - 1 arcmin).

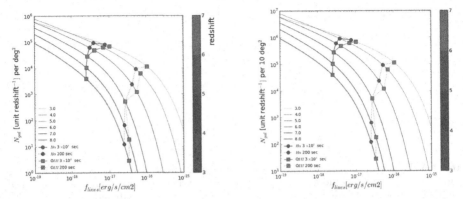

Figure 5. Expected number of spectroscopically detected galaxies at $3 < z < 10$ estimated by translating Far-UV luminosity functions into Hα luminosity function for the 1deg^2 Deep Spectroscopic Survey (left) and the 10 deg^2 Wide Spectroscopic Survey (right).

At very high redshifts, WISH photometric surveys will provide 10^3 - 10^4 galaxies at $z \sim 12$ and 10^2 at $z \sim 15$ (depending on assumed evolution) at AB $\sim$ 27 - 28 from the 100-deg^2 UDS survey. Following up these galaxies spectroscopically requires ELTs. The E-ELT limiting magnitudes in spectroscopy reach about AB = 27 - 28 for the longest exposure times. This means that fainter (e.g., JWST) objects might be difficult to follow up even with ELTs while WISH ones would be more easily observed with the E-ELT spectrographs.

In the range $3 < z < 8$ WISH-Spec will also build large samples of galaxies over non contiguous fields of 1deg^2 and 10deg^2 observed spectroscopically with R = 1000. We will be able to study the physics of the galaxies and their very early phases of formation around the epoch of reionization.

Acknowledgements

A support from the French Centre National d'Etudes Spatiales (CNES), the Programme National Cosmologie et Galaxies and the LABEX OCEVU (Origines, Constituants & EVolution de lUnivers) is acknowledged.

References

Bouwens, R J., Illingworth, G. D., Franx, M., *et al.* 2009, *ApJ* 705, 936
Burgarella D., *et al.*, 2013, *A&A* 554, 70
Burgarella, D., Buat, V., & Iglesias-P/'aramo, J., 2005, *MNRAS* 360, 1413
Cucciati, O., Tresse, L., Ilbert, O., *et al.* 2012, *A&A* 539, 31
Riechers, D., *et al.* 2013, *Nature* 496, 329
Takeuchi, T. T., Buat, V., & Burgarella, D., 2005, *A&A* 440, L17

Galaxies in 3D across the Universe
Proceedings IAU Symposium No. 309, 2014
B. L. Ziegler, F. Combes, H. Dannerbauer, M. Verdugo, eds.

© International Astronomical Union 2015
doi:10.1017/S1743921314009247

The Hector Survey: integral field spectroscopy of 100,000 galaxies

J. Bland-Hawthorn

Sydney Institute for Astronomy (SIfA), School of Physics, The University of Sydney, NSW
2006, Australia
email: jbh@physics.usyd.edu.au

Abstract. In March 2013, the Sydney–AAO Multi-object Integral field spectrograph (SAMI) began a major survey of 3400 galaxies at the AAT, the largest of its kind to date. At the time of writing, over a third of the targets have been observed and the scientific impact has been immediate. The Manga galaxy survey has now started at the SDSS telescope and will target an even larger sample of nearby galaxies. In Australia, the community is now gearing up to deliver a major new facility called Hector that will allow integral field spectroscopy of 100 galaxies observed simultaneously. By the close of the decade, it will be possible to obtain integral field spectroscopy of 100,000 galaxies over 3000 square degrees of sky down to $r = 17$ (median). Many of these objects will have HI imaging from the new ASKAP radio surveys. We discuss the motivation for such a survey and the use of new cosmological simulations that are properly matched to the integral field observations. The Hector survey will open up a new and unique parameter space for galaxy evolution studies.

Keywords. galaxies: evolution – galaxies: kinematics and dynamics – galaxies: structure – techniques: imaging spectroscopy

1. Introduction

A chain of argument is only as strong as its weakest link. In this respect, our understanding of how galaxies evolve from the first seeds to the present day is seriously incomplete. We have a reasonable picture of how the first dark matter structures came together out of the initial matter perturbations. But just how gas settled into these structures to form the first stars and subsequent generations remains an extremely difficult problem. From an observational perspective, the main advances in evolutionary studies have come from spatially resolved imaging and spectroscopy across many wavebands, from the development of optical and imaging spectroscopy since the 1960s, the HI and radio continuum surveys of the 1970s, improvements in infrared imaging and spectroscopy in the 1980s, and the emergence of mid-infrared, UV, x-ray and gamma-ray satellites in the 1990s to the present day. These advances have come from a combination of detailed studies of individual objects and large surveys of global parameters.

Over the past twenty years, imaging surveys from the Hubble Space Telescope (far field) and the Sloan Digital Sky Survey (near field) have been particularly effective in identifying evolution of galaxy parameters with cosmic time and with environment across large-scale structure. This has been matched by large galaxy surveys using multi-object spectroscopy (MOS), most notably at the AAT, SDSS, VLT, Magellan and Keck telescopes (e.g. York *et al.* 2000; Colless *et al.* 2001; Driver *et al.* 2011). MOS instruments provide a single spectrum within a fixed fibre aperture at the centre of each galaxy; spatial information must be drawn from multi-wavelength broadband images. While this approach has its place, the question always remained as to whether imaging spectroscopy was feasible for a cosmologically significant sample.

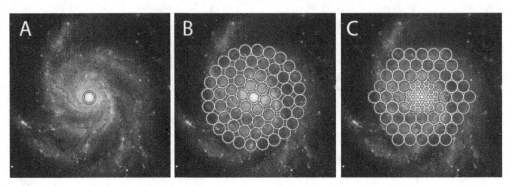

Figure 1. (a) Single fibre apertures are the basis for galaxy surveys to date; (b) the 61-fibre configuration of the SAMI hexabundle; (c) the 85-fibre (54 large, 31 small) configuration of the Hector hexabundle under development at the University of Sydney.

In the 1980s, the scientific potential for imaging spectroscopy was clear from early Fabry-Perot (Bland *et al.* 1987; Cecil 1988; Amram *et al.* 1992) and tuneable filter studies (Jones & Bland-Hawthorn 2001). After the pioneering work of Courtès *et al.*(1988), integral field spectroscopy (IFS) soon exploited the plunging costs of large-area detectors to dominate extragalactic studies today (Hill 2014). Recent integral field surveys involving hundreds of nearby galaxies include ATLAS-3D (Cappellari *et al.* 2011) and CALIFA (Sánchez *et al.* 2012). A useful summary of recent scientific results is given by Glazebrook (2013).

So how do we combine the power of the MOS and IFS techniques, the 'missing link' in astronomical instrumentation? VLT KMOS provides an elegant solution involving small image slicers: this instrument employs 24 configurable arms that position pickoff mirrors at specified locations in the Nasmyth focal plane. But this expensive technology would be prohibitively costly to adapt for the degree-scale fields of cosmologically motivated MOS instruments. Seven years ago, we began to look at compact fibre bundles (hexabundles) that would work with existing robotic positioners (Bland-Hawthorn *et al.* 2011; Bryant *et al.* 2011). This led to the Sydney–AAO Multi-object Integral field spectrograph (SAMI), the first instrument of its kind. SAMI deploys 13 IFUs, each with a field of view of 15″across a 1-degree patrol field (Croom *et al.* 2012). Each IFU consists of a bundle of 61 optical fibres lightly fused to have a high ($\sim$75%) filling factor (Bryant *et al.* 2014a). SAMI is installed on the 3.9-m Anglo-Australian Telescope (AAT), feeding the existing AAOmega spectrograph. For the first time, SAMI allows very large samples of IFS observations to be obtained in a short period of time.

2. The SAMI Galaxy Survey: early results

The SAMI Galaxy Survey is an ongoing project to obtain integral field spectroscopic data for $\sim$3400 galaxies, with an expected completion date of mid 2016. The bundles are dithered with respect to sky to wash out the footprint (Fig. 1(b)). An example of the data quality is shown in Fig. 2. At the time of writing, roughly 1000 galaxies have been observed, including the pilot survey of $\sim$100 galaxies. The updated instrument performance, target selection, survey parameters and reduction procedures are discussed elsewhere (Bryant *et al.* 2014b; Sharp *et al.* 2014). The Galaxy Survey, which includes both cluster and field objects selected from the GAMA survey, had an early data release in 2014 (Allen *et al.* 2014). The use of GAMA fields means that the SAMI data are complemented by many surveys, including GALEX MIS, VST KiDS, VISTA VIKING,

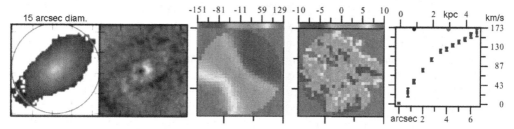

Figure 2. SAMI results on a nearby galaxy: (a) red stellar continuum; (b) residual after disk model subtracted to reveal spiral arms; (c) gas velocity field in km s^{-1}; (d) <7% residuals after rotation model subtracted; (e) gas rotation as a function of radius (with errors shown). SAMI also provides stellar velocities and emission-line ratios giving information on gas chemistry and ionization conditions. The circular bundle in (a) is dithered using small shifts on sky to remove the footprint (Fig. 1) and fill out the field with complete data.

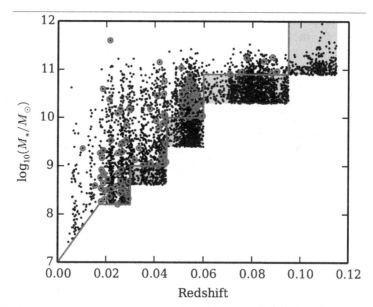

Figure 3. Stellar mass and redshift for all galaxies in the GAMA field regions of the SAMI Galaxy Survey (black points); the early data release sample are also shown (red circles). The blue boundaries indicate the primary selection criteria, while the shaded regions indicate lower-priority targets. Large-scale structure within the GAMA regions is seen in the overdensities of galaxies at particular redshifts (Allen *et al.* 2014).

WISE, Herschel-ATLAS, GMRT and ASKAP (Driver *et al.* 2011). Early science results include studies of galactic winds (Fogarty *et al.* 2012; Ho *et al.* 2014), a universal dynamical relation for all galaxy types (Cortese *et al.* 2014), the kinematic morphology–density relation for early-type galaxies (Fogarty *et al.* 2014), bar streaming in spiral galaxies (Cecil *et al.* 2014), and star formation in dwarf galaxies (Richards *et al.* 2014); see the related contributions in this volume.

3. Hector instrument

The main limitation of SAMI is the need to plug each bundle into pre-cut metal plates. A. Bauer has made a short movie – www.youtube.com/watch?v=MB5H-5XZ9ZE – of this labour intensive process, something we would rather avoid with 100,000 targets in mind. The Hector concept (Lawrence *et al.*2012) will exploit the full two-degree field of the AAT prime focus using robotically positioned hexabundles. The AAO has invested several decades in robotic technology for positioning fibres over large focal planes. Their most recent development is the semi-autonomous "starbug" that positions fibres (or bundles) with a piezo-electric tube that walks over a glass plate. For a movie of how this works in practice, see here: www.youtube.com/watch?v=YvZDF54Si5w. Recent tests by S. Richards and C. Betters show that this technology is ideal for moving the bundles around. An update on the performance statistics for starbugs is given in Goodwin *et al.* (2010).

Much like HETDEX and VLT MUSE before us, we will need to replicate cheap spectrographs to handle 8500 fibres from the 100 hexabundles. Replicating spectrographs at $R = 2000$ is relatively inexpensive using 4000-pixel Andor detectors, for example. We are exploring simple designs involving curved VPH gratings but it is unclear if this technology can reach our goal of $R = \lambda/\delta\lambda \approx 4000$ across the optical band (370-900 nm). A resolution and rough costings should be ready by early 2015. The modular design of Hector means that the first devices will be ready for use by the end of the SAMI Galaxy Survey in mid 2016.

The need for a new hexabundle with a supersampled inner bundle (Fig. 1(c)) is to minimise loss of spatial resolution over the core regions of galaxies. Essentially all SAMI galaxies are brightest over the inner scale length where the gravitational potential is changing most rapidly. The benefits of proper sampling includes more accurate modelling of the inner potential (e.g. bars), sensitivity to kinematic substructure, higher order moments, and so forth (Naab *et al.* 2013). The new design allows for excellent sampling within an effective radius (R_e) for most galaxies while allowing us to reach $2R_e$ with the larger outer fibre apertures. This design also has the added benefit of providing better intrinsic sampling of more distant and compact targets.

4. Hector science

4.1. *Big questions*

Fossil signatures. The formation history of galaxies leaves its imprint on the gas and stellar kinematic properties of present day galaxies. Epochs dominated by gas dissipation will result in the formation of flattened stellar distributions (disks), supported by rotation. During merger dominated phases the stellar systems experience stripping and violent relaxation, existing cold gas might be driven to the central regions causing starbursts, trigger the formation and growth of supermassive black holes or be expelled form the systems in a galactic wind. This, in turn, will impact the distribution of cold gas and the kinematics of forming stars.

Using cosmological 'zoom' simulations of representative synthetic galaxies, Naab *et al.* (2013) demonstrate that gas dissipation and merging result in observable features (at the present day) in the 2D kinematic properties of galaxies, which are clear signatures of distinct formation processes. Dissipation favours the formation of fast rotating systems and line-of-sight velocity distributions with steep leading wings, a property that can be directly traced back to the orbital composition of the systems (Röttgers *et al.* 2014; Sharma *et al.* 2012). Merging and accretion can result in fast or slowly rotating systems with counter-rotating cores, cold nuclear or extended (sometimes counter-rotating) disks showing dumbbell-like features - all observed in real galaxies. The strength of these kinematic signatures will be influenced by feedback from supernovae and AGN, but also by the mass of the galaxies and their environment.

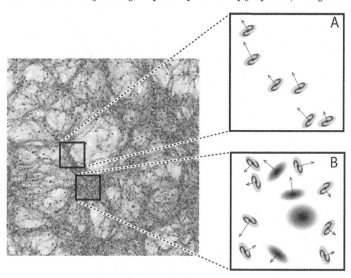

Figure 4. Cosmological simulation (GIF consortium) of the universe today within a 30 Mpc volume. Blue, yellow and red dots are young, middle-aged and old galaxies; the grey is unprocessed gas. The most detailed simulations predict that (A) low-mass galaxy spins align along filaments, and (B) spins are more randomised in dense regions (e.g Dubois *et al.*2014).

Low mass star forming disk galaxies favor low-density environments, predominantly grown by accretion of gas and subsequent in-situ star formation, and are affected by stellar feedback. Higher mass early-type galaxies form in high-density environments - possibly affected by feedback from accreting super-massive black holes - and their late assembly involves merging with other galaxies, which might also be of an early type. Before now, it was not possible to perform a statistically meaningful comparison of kinematic properties of galaxy populations to observed population properties, like the observed increasing fraction of slow versus fast rotators for early-type galaxies as a function of environmental density (Cappellari *et al.* 2012). With the new Eagle simulations (Schaye *et al.* 2014), we can tackle these questions for the first time using the large simulated volume and the higher spatial resolution.

With the Hector instrument, the 2D kinematic (gas, stars) and abundance patterns will be compared over different environs. Preliminary work will be possible with SAMI and Manga, but the higher target density across large-scale structure will allow a more detailed statistical connection to investigate trends with environment. We will attempt to (i) identify characteristic formation histories, (ii) identify properties of progenitor galaxy populations, (iii) assess the impact of the major feedback mechanisms (from massive stars and AGN) on the kinematic properties of high and low mass galaxies. This study that will be supported by higher resolution cosmological zoom simulations for characteristic cases.

The origin of angular momentum. An interesting future area of study is the build-up of angular momentum in galaxies and across different environs (Fig. 4). Codis *et al.* (2012) exploit the *Horizon* simulations ($z > 1.3$) and find an interesting interplay between the gas and dark matter during accretion onto galaxies. The gas and dark matter phases separate when sheets and filaments form between expanding voids of different sizes. Material flows out of voids at a rate that depends on the size of the void. The dark matter collapses to form asymmetric sheet or filament but the gas does not. The gas

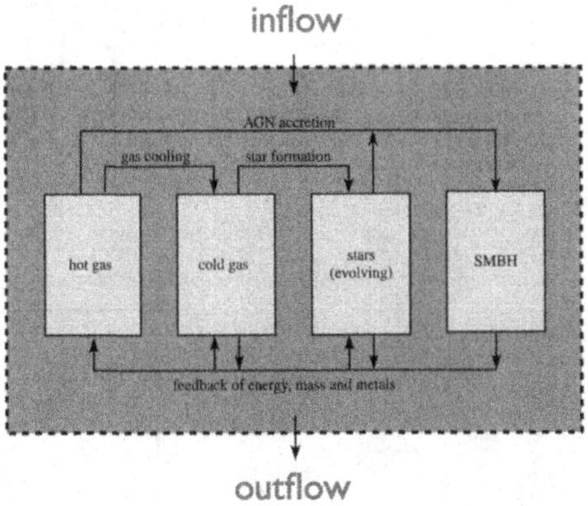

J. Bland-Hawthorn

inflow

outflow

Figure 5. A schematic of the inner workings of a galaxy (adapted from Mo *et al.* 2010), akin to a 4-stroke engine fuelled by gas with a complex exhaust system. Galaxy evolution is an extremely complex process that *demands* much larger 3D spectroscopic surveys and the measurement of new parameters.

undergoes shocks and compresses to form a medium whose main concentration is offset from the dark matter the protocloud. The protocloud oscillates slowly in the gravity field of the collapsed dark matter sheet and rains onto galaxies as a cold or a warm flow. This idea has far-reaching consequences. The wider the separation of gas and dark matter, the longer the delay in accreting to the outer disk leading to galaxies with different spin properties and star formation histories. The *Horizon* team find that low mass gas-accreting galaxies develop spins aligned with filaments (Fig. 4) whereas the high mass galaxies are anti-aligned due to major collisions Dubois *et al.* (2014). The signal is expected in the angular momentum vector field (spin direction and magnitude) when averaged over many thousands of galaxies.

To see this weak signature at $z \sim 0.1$, we need to observe of order 60,000 galaxies in a contiguous volume (Trowland *et al.* 2013) although a larger sample may be required (Dubois *et al.* 2014). This requires mass and rotation estimates for stars and/or gas in each galaxy, which Hector is uniquely situated to do. In this paradigm, the collective properties of galaxies in sheets and filaments may be different in character across the hierarchy. Once again, galaxies within a single filament, for example, are expected to have correlated properties but the properties can vary between different filaments. If this is correct, it will radically overhaul our ideas on how galaxies form and evolve because now the disk properties of galaxies are linked to the large-scale environment. In the earlier classical paradigm, shock heating removes any vestige of where the gas came from, thereby removing any large-scale correlations; the hot gas simply cools and collapses to a disk within the halo (Fall & Efstathiou 1980).

4.2. *Matched data from GAMA and HI surveys*

The GAMA survey is the ideal input catalogue for the SAMI survey because of its supporting data. As we have shown, galaxies are observed in volume-limited regions at several different redshifts (Fig. 3) with a Petrosian limit of about $r \approx 17 - 17.5$ for the faintest sources. The GAMA field selection means that our survey data are complemented by many surveys from UV to radio wavelengths (Driver *et al.* 2011). Ideally, for the Hector survey, we would require the GAMA survey to be extended about a magnitude deeper and over a larger region of sky. By the end of the decade, the Large Synoptic Survey Telescope (LSST) will provide very deep multiband photometry across the Southern Sky.

The GAMA depth and selection complements the all-sky HI surveys soon to be carried out by ASKAP in Australia (Duffy *et al.* 2012). Some fraction of the envisaged Hector sources will have spatially resolved ($10''$ FWHM) HI imaging and spectroscopy over roughly $\sim 1'$ diameter. There are to be two major HI surveys: (1) the Widefield ASKAP L-band Legacy All-sky Blind surveY (WALLABY) is a shallow 3π sr survey ($z < 0.26$) which will survey the mass and dynamics of more than 6×10^5 galaxies; (2) a deeper small-area HI survey, called Deep Investigation of Neutral Gas Origins (DINGO), will trace the evolution of HI ($z < 0.43$) over a cosmological volume of 4×10^7 Mpc^3 detecting up to 10^5 galaxies.

4.3. *Survey design*

All large surveys, both stellar and extragalactic, struggle to defend the survey size with any statistical rigour. This is especially true for galaxy evolution studies since many free parameters are needed to capture the microphysics in galaxies (Fig. 5) and the relative importance of different microphysics may vary with the underlying density field and the galaxy's mass.

Our initial study indicates that 10^5 galaxies over a contiguous volume is of order the size we need to probe (i) variations on local overdensity; (ii) different redshifts; (iii) a range of galaxy masses at each redshift; (iv) galaxy orientations; (v) sufficient cell division in the galaxy colour-magnitude diagram. We also need sufficient numbers of objects per bin to counteract complexity and fine structure in galaxies. A key uncertainty is the association of galaxies with local overdensity. Statistical 'crowding' techniques (e.g. five nearest neighbours) are intrinsically uncertain for measuring local density and indeed there are many methods on offer (Muldrew *et al.* 2012). The target density needs to be sufficiently high, as for the SAMI Galaxy Survey, to ensure groups and clusters are well sampled (Bryant *et al.* 2014b).

To examine survey parameters more qualitatively needs high-resolution cosmological simulations where individual galaxy properties are resolved. The necessary simulations are only now becoming available for the first time (e.g. Schaye *et al.* 2014). A proper survey estimate depends ultimately on what is being measured. Broad categories like fast vs. slow rotators may require smaller samples whereas weak measurables (e.g. Gauss-Hermite parameters h_3, h_4), or detailed properties defined by many parameters, push us to larger samples.

We have begun a study of the Eagle simulations through synthetic IFS observations of the stars and gas in each evolving galaxy (cf. Naab *et al.* 2013). There are basic properties that we would like to have, e.g. fractional variations in slow rotators with respect to fast rotators as a function of environment. Fogarty *et al.* (2014) have shown that, in the SAMI survey, the ratio of slow to fast rotators may vary across cluster environments. Importantly, Naab *et al.* (2013) argue the case for good spatial sampling to measure high order moments (h_3, h_4) of the stellar kinematics which may also vary in interesting ways over large-scale structure due to preserved kinematic subtructure.

In summary, the Hector survey will allow us to explore how galaxy properties vary across a contiguous volume of the universe. The huge database will provide information on galaxy properties, in particular, stellar and gas kinematics, the presence of nuclear activity (starburst vs. black hole), the star formation distribution, the spread of chemical abundance in both stars and gas, and so forth. The Vienna conference has underscored the broad community support for the the SAMI and Manga surveys which in turn build on the early successes of the ATLAS-3D and CALIFA surveys. Future galaxy studies will be dominated by IFS observations of cosmologically significant numbers of galaxies supported by large HI surveys.

Acknowledgements

JBH acknowledges an ARC Federation Fellowship and LIEF grant for the SAMI instrument and an ARC Australian Laureate Fellowship for the ongoing development of the Hector instrument. JBH acknowledges insights from members of the SAMI team, including S. Croom, J. Bryant, J. Allen, L. Fogarty, S. Richards and R. Sharp. We are indebted to J. Lawrence, the Anglo-Australian Observatory and the Anglo-Australian Telescope for their role in realising the SAMI project. My thanks to G. Cecil for producing Fig. 2 and to T. Naab for insights on where the EAGLE simulations can take us.

References

Allen, J. T., *et al.* 2014, *MNRAS* submitted, arXiv:1407.6068
Amram, P., Le Coarer, E., Marcelin, M., *et al.* 1992, *A&A Suppl.*, 94, 175
Bland, J., Taylor, K., & Atherton, P. D. 1987, *MNRAS*, 228, 595
Bland-Hawthorn, J., Bryant, J., Robertson, G., *et al.* 2011, *Optics Express*, 19, 2649
Bryant, J. J., *et al.* 2011, *MNRAS*, 415, 2173
Bryant, J. J., *et al.* 2014a, *MNRAS*, 438, 869
Bryant, J. J., Owers, M. S., Robotham, A. S. G., *et al.* 2014b, arXiv:1407.7335
Cappellari, M., *et al.* 2011, *MNRAS*, 416, 1680
Cappellari, M., McDermid, R. M., Alatalo, K., *et al.* 2012, *Nature*, 484, 485
Cecil, G. 1988, *ApJ*, 329, 38
Cecil, G., Fogarty, L., Bland-Hawthorn, J., *et al.* 2014, *ApJ*, submitted
Codis, S., Pichon, C., Devriendt, J., *et al.* 2012, *MNRAS*, 427, 3320
Colless, M., Dalton, G., Maddox, S., *et al.* 2001, *MNRAS*, 328, 1039
Cortese, L., Fogarty, L., Ho, I.-T., *et al.* 2014, *ApJ*, submitted
Croom, S. M., *et al.* 2012, *MNRAS*, 421, 872
Driver, S. P., *et al.* 2011, *MNRAS*, 413, 971
Dubois, Y., Pichon, C., Welker, C., *et al.* 2014, *MNRAS*, 444, 1453
Duffy, A. R., Meyer, M. J., Staveley-Smith, L., *et al.* 2012, *MNRAS*, 426, 3385
Fall, S. M. & Efstathiou, G. 1980, *MNRAS*, 193, 189
Fogarty, L. M. R., *et al.* 2012, *ApJ*, 761, 169
Fogarty, L. M. R., *et al.* 2014, *MNRAS*, 443, 485
Glazebrook, K. 2013, *PASA*, 30, 56
Goodwin, M., Heijmans, J., Saunders, I., *et al.* 2010, *SPIE*, 7739
Hill, G. J. 2014, *Advanced Optical Technologies*, 3, 265
Ho, I., Kewley, L. J., Dopita, M. A., *et al.* 2014, arXiv:1407.2411
Jones, D. H. & Bland-Hawthorn, J. 2001, *ApJ*, 550, 593
Lawrence, J., Bland-Hawthorn, J., Bryant, J., *et al.* 2012, *SPIE*, 8446
Mo, H., van den Bosch, F. C., & White, S. 2010, *Galaxy Formation and Evolution*, CUP
Muldrew, S. I., Croton, D. J., Skibba, R. A., *et al.* 2012, *MNRAS*, 419, 2670
Naab, T., Oser, L., Emsellem, E., *et al.* 2013, arXiv:1311.0284
Richards, S. N., Schaefer, A. L., Lopez-Sanchez, A. R., *et al.* 2014, arXiv:1409.4495
Röttgers, B., Naab, T., & Oser, L. 2014, arXiv:1406.6696
Sánchez S. F., *et al.* 2012, *A&A*, 538, A8
Schaye, J., Crain, R. A., Bower, R. G., *et al.* 2014, arXiv:1407.7040
Sharma, S., Steinmetz, M., & Bland-Hawthorn, J. 2012, *ApJ*, 750, 107
Sharp R., *et al.* 2014, *MNRAS* submitted, arXiv:1407.5237
Trowland, H. E., Lewis, G. F., & Bland-Hawthorn, J. 2013, *ApJ*, 762, 72
York D. G., *et al.* 2000, *AJ*, 120, 1579

Galaxies in 3D across the Universe
Proceedings IAU Symposium No. 309, 2014
B. L. Ziegler, F. Combes, H. Dannerbauer, M. Verdugo, eds.

© International Astronomical Union 2015
doi:10.1017/S1743921314009259

J-PAS : Low-resolution ($R \sim 50$) spectroscopy covering 8000 deg^2

C. López-Sanjuan, A. J. Cenarro, L. A. Díaz-García, D. J. Muniesa, I. San Roman, J. Varela, K. Viironen, and the J-PAS collaboration

Centro de Estudios de Física del Cosmos de Aragón, Plaza San Juan 1, 44001 Teruel, Spain
email: clsj@cefca.es

Abstract. We present the ambitious project J-PAS, that will cover 8000 deg^2 of the northern sky with 54 narrow-band ($\sim 145\mathring{A}$) contiguous filters, all of them in the optical range ($3700\mathring{A} - 9200\mathring{A}$). J-PAS will provide a low resolution spectra ($R \sim 50$) in every pixel of the northern sky by 2020, leading to excellent photometric redshifts (0.3% uncertainty) of 100 million sources. J-PAS will permit the study of the 2D properties of nearby galaxies with unprecedented statistics. Some viable studies are the distribution of the star formation rate traced by $H\alpha$, the stellar populations gradients in elliptical galaxies up to a few effective radii, or the impact of environment in galaxy properties. In summary, J-PAS will bring a superb data set for 3D analysis in the local Universe.

Keywords. galaxies: evolution - galaxies: formation

1. The J-PAS

The J-PAS† (Javalambre - PAU Astronomical Survey, Benítez *et al.* 2014) will map 8000 deg^2 of the northern hemisphere with 54 narrow-band ($\sim 145\mathring{A}$) filters from $3700\mathring{A}$ to $9200\mathring{A}$ and two extra broad-band filters at the blue and red ends. The depth in these 56 optical bands will be $\sim 22 - 23$ (5σ AB), with a detection limit of $r \sim 23.5$ (5σ AB) performed in a deep broad-band r image. The J-PAS will provide low-resolution ($R \sim 50$) spectra of a hundred million sources from the local Universe up to z = 3, leading to excellent photometric redshifts with a precision of $\Delta z/(1 + z) \sim 0.3\%$.

The J-PAS will be carried out with the 1.2 Gpixel camera JPCam (Taylor *et al.* 2014), mounted at the 2.55m Javalambre Survey Telescope (JST/T250) at the OAJ (Observatorio Astrofsico de Javalambre, Cenarro *et al.* 2012), a new astronomical facility located in the Pico del Buitre in Teruel, Spain. All the data produced by the OAJ will be reduced, processed, and stored at CEFCA with the UPAD (Unidad de Procesado y Archivo de Datos, Cristóbal-Hornillos *et al.* 2012). The J-PAS final catalogue will comprise 56 bands photometry of a hundred million sources.

J-PAS will address cosmological and galaxy evolution topics, and will complement integral field surveys (IFSs) in the local Universe, such as SAURON (de Zeeuw *et al.* 2002), CALIFA (Sánchez *et al.* 2012), or MaNGA (P. I.: K. Bundy).

2. The 2D study of galaxy properties with J-PAS

The main goals of J-PAS in the local Universe are (i) the 2D study of stellar populations in nearby elliptical galaxies (Fig. 1), (ii) explore the 2D star-formation properties of emission lines galaxies (Fig. 2), and (iii) constrain the impact of environment in the previous items (Fig. 3).

---

† www.j-pas.org

30 C. López-Sanjuan *et al.*

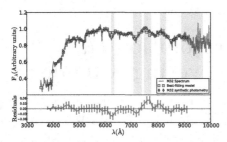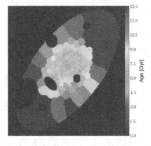

Figure 1. *Left panel*: M32 through the J-PAS glasses (dots) and the best fitting with two stellar populations (squares). The bands mark telluric and sky lines. *Right panel*: The 2D age of a local elliptical galaxy ($z = 0.075$) measured with the ALHAMBRA survey (Moles *et al.* 2008). The outskirts of the galaxy are older than the inner parts.

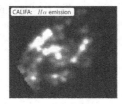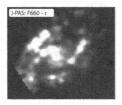

Figure 2. SDSS (Sloan Digital Sky Survey, Aihara *et al.* 2011) colour image of UGC09476 (*left panel*). The $H\alpha$ emission of UGC09476 shown by CALIFA (*central panel*) and the expectation for J-PAS with a narrow vs broad–band image (*right panel*). Relevant information about $H\alpha$, $H\beta$, [OIII], and [OII] lines will be present in the J-PAS photo–spectra.

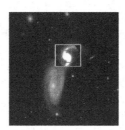

Figure 3. J-PAS will permit the detailed study of environmental effects in galaxy properties. As example, we shown three CALIFA galaxies with close companions. The red squares mark the field-of-view of the SAURON instrument. The coloured images are from SDSS.

Acknowledgements

This work has been mainly funding by the FITE (Fondos de Inversiones de Teruel) and the projects AYA2012-30789, AYA2006-14056, and CSD2007-00060.

References

Aihara, H., Allende Prieto, C., An, D., *et al.* 2011, *ApJS*, 193, 29
Benítez, N., Dupke, R., Moles, M., *et al.* 2014, [ArXiv:1403.5237]
Cenarro, A. J., Moles, M., Cristóbal-Hornillos, D., *et al.* 2012, in SPIE CS, Vol. 8448
Cristóbal-Hornillos, D., Gruel, N., Varela, J., *et al.* 2012, in SPIE CS, Vol. 8451
de Zeeuw, P. T., Bureau, M., Emsellem, E., *et al.* 2002, *MNRAS*, 329, 513
Moles, M., Benítez, N., Aguerri, J. A. L., *et al.* 2008, *AJ*, 136, 1325
Sánchez, S. F., Kennicutt, R. C., Gil de Paz, A., *et al.* 2012, *A&A*, 538, A8
Taylor, K., Marín-Franch, A., Laporte, R., *et al.* 2014, JAI, 3, 50010

Galaxies in 3D across the Universe
Proceedings IAU Symposium No. 309, 2014
B. L. Ziegler, F. Combes, H. Dannerbauer, M. Verdugo, eds.
© International Astronomical Union 2015
doi:10.1017/S1743921314009260

On Schmidt's Conjecture and Star Formation Scaling Laws

Charles J. Lada

Harvard-Smithsonian Center for Astrophysics,
60 Garden Street, Cambridge, MA 02138 USA
email: clada@cfa.harvard.edu

Abstract. Ever since the pioneering work of Schmidt a half-century ago there has been great interest in finding an appropriate empirical relation that would directly link some property of interstellar gas with the process of star formation within it. Schmidt conjectured that this might take the form of a power-law relation between the rate of star formation (SFR) and the surface density of interstellar gas. However recent observations suggest that a linear scaling relation between the total SFR and the amount of dense gas within molecular clouds appears to be the underlying physical relation that most directly connects star formation with interstellar gas from scales of individual GMCs to those encompassing entire galaxies both near and far. Although Schmidt relations are found to exist within local GMCs, there is no Schmidt relation observed between GMCs. The implications of these results for interpreting and understanding the Kennicutt-Schmidt scaling law for galaxies are discussed.

Keywords. galaxies: star formation – stars: formation

1. Introduction

One of the fundamental challenges confronting modern day astrophysics is the determination of the physical processes that control the conversion of interstellar gas into stars. Understanding of this process is a key to deciphering both star formation and galaxy evolution across cosmic time. A little more than half a century ago Schmidt (1959) conjectured that "It would seem probable that the rate of star formation depends on the gas density and...varies with a power-law of the gas density." Or as later parameterized by Kennicutt (1998): $\Sigma_{SFR} = \kappa \Sigma_{gas}^n$, where Σ_{SFR} and Σ_{gas} are *surface* densities of the respective quantities. Based on a number considerations, including the local distributions over z of B stars and HI gas, Schmidt argued that such a scaling relation existed in the solar neighborhood and that n $\approx$ 2. He also noted that "It is rather tempting to try to estimate the effects of star formation...in galaxies as a whole." However, it would take more than four decades before astronomers could adequately test Schmidt's idea in galaxies. It turns out that it was not so easy to determine accurate star formation rates for whole galaxies when what one actually measures is a flux of the galaxy at one or perhaps a number of different wavelengths. The solution to this problem required the development of population synthesis modeling in order to reasonably predict star formation rates from observed fluxes. It wasn't until the late 1970s and early 1980s that technology and astrophysical knowledge of such issues as stellar evolution, stellar spectra, stellar IMFs, dust, etc. reached the necessary level to enable meaningful determinations of star formation rates in galaxies using population synthesis calculations. Moreover, it wasn't until the late 1970s that the molecular gas content in galaxies could be measured using CO and thus provide a complete census of both HI and H_2 gas surface densities for comparison with the galaxy-wide SFRs. By 1989, nearly 30 years after Schmidt first published his conjecture, observations of blue stars, HII regions and gas surface densities

C. J. Lada

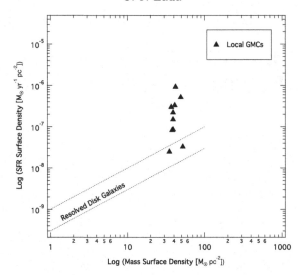

Figure 1. The Kennicutt-Schmidt diagram for nearby GMCs. There is no Schmidt scaling relation between local GMCs. The parallel dotted lines represent the approximate locus of (resolved) disk galaxies which can be described by a linear scaling law in the Kennicutt-Schmidt plane (Schruba *et al.* 2011). GMC data from Lada *et al.* (2010). See text.

confirmed the existence of Schmidt-like relations for disk galaxies but with a large dispersion in the value of the power-law index, $0 \leqslant n \leqslant 4$ (Kennicutt 1989 and references therein). Another decade later it became possible for Kennicutt (1998) to put together existing observations and models and provide a more robust test of Schmidt's conjecture for whole galaxies. In his classic paper he showed that a power-law scaling relation (now known as the Kennicutt-Schmidt law) exists between star forming galaxies and is furthermore characterized by an index, n = 1.6. Twenty years later observational capabilities improved to the point where it was possible to investigate the Kennicutt-Schmidt law *within* individual galaxies on kpc spatial scales. In an influential paper, Bigiel *et al.* (2008) produced a Kennicutt-Schmidt diagram that combined resolved observations of local disk galaxies with more global observations of local starburst galaxies. They found a more complex scaling between Σ_{SFR} and Σ_{gas} that could not simply be described by a single power-law (see Figure 5). Given this development it is interesting to consider what the Schmidt relation would be if we could make measurements on smaller, sub-kpc scales, indeed, even on the spatial scales of giant molecular clouds (GMCs) themselves.

2. The Schmidt Law on GMC scales

Over the last decade wide-field infrared surveys from the ground and space have made it possible to obtain both accurate masses of local molecular clouds (from infrared extinction measurements) and nearly complete inventories of the populations of young stellar objects (YSOs) contained within these clouds (from both ground and space-based infrared, optical and x-ray observations). Furthermore, ages and masses of these YSOs can be derived from their positions on the HR diagram. And, for clouds within 0.5 kpc of the sun, the stellar IMF is sampled over its entire range, from OB stars down to the hydrogen burning limit. Using these data fairly accurate determinations of cloud sizes, masses and star formation rates (SFRs) are possible. Combining these measurements, the individual GMCs can all be placed on the Kennicutt-Schmidt (KS) diagram. Figure 1 shows the result of this exercise for a nearly complete sample of molecular clouds within 0.5 kpc

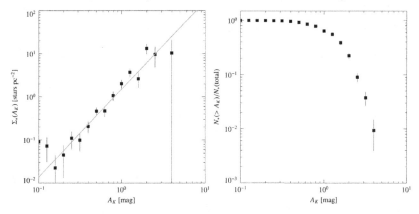

Figure 2. The left panel shows that a Schmidt scaling relation does exist within a single GMC, e.g., the Orion A cloud. Here $\Sigma_* \propto A_V{}^{2.0}$, where A_V is the dust surface density, a proxy for Σ_{H_2}. The right panel shows the cumulative protostellar fraction (or fractional yield of star formation) in the cloud as a function of A_V. From Lada *et al.* (2013).

of the sun (Lada *et al.* 2013). Clearly, there is *no* Schmidt scaling relation between local GMCs. This result is perhaps surprising at first look, but it is in fact a consequence of the well know scaling relation between mass and radius ($\mathrm{M_{GMC}} \propto \mathrm{R^2}$) of Milky Way (MW) GMCs, discovered over three decades ago by Larson (1981). Yet, in the context of the observed KS relation for disk galaxies (e.g., Schruba *et al.*, 2011), the lack of a Schmidt relation for GMCs is nonetheless intriguing.

3. The Schmidt Law on sub-GMC scales

Because local GMCs are so well resolved by observations, it is interesting to next consider whether a Schmidt scaling relation exists *within* clouds (i.e., on sub-GMC or pc scales). Indeed, a series of recent observational studies have demonstrated that a Schmidt scaling relation does exist within local clouds (Heidermann *et al.* 2010, Gutermuth *et al.* 2011, Lombardi *et al.* 2013, Lada *et al.* 2013, Evans *et al.* 2014). As an example, Figure 2 (left panel) shows the Schmidt relation for the Orion A GMC derived by Lada *et al.* (2013) from highly resolved observations of the protostellar surface density, Σ_*, and extinction, A_V (i.e., dust surface density). These two quantities are directly proportional to Σ_{SFR} and Σ_{H_2}, respectively. Here a fit to the data gives $\Sigma_* \propto A_V{}^{2.0}$. The unambiguous presence of a Schmidt scaling relation on sub-GMC scales sharply contrasts with the unambiguous lack of such a relation on GMC scales. How can one reconcile these different results derived using the same set of observations?

Protostars are the youngest, least evolved YSOs and can be thought of as the instantaneous product of the star formation process. The right panel of Figure 2 shows the (cumulative) fractional number of protostars plotted as a function of extinction in the Orion GMC. Despite the fact that Σ_* rises so steeply with extinction, the actual number of protostars produced by the cloud sharply decreases with extinction. This is equivalent to the statement that although Σ_{SFR} increases non-linearly with extinction, the overall SFR decreases sharply with extinction. This seemingly paradoxical situation is resolved when one considers that, 1) the total star formation rate is given by: SFR = $\int \Sigma_{SFR} dS$, the integral of the surface density over the surface area, S, and 2) the surface area, $S(A_K)$, of a cloud sharply decreases with increasing extinction (Lada *et al.* 2013). Indeed, it is the coupling of these two functions, Σ_{SFR} (A_K) and $S(A_K)$ that sets the global star formation rate in a cloud. Clearly, cloud structure plays a fundamental role in

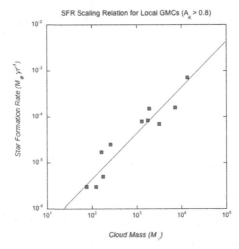

Figure 3. The scaling relation between (integrated) SFR and high extinction, presumably dense, gas mass for local GMCs. Adapted from Lada *et al.* (2010). See text.

determination of its star formation rate. In this context the Schmidt scaling relation can be considered a differential scaling relation and is, by itself, not a reliable predictor of the level of star formation in a cloud. Two clouds of the same mass and size and characterized by the same Schmidt scaling relation will not necessarily produce similar levels of star formation. Since most clouds seem to be characterized by Schmidt relations with similar power-law indices (n $\approx$ 2; Lada *et al.* 2013) observed differences in their SFRs (Figure 1) are primarily the result of differences in their surface area distribution functions, $S(A_K)$.

4. Dense Gas and Integral Scaling Laws

Close inspection of the right panel in Figure 2 shows that although the fractional number of protostars sharply decreases with extinction, 80% of all protostars are found at extinctions in excess of $A_K \approx 0.8$ magnitudes. More recent observations using higher angular resolution Herschel observations (Lombardi *et al.* 2014) to map dust column density show that 90% of all protostars in Orion lie above this extinction level. This suggests that star formation is intimately related to high opacity and presumably high (volume) density material. Indeed, for local clouds there is a relatively tight scaling relation between the mass of high opacity gas and the global SFR (Lada *et al.* 2010). This (integral) scaling relation is shown in Figure 3, where the global SFR is plotted against the mass of the cloud contained above an extinction threshold of $A_K = 0.8$ magnitudes. Unlike typical Schmidt relations, this scaling follows a linear relation, i.e., n = 1. This has been interpreted to mean that the SFR in a cloud is simply controlled by the amount of high extinction (and presumably dense) material contained within it. A similar linear scaling between the total gaseous mass in a cloud and its global SFR has also been found to exist (Lada *et al.* 2012), but it is characterized by a significantly higher dispersion (an order of magnitude in SFR). The increased dispersion in the latter relation compared to that of the dense gas scaling relation has been shown to be entirely due to the differing dense gas fractions in the clouds.

The linear scaling between dense gas mass and the SFR in local clouds is reminiscent of a similar global relation found for galaxies by Gao and Solomon (2004) who showed that FIR luminosities of galaxies (including disk galaxies and nuclear starbursts) were linearly correlated with the luminosities of HCN molecular-line emission, a tracer of dense (n_{H_2}

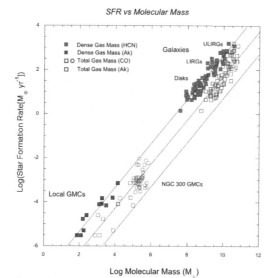

Figure 4. The integral scaling relation (SFR vs M(H_2)) for GMCs and galaxies for both dense gas (filled symbols) and total gas (open symbols) masses. The dotted parallel lines represent lines of constant depletion times corresponding to 22 Myr, 220 Myr and 2.2 Gyr, from top to bottom, respectively, or alternatively, constant dense gas fractions of 100%, 10% and 1%, respectively. See text.

$> 10^4$ cm^{-3}) molecular gas. Subsequent observations by Wu *et al.* (2005) comparing FIR and HCN luminosities of massive GMCs in the Milky Way also showed a linear correlation between the two quantities and this relation was found to extrapolate smoothly to that found by Gao and Solomon for galaxies, spanning a range in scale of over nine orders of magnitude. This can be seen in Figure 4 where the integral scaling relations (i.e., SFR vs M(H_2)) for both dense gas and total mass measurements of MW GMCs (Lada *et al.*2010) and galaxies (Gao & Solomon 2004) are presented on the same plot. Also shown are recent observations of GMCs in the nearby galaxy NGC 300, obtained on a $\sim$ 200 pc spatial scale (Faesi *et al.* 2014). Gas masses were obtained from extinction measurements for MW GMCs and from CO (total) and HCN (dense) gas measurements for the extragalactic objects. SFRs were determined from direct counting of YSOs in local GMCs and from population synthesis analyses for the extragalactic sources with appropriate calibrations (Chomiuk & Povich 2011, Lada *et al.* 2012, Faesi *et al.* 2014).

Qualitatively, the relationship between SFR scaling relations for dense and total gas masses found in local GMCs also appears to characterize the observations of galaxies. This similarity suggests that we are observing a similar physical process in star forming environments across all spatial scales. In particular, the linear scaling between SFR and dense gas mass suggests that the rate of star formation is directly controlled by the amount of dense gas that can be assembled in a star forming region.

The scaling relation between SFR and total cloud masses is also seemingly linear from GMCs to disk galaxies but appears to depart from this linearity at the highest SFRs, in the region occupied by active starbursts (i.e., ULIRGs). Lada *et al.* (2012) argued that this departure from a single linear relation was the result of the fact that the dense gas fractions in these starburst galaxies are systematically higher than those that characterize GMCs and disk galaxies. They postulated the existence of a family of linear scaling laws between SFRs and total gas masses that are parameterized by the dense gas fraction in the star forming material. In other words, the star formation scaling law is always

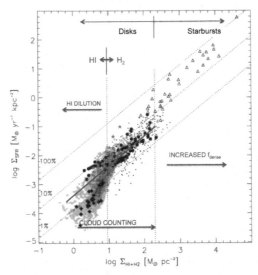

Figure 5. Interpreting the Kennicutt-Schmidt diagram for nearby disk and starburst galaxies. Figure adapted from Bigiel *et al.* (2008). The solid brown line approximately represents the linear relation between Σ_{SFR} and Σ_{H_2} from $\sim 1\ M_\odot\ \mathrm{pc}^{-2}$ to $\sim 100\ M_\odot\ \mathrm{pc}^{-2}$ found by Schruba *et al.* (2011). See text for discussion.

linear for systems with the same dense gas fraction. It is interesting to note that even galaxies at high z (> 2-4) are characterized by similar scaling relations when FIR and CO luminosities are compared (Genzel *et al.* 2010, Shapley 2011) further suggesting that the physical process of star formation in distant galaxies and even through much of cosmic time may be similar in some fundamental respect to that we are observing today in local GMCs.

5. The Nature of the Kennicutt-Schmidt Law for Galaxies

Determinations of the KS relation for disk galaxies likely consist of measurements of star forming GMCs and GMC complexes on spatial scales from 1 - 10 kpc. These measurements clearly display power-law scaling relations. How is this reconciled with the observations in Figure 1 that show no Schmidt relation for MW GMCs? Figure 5 primarily shows resolved observations of disk galaxies in the KS plane (Σ_{SFR} vs Σ_{gas}) from Bigiel *et al.* (2008). Unlike the situation for global measurements of such galaxies (Kennicutt 1998), a single power-law relation cannot adequately match the data. The relation is more complex and perhaps could be characterized by multiple (3) power-law relations. In order to interpret this diagram, we note that for $\Sigma_{gas} < 10\ M_\odot\ \mathrm{pc}^{-2}$, the gas in the disks is predominately atomic (HI) in form, and above this threshold the gas is almost exclusively in molecular (H$_2$) in form. Star formation occurs exclusively in molecular gas, thus HI gas is inert to star formation. The combination of HI and H$_2$ gas in the measurement of Σ_{gas} tends to dilute the star formation signal in disks dominated by HI, (with $\Sigma_{gas} < 10\ \mathrm{pc}^{-2}$). This artificially steepens the slope between Σ_{SFR} and Σ_{gas} . The functional form of the scaling relation in this part of the diagram provides no insight into the actual physical process of star formation operating in the molecular gas.

For $\Sigma_{gas} > 10\ M_\odot\ \mathrm{pc}^{-2}$, the scaling relation appears to be linear up to $\approx 100\ M_\odot$ pc^{-2}. Using an improved analysis of the data in which the molecular observations of the galaxies were stacked to improve the sensitivity, Schruba *et al.* (2011) showed that if only CO observations are considered, this linear relation extends down to $\Sigma_{H_2} \approx 1\ M_\odot\ \mathrm{pc}^{-2}$.

However, these observations are not consistent with the fact that GMCs are thought to be characterized by a constant column density of $\sim$ 40 - 50 $M_\odot$ pc^{-2} (Figure 1 and Larson 1981). Moreover, a molecular cloud cannot exist if $\Sigma_{H_2} < 16$ $M_\odot$ pc^{-2}, the threshold surface density required for self-shielding against UV photo-dissociation. Consequently, the vast majority of the (CO derived) measurements of Σ_{H_2} likely represent beam diluted observations. Consider that in a 1 kpc square pixel a single GMC would be characterized by a filling factor, f $\sim$ 0.002. Consequently, the observed linear scaling relation between $1 - 100$ $M_\odot$ pc^{-2} likely represents a cloud-counting sequence which measures the surface density of GMCs rather than the surface density of gas within them. As a result, the linear KS relation between 1 - 100 $M_\odot$ pc^{-2} reveals little about physics of the underlying process of star formation within the molecular gas.

In the portion of the KS diagram for which $\Sigma_{gas} > 100$ $M_\odot$ pc^{-2} there appears to be a distinct break and jump to higher Σ_{SFR} followed by a more or less linear scaling between Σ_{SFR} and Σ_{gas}. These observations represent global observations of primarily starburst galaxies. We interpret this shift as being a consequence of an increased dense gas fraction in these objects. Consider that in typical starburst nuclei, the mass of molecular gas can be similar to or larger than that contained in the entire disk of a normal star forming spiral galaxy (see Figure 4), but confined to a region only $\sim$ 1 kpc in size. The geometric decrease in surface area alone would be enough to significantly increase Σ_{SFR} and Σ_{gas} compared to normal galaxies. In addition the mean volume density of the material must also be significantly enhanced which in turn would lead to increases in the global SFRs for the molecular gas in these systems compared to that in GMCs. Interestingly, observations suggest that the molecular gas is still beam diluted by observations on these scales, but this likely cannot account fully for the linear nature of the KS scaling in this regime. The molecular gas in starbursts is likely not organized in units similar to galactic GMCs, and the basic physical properties and nature of this material are likely much more complex than those of local GMCs. Consequently, the nature of the Σ_{SFR} - Σ_{gas} relation in this part of the KS diagram still awaits an explanation. Nonetheless, the considerations discussed in the previous section do suggest that even in these extreme conditions the role dense gas may still be the critical factor in controlling the SFR.

However, a discrepancy remains between the local cloud and extragalactic observations. Namely, the gas depletion times ($\sim$ 220 - 250 Myrs) derived for local GMCs (and GMCs in NGC 300) differ significantly from those derived on 1 kpc scales of local disk galaxies (2-3 Gyr, e.g., Biegiel *et al.*2008). The depletion times derived for local GMCs are likely robust suggesting that either: 1) the extragalactic SFR calibrations underestimate the actual SFRs or 2) in addition to discrete GMCs, the extragalactic observations include diffuse molecular gas that is not participating in star formation or 3), some combination of these two effects. This issue clearly warrants more attention.

6. Conclusions

Comprehensive observations of nearby GMCs indicate that a Schmidt scaling relation of the form $\Sigma_{SFR} \propto \Sigma_{H_2}^n$, with n $\approx$ 2, exists *within* local molecular clouds. However, the observations also indicate that this relation alone does not provide a complete description of a cloud's star forming activity. Instead, the structure of a cloud is found to play a pivotal role in setting its global SFR. Moreover, observations clearly indicate that there is (and can be) *no* Schmidt relation connecting Σ_{SFR} and Σ_{gas} between GMCs, contrary to the case for galaxies. However, the integrated SFR is found to scale *linearly* with, and is most reliably traced by, the mass of *dense* gas in star forming GMCs. Moreover, this scaling relation appears to smoothly extend to global measurements of galaxies suggesting that the amount of dense gas in star forming regions sets the SFR in systems ranging from individual GMCs to entire galaxies. Deconstruction of the (resolved) KS law for local disk galaxies indicates that its functional form is largely a result of two factors: the unresolved observations of GMCs and the diluting effects of the inclusion of "inert" HI gas in measurements of Σ_{gas}. Consequently, the KS relation provides little physical insight into the underlying process of star formation characterizing the molecular gas within galaxies. Nonetheless, indications are that the physics of star formation in distant galaxies and even through much of cosmic time may be similar in some fundamental respects to that we are witnessing today in local GMCs.

Acknowledgements

I acknowledge the dedicated efforts of Joao Alves, Chris Faesi, Jan Forbrich, Marco Lombardi and Carlos Roman-Zuniga who made significant contributions to much of the research reviewed here.

References

Bigiel, F., Leroy, A., Walter, F., *et al.* 2008, *AJ*, 136, 2846
Chomiuk, L. & Povich, M. S. 2011, *AJ*, 142, 197
Evans, N. J., II, Heiderman, A., & Vutisalchavakul, N. 2014, *ApJ*, 782, 114
Faesi, C. M., Lada, C. J., Forbrich, J., Menten, K. M., & Bouy, H. 2014, *ApJ*, 789, 81
Gao, Y. & Solomon, P. M. 2004, *ApJ*, 606, 271
Genzel, R., Tacconi, L. J., Gracia-Carpio, J., *et al.* 2010, *MNRAS*, 407, 2091
Gutermuth, R. A., Pipher, J. L., Megeath, S. T., *et al.* 2011, *ApJ*, 739, 84
Heiderman, A., Evans, N. J., II, Allen, L. E., Huard, T., & Heyer, M. 2010, *ApJ*, 723, 1019
Kennicutt, R. C., Jr. 1989, *ApJ*, 344, 685
Kennicutt, R. C., Jr. 1998, *ARA&A*, 36, 189.
Lada, C. J., Lombardi, M., & Alves, J. F. 2010, *ApJ*, 724, 687
Lada, C. J., Forbrich, J., Lombardi, M., & Alves, J. F. 2012, *ApJ*, 745, 190
Larson, R. B. 1981, *MNRAS*, 194, 809
Lombardi, M., Lada, C. J., & Alves, J. 2013, *A&A*, 559, A90
Lombardi, M., Bouy, H., Alves, J., & Lada, C. J. 2014, *A&A*, 566, A45
Schmidt, M. 1959, *ApJ*, 129, 243
Schruba, A., Leroy, A. K., Walter, F., *et al.* 2011, *AJ*, 142, 37
Shapley, A. E. 2011, *ARA&A*, 49, 525
Wu, J., Evans, N. J., Gao, Y., Solomon, P. M., Shirle, Y. L., & Vanden Bout, P. A. 2005, *ApJ*, 635, L173

Galaxies in 3D across the Universe
Proceedings IAU Symposium No. 309, 2014
B. L. Ziegler, F. Combes, H. Dannerbauer, M. Verdugo, eds.

© International Astronomical Union 2015
doi:10.1017/S1743921314009272

The Local Universe: Galaxies in 3D

Bärbel S. Koribalski

CSIRO Astronomy and Space Science, Australia Telescope National Facility,
P. O. Box 76, Epping, NSW 1710, Australia
email: Baerbel.Koribalski@csiro.au

Abstract. Here I present results from individual galaxy studies and galaxy surveys in the Local Universe with particular emphasis on the spatially resolved properties of neutral hydrogen gas. The 3D nature of the data allows detailed studies of the galaxy morphology and kinematics, their relation to local and global star formation as well as galaxy environments. I use new 3D visualisation tools to present multi-wavelength data, aided by tilted-ring models of the warped galaxy disks. Many of the algorithms and tools currently under development are essential for the exploration of upcoming large survey data, but are also highly beneficial for the analysis of current galaxy surveys.

Keywords. galaxies: individual (M 83), evolution & formation, radio surveys — technology: interferometry, wide-field phased array feeds

1. Introduction

Analysing the well-resolved stellar and gas kinematics of disk galaxies provides insights into their rotational and non-rotational components, both of which can be measured as a function of radius and disk height. The resulting rotation curves reflect the overall mass distribution of galaxies (eg., Bosma 2004), including their very large dark matter halos, which extend well beyond the observed disks (e.g., Jones *et al.* 1999; Warren *et al.* 2004; Kreckel *et al.* 2011; Koribalski & Lopez-Sanchez 2009; Westmeier *et al.* 2011, 2013). High-resolution spectroscopic data cubes (targeting, eg., the H I, CO, and Hα spectral lines) also allow us to determine the 3D shape of galaxies, including their warped disks. The various shapes of 3D models derived with TiRiFiC, the *Tilted Ring Fitting Code*, are nicely illustrated by Józsa (2007). I then construct 3D visualisations of the observed galaxies based on multi-wavelength imaging and spectral line cubes at various angular resolutions, allowing me to analyse and improve the 3D representation of each galaxy until it reflects the observed data and knowledge derived from the data. Another benefit of the re-constructed 3D particle model is the ability to view each galaxy from any angle and produce fly-through movies (as presented for the galaxy M 83). In this first attempt, I used high-resolution optical and ultraviolet images as well as H I spectral line cubes from the Australia Telescope Compact Array (ATCA).

Optical galaxy surveys with ever more powerful integral fields units (IFUs), such as ATLAS-3D (Cappellari *et al.* 2011), CALIFA (Sánchez *et al.* 2012) and SAMI (Allen *et al.* 2014), are now delivering high-resolution data cubes. Radio synthesis telescopes like the ATCA, which consists of six 22-m dishes, have been recording large spectral line data cubes for over two decades. The ATCA primary beam is ∼0.5 degr at 21-cm; H I emission from nearby galaxies within this large field of view can be mapped with angular resolutions of up to ∼10 arcsec and velocity resolution of less than 1 km s^{-1}. Much higher resolutions are typically used when observing molecular lines (ATCA receivers cover frequencies from 1 to 105 GHz observable with two 2 GHz-wide bands; Wilson *et al.* 2011).

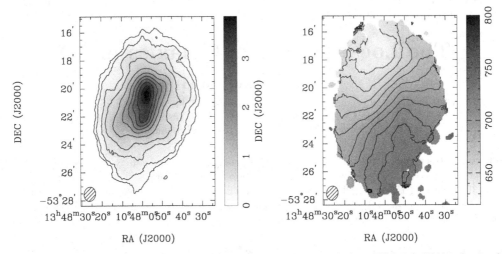

Figure 1. ATCA H I moment maps of the low-surface brightness galaxy ESO174-G?001 obtained as part of the *Local Volume HI Survey* (LVHIS) project (Koribalski 2010). The H I distribution (left) extends far beyond the stellar disk, while the mean H I velocity field (right) shows a regular, but rather twisted rotation pattern.

ASKAP, the Australian Square Kilometre Array Pathfinder (Johnston *et al.* 2008), which will be fully equipped with novel wide-field phased array feeds (PAFs), is just starting to deliver the first well-resolved H I maps of nearby galaxies and groups. We are currently observing with six of the 36 antennas, forming nine beams (each with FWHM ∼1 degr) which can be placed anywhere within the 30 sq degr field-of-view. After successful ASKAP H I mapping the the Sculptor Group galaxies NGC 253 and NGC 247 plus the area in between this wide pair, we are now imaging the nearby gas-rich galaxy group IC 1459 and its environment (Serra *et al.* 2015). The recording bandwidth is 300 MHz divided into around 17 000 channels giving a velocity resolution of 4 km s^{-1}. The angular resolution of the current 6-antenna array, known as the Boolardy Engineering Test Array (BETA; Hotan *et al.* 2014) is ∼1 arcmin; the frequency coverage is 700 to 1800 MHz.

ASKAP is a powerful 21-cm survey machine and will — once fully equipped with the equivalent of wide-field IFUs — be mapping the neutral hydrogen line in emission and absorption over the whole southern sky and a good fraction of the northern sky. Several large ASKAP H I surveys are planned. Here I briefly introduce WALLABY, a 21-cm survey of the sky ($\delta < +30°$; $z < 0.26$) which in about one year of observing time will detect more than 500 000 galaxies in the H I spectral line (Duffy *et al.* 2012, Koribalski 2012b), ie. a factor ∼20 more than currently catalogued. Novel Chequerboard PAFs are providing 30 sq degr instantaneous field-of-view. In WALLABY ∼1000 galaxies will have H I diameters larger than 5 arcmin (>10 beams) and ∼5000 galaxies will have major-axis H I diameters greater than 2.5 arcmin (>5 beams), allowing us to study in detail their morphology, kinematics and mass distribution. The number would rise to 1.6×10^5 galaxies if all 36 ASKAP antennas could be used; the additional six antennas provide baselines up to 6 km, resulting in an angular resolution of 10″. Creating highly reliable and complete source catalogs requires sophisticated source-finding algorithms as well as accurate source parametrisation. We are aiming to achieve this with our new *Source Finding Application* (SoFiA; Serra *et al.* 2014) and H I profile fitting with the versatile "Busy Function" (Westmeier *et al.* 2014). For an overview on continuum and spectral line source finding see the *PASA Special Issue* (Koribalski 2012a).

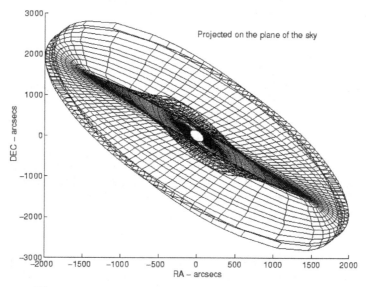

Figure 2. Example 3D model of a warped spiral galaxy.

2. Kinematic Modelling

A common way of analysing a galaxy's H I velocity field is to carefully fit tilted ring models (de Blok *et al.* 2008, Oh *et al.* 2010). The resulting residual velocity field is a good indicator of the goodness of fit and any remaining peculiar (non-rotational) structures. Tilted ring fits retrieve structural parameters, such as position angle, inclination and rotational velocity, as a function of radius when a galaxy is moderately well resolved. Several well-tested algorithms/programs exist, most of which are applied to the 2D velocity field. In some cases, for example edge-on disks and large well-resolved H I disks of galaxies, 3D modelling of the H I data cubes is of advantage (eg., Kamphuis *et al.* 2013).

The WALLABY kinematics working group (led by Kristine Spekkens) has started to create a modular package that allows to input spectral data cubes of individual galaxies, calculate H I velocity fields (eg., as shown in Fig. 1) and fit them with one or more of the algorithms described below. We base this approach on our successful SoFiA package (Serra *et al.* 2014). Assuming circular symmetry, the fitted rings allow an accurate description of warped galaxy disks where both inclination, $i(r)$, and position angle, $PA(r)$, vary with radius r. This is necessary to correctly deproject the measured rotational velocities to derive the galaxy rotation curve, $v_{rot}^i(r)$. Rogstad *et al.* (1974) introduced the tilted-ring model fitting, which is widely used through the *Rotcur* program and applied to 2D velocity fields. *TiRiFiC* uses spectroscopic data cubes to obtain tilted ring model fits of galaxies; it is now available as a stand-alone routine. This approach is essential for galaxy disks seen edge-on and allows sophisticated modelling of the disk thickness and surface brightness distribution (Kamphuis *et al.* 2013, 2014). *DiskFit* was created for the kinematic modeling of barred spiral galaxies. It does not allow the fitting of warped disk where the inclination and position angles vary with radius. DiskFit buids on Velfit (Spekkens & Sellwood 2007). The *Kinemetry* technique (Krajnovic *et al.* 2006) was devised to model the stellar velocity fields of early-type galaxies.

3. The Spiral Galaxy M 83

In the following I will use the grand-design spiral galaxy as an example of 3D H I mapping. The H I envelope of M 83 is much more extended than the well-known stellar disk. A prominent tidal arms hints at gravitational interactions with its dwarf galaxy neighbours while the western edge of the gas appears compressed, possibly affected by ram pressure stripping. The H I velocity field shows both the rotation and the changing orientation of the gaseous disk. By modelling the disk we are able to reconstruct the true 3D shape of a galaxy and place it within the group volume. In Section 5 I give more details on the 3D visualisation of stars and gas using the ray-tracing software Splotch. Currently, we can model several hundred galaxies in this way. Future large-scale ASKAP H I surveys will allow us to do this for several thousand galaxies, obtaining a true 3D dynamic picture of the nearby Universe.

Our large-scale H I mosaic of the beautiful, grand design spiral galaxy M 83 (NGC 5236, HIPASS J1337–29), obtained by combining ATCA interferometric and Parkes single-dish data (see Fig. 3), reveals an enormous H I disk with a diameter of $\sim$80 kpc, several times larger than M 83's optical Holmberg diameter. Here we adopt a distance of $D = 4.5$ Mpc. While the inner disk of M 83 rotates remarkably regular, the H I gas dynamics appear increasingly peculiar towards the outer regions which show clear signs of tidal disruption. The most prominent tidal features of M 83 are the one-sided outer H I arm which can be traced over 180 degr from the western to the eastern side, and the spectacular stellar stream, consisting of mainly old stars, to the north; their origin and possible relation are explored. M 83 is surrounded by numerous dwarf galaxies and, given its dynamical mass of about 5×10^{11} M$_\odot$, is likely to attract and accrete them in regular intervals.

From a very deep optical image of M 83 by Malin & Hadley (1997) we estimate a stellar diameter of $\sim 22' \times 19'$ (ie, 29 kpc $\times$ 25 kpc) for the disk at a B-band surface brightness of $\sim$28 mag arcsec^{-2}. For comparison, the optical B_{25} diameter of M 83 is $12.9' \times 11.5'$ (de Vaucouleurs et $al.$ 1991). Beyond the faint stellar disk, the deep optical image reveals three peculiar features: a prominent north-western $stellar$ $stream$ at a distance of $\sim 18'$ from the center of M 83, a $small$ arc towards the north-east at a distance of $\sim 11'$, and a $southern$ $ridge$. GALEX FUV emission is clearly detected in the $southern$ $ridge$ as well as the western part of the $small$ arc which we identified in the deep optical and H I images of M 83. (see Fig. 3). The mean B-band surface brightness of the extended stellar stream to the north of M 83 is around 27 mag arcsec^{-2}. It covers an area of about 10 arcmin^2 or about 17 kpc^2. The H I distribution (0. moment) of M 83 is most remarkable. No longer does this grand-design spiral look regular and undisturbed. The H I maps show streamers, irregular enhancements, an asymmetric tidal arm, diffuse emission, and a highly twisted velocity field, much in contrast to its regular appearance in short-exposure optical images. M 83's H I mass (M$_{HI}$ = 7.8 ($\pm$0.5) $\times 10^9$ M$_\odot$) is more than twice that of an M$_{HI}^*$ galaxy. The H I appearance of M 83 clearly suggests that it has been and possibly still is interacting with neighbouring dwarf galaxies. The effect of this interaction on the dwarfs can of course be rather devastating; it is quite likely that M 83 has accreted dwarf galaxies in the past as the stellar and gaseous streams/tails.

The eastern-most H I emission of M 83 which forms part of its peculiar, outer arm lies $\sim 34.5'$ or 45 kpc away from the center of M 83. And the the dwarf irregular galaxy NGC 5264 lies at a projected distance of only 25.5' or 33 kpc from the eastern H I edge of M 83. For NGC 5264 we measure M$_{HI}$ = 6×10^7 M$_\odot$ and calculate a total dynamical mass of $\sim 1.2 \times 10^9$ M$_\odot$. We also detected the dwarf irregular galaxy UGCA 365 to the north of M 83, just outside its large H I envelope. We measure M$_{HI}$ = 1.5×10^7 M$_\odot$ and a total dynamical mass of $\sim 1.2 \times 10^9$ M$_\odot$.

Figure 3. Multi-wavelength image of the spiral galaxy M 83. The large neutral hydrogen (H I) disk, shown in blue, was mapped with the ATCA and extends well beyond the stellar disk. For the color composition we also used GALEX NUV+FUV (light blue), DSS *R*-band (green), and 2MASS *J*-band (red) images. (Image credit: Ángel R. López-Sánchez).

Figure 4. — (Left) Three of the 36 ASKAP antennas at the Murchison Radio astronomy observatory (MRO) in Western Australia. — (Right) The first Mk II PAF was installed on ASKAP antenna 29 in September 2014. (Image credits: Ant Schinckel, CSIRO).

For comparison, the H I disk of the Circinus galaxy has a diameter of $\sim 80'$ or ~ 100 kpc at an adopted distance of 4.2 Mpc (For, Koribalski & Jarrett 2012), $\sim 5\times$ larger than the extrapolated Holmberg diameter of $\sim 17'$ (Freeman *et al.* 1977).

4. SKA Pathfinder HI Surveys

We have come a long way since the discovery of the 21-cm spectral line by Ewen & Purcell in 1951. Nevertheless, detecting H I emission in a Milky Way-like galaxy at redshift $z = 1$ will require the Square Kilometre Array (SKA; Rawlings ?; Obreschkow *et al.* 2011). Several SKA pathfinder and precursor telescopes are currently testing new technologies and used to develop reliable data processing pipelines. Much of the data is too large to store long-term, so data processing, quality control, imaging and source finding has to be done as the raw data is being recorded.

Several tens of thousands galaxies have so far been detected in the 21-cm H I line, the vast majority with single dish radio telescopes (e.g., Meyer *et al.* 2004, Springob *et al.* 2005, Wong *et al.* 2006, Haynes *et al.* 2011). The intrinsic faintness of the electron spin-flip transition of neutral atomic hydrogen (rest frequency 1.42 GHz) makes it difficult to detect H I emission from individual galaxies at large distances. To study the H I content of galaxies and diffuse H I filaments between galaxies, we need radio synthesis telescopes with large collecting areas, low-noise receivers and large fields of view.

ASKAP consists of 36×12-m antennas and is located in the Murchison Shire of Western Australia (see Fig. 4). Of the 36 antennas, 30 are located within a circle of ~ 2 km diameter, while six antennas are at larger distances providing baselines up to 6 km. Six ASKAP dishes are currently equipped with first-generation (Mk I) Chequerboard PAFs. The instantaneous field-of-view of the ASKAP PAFs is 5.5 deg $\times$ 5.5 deg, ie. 30 square degrees, making ASKAP a 21-cm survey machine. The WSRT APERTIF upgrade employs Vivaldi PAFs, delivering a field-of-view of 8 square degrees (Verheijen *et al.* 2008).

WALLABY, the *Widefield ASKAP L-band Legacy All-sky Blind surveY* (led by me and Lister Staveley-Smith; see Koribalski *et al.* 2009), will cover 75% of the sky ($-90° < \delta < +30°$) over a frequency range from 1.13 to 1.43 GHz (corresponding to $-2000 < cz < 77,000$ km s^{-1}) at resolutions of $30''$ and 4 km s^{-1}. WALLABY will be carried out using the inner 30 antennas of ASKAP, which provide excellent uv-coverage and baselines up to 2 km. High-resolution ($10''$) ASKAP H I observations using the full 36-antenna array will require further computing upgrades. **WNSHS**, the *Westerbork Northern Sky HI Survey* (led by Guyla Józsa), will cover a large fraction of the northern sky ($\delta > +27°$) with APERTIF over the same frequency range as WALLABY with ASKAP. Both H I surveys combined will achieve a true all-sky survey with unprecedented resolution and depth. The science goals of both surveys are well developed and complement, as well as enhance, each other. For a summary see Koribalski (2012b). **WALLABY** and **WNSHS** are made possible by the development of phased array feeds, delivering a much larger field-of-view than single feed horns or multi-beam systems. WALLABY will take approximately one year (ie 8 hours per pointing) and deliver an rms noise of 1.6 mJy beam^{-1} per 4 km s^{-1} channel. WALLABY is a precursor for future H I surveys with SKA Phase I and II, exploring the role of atomic hydrogen in galaxy formation and evolution. Using the H I-detected galaxies over two thirds of sky we will be able to unveil their large-scale structures and cosmological parameters. For nearby galaxies, we can detect their extended, low-surface brightness disks as well as gas streams and filaments between galaxies.

5. 3D Visualisation of gas and stars in galaxies

Animations of several thousand HIPASS galaxies show their 3D distribution in the nearby Universe ($z < 0.03$), based on their measured positions, velocities / distances and H I masses (created by Mark Calabretta and available on-line). Large-scale structures such as the Supergalactic Plane and the Local Void (see Koribalski *et al.* 2004) are clearly visible. The resolution of HIPASS, the H I Parkes All Sky Survey, is 15.5 arcmin and 18 km s^{-1}. Once WALLABY data are in hand (resolution 30 arcsec and 4 km s^{-1}), we should be able to create animations of 500 000 galaxies as well as detailed 3D models of ∼5000 galaxies (ie, all HIPASS galaxies) by a 3D multi-wavelength rendering of the respective galaxy. Multi-wavelength images and spectral line data cubes of galaxies allow us to measure their stellar, gas and dark matter properties. Visualisation packages such as KARMA provide a range of tools to interactively view 2D and 3D data sets as well as apply mathematical operations. This allows not only the quick inspection and evaluation of multiple images, spectra and cubes, but also the production of beautiful multi-color images and animations. To improve our understanding of galaxy formation and evolution, we need to include models and theoretical knowledge together with observations of galaxy disks and halos. By fitting and modelling the observed gas distribution and kinematics of extended galaxy disks, we derive their 3D shapes and rotational velocities. Visualisation can then be employed to combine the actual data with our derived knowledge to re-construct the most likely 3D representation of each galaxy. By adding time as the fourth dimension one can also visualize the evolution of galaxies and the Universe (e.g., see the *4D Universe* visualisation by Dolag *et al.* 2008). SPLOTCH is a powerful and very flexible ray-tracer software tool which supports the visualisation of large-scale cosmological simulation data (Dolag *et al.* 2008, 2011; Jin *et al.* 2010). It is publicly available and continues to be enhanced. A small team is currently working on supporting the visualisation of multi-frequency observational data to achieve realistic 3D views and fly-throughs of nearby galaxies and galaxy groups.

Acknowledgements

I like to thank the conference organisers for inviting me to participate in such a fantastic conference in the beautiful city of Vienna. Furthermore, I like to thank my collaborators, in particular Claudio Gheller and Klaus Dolag from the 3D visualisation team, my wonderful WALLABY team, and my colleagues Peter Kamphuis and Tiffany Day who continue to improve the H I models of M 83. I gratefully acknowledge financial support from the IAU.

References

Allen, J. T., *et al.* 2014, *MNRAS*, in press
Bosma, A. 2004, in *Dark Matter in Galaxies*, IAU Symposium 220, p. 39
Cappellari, M., *et al.* 2011, *MNRAS*, 413, 813
de Blok, W. J. G., *et al.* 2008, *AJ*136, 2648
Duffy, A., *et al.* 2012, *MNRAS*, 426, 3385
For, B.-Q., Koribalski, B. S., & Jarrett, T. H. 2012, *MNRAS*, 425, 1934
Freeman, K., *et al.* 1977,
Hotan, A., *et al.* 2014, *PASA*, in press (arXiv:1409.1325)
Johnston, S., *et al.* 2008, *Experimental Astronomy* 22, 151
Jones, K. L., Koribalski, B. S., Elmouttie, M., & Haynes, R. F. 1999, *MNRAS*, 302, 649
Józsa, G. I. G. 2007, *A&A*, 468, 903
Kamphuis, P., *et al.* 2013,
Kamphuis, P., *et al.* 2014,
Koribalski, B. S., *et al.* 2004, *AJ*, 128, 16
Koribalski, B. S. & López-Sánchez, Á. R. 2009, *MNRAS*, 400, 1749
Koribalski, B. S. 2008, in *Galaxies in the Local Volume*, Astrophysics and Space Science, eds. B. S. Koribalski, and H. Jerjen, p. 41
Koribalski, B. S., *et al.* 2009, ASKAP proposal
Koribalski, B. S. 2012a, *PASA* 29, 213
Koribalski, B. S. 2012b, *PASA* 29, 359
Kreckel, K., *et al.* 2011, *AJ*, 141, 204
Obreschkow, D. *et al.* 2011, ...
Oh, Se-Heon, *et al.* 2010, ...
Rogstad, *et al.* 1974, ...
Sánchez, *et al.* 2012, A&A, 538, 8
Serra, P., Westmeier, T., *et al.* 2014, *MNRAS*, submitted
Serra, P., Koribalski, B. S., Kilborn, V. A., *et al.* 2015, *MNRAS*, submitted
Spekkens, K. & Sellwood, J. A. 2007, *ApJ*, 664, 204
Krajnovic, *et al.* 2006,
Verheijen, M., *et al.* 2008, *AIP Conf. Proc.* Vol. 1035, p. 265
Warren, B. E., Jerjen, H., & Koribalski, B. S. 2004, *AJ*, 128, 1152
Westmeier, T., Braun, R., & Koribalski, B. S. 2011, *MNRAS*, 410, 2217
Westmeier, T., Koribalski, B. S., & Braun, R. 2013, *MNRAS*, 434, 3511
Westmeier, T., Jurek, R., Obreschkow, D., Koribalski, B. S., & Staveley-Smith, L. 2014, *MNRAS*, 438, 1176
Wilson, W. E., *et al.* 2011, *MNRAS*, 416, 832

Galaxies in 3D across the Universe
Proceedings IAU Symposium No. 309, 2014
B. L. Ziegler, F. Combes, H. Dannerbauer, M. Verdugo, eds.

© International Astronomical Union 2015
doi:10.1017/S1743921314009284

The Recent Evolution of Early-Type Galaxies as Seen in their Cold Gas

Lisa M. Young

Physics Department, New Mexico Tech, 801 Leroy Place, Socorro NM, 87801, USA
email: lyoung@physics.nmt.edu

Abstract. I present an overview of new observations of atomic and molecular gas in early-type galaxies, focusing on the ATLAS^{3D} project. Our data on stellar kinematics, age and metallicity, and ionized gas kinematics allow us to place the cold gas into the broader context of early-type galaxy assembly and star formation history. The cold gas data also provide valuable constraints for numerical simulations of early-type galaxies.

Keywords. galaxies: elliptical and lenticular, cD - galaxies: evolution - galaxies: formation

1. Introduction

Two of the major interrelated challenges of galaxy formation are understanding the basic origin of the Hubble sequence (spiral vs. elliptical and lenticular galaxies) and the present-day dichotomy in galaxy colors (the red sequence vs. the blue cloud or star-forming sequence). Most early-type (elliptical and lenticular) galaxies are poor in cold gas; yet their stars did form, albeit a long time ago, in a precursor that was probably gas-rich and blue. In this context, then, the story of what happened to the cold gas in those precursors is a crucial part of the evolution of early-type galaxies. The current gas contents of early-type galaxies can offer valuable quantitative benchmarks for numerical simulations of galaxy evolution.

Early-type galaxies probably moved onto the red sequence as they lost their cold gas to consumption (star formation), AGN and star formation feedback, and environmental processing in groups and clusters. On the other hand, they could also have acquired cold gas through cooling out of a hot medium and through mergers. Their cold gas supports some current star formation activity, and the rates and efficiencies of that process (in comparison to the behaviors of spirals) can offer some useful constraints on star formation in general. These are some of the issues addressed through observations of nearby early-type galaxies in the ATLAS^{3D} project.

2. Sample and Observations

The ATLAS^{3D} sample is a complete volume-limited sample of early-type galaxies nearer than 42 Mpc and brighter than $M_K = -21.5$, i.e. having stellar masses greater than $M_\star = 10^{9.9}$ M$_\odot$ (Cappellari *et al.* 2011a). The early-type galaxy sample is actually drawn from a parent sample which has no color or morphological selection, and the 260 galaxies lacking spiral structure form the basis of the project. The core of the observations consists of integral-field optical spectroscopy over a field of at least $33'' \times 41''$. For these nearby galaxies, 80% of the sample has spectroscopic coverage out to $0.8R_e$ or farther (Emsellem *et al.* 2011). While the wavelength coverage of the spectra is not large by current standards, the data do offer stellar kinematics, [O III], Hβ and occasionally other emission lines, and line indices sensitive to age, metallicity, and α enhancement.

This is all useful for inferring both the star formation and the assembly histories of the galaxies, in addition to the ionized gas excitation and kinematics.

ATLAS^{3D} galaxies primarily inhabit the red sequence, as expected, though a small percentage are blue, especially at masses $M_\star < 10^{10.5}$ M$_\odot$ (Cappellari et al. 2011a; Young et al. 2014). A discussion of their morphological characteristics is found in Cappellari et al. (2011b). Basically, the majority of early-type galaxies are disky, fast rotating, oblate galaxies, and relatively few, primarily at high masses, are spherical or triaxial and slowly rotating (Krajnović et al. 2011; Emsellem et al. 2011; Weijmans et al. 2014). The two-dimensional kinematic data highlight the incidence of multiple kinematic subcomponents, decoupled or even counterrotating cores, and kinematic twists within the galaxies (see especially Krajnović et al. 2011). Ongoing work also involves deep optical imaging (Duc et al. 2011) to give better limits on recent interactions.

An initial CO survey with IRAM 30m telescope probed the molecular content in the central 20″ of the ATLAS^{3D} galaxies, giving detection limits in the range 10^7 to 10^8 M$_\odot$ (Young et al. 2011). Some more exotic high density molecular tracers have also been observed in a few cases (Bayet et al. 2012; Crocker et al. 2012; Bayet et al. 2013; Davis et al. 2013). Alatalo et al. (2013a) present interferometric CO observations of most of the CO detections, at a typical spatial resolution of 5″ (300 pc to 1 kpc) over a 1′ (5 – 11 kpc) field of view. Serra et al. (2012) also present interferometric H I observations of 170 of the ATLAS^{3D} galaxies from the Westerbork array; those data cover a 30′ field and a velocity range of 4000 km s^{-1} at typical resolutions of 35″, and they have the sensitivity to detect H I column densities of a few 10^{19} cm^{-2}. For a modest-sized cloud of gas this corresponds to mass limits of 10^6 to 10^8 M$_\odot$.

3. Abundant Cold Gas in Early-Type Galaxies

From the observations noted above, the detection rate of molecular gas in massive early-type galaxies is 22% ±3% (Young et al. 2011). To place this value in context it is important to remember that the sample selection is not biased by far-IR detections or even by selection in the B band, as were many previous surveys in early-type galaxies. The H$_2$ masses fall in the range $10^{7.1}$ to $10^{9.3}$ M$_\odot$, or log M(H$_2$)/M$_\star$ in the range -3.5 to -1.1. For the galaxies with CO maps, the molecular gas is usually concentrated in the central kpc or so (Davis et al. 2013) in disks, rings, bars, and irregular configurations (Alatalo et al. 2013a). The detection rate of H I is 40% outside of the Virgo Cluster; H I masses range from $10^{6.3}$ M$_\odot$ to 10^{10} M$_\odot$, and the atomic gas is found in isolated clouds, regular disks (sometimes many tens of kpc in diameter) and disturbed configurations (Serra et al. 2012).

The ATLAS^{3D} project thus has the largest set of cold gas maps ever assembled for early-type galaxies. Atomic and/or molecular gas in quantities of 10^7 to 10^9 M$_\odot$ is surprisingly common in these galaxies, being found in about 50% of them. Often the gas-rich galaxies are detected in both phases, but not always; and the galaxies have values of M(H$_2$)/M(H I) that range from 10^{-2} to 10^2, even considering just the ones detected in both phases.

4. Which kinds of ETGs have cold gas?

Figure 1 shows a color-magnitude diagram of the ATLAS^{3D} galaxies, and it demonstrates that molecular gas detection rates are still strong among red sequence galaxies, particularly at high stellar masses where the star formation rates are not large enough to move the galaxies off the red sequence in integrated colors. Young et al. (2014) have also discussed the reddening effects of dust, which are modest. Careful statistical analysis

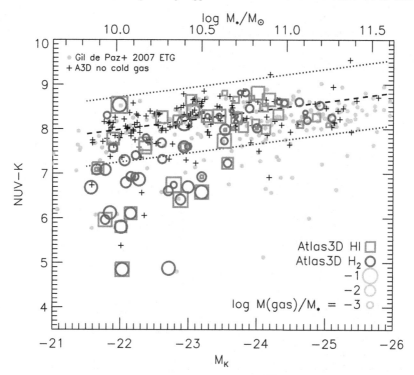

Figure 1. NUV-K color-magnitude diagram for the ATLAS^{3D} sample. The dashed line is the red sequence ridge line and the dotted lines are 2σ redder and bluer, as described in Young *et al.* (2014). H I and H$_2$ masses are indicated by the symbol sizes.

shows that the molecular mass detection rate is not a function of stellar mass over the observed ranges, nor is the H$_2$ mass distribution function (Young *et al.* 2011).

While the molecular masses of early-type galaxies do not correlate with stellar mass, they do correlate with structural parameters. Cappellari *et al.* (2013b) show that, at a fixed stellar mass, the early-type galaxies with low stellar velocity dispersion and large effective radius (i.e. with more dominant disky stellar components rather than spheroids) have larger molecular masses. It is tempting to assume that this might be because the molecular gas disk gives birth to a young, disky stellar component. And while that is partly true, the currently observed molecular masses are usually only 1% of the stellar masses, or less, so the young stellar disks are probably too small to drive the whole-galaxy bulge/disk ratios. They must be driven by older stellar disks. In addition, as we will discuss below, much of the molecular gas is kinematically misaligned with respect to the stars.

There are some interesting differences in the atomic and molecular contents of early-type galaxies. Krajnović *et al.* (2011) show that about 86% of early-type galaxies are the oblate fast rotators. These galaxies can grow through gas-rich minor mergers and associated *in situ* star formation, or through recent major mergers with the appropriate geometry to preserve a large net angular momentum (Naab *et al.* 2013). They are detected at about the same rate in both CO and H I (Serra *et al.* 2012). More rare are the slow rotators, which can be formed from recent mergers with special geometries ensuring near-cancellation of the net angular momentum, and the spherical non-rotators, which are proposed to grow through a large number of isotropically oriented minor mergers (Naab *et al.* 2013). Slow- and non-rotators are rarely detected in CO but are detected in

H I at about the same rate as the fast rotators are detected in H I (Serra *et al.* 2012). This information suggests that at least in some cases the origin of the atomic gas is different from that of the molecular gas, or perhaps the galaxy structure affects the formation or longevity of the molecular phase. Sarzi *et al.* (2013) also discusses some effects of the galaxy structure on the hot gas content of the ATLAS^{3D} galaxies.

We conclude that the quenching of a galaxy and the approach to the red sequence do not always involve the loss of all, or even most, of the cold gas from an early-type galaxy. Alternatively, perhaps some cold gas is reacquired during or after the approach to the red sequence. The presence of molecular gas in an early-type galaxy is strongly connected to the presence of a dynamically cold, disky stellar component, and the connection may be partly (but not entirely) due to the formation of a young stellar disk out of said molecular gas. Contrary to this, atomic gas is more indiscriminate, being found in approximately equal quantities in both the disky fast rotators and the bulge-dominated slow rotators, and often not associated with any notable star formation.

5. Angular Momenta as clues to the history of the gas

Early-type galaxies have much more interesting and diverse stellar kinematics in their inner regions than spirals do — as is abundantly clear from the examples shown in Krajnović *et al.* (2011). The comparisons between gas and stellar kinematics are also important clues to the evolution of these galaxies. For example, Davis *et al.* (2011) plot the misalignment angle between stellar and ionized or molecular gas kinematics; the distribution peaks at zero but has a long flat tail all the way to 180° (retrograde gas). Stellar–gas kinematic misalignments are common in molecular and ionized gas, occurring in at least 35% of early-type fast rotators (see also Sarzi *et al.* 2010). Similarly, half of all H I-detected fast rotators have that H I kinematically misaligned with respect to the stars (Serra *et al.* 2014). These papers argue that gas at misalignments > 30° cannot be simply recycled material from internal stellar mass loss, because it has the wrong kinematics. Perhaps it could be brought in to the galaxy in an accreted satellite, or perhaps (as discussed below) it cooled out of hot gas at large radii in the galaxy halo, where the hot gas is vulnerable to many external torques. If these arguments apply to all of the misaligned gas plus some portion of the aligned gas as well, they apply to something like $\gtrsim 50\%$ of the cold gas in nearby early-type galaxies.

There is also some evidence for external gas accretion from the metallicity measurements. The ATLAS^{3D} sample galaxies follow a well-established stellar mass–metallicity relation, with a few notable outliers at markedly low metallicity (but enhanced α-element abundances) for their mass. These also tend to be blue and rich in molecular gas. McDermid *et al.* (2014) postulate that these are cases in which low metallicity gas was accreted, so the population of younger stars with lower metallicity is biasing the "average" metallicity estimate.

6. Comparison to simulations

Past work on simulating the cold gas content of galaxies has mainly focused on the gas-rich spiral galaxies (e.g. Lagos *et al.* 2011). The simulations of early-type galaxies are more challenging if for no other reason than the quantities of cold gas are generally smaller, requiring higher fidelity. Serra *et al.* (2014) made high resolution re-simulations of dark matter halos extracted from a larger λCDM simulation, implementing gas cooling, star formation and supernova feedback. Cooling is not carried out to molecular gas temperatures, but H I is converted to H_2 by fiat at surface densities of 10 $M_\odot pc^{-2}$. Not surprisingly, the details of the feedback prescription have strong influence on the final gas distribution and kinematics, and even in some cases on the stellar kinematics. There is significant tension between these simulations and the observations in the sense that the simulations produce too much prograde cold gas and not enough misaligned cold gas.

Lagos *et al.* (2014) describe a semi-analytical model which follows the quantities of cold gas that cooled out of the hot halo, were brought in by mergers, returned to the ISM from internal stellar mass loss, and consumed by star formation. They stress the co-evolution of the hot and cold gas, emphasizing that an improved treatment of gentle ram pressure stripping of the hot halo is necessary to reproduce the observed cold gas contents of early-type galaxies. A work in preparation also addresses the kinematic misalignments of the cold gas in early-type galaxies; there are not enough minor mergers to explain all of the misaligned gas, but cooling out of a hot halo might contribute misaligned cold gas if the hot halo has been torqued so that it is also misaligned with the stellar body of the galaxy (Lagos 2014, private communication).

7. Star formation and galaxy dynamics

For some years it has been suspected that the empirical star formation "laws" (predicting the star formation rate produced by a given quantity of cold gas) might be somewhat different in early-type galaxies than in spirals, due to the systematically different gravitational potentials (e.g. Kawata *et al.* 2007; Martig *et al.* 2009). Indeed, hydrodynamical simulations of Martig *et al.* (2013) found that the molecular gas in spherical galaxies should not reach the same high local densities as the molecular gas in spirals, and they predicted something like factors of 2 lower star formation efficiencies in early-type galaxies. These kinds of factors of two could prove to be important as simulations of galaxy evolution move into precision mode.

Interestingly, Davis *et al.* (2014) observe that the star formation efficiency of early-type galaxies correlates with a measurement of whether most of the gas is in the rising or the flat part of the rotation curve. Specifically, where most of the molecular gas is in the rising part, the efficiency is lower by factors of a few. These types of observations offer the opportunity to make empirical tests of the star formation "laws" beyond the spiral galaxies where they were developed, so they should help refine the prescriptions used in simulations.

8. Summary

There is more cold gas in early-type galaxies than most of us expected. Approximately 50% of massive early-type galaxies ($M_\star \gtrsim 10^{10}\ M_\odot$) contain 10^7 to $10^9\ M_\odot$ of H I and/or H_2. The atomic and molecular phases are not always found together; for example, Virgo Cluster members are more likely to be detected in H_2 but not in H I.

Because many of these cold-gas-rich early-type galaxies are found on the red sequence, and the red colors are *not* solely the effect of dust associated with the cold gas, we infer that 'red sequence' and 'cold gas' are not mutually exclusive properties; the approach to the red sequence does not always involve the loss of all cold gas. Alternatively, some red sequence galaxies may reacquire gas after going onto the red sequence.

Much of the cold gas in early-type galaxies is kinematically misaligned with respect to the stars. Because of the misalignment it is apparently not a simple recycling of gas from stellar mass loss directly back into a cold gas phase. It remains to be seen — and would be fascinating to know — how the cold gas is aligned (or not) with respect to the hot gas, as it becomes increasingly obvious that hot and cold phases probably do interact with each other. We have posited that a significant portion of the cold gas in early-type galaxies is accreted either with incoming satellite galaxies in minor mergers, or has cooled from the intergalactic medium or a hot halo.

The cold gas contents and kinematics of early-type galaxies are thus shown to be sensitive probes of their evolution. The diversity in the stellar kinematics, gas contents, and gas kinematics of early-type galaxies emphasizes the diversity in their evolutionary paths. Even the diversity in the star formation properties of early-type galaxies can help improve our understanding of the star formation process. These features are now beginning to give useful constraints on numerical simulations.

Additional work on the hot gas in early-type galaxies is needed, especially regarding the thermal and dynamical interactions between hot and cold gas. More insights to the evolution of early-type galaxies should also come from the metallicities of all the gas phases.

Acknowledgements

I acknowledge support from NSF AST 1109803 and travel support from the IAU; I also thank Dr Paul T. P. Ho for the invitation to spend a sabbatical year at ASIAA.

References

Alatalo, K., Davis, T. A., Bureau, M., et al. 2013, MNRAS, 432, 1796
Bayet, E., Davis, T. A., Bell, T. A., Viti, S., 2012, MNRAS, 424, 2646
Bayet, E., et al. 2013, MNRAS, 432, 1742
Cappellari, M., et al. 2011, MNRAS, 413, 813 (Paper I)
Cappellari, M., et al. 2011, MNRAS, 416, 1680 (Paper VII)
Cappellari, M., McDermid, R. M., et al. 2013, MNRAS, 432, 1862
Crocker, A. F., Krips, M., Bureau, M., et al. 2012, MNRAS, 421, 1298
Davis, T. A., Alatalo, K., Sarzi, M., et al. 2011, MNRAS, 417, 882
Davis, T. A., Bureau, M., et al. 2013, MNRAS, 429, 534
Davis, T. A., Bayet, E., Crocker, A., Topal, S., Bureau, M., 2013, MNRAS, 433, 1659
Davis, T. A., Young, L. M., et al. 2014, MNRAS, 444, 3427
Duc, P.-A., Cuillandre, J.-C., Serra, P., et al. 2011, MNRAS, 417, 863
Emsellem, E., et al. 2011, MNRAS, 414, 888
Kawata, D., Cen, R., Ho, L. C. 2007, ApJ, 669, 232
Krajnović, D., et al. 2011, MNRAS, 414, 2923
Lagos, C. d. P., Baugh, C. M., Lacey, C. G., et al. 2011, MNRAS, 418, 1649
Lagos, C. d. P., Davis, T. A., Lacey, C. G., et al. 2014, MNRAS, 443, 1002
Martig, M., Bournaud, F., Teyssier, R., & Dekel, A., 2009, ApJ, 707, 250
Martig, M., Crocker, A. F., Bournaud, F., et al. 2013, MNRAS, 432, 1914
McDermid, R. M., et al. 2014, MNRAS, submitted
Naab, T., Oser, L., Emsellem, E., et al. 2013, MNRAS, 444, 3357
Sarzi, M., et al. 2010, MNRAS, 402, 2187
Sarzi, M., et al. 2013, MNRAS, 432, 1845
Serra, P., et al. 2012, MNRAS, 422, 1835
Serra, P., Oser, L., Krajnović, D., et al. 2014, MNRAS, 444, 3388
Weijmans, A.-M., et al. 2014, MNRAS, 444, 3340
Young, L. M., et al. 2011, MNRAS, 414, 940
Young, L. M., et al. 2014, MNRAS, 444, 3408

Galaxies in 3D across the Universe
Proceedings IAU Symposium No. 309, 2014
B. L. Ziegler, F. Combes, H. Dannerbauer, M. Verdugo, eds.

© International Astronomical Union 2015
doi:10.1017/S1743921314009296

Inner polar ionized-gas disks and properties of their host galaxies

Olga K. Sil'chenko

Sternberg Astronomical Institute, Lomonosov Moscow State University,
University av. 13, 119991 Moscow, Russia
email: olga@sai.msu.su

Abstract. I have analyzed line-of-sight velocity fields of the stellar and ionized-gas components for the volume-limited sample of nearby lenticular galaxies by using the raw data of
the ATLAS-3D survey undertaken with the integral-field spectrograph SAURON. Among 200
nearby lenticular galaxies, I distinguish 20 cases of nearly orthogonal rotation of the inner ionized gas with respect to the central stellar components; so I estimate a frequency of the inner
polar disks in nearby S0 galaxies as 10%. Properties of the central stellar populations – mean
ages, metallicities, magnesium-to-iron ratios – are derived through the Lick indices. The typical
stellar population properties of the polar-disk host galaxies are exactly the same as the stellar
population properties of the complete sample.

Keywords. galaxies: elliptical and lenticular, cD - galaxies: evolution - galaxies: formation

1. Introduction: Inner polar rings, what is it?

Everybody knows well such rare but spectacular phenomenon as large-scale polar rings,
when the outer gas, and sometimes young stars, rotates in the plane highly inclined to
the main galactic disks. However starting from 1990ties, inner analogs of the large-scale
polar rings have become to be found. Inner polar rings, or disks, represent circumnuclear,
regularly rotating ionized gas which rotation axis is strongly inclined to the rotation axis
of the central stellar component – by more than 50 degrees. Recently Moiseev (2012) has
collected a list of the inner polar disks found to date and has analyzed their properties.
From the statistics of 47 galaxies, it becomes clear that the inner polar disks are indeed
polar: even taken over the range of the inclinations to the stellar disks > 50 degrees,
they tend to concentrate to the value of $\Delta i = 90°$. It is not a surprise because from the
dynamical point of view the polar plane is stable against precession. The typical radii of
the inner polar disks are 0.2–2 kpc, the outer border being real and the inner one being
an artifact of the finite spatial resolution of our observations. The inner polar disks are
met mostly in early-type disk galaxies, S0s-Sbs, but a few are found in very late-type
dwarfs.

A main method by which we can identify inner polar disks is the analysis of two-
dimensional velocity fields both for the ionized-gas and stellar components, to derive the
spatial orientations of their rotation planes by applying tilted-ring approach. It requires
the data of 3D spectroscopy. After the Moiseev's review publication, I have undertaken
the further search of the inner polar disks by using the rich raw data of the IFU SAURON
collected in the frame of the survey ATLAS-3D (Cappellari *et al.* 2011) and stored in the
open ING archive of the CASU Astronomical Data Centre at the Institute of Astronomy,
Cambridge. About 150 nearby S0 galaxies of this survey have been analyzed. In this
contribution I present the results of my analysis.

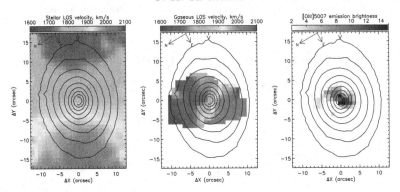

Figure 1. Three maps for NGC 5507 derived from the SAURON data: stellar LOS velocity field (*left*), ionized-gas LOS velocity field (*center*), and surface brightness distribution for the gas emission line [OIII]λ5007 (*right*).

2. Inner polar disks in the S0s of the ATLAS-3D survey

By reducing the raw SAURON data for 149 S0 galaxies observed in 2007–2008 in the frame of the ATLAS-3D survey (Cappellari *et al.* 2011), I have calculated line-of-sight (LOS) velocity fields for the stellar and ionized-gas (when present) components. These 2D LOS-velocity distribution have been analyzed by a tilted-ring approach in a particular modification by Alexei Moiseev (Moiseev *et al.* 2004) allowing to estimate parameters of the spatial orientation of the rotation planes for both components, inclinations $i_{s,g}$ and position angles of the lines of nodes, $PA_{s,g}$. After that, the mutual inclinations of the stellar and gaseous rotation planes have been calculated by using the following formula (Moiseev 2012):

$$\cos \Delta i = \pm \cos(PA_s - PA_g) \sin i_s \sin i_g + \cos i_s \cos i_g.$$

The equation has two solutions because typically we do not know what side of a galactic disk is closer to us. By making these calculations, among 149 S0s of the ATLAS-3D survey, I have found 6 new cases of the nearly polar inner ionized-gas disks with $\Delta i > 50°$. These are NGC 2962 ($\Delta i = 50$ or 114 deg), NGC 3499 ($\Delta i = 61$ or 68 deg), NGC 3648 ($\Delta i = 68$ or 88 deg), NGC 4690 ($\Delta i = 64$ or 68 deg), NGC 1121 and NGC 5507 (edge-on?). As an illustration of the data quality, for the latter galaxy, NGC 5507, I give here three maps (Fig. 1): stellar LOS velocity field, ionized-gas LOS velocity field, and surface brightness of the gas emission line [OIII]λ5007 which demonstrates clearly that we deal with the gaseous disk edge-on, in full agreement with the gas kinematical major axis orientation. Also the inner polar rings have been noted in the isolated S0 UGC 9519 through the central dust distribution though the ionized-gas velocity field is rather irregular in this galaxy (Katkov *et al.* 2014), and in NGC 4684 by Davis *et al.* (2011) who found the kinematical major axes stars/ionized-gas misalignment of about 90 deg in this edge-on galaxy. Taking these 8 galaxies together with those listed earlier by Moiseev (2012), and considering the full ATLAS-3D sample including the galaxies observed earlier than 2007, we obtain 20 inner polar ionized-gas disks in the sample of 200 nearby S0 galaxies. Since the sample of the early-type galaxies in the ATLAS-3D survey is complete within the volume limited by $D < 42$ Mpc and above the luminosity $M_K < -21.5$ (Cappellari *et al.* 2011), I conclude that the frequency of the inner polar disks among the nearby S0 galaxies is about 10%. So the phenomenon of the *inner* polar disks is much more common than the *large-scale* polar rings (Moiseev *et al.* 2011).

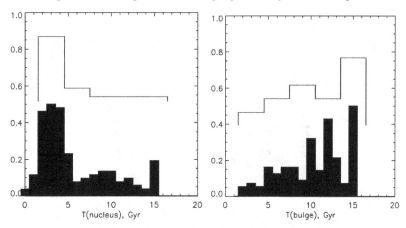

Figure 2. The distributions of the SSP-equivalent ages for the nuclei (*left*) and bulges (*right*) of the ATLAS-3D S0s: filled histograms present the total sample, while the blue line – the inner polar-disk hosts.

3. S0 hosts of the inner polar disks: are they special?

I have also analyzed properties of the stellar populations in the nuclei and the bulges of 149 S0 galaxies of the ATLAS-3D survey. To get them, I have calculated Lick indices (Worthey *et al.* 1994) Hβ, Mgb, and Fe5270 in every spaxel for every galaxy, then I have averaged them for the nuclei, $R < 1''$, and for the bulges, taken in the rings of $4'' < R < 7''$ for galaxies close to face-on and along the minor axes in $4'' - 6''$ from the centers for galaxies close to edge-on. I have confronted the obtained mean indices to the Simple Stellar Populations (SSP) models by Thomas *et al.* (2003) which allowed to take into account different abundance ratios Mg/Fe and to obtain SSP-equivalent (luminosity-averaged) ages and metallicities. With these data in hands, I have constructed the distributions of the ages and metallicities, both for the nuclei and the bulges. I wanted to search for particular ecolutionary paths of the hosts of the inner polar disks. The comparison of the distributions for the total sample with those for the hosts of the inner polar disks is presented in Fig. 2 and Fig. 3. Curiously, the polar-disk host galaxies do not differ in any sense from the other lenticular galaxies. Even a rather exotic bimodal age distribution of the stellar nuclei where the majority is young, of 2–5 Gyr old, but a significant fraction of very old stellar nuclei is also present, is perfectly reproduced by the inner polar-disk hosts.

Further on, the hosts of the inner polar disks fill up the full range of galaxy masses which was probed by the ATLAS-3D survey, and take part in all known correlations. Interestingly, the well-known correlation of the stellar population ages with the stellar velocity dispersion for the nuclei holds only for the rather young nuclei while the old nuclei are homogeneously distributed among the galaxies of various masses.

4. Conclusions

By analyzing the kinematical maps for the full lenticular-galaxy sample of the ATLAS-3D survey, I have found a few new cases of the nearly polar rotation of the circumnuclear ionized gas and have estimated the frequency of the inner polar ionized-gas disks as 10%. Inner polar disks can be met in lenticular galaxies of any mass and over the environments of any density, including some cases in completely isolated galaxies. The properties of the stellar populations of the nuclei and circumnuclear bulges derived through the Lick index

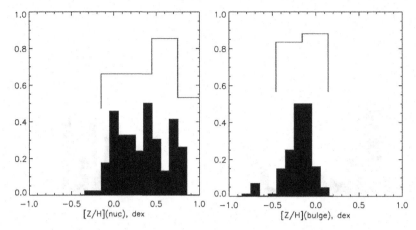

Figure 3. The distributions of the stellar metallicities for the nuclei (*left*) and bulges (*right*) of the ATLAS-3D S0s: filled histograms present the total sample, while the blue line – the inner polar-disk hosts.

modelling have appeared to be the same in the host galaxies of the inner polar disks and over the full sample, including a large fraction of young stellar nuclei, of 2–5 Gyr old.

Acknowledgements

I acknowledge financial support from the IAU to attend this Symposium. The present contribution makes use of data obtained from the Isaac Newton Group Archive which is maintained as part of the CASU Astronomical Data Centre at the Institute of Astronomy, Cambridge.

References

Cappellari, M., Emsellem, E., Krajnovic, D., McDermid, R. M., Scott, N., *et al.* 2011, *MNRAS*, 413, 813

Davis, T. A., Alatalo, K., Sarzi, M., Bureau, M., Young, L. M., *et al.* 2011, *MNRAS*, 417, 882

Katkov, I. Yu., Sil'chenko, O. K., & Afanasiev, V. L. 2014, *MNRAS*, 438, 2798

Moiseev, A. V., Valdés, J. R., & Chavushyan, V. H. 2004, *A&A*, 421, 433

Moiseev, A. V., Smirnova, K. I., Smirnova, A. A., & Reshetnikov, V. P. 2011, *MNRAS*, 418, 244

Moiseev, A. V. 2012, *Astrophys. Bull.*, 67, 147

Thomas, D., Maraston, C., & Bender, R. 2003, *MNRAS*, 339, 897

Worthey, G., Faber, S. M., González, J. J., & Burstein, D. 1994, *ApJS*, 94, 687

Galaxies in 3D across the Universe
Proceedings IAU Symposium No. 309, 2014
B. L. Ziegler, F. Combes, H. Dannerbauer, M. Verdugo, eds.

© International Astronomical Union 2015
doi:10.1017/S1743921314009302

The Spatially-Resolved Star Formation History of the M31 Disk from Resolved Stellar Populations

Alexia R. Lewis[1], Julianne J. Dalcanton[1], Andrew E. Dolphin[2], Daniel R. Weisz[1,3,4] and Benjamin F. Williams[1], PHAT

[1]Department of Astronomy, University of Washington, Box 351580, Seattle, WA 98195
email: arlewis@astro.washington.edu
[2]Raytheon Company, 1151 East Hermans Road, Tucson, AZ 85756
[3]University of California Observatories/Lick Observatory, University of California, 1156 High St., Santa Cruz, CA 95064
[4]Hubble Fellow

Abstract. The Panchromatic Hubble Andromeda Treasury (PHAT) is an HST multi-cycle treasury program that has mapped the resolved stellar populations of ∼1/3 of the disk of M31 from the UV through the near-IR. This data provides color and luminosity information for more than 150 million stars. Using stellar evolution models, we model the optical color-magnitude diagram to derive spatially-resolved recent star formation histories (SFHs) over large areas of M31 with 100 pc resolution. These include individual star-forming regions as well as quiescent portions of the disk. With these gridded SFHs, we create movies of star formation activity to study the evolution of individual star-forming events across the disk. We analyze the structure of star formation and examine the relation between star formation and gas throughout the disk and particularly in the 10-kpc star-forming ring. We find that the ring has been continuously forming stars for at least 500 Myr. As the only large disk galaxy that is close enough to obtain the photometry for this type of spatially-resolved SFH mapping, M31 plays an important role in our understanding of the evolution of an L* galaxy.

Keywords. galaxies: evolution - galaxies: individual (M31) - galaxies: spiral - galaxies: structure

1. Introduction

The star formation history (SFH) of a galaxy encodes much of the physics of the evolution of that galaxy. It tells us about the evolution of the star formation rate (SFR) throughout the lifetime of the galaxy as well as the evolution of the mass and metallicity distributions.

The SFH can be inferred for bulk populations by fitting the spectral energy distributions (SEDs) of galaxies. The evolution of large samples of galaxies can then be studied by linking local universe galaxies to their likely high redshift progenitors and examining the evolution of the physical properties of the galaxies over time (e.g. Papovich *et al.* 2011). This technique is advantageous in that it enables a statistically complete analysis of galaxy evolution due to the large number of galaxies in the sample. However, changes to the approximations it requires approximations (SFR as a function of time, dust model, etc.) can have a significant affect on the derived ages and SFRs. It also sacrifices detail for lookback time.

We can perform a more detailed analysis of the evolution of the structure and populations of local universe galaxies by using resolved stellar populations in color-magnitude diagram (CMD) analysis. The CMD contains the history of star formation (SF) and the metallicity distribution of the galaxy. Using individually resolved stars we can probe look-back time, allowing us to self-consistently study the evolution of a single galaxy based on its stellar population. With this method, though, we can only study the local universe, which sacrifices lookback time for increased detail.

Using recent observations of the Andromeda Galaxy (M31) as part of the Panchromatic Hubble Andromeda Treasury (PHAT; Dalcanton *et al.* 2012) we examine the spatially-resolved SFH of M31 using this 'gold standard' technique. Derivation of the SFH on small scales enables a more detailed comparison of SF with the interstellar medium of a galaxy, helping us uncover clues about the specific conditions that promote SF (e.g. Kennicutt 1998). Historically, this method has been used only to examine the SFHs of low-mass galaxies like the LMC and SMC (Harris & Zaritsky 2004, 2009). This is the first time it has been used to probe the SFH of an L_* galaxy.

2. Star Formation History Recovery Method

We use the CMD fitting code MATCH (Dolphin 2002) to derive the SFHs on $\sim$100 pc (projected) scales in the disk of M31. The user specifies ranges to search in age, metallicity, distance, and extinction, as well as an initial mass function and a binary fraction. The code then populates synthetic CMDs, including photometric and completeness errors determined by artificial star tests, and linearly combines them to form many different SFHs. Each of these complex CMDs is compared with the observed CMD and the one that provides the best fit, as determined by a Poisson maximum likelihood technique, is taken to be the model SFH that best describes the data.

To lessen some of the effects of dust on the SFHs, we simplify the fitting process by excluding the redder portions of the CMD from the fit. We model only the main sequence. As a result, we determine only the recent SFH in each region. Our SFHs are reliable back to $\sim$600 Myr ago. This process is repeated in each of our 100 pc regions independently of all other regions. We combine the final results to create maps of SFH in M31.

3. Results

We combine the single best fit SFHs in each region into maps of SFR as a function of time. As an ensemble, these reveal the recent SFH of M31 in the PHAT footprint over the last 630 Myr. This SFH is shown in Figure 1. The SFHs reveal large-scale, long-lasting, coherent structure in the M31 disk. There are three star-forming rings; a modest inner ring at $\sim$5 kpc, the well-known 10-kpc ring, and an outer, low-intensity ring at $\sim$15 kpc that partially merges with the 10-kpc ring due to a combination of projection effects and a possible warp as visible in HI (e.g. Brinks & Shane 1984). The 10-kpc ring is visible and actively forming stars throughout the past 600 Myr. The ring features at 5 kpc peaked in SF $\sim$100 Myr ago but has largely dispersed today. SF in the 15-kpc feature has increased to the present day, aided in large part by OB 102 which is clearly visible in the maps.

We generate the overall recent SFH within the PHAT footprint by integrating the SFHs of all regions. On the left side of Figure 2 we show the total SFR per time bin over the survey area. The SFR has gradually declined over the last 600 Myr, but we do see a peak at $\sim$50 Myr. Further examination reveals that this peak is not a galaxy-wide feature; instead it is likely due to stellar clusters and OB associations in the rings. On the right side of Figure 2, we plot the average SFR per time bin in regions that fall within the 10-kpc ring and in all other regions. It is very clear the SF in the ring drives the overall SFH of the disk. The regions outside of the ring show a very low and mostly constant SFR over the last 500 Myr. A more detailed analysis and discussion of the results can be found in Lewis *et al.* (submitted).

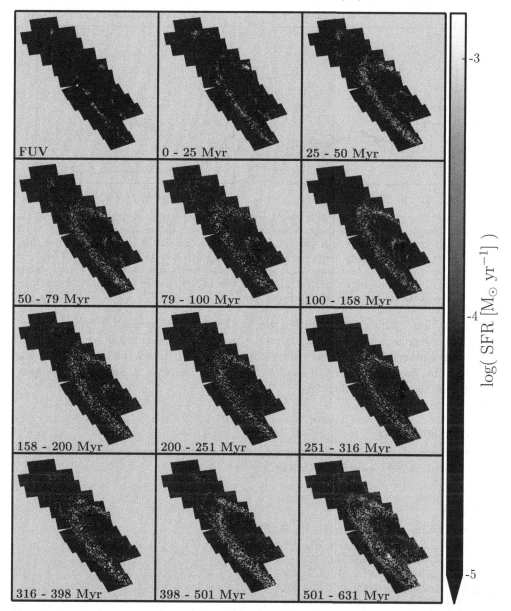

Figure 1. Map of the SFH of M31 covered by the PHAT survey. The pixel value is proportional to the SFR in that region. We have applied a lower cut at a SFR of 10^{-5} $M_\odot$ yr^{-1} in order to highlight structure in the image. The time range is shown in the lower left of each plot. The plot in the top left shows the *GALEX* FUV image, convolved to the same physical scale as our SFHs for better comparison. The maps are oriented such that north is up and east is left.

4. Conclusions

We have measured the spatially-resolved SFH of ∼1/3 of the star-forming disk of M31 from HST images as part of the PHAT survey. SF in M31 is largely confined to three ring features, including the well-known 10-kpc ring. These features show elevated levels of SF over the entire reliable lifetime of the SFHs, approximately 600 Myr. The 10-kpc ring is long-lived and stationary, producing stars over at least the past 600 Myr. It has also

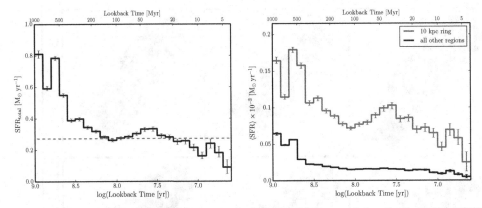

Figure 2. Left:The global SFH of M31 over the last 630 Myr, combining the individual SFHs of each field. The dashed line (blue) shows the average SFR over the most recent 100 Myr ($\sim$0.3 $M_\odot$ yr^{-1}). Right: The average SFR per time bin in regions in the 10-kpc ring (top red line) and in all other regions (bottom black line). SF in the ring drives to the recent SFH of M31. In both plots, the error bars are a combination of the random uncertainties and the uncertainties in A_V and dA_V. The time bins are $\Delta(\log t) = 0.1$. The region in gray indicates the timescale over which the SFHs are not reliable in all regions.

shown little or no significant propagation over the lifetime of these SFHs. Over this time, SF in the ring has been elevated relative to the surrounding regions and as a result, the ring drives the overall SFH. We have also found that the SFH over the last 500 Myr has been relatively constant, showing a shallow but steady decrease in SFR to the present day with a small burst $\sim$50 Myr ago.

Acknowledgements

This project made extensive use of the Stampede supercomputing system at TACC/UT Austin, funded by NSF award OCI-1134872.

References

Brinks, E. & Shane, W. W. 1984, *A&AS*, 55, 179

Dalcanton, J. J., *et al.* 2012, *ApJS*, 200, 18

Dolphin, A. E. 2002, *MNRAS*, 332, 91

Harris, J. & Zaritsky, D. 2004, *AJ*, 127, 1531

Harris, J. & Zaritsky, D. 2009, *AJ*, 138, 1243

Kennicutt, R. C., Jr. 1998, *ApJ*, 498, 541

Papovich, C., Finkelstein, S. L., Ferguson, H. C., Lotz, J. M., & Giavalisco, M. 2011, *MNRAS*, 412, 1123

Galaxies in 3D across the Universe
Proceedings IAU Symposium No. 309, 2014
B. L. Ziegler, F. Combes, H. Dannerbauer, M. Verdugo, eds.

© International Astronomical Union 2015
doi:10.1017/S1743921314009314

ALMA Explorations of Warm Dense Molecular Gas in Nearby LIRGs

C. Kevin Xu

Infrared Processing and Analysis Center,
MS 100-22, California Institute of Technology,
Pasadena, CA 91125
email: cxu@ipac.caltech.edu

Abstract. We present results of ALMA (Cycle-0) observations of the CO (6-5) line emission and the $435\mu m$ continuum of two nearby luminous infrared galaxies (LIRGs) NGC 34 (a major merger with an AGN) and NGC 1614 (a minor merger with a circum-nuclear starburst). Using receivers in the highest frequency ALMA band available (Band-9), these observations achieved the best angular resolutions ($\sim 0''.25$) for ALMA Cycle-0 observations and resolved for the first time distributions of warm dense molecular gas ($n > 10^5$ cm^{-3}, T > 100 K) in LIRGs with spatial resolutions better than 100 pc. Our ALMA data show a very tight correlation between the CO (6-5) line emission and the $435\mu m$ dust continuum emission, suggesting the warm dense molecular gas dominates the ISM in the central kpc of LIRGs, and gas heating and dust heating in the warm dense gas cores are strongly coupled. On the other hand, we saw very different spatial distributions and kinematic properties of warm dense gas in the two LIRGs, indicating that physical conditions in the ISM can be very different in different LIRGs.

Keywords. galaxies: interactions - galaxies: evolution - galaxies: starburst - galaxies: general - galaxies: active - submillimeter: galaxies

1. Introduction

Luminous infrared galaxies (LIRGs) with $L_{IR} [8-1000\mu m] > 10^{11.5}$ L$_\odot$, including ultra-luminous infrared galaxies (ULIRGs: $L_{IR} > 10^{12}$ L$_\odot$) are mostly advanced mergers (?). Extensive surveys of CO rotation lines in low J transitions such as CO (1-0) at 2.6 mm and CO (2-1) at 1.3 mm have found very large amounts of molecular gas (up to a few times 10^{10} M$_\odot$) in the central kpc of (U)LIRGs (Scoville *et al.* 1989; Sanders *et al.* 1991). This gas, funneled into the nuclear region by the gravitational torque in a merger (Barnes & Hernquist 1996; Hopkins *et al.* 2009), provides fuel for the nuclear starburst and/or AGN. However, due to the heavy dust extinction for the UV/optical/NIR observations and the lack of high angular resolution FIR/sub-mm/mm observations, it is still not very clear how the different constituents (i.e., gas, dust, stars, and black holes) in (U)LIRG nuclei interplay with each other. Some studies (Scoville *et al.* 1997; Downes & Solomon 1998; Bryant & Scoville 1999; Gao *et al.* 2001) suggest that much of the low J CO luminosities may be due to the emission of diffuse gas not closely related to the active star formation regions. Indeed, single dish and interferometry mm and submm observations have found that the intensities and spatial distributions of star formation in (U)LIRGs correlate significantly stronger with those of higher J CO lines (with upper level J $\geqslant$ 3), which probe warmer and denser gas than low J lines (Yao *et al.* 2003; Iono *et al.* 2004; Wang *et al.* 2004; Wilson *et al.* 2008; Iono *et al.* 2009; Sakamoto *et al.* 2008, 2013; Xu *et al.* 2014; Lu *et al.* 2014).

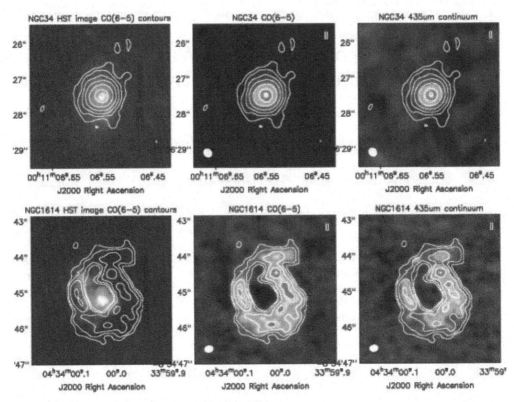

Figure 1. *Upper-left:* The HST V-band image of NGC 34 overlaid by contours of the integrated CO (6-5) line emission. The contour levels are [1, 2, 4, 8, 16, 32, 50]×3.6 Jy km s^{-1} beam^{-1}. *Upper-mid:* Image and contours of the integrated CO (6-5) line emission of NGC 34. *Upper-right:* Image of the continuum of NGC 34 overlaid by contours of the integrated CO (6-5) line emission. *Lower-left:* The HST V-band image of NGC 1614 overlaid by contours of the integrated CO (6-5) line emission. The contour levels are [1, 2, 4, 8]×2.7 Jy km s^{-1} beam^{-1}. *Lower-mid:* Image and contours of the integrated CO (6-5) line emission of NGC 1614. *Lower-right:* Image of the continuum of NGC 1614 overlaid by contours of the integrated CO (6-5) line emission.

2. NGC 1614 and NGC 34: A Tale of Two LIRGs

In order to study the warm dense gas in nuclear regions of (U)LIRGs, we observed the CO (6-5) line emission (rest-frame frequency = 691.473 GHz) and associated dust continuum emission in two nearby (U)LIRGs (NGC 34 and NGC 1614) using the Band 9 receivers of the Atacama Large Millimeter Array (ALMA; Wootten & Thompson 2009). Both NGC 34 and NGC 1614 were chosen for these early ALMA observations, among the complete sample of 202 LIRGs of the Great Observatories All-sky LIRG Survey (GOALS; Armus *et al.* 2009), because of their close proximity (D < 85 pc) and bright CO (6-5) line flux (f$_{\text{CO (6-5)}}$ ≳ 1000 Jy km s^{-1}) observed in the Herschel SPIRE FTS survey of GOALS galaxies (angular resolution: ~ 30″; van der Werf *et al.* 2010; Lu *et al.* 2014). This enables high signal-to-noise-ratio ALMA observations of warm gas structures with linear resolutions of ≲ 100 pc for the given angular resolutions of ~ 0″.25. Further, both LIRGs have declination angles close to the latitude of the ALMA site, therefore the Band 9 observations are affected by minimal atmospheric absorption when being carried out near transit.

Table 1. Results of ALMA Observations

name	beam	$f_{CO(6-5)}$	$f_{CO(6-5)}^{ALMA}/f_{CO(6-5)}^{Hsch}$	$f_{435\mu m}$	$f_{435\mu m}^{ALMA}/f_{435\mu m}^{Hsch}$
		(Jy km s^{-1})		(Jy)	
NGC 34	$0''.26 \times 0''.23$	1004 ± 151	1.07 ± 0.18	275 ± 41	0.53 ± 0.08
NGC 1614	$0''.26 \times 0''.20$	898 ± 153	0.63 ± 0.12	269 ± 46	0.32 ± 0.06

In Fig 1 we compare the ALMA images with the HST V-band images (Malken *et al.* 1998) of the two LIRGs. We obtained the following results from the ALMA observations:

• The CO (6-5) line emission and the $435\mu m$ continuum in the central kpc of both LIRGs are well detected and resolved by the ALMA observations.

• In both LIRGs, the CO (6-5) line emission and the continuum correlate tightly with each other. This suggests that the warm dense molecular gas (probed by the CO (6-5) line emission) dominates the ISM in the central kpc of LIRGs, and gas heating and dust heating in the warm dense gas cores are strongly coupled.

• The CO (6-5) line emission shows very different morphologies in the central kpc of the two LIRGs. In NGC 34, a major-merger with a weak AGN, it is concentrated in a nuclear disk of r ∼ 100 pc, which is ∼ 6 times smaller than the CO (1-0) disk (Fernandez *et al.* 2014). In NGC 1614, a minor-merger without any detectable AGN, the CO (6-5) line emission is in a circum-nuclear ring ($100 \lesssim r \lesssim 350$ pc) with a central hole.

• In the CO (6-5) line profile of NGC 34, a significant emission feature is detected on the redshifted wing, coincident with the frequency of the H^{13}CN (8-7) line emission which is a very high critical density isotopic molecular line. However, it cannot be ruled out that the feature is due to an outflow of warm dense gas with a mean velocity of 400 km s^{-1}.

• The non-detections of the nucleus in both the CO (6-5) and the 435 μm continuum rule out, with relatively high confidence, a Compton-thick AGN in NGC 1614.

• Comparisons between the ALMA maps and high resolution radio continuum and Pa α maps of NGC 1614 show that the local correlation (on the linear scale of ∼ 100 pc) between Σ_{Gas} and Σ_{SFR} (i.e. the Kennicutt-Schmidt law) is severely disturbed. In particular, the nucleus has a lower-limit of the Σ_{SFR}-to-Σ_{Gas} ratio about an order of magnitude above the nominal value in the standard Kennicutt-Schmidt law.

Acknowledgements

This paper makes use of the following ALMA data: ADS/JAO.ALMA#2011.0.00182.S. ALMA is a partnership of ESO (representing its member states), NSF (USA) and NINS (Japan), together with NRC (Canada) and NSC and ASIAA (Taiwan), in cooperation with the Republic of Chile. The Joint ALMA Observatory is operated by ESO, AUI/NRAO and NAOJ.

References

Armus, L., *et al.* 2009, *PASP*, 121, 599
Barnes, J. & Hernquist, L. 1996, *ApJ*, 471, 115
Bryant, P. M. & Scoville, N. Z. 1999, *AJ*, 117, 2632
Downes, D. & Solomon, P. M. 1998, *ApJ*, 507, 615
Fernandez, X., Petric, A. O., Schweizer, F., & van Gorkom, J. H. 2014, arXiv: 1401.1821

Gao, Y., Lo, K. Y., Lee, S.-W., & Lee, T.-H. 2001, *ApJ*, 548, 172

Hopkins, P. F., *et al.* 2009, *ApJ*, 691, 1186

Iono, D., Ho, P. T. P., Yun, M. S., *et al.* 2004, *ApJ*, 616, L63

Iono, D., Wilson, C. D., Yun, M. S., *et al.* 2009, *ApJ*, 695, 1537

Lu, N., Zhao, Y., Xu, K. C., *et al.* 2014, *ApJ*, 787, L23

Malken, M. A., *et al.* 1998, *ApJS*, 117, 25

Sakamoto, K., Aalto, S., Costagliola, F., *et al.* 2013, *ApJ*, 764, 42

Sakamoto, K., Wang, J., Wiedner, M. C., *et al.* 2008, *ApJ*, 684, 957

Sanders, D. B., Scoville, N. Z., & Soifer, B. T. 1991, *ApJ*, 370, 158

Scoville, N. Z., Sanders, D. B., Sargent, A. I., *et al.* 1989, *ApJ*, 345, L25

Scoville, N. Z., Yun, M. S., & Bryant, P. M. 1997, *ApJ*, 484, 702

van der Werf, P. P., Isaak, K. G., Meijerink, R., *et al.* 2010, *A&A*, 518, L42

Wang, J., Zhang, Q., Wang, Z., *et al.* 2004, *ApJ*, 616, L67

Wilson, C. D., Petitpas, G. R., Iono, D., *et al.* 2008, *ApJS*, 178, 189

Wootten, A. & Thompson, A. R. 2009, *IEEE*, 97, 1463

Xu, C. K., Cao, C., Lu, N., *et al.* 2014, *ApJ*, 787, 48

Yao, L., Seaquist, E. R., Kuno, N., & Dunne, L. 2003, *ApJ*, 588, 771

Galaxies in 3D across the Universe
Proceedings IAU Symposium No. 309, 2014
B. L. Ziegler, F. Combes, H. Dannerbauer, M. Verdugo, eds.
© International Astronomical Union 2015
doi:10.1017/S1743921314009326

Ionized and neutral gas in the XUV discs of nearby spiral galaxies

López-Sánchez[1,2], B. S. Koribalski[3], T. Westmeier[4] and C. Esteban[5,6]

[1] Australian Astronomical Observatory, PO Box 915, North Ryde, NSW 1670, Australia
email: Angel.Lopez-Sanchez@aao.gov.au
[2] Department of Physics and Astronomy, Macquarie University, NSW 2109, Australia
[3] CSIRO Astronomy and Space Science, Australia Telescope National Facility, PO BOX 76,
Epping, NSW 1710, Australia
[4] International Centre for Radio Astronomy Research, The University of Western Australia, 35
Stirling Hwy, Crawley, WA 6009, Australia
[5] Instituto de Astrofísica de Canarias, E-38200 La Laguna, Tenerife, Spain
[6] Departamento de Astrofísica, Universidad de La Laguna, E-38205 La Laguna, Tenerife, Spain

Abstract. We are conducting a multiwavelength study of XUV discs in nearby, gas-rich spiral galaxies combining the available UV (GALEX) observations with H I data obtained at the ATCA as part of the *Local Volume HI Survey* (LVHIS) project and multi-object fibre spectroscopy obtained using the 2dF/AAOmega instrument at the 3.9m AAT. Here we present the results of the multiwavelength analysis of the galaxy pair NGC 1512/1510. The H I distribution of NGC 1512 is very extended with two pronounced spiral/tidal arms. Hundreds of independent UV-bright regions are associated with dense H I clouds in the galaxy outskirts. We confirm the detection of ionized gas in the majority of them and characterize their physical properties, chemical abundances and kinematics. Both the gas distribution andthe distribution of the star-forming regions are affected by gravitational interactionwith the neighbouring blue compact dwarf galaxy NGC 1510. Our multiwavelength analysis provides new clues about local star-formation processes, the metal redistribution in the outer gaseous discs of spiral galaxies, the importance of galaxy interactions, the fate of the neutral gas and the chemical evolution in nearby galaxies.

Keywords. galaxies: ISM, abundances, kinematics and dynamics, evolution, interactions, dwarf, individual: NGC 1510, NGC 1512

1. Introduction

The study of star-formation processes in galaxy outskirts gives key clues about the formation and evolution of galaxies, not only because these regions may probe physical conditions similar to those present of the early Universe, but also because they test the *inside-out* scenario of galaxy disc formation. The discovery of UV-bright complexes in the outskirts (beyond their Hα or B_{25} radius) of nearby spiral galaxies (Thilker *et al.* 2005; Gil de Paz *et al.* 2005) has provided an excellent opportunity to investigate both how stars are formed in regions of low gas density and the chemical enrichment of the outer parts of galaxies. When combining the UV data with optical spectroscopy and 21-cm H I observations we get important constraints to the star-formation activity and star-formation history of galaxies, the relationships between gas and stars, as well as the physics behind them. We are conducting such study of XUV discs in nearby, H I -rich galaxies included in the *Local Volume HI Survey* (LVHIS) project (Koribalski 2008, Koribalski *et al.* in prep.) using the multi-object fiber-positioner instrument 2dF and the AAOmega spectrograph installed at the 3.9m Anglo-Australian Telescope (AAT, Siding Spring Observatory, NSW, Australia).

Figure 1. Multi-wavelength color-composite image of the galaxy pair NGC 1512/1510 obtained usingthe DSS *R*-band image (red), the ATCA H I distribution (green) and the GALEX NUV -band image (blue).The Spitzer 24μm image was overlaid just in the center of the two galaxies. We note that in the outer diskthe UV emission traces the regions of highest H I column density.

Here we present the results of our analysis of the XUV emission in the galaxy pair NGC 1512/1510. At the adopted distance of 9.5 Mpc, 1′ corresponds to 2.49 kpc. This system hosts hundreds of independent UV-bright regions associated with dense H I clouds in the outskirts, the blue compact dwarf galaxy (BCDG) NGC 1510 (which is located at just 13.8 kpc from the center of NGC 1512), a central star-forming ring and two tidal dwarf galaxy (TDG) candidates at very large distance (projected radius of 83 kpc) identified by their H I emission.

2. Neutral gas distribution in NGC 1512/1510

The analysis of the neutral gas distribution in NGC 1512/1510 was presented in Koribalski & López-Sánchez (2009). Our deep ATCA H I mosaic revealed a very extended (∼ 4× its optical diameter) H I distribution (see Fig. 1), spanning a diameter of ∼40 (or 110kpc). Two prominent spiral arms, which appear to wrap around ∼1.5 times, are among the mostremarkable H I features. The brightness and width of both H I arms varies with radius. The southern arm splits into three branches, followed by a broad region of H I debris towards the west before continuing on as a single feature towards the north. Thesedisturbances in the outer disk of NGC 1512 are likely caused by tidal interaction with andaccretion of the dwarf companion, NGC 1510. Individual H I clouds are found out to projectedradii of 30 (∼83 kpc). The velocity gradient detected within

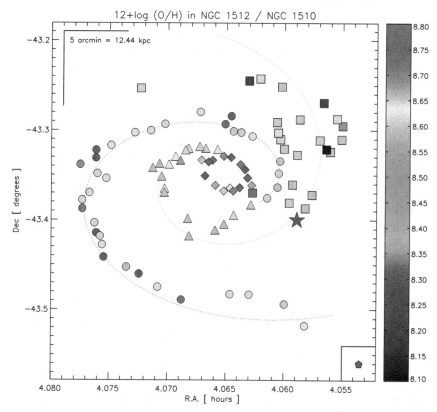

Figure 2. Map of the oxygen abundance in the NGC 1512/1510 system, as it is provided via the analysis of the emission lines detected in the UV-rich regions. The four areas identified in the system are shown using different symbols: Area 1 = Ring (Diamonds), Area 2 = Arm 1 (circles), Area 3 = Internal Arm (triangles), Area 4 = External Debris (squares). The position of the BCDG NGC 1510 is shown with a star, while the TDG 1 (which actually lies far from the system, at 78 kpc from the center of NGC 1512) is plotted using a pentagon. The scale included in the top left corner indicates the length of 5 arcmin, which corresponds to 12.44 kpc at the distance of NGC 1512. The value for R_{25} in NGC 1512 is 11.76 kpc (4.26 arcmin). The two H I tidal/spiral arms are sketched following a blue (Arm 1) or green (Arm 2) dotted line.

the extended clumps agrees withthat of the neighbouring spiral arms, suggesting that they are condensations within theoutermost parts of the disk. We detect star-formation activity in the two most distant H I clumps (TDG candidates). The large majority of the observed UV-complexes lie in regions where the H I column density isabove 2×10^{21} atoms cm^{-2}. Hence, XUV-discs should be embedded in larger H I envelopes –a 2X- H I disc , which are providing the fuel for their star-formation activity.

3. Ionized gas in the XUV regions of NGC 1512/1510

Our 2dF/AAOmega observations (López-Sánchez *et al.* in revision) provided deep, intermediate-resolution optical spectroscopy of 136 genuine UV-bright regions located in both the spiral and the XUV discs of the NGC 1512/1510 system. We detect ionized gas in the huge majority of them and confirm that in all cases except the centre of NGC 1512 photoionization by massive stars is the main excitation mechanism of the gas. In the XUV complexes, only few (1-5) O7V stars are responsible for this.

Using a comprehensive analysis of the emission lines of [O II] λ3727, Hγ, Hβ, [O III] λ5007, Hα, and [N II] λ6583 we computed the oxygen abundance and the N/O ratio for the majority of the detected UV-bright regions. The obtained chemical abundance map (see Fig. 2) reaches a projected distance of 78 kpc (6.6 R_{25}) from the center of NGC 1512. We found significant differences between regions along the Arm 1 at the east, which have oxygen abundances 8.25$\lesssim$12+log(O/H)$\lesssim$8.45, and knots located in the external debris of Arm 2, which typically have 8.40$\lesssim$12+log(O/H)$\lesssim$8.60. Considering just a radial gradient involving all complexes it is difficult to see a radial decrease of metallicity from the center of NGC 1512 to the outer regions. However, assuming both a radial and an azimuthal gradient following the spiral arms we are able to clearly distinguish that Arm 1 has an almost flat oxygen abundance –12+log(O/H)$\sim$8.35– and flat N/O ratio –log(N/O)$\sim$ –1.4– while regions located in the disrupted Arm 2 (inner arm and external debris), show a large dispersion in oxygen abundances –8.1$\lesssim$12+log(O/H)$\lesssim$8.6– and N/O ratios. Arm 2 has experienced an enhancement in star-formation because of the interaction with NGC 1510, flattening the radial metallicity at large radii, while Arm 1 still shows the original and poorly-disturbed radial distribution (gradient+flat in the outskirts).

Using the Hα emission line profile we trace the kinematics of the system out to 78 kpc, which generally matches well with that provided by the H I kinematics (velocity differences of less than 15 km s^{-1}). Some small velocity discrepancies (differences between 20 and 40 km s^{-1}) are found in some clusters of the external debris and in particular areas of Arm 1. We locate a region in the internal arm for which its Hα radial velocity differs by 136 km s^{-1} with respect its H I radial velocity. The upper limit to the metallicity of this knot is 12+log(O/H)=8.1, which is 0.4 dex lower than that found in nearby complexes. We suggest that this region actually is an independent dwarf galaxy or its remnant.

Although the XUV complexes are very likely induced by the NGC 1512/1510 interaction, they cannot be strictly defined as TDGs, as they have not been formed from material stripped from the galaxies, but from the diffuse gas already existing at large galactocentric radii. We propose to define these UV-bright, young, relatively low-metallicity, gas-rich entities to be "tidally-induced star-forming clusters" (TSFCs) in the galaxy outskirts.

4. Conclusions of the multiwavelength analysis of NGC 1512/1510

Our data suggest that the gas in the outer regions of NGC 1512 already had a metallicity of 12+log(O/H)$\sim$8.1 about 400 Myr ago, before the interaction with NGC 1510 started. The metals within the diffuse H I gas are very likely not coming from the inner regions of NGC 1512, but probably from material accreted during minor merger events during the life of the galaxy. We indeed found a probable remnant of an independent dwarf galaxy and three knots which may also be independent systems, all of them having 12+log(O/H)$\lesssim$8.1. This hypothesis agrees with the finding around other nearby large galaxies (e.g. Bresolin *et al.* 2009; Bresolin, Kennicutt & Ryan-Weber 2012), as well as constraints chemical and dynamical models of galaxy evolution.

References

Bresolin, F., *et al.* 2009, *ApJ*, 700, 309
Bresolin, F., Kennicutt, R. C., & Ryan-Weber, E., 2012, *ApJ*, 750, 122
Gil de Paz, A., *et al.* 2005, *ApJL*, 627, L29
Koribalski, B. S., 2008, *The Local Volume* H I *Survey*, Koribalski B. S., Jerjen H., eds. p. 41
Koribalski, B. S. & López-Sánchez, Á. R., 2009, *MNRAS*, 400,1749
Thilker, D. A., *et al.* 2005, *ApJL*, 619, L79

Galaxies in 3D across the Universe
Proceedings IAU Symposium No. 309, 2014
B. L. Ziegler, F. Combes, H. Dannerbauer, M. Verdugo, eds.

© International Astronomical Union 2015
doi:10.1017/S1743921314009338

The Wsrt Halogas Survey

George Heald[1,2] and the HALOGAS Team

[1] ASTRON, Postbus 2, 7990 AA Dwingeloo, The Netherlands
[2] Kapteyn Astronomical Institute, Postbus 800, 9700 AV, Groningen, The Netherlands
email: `heald@astron.nl`

Abstract. We present an overview of the HALOGAS (Hydrogen Accretion in LOcal GAlaxieS) Survey, which is the deepest systematic investigation of cold gas accretion in nearby spiral galaxies to date. Using the deep H I data that form the core of the survey, we are able to detect neutral hydrogen down to a typical column density limit of about 10^{19}cm^{-2} and thereby characterize the low surface brightness extra-planar and anomalous-velocity neutral gas in nearby galaxies with excellent spatial and velocity resolution. Through comparison with sophisticated kinematic modeling, our 3D HALOGAS data also allow us to investigate the disk structure and dynamics in unprecedented detail for a sample of this size. Key scientific results from HALOGAS include new insight into the connection between the star formation properties of galaxies and their extended gaseous media, while the developing HALOGAS catalogue of cold gas clouds and streams provides important insight into the accretion history of nearby spirals. We conclude by motivating some of the unresolved questions to be addressed using forthcoming 3D surveys with the modern generation of radio telescopes.

Keywords. galaxies: spiral — galaxies: evolution — galaxies: kinematics and dynamics

1. Background

Spiral galaxies require a continuous supply of gas to maintain star formation over long timescales and to explain their metallicity distribution and evolution (see Sancisi *et al.* 2008, and references therein). What form this gas supply takes has not been observationally established, but it may be that the dominant form of supply is cold gas accretion (e.g., Kereš *et al.* 2009). In the Milky Way (MW) and the nearest galaxies high velocity clouds (HVCs) are observed (Wakker & van Woerden 1997; Thilker *et al.* 2004), some with low metallicity (Wakker *et al.* 1999; Collins *et al.* 2007). The HVCs may contribute a small fraction of the fuel needed for star formation, but a more complete census is required to be able to draw general conclusions. While distinct H I clouds of this nature are not commonly detected in the outskirts of galaxies (e.g., Giovanelli *et al.* 2007; Putman *et al.* 2012), another morphological feature – thick gas disks – is becoming increasingly recognized as a typical structural feature of spiral galaxies (e.g., Oosterloo *et al.* 2007; Putman *et al.* 2012). These gas disks can contribute up to $\approx 10 - 20\%$ of the total H I mass of their host galaxies, and have distinct kinematics (e.g., Fraternali *et al.* 2002). These slowly rotating thick disks may be connected to the accretion history (Marinacci *et al.* 2010) and provide an observational tracer of coronal material feeding the disk.

Searching for clouds and for thick H I disks requires specialized deep observations. The HALOGAS (Hydrogen Accretion in Local Galaxies; Heald *et al.* 2011) Survey was undertaken to provide the required capability for a large enough sample to begin to draw statistical conclusions. The core of the survey are Westerbork Synthesis Radio Telescope (WSRT) observations of 22 edge-on and moderately-inclined nearby galaxies, each observed for 120 hours in the 21-cm H I line. The typical 5σ column density sensitivity is $N_{\rm HI} \lesssim 10^{19}$cm^{-2} for typical linewidth of $\Delta v = 12$kms^{-1}, making HALOGAS the deepest interferometric galaxy H I survey available to date. HALOGAS has an optimal angular resolution of $\approx 30''$ (1.5 kpc at a typical distance of 10 Mpc), unsuitable for detailed high-resolution studies of the gas-star formation connection like THINGS (The H I Nearby Galaxy Survey; Walter *et al.* 2008), but ideal for faint diffuse emission. The typical HALOGAS mass sensitivity for unresolved clouds with the same linewidth allows the detection of HVC analogues in the outskirts of the survey galaxies,

$$M_{HI} = 2.7 \times 10^5 \left(\frac{D}{10\text{Mpc}} \right)^2 M_\odot. \qquad (1.1)$$

The primary HALOGAS observations were completed as of early 2013. A number of ancillary data products have been collected to supplement our deep H I observations. Deep multi-band optical imagery have been obtained at the *Isaac Newton Telescope* (INT) and Kitt Peak National Observatory (KPNO); see e.g. de Blok *et al.* (2014). Sensitive *GALEX* UV data have also been collected to supplement deep survey data already available for some HALOGAS targets. Together with the H I line observations, broadband full-polarization continuum data are also available for many of the survey galaxies, allowing the detailed investigation of magnetic fields and cosmic rays in the same galaxies. The interferometric HALOGAS data are being supplemented with single dish data from Effelsberg and the the Green Bank Telescope (GBT) — useful not only to provide a missing short-spacing correction, but to enhance the search for clouds and streams in the outer parts of HALOGAS galaxies.

2. Overview of HALOGAS results

Sophisticated data analysis techniques are required in order to achieve reliable structural parameters from the sensitive H I data cubes provided by HALOGAS. Isolating thick H I disks in particular requires detailed modeling of the 3D gas distribution and kinematics. This modeling work is mostly performed using the TiRiFiC (Tilted Ring Fitting Code; Józsa *et al.* 2007) software.

Two examples that illustrate the need for detailed tilted ring modeling are shown in Figure 1. In the case of UGC 7774, the strong warping already apparent in the plane of the sky (see also García-Ruiz *et al.* 2002) also causes the vertical thickening. In the other galaxy, NGC 1003, the disk structure recovered from a detailed 3D analysis shows a substantial warp ($\approx 15-20°$) along the line of sight, causing the disk to appear thicker in projection than the actual vertical distribution of the H I layer.

Another interesting example, NGC 5023 (not shown), illustrates an extreme case: recovery of spiral structure in an edge-on disk (see Kamphuis *et al.* 2013). The modeling works best when these spiral structures are present and are required to fully understand the kinematics, emphasizing that this level of detail is essential to properly describe the 3D structure of the H I in galaxies.

HALOGAS galaxies are both found to host thick H I disks (Gentile *et al.* 2013; de Blok *et al.* 2014) and to have little extraplanar H I (Zschaechner *et al.* 2011, 2012). This dichotomy provides powerful leverage on the origin of H I thick disks in galaxies.

2.1. *Star formation and the origin of thick* H I *disks*

The entire HALOGAS sample has been inspected in order to characterize the presence and characteristics of extraplanar H I. We use both the 3D modeling mentioned in §2, as well as an approximate disk-halo separation of the type presented by Fraternali *et al.* (2002). On this basis we have searched for evidence of a common origin of the extraplanar H I by comparing with the general properties of the sample galaxies. From the HALOGAS

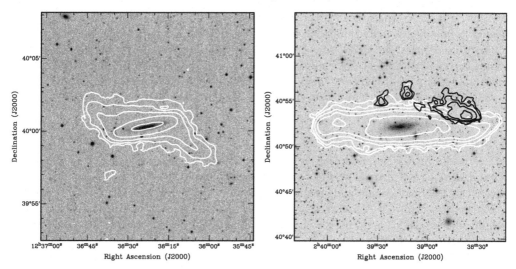

Figure 1. Two HALOGAS sample galaxies in optical and H I. *Left*: DSS2 R-band image of UGC 7774 displayed on a logarithmic stretch and overlaid with HALOGAS H I total intensity contours, starting at $N_{\rm HI} = 9 \times 10^{18}\,{\rm cm}^{-2}$ and increasing by multiples of 2. The angular resolution of the H I data is $36'' \times 33''$. *Right*: DSS2 R-band image of NGC 1003 displayed on a logarithmic stretch and overlaid with HALOGAS H I total intensity contours. The H I from the entire galaxy is shown in white contours, starting at $N_{\rm HI} = 5 \times 10^{18}\,{\rm cm}^{-2}$ and increasing by multiples of 4. The black contours show the HVC analogues discussed in the text (§2.2), starting at about the same column density and increasing by multiples of 2. The resolution of the H I data is $39'' \times 34''$.

sample we find a tentative connection between the star formation rate density and the presence of thick H I disks, such that thick H I disks appear to be present in galaxies above a threshold SF energy injection. We note that a similar relationship has been identified for radio continuum halos (Dahlem *et al.* 2006). In the case of the H I, there appears to be a strong dependence on the star formation rate density, but a weaker dependence on the stellar mass density that traces the strength of the gravitational potential.

2.2. *Cold gas accretion in galaxies*

Another key result of the HALOGAS Survey is the identification of H I clouds and streams in the vicinity of the sample galaxies that could originate through the cold gas accretion process, and to determine the impact on the star formation history of the sample. As previously noted, the occurrence of gas clouds in the outskirts of galaxies is low. The HALOGAS Survey, despite its high sensitivity, does not substantially change this picture. The HALOGAS catalog of isolated gas features is still in preparation (as we work toward a statistically sound census across the full sample), but we can already state qualitatively that there is a low incidence of clouds and streams. The contribution of visible H I accretion to the star formation fueling process in nearby galaxies appears to be minimal.

As a representative example, we highlight NGC 1003 as shown in Figure 1. A small number of morphologically and kinematically distinct gas features have been identified from an inspection of the 3D H I dataset; these are shown with black contours in Fig. 1. These features do not appear to have stellar counterparts, even in our deep optical images. The H I masses and distances from NGC 1003 of these clouds are similar to the properties of the HVCs around the MW. The mass of these HVC analogues adds up to only $M_{\rm HI} = 4 \times 10^6 M_\odot$, which over a dynamical time ($\tau_{\rm dyn} \approx 500{\rm Myr}$) contributes only 2% of the star formation rate of NGC 1003 (SFR $= 0.40 M_\odot {\rm yr}^{-1}$; Heald *et al.* 2012). Alternatively, if the gas contained in these clouds were used to completely fuel the current SFR, they would need to be replenished after only $\tau_{\text{repl}} \approx 10\text{Myr}$. These rates neglect a possible contribution from an ionized gas component to which we do not have observational constraints. We note however that the contribution of ionized gas to the total mass of MW HVCs can be substantial (e.g., Lehner & Howk 2011; Fox *et al.* 2014); QSO absorption line studies of low column density H I probed by HALOGAS where possible would be very useful to constrain this aspect of the mass budget.

3. Future prospects

The scientific themes addressed by the HALOGAS project will help to focus the groundbreaking science questions to be addressed with the Square Kilometre Array (SKA) and its pathfinder and precursor projects. For descriptions of surveys targeting deeper observations of larger nearby galaxy samples, as well as investigations of low column density IGM material, see the discussions by de Blok *et al.* (2015) and Popping *et al.* (2015), respectively.

Acknowledgements

The Westerbork Synthesis Radio Telescope is operated by ASTRON (Netherlands Institute for Radio Astronomy) with support from the Netherlands Foundation for Scientific Research (NWO).

References

de Blok, W. J. G., Józsa, G. I. G., Patterson, M., *et al.* 2014, *A&A*, 566, A80
de Blok, W. J. G., *et al.* 2015, in *"Advancing Astrophysics with the Square Kilometre Array"*
Collins, J. A., Shull, J. M., & Giroux, M. L. 2007, *ApJ*, 657, 271
Dahlem, M., Lisenfeld, U., & Rossa, J. 2006, *A&A*, 457, 121
Fox, A. J., Wakker, B. P., Barger, K. A., *et al.* 2014, *ApJ*, 787, 147
Fraternali, F., van Moorsel, G., Sancisi, R., & Oosterloo, T. 2002, *AJ*, 123, 3124
García-Ruiz, I., Sancisi, R., & Kuijken, K. 2002, *A&A*, 394, 769
Gentile, G., Józsa, G. I. G., Serra, P., *et al.* 2013, *A&A*, 554, A125
Giovanelli, R., Haynes, M. P., Kent, B. R., *et al.* 2007, *AJ*, 133, 2569
Heald, G., Józsa, G., Serra, P., *et al.* 2011, *A&A*, 526, A118
Heald, G., Józsa, G., Serra, P., *et al.* 2012, *A&A*, 544, 1
Heald, G., *et al.* 2015, in prep.
Józsa, G. I. G., Kenn, F., Klein, U., & Oosterloo, T. A. 2007, *A&A*, 468, 731
Jütte, E., *et al.* 2015, in prep.
Kamphuis, P., Rand, R. J., Józsa, G. I. G., *et al.* 2013, *MNRAS*, 434, 2069
Kereš, D., Katz, N., Fardal, M., Davé, R., & Weinberg, D. H. 2009, *MNRAS*, 395, 160
Lehner, N. & Howk, J. C. 2011, *Science*, 334, 955
Marinacci, F., Binney, J., Fraternali, F., *et al.* 2010, *MNRAS*, 404, 1464
Oosterloo, T., Fraternali, F., & Sancisi, R. 2007, *AJ*, 134, 1019
Popping, A., *et al.* 2015, in *"Advancing Astrophysics with the Square Kilometre Array"*
Putman, M. E., Peek, J. E. G., & Joung, M. R. 2012, *ARA&A*, 50, 491
Sancisi, R., Fraternali, F., Oosterloo, T., & van der Hulst, T. 2008, *A&A Rev.*, 15, 189
Thilker, D. A., Braun, R., Walterbos, R. A. M., *et al.* 2004, *ApJ*, 601, L39
Wakker, B. P., Howk, J. C., Savage, B. D., *et al.* 1999, *Nature*, 402, 388
Wakker, B. P. & van Woerden, H. 1997, *ARA&A*, 35, 217
Walter, F., Brinks, E., de Blok, W. J. G., *et al.* 2008, *AJ*, 136, 2563
Zschaechner, L. K., Rand, R. J., Heald, G. H., Gentile, G., & Kamphuis, P. 2011, *ApJ*, 740, 35
Zschaechner, L. K., Rand, R. J., Heald, G. H., Gentile, G., & Józsa, G. 2012, *ApJ*, 760, 37

Galaxies in 3D across the Universe
Proceedings IAU Symposium No. 309, 2014   © International Astronomical Union 2015
B. L. Ziegler, F. Combes, H. Dannerbauer, M. Verdugo, eds.   doi:10.1017/S174392131400934X

Ionized gas outflows and global kinematics of low-z luminous star-forming galaxies

S. Arribas[1], L. Colina[1], E. Bellocchi[1], R. Maiolino[2] and M. Villar-Martín[1]

[1] CSIC - Departamento de Astrofísica-Centro de Astrobiología (CSIC-INTA),Torrejón de Ardoz, Madrid, Spain
email: `arribas@cab.inta-csic.es`
[2] Cavendish Laboratory, University of Cambridge 19 J. J. Thomson Avenue, Cambridge CB3 0HE, UK **and** Kavli Institute for Cosmology, University of Cambridge, Madingley Road, Cambridge CB3 0HA, UK

Abstract. We have studied the kinematic properties of the ionized gas outflows and ambient interstellar medium (ISM) in a large and representative sample of local U/LIRGs (58 systems, 75 galaxies) at galactic and sub-galactic (i.e.,star-forming clumps) scales, thanks to integral field spectroscopy (IFS)-based high signal-to-noise integrated spectra. The main results have been recently presented in Arribas *et al.* (2014), and are briefly summarized here.

Keywords. - galaxies: evolution - galaxies: ISM -galaxies: Outflows

1. Introduction

The role of outflows governing galaxy evolution is believed to be crucial, as they can regulate and quench both SF and black hole activity, being also the primary mechanism by which dust and metals are redistributed over large scales within the galaxy, or even expelled into the intergalactic medium (IGM, see Veilleux, Cecil, & Bland-Hawthorn 2005 for a review). Hence most of the models of galaxy formation require energetic outflows to reproduce the observed properties (e.g., Silk and Rees 1998; Di Matteo *et al.* 2005; Hopkins *et al.*2012). Although the general theoretical framework seems reasonably well established, it is not well constrained by observations as it is unclear how the dynamical status of the ISM and the properties of the outflows depend on the galaxy mass, large-scale tidal forces, SFR, and/or type and luminosity of the AGN.

We have studied the main integrated kinematic properties of the ionized interstellar medium in the largest sample of local (U)LIRGs observed via IFS (Arribas *et al.* 2014). Specifically, we have focussed the study on i) the global dynamical status of the ambient ionized gas via the velocity dispersion of the ISM, and ii) the properties of the ionized gas outflows, and how they relate to the characteristics of the system like the global SFR, SF density, morphological merging phase, and the presence of AGNs. For the present study, we follow a methodology often used at high-z, which consists in integrating the IFS spectra for deriving the (flux weighted) average properties of the system. This allow us a direct comparison with high-z samples. The properties of a sample of luminous SF clumps have also been investigated.

2. Main results and conclusions

A summary of the main results are given below (see Arribas *et al.* 2014 for details).
(*a*) Global dynamical status of the ambient ionized gas

- *The role of star formation regulating* σ. Isolated non-AGN LIRG discs have velocity dispersions of the ambient ionized gas (45 ± 4 kms^{-1}) about a factor 2 larger than in less active spirals, likely as a consequence of the increased star formation. However, the heating effects of the star formation tend to saturate in the U/LIRGs luminosity range as indicated by the weak dependency of the velocity dispersion on the IR-derived SFR ($\sigma \propto$ SFR$(L_{IR})^{+0.12}$). The velocity dispersion also has a very slight dependency on the Hα star formation surface density ($\sigma \propto \Sigma_{SFR}^{0.06}$), departing significantly from the expected behaviour if energy release by the starburst were regulating σ (i.e., $\sigma \propto \Sigma_{SFR}^{0.5}$) (see Fig.1). The relatively small role of the star formation driving the dynamical status of the ambient ionized gas in U/LIRGs is reinforced by the fact that the clumps of strong SF do not have an increase in velocity dispersion associated, as we find in the corresponding IFS maps.
- *Heating by tidal forces.* The σ of the ionized gas follows a sequence of increasing values with the dynamical phase of the system, from isolated discs to interacting pairs and mergers ($< \sigma >$ of 45, 59, and 73 km s^{-1}, respectively). This evolution towards dynamically hotter systems seems to be mainly driven by the gravitational energy release associated with the interacting/merging process, rather than with SF.
- *The role of AGNs.* The impact of AGNs in the dynamical status of the ionized gas is significant in ULIRGs, with a net increase in the average velocity dispersion from 74 ± 6 km s^{-1} for non-AGNs to 109 ± 14 km s^{-1} for AGNs, i.e., by a factor 1.5. For LIRGs the impact is significantly lower (i.e., a factor of 1.2).
- *Local versus high-z SF galaxies.* The low-z U/LIRGs cover a range in σ and Σ_{SFR} similar to that of high-z SFGs. Moreover, the observed weak dependency of σ of the ionized gas on the compactness of the SF for local U/LIRGs ($\sigma \propto \Sigma_{SFR}^{n}$, with $n = 0.06 \pm 0.03$) is in very good agreement with the dependency reported by Genzel *et al.* (2011) at high-z ($n = 0.07 \pm 0.025$) (Fig.1).
- *SF clumps properties and scaling laws.* The sizes, luminosities and velocity dispersions of the clumps are in general consistent with the mean scaling relations derived for (mainly) giant HII regions in spirals, although the smaller clumps (diameter less than 1 kpc) appear more luminous, at a given size. When compared with high-z clumps, they are in general smaller, less luminous and have lower σ. However, part of these differences could be due to systematic observational effects.

(*b*) Ionized gas outflows

- *Detection rate.* The presence of ionized gas outflows in U/LIRGs seems to be universal based on the detection of a secondary broad, usually blueshifted, kinematic component in the Hα emission line.
- *The relative role of SF and AGN.* In non-AGN U/LIRGs, the maximum velocity of the outflow (V$_{max}$) shows a dependency on the SFR of the type V$_{max}(non - AGN) \propto SFR^{+0.24}$. In U/LIRGs with an AGN, the velocity increases by a factor of $\sim$2 with respect to non-AGNs, independent of the global SFR (i.e., V$_{max}(AGN) \propto SFR^{+0.23}$). Moreover, the fraction of Hα flux associated with the outflow increases by a factor of ~ 1.4 in U/LIRGs with an AGN. Thus, AGNs in U/LIRGs are able to generate faster and more massive ionized gas outflows than pure starbursts.
- *Dependency on merger evolution.* The observed differences in the properties of the outflows for isolated discs, interacting systems, and mergers are consistent with their associated change in SF. Therefore, the release of gravitational potential along the merger evolution does not seem to affect the properties of the outflow, except for its effects on the SF.
- *Outflowing mass and mass loading factors.* The outflowing mass rate covers a wide range from about 0.1 to 100 M$_\odot yr^{-1}$ with a close to linear dependency with the

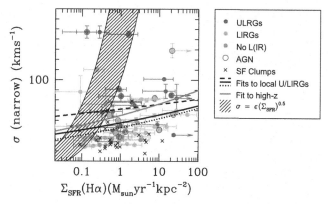

Figure 1. Velocity dispersion of the ISM (narrow component) as a function of the star formation density, Σ_{SFR}, derived considering the extinction corrected Hα fluxes (see text), and Hα half-light radii (Arribas *et al.* 2012). The symbol and line codes are indicated in the right panel. The black continuous line corresponds to an unweighted regression to all the U/LIRGs ($\sigma \propto \Sigma_{SFR}^{n}$, $n = 0.06 \pm 0.03$, r = 0.3), while the dotted and dashed lines are for non-AGNs (n = 0.06, r = 0.4) and AGNs (n = 0.03, r = 0.1), respectively. The shaded region corresponds to a behaviour of the type σ (kms^{-1})=$\epsilon\Sigma_{SFR}^{0.5}$ for efficiencies (ϵ) between 100 and 240, which is expected for SF driven motions (Lehnert *et al.* 2009). The magenta line corresponds to the unweighted) fit, found by Genzel *et al.* (2011), to a large compilation of high-z galaxies and clumps (see text)

IR-derived SFR ($\dot{M}_g \propto SFR^{+1.11}$). Mass loading factors (η) are in general below one, with only a few AGNs above this limit. The dependency of η on the dynamical mass of the system agrees well with recent simulations (Hopkins *et al.* 2012). However, the observed absolute values for η, as well as its trend with star formation surface density (($\eta \propto \Sigma_{SFR}^{+0.21}$, for all non-AGN U/LIRGs) disagree with these models.

• *Escape fractions and IGM metal enrichment.* Only less massive (log(M$_{dyn}$/M$_{\odot}$) < 10.4) U/LIRGs have outflowing maximum velocities (V$_{max}$) higher than their escape velocities, and therefore part of their ejected metal-rich gas could escape the gravitational potential of the host galaxy. More massive galaxies (log(M$_{dyn}$/M$_{\odot}$) > 10.4) would retain all the gas, even if they host an AGN. These results are consistent with chemical evolutionary models (Fig.2).

• *Outflows in low- and high-z star-forming galaxies.* The observed average properties (line width, velocity shift, and broad-to-narrow line flux ratio) are similar in low-z U/LIRGs and high-z star-forming galaxies, with hints that the fraction of Hα flux in the broad component could somehow be larger in low-z U/LIRGs. While high-z outflows show mass loading factors above one for galaxies with star formation surface densities above 1 $M_{\odot}yr^{-1}kpc^{-2}$, such a threshold is not observed in low-z U/LIRGs even after considering the different gas fraction.

• *Outflows in luminous star-forming clumps.* Ionized gas outflows appear to be very common (detection rate over 80%) in the bright SF clumps found in LIRGs, which are characterized by a median star formation surface density of 0.13 $M_{\odot}yr^{-1}kpc^{-2}$ and half-light radius of 0.49 kpc. Their properties (line width, velocity shift, and fraction of Hα flux in the broad component) appear to be less extreme than those associated with the galaxy, and therefore the energy involved in the outflows associated with individual clumps is (in relative terms) more modest than that for the entire galaxy.

• *Outflows in low- and high-z SF clumps.* Ionized gas outflows generated in SF clumps at all redshifts appear to carry a similar fraction of their Hα luminosity. However, high-z SF clumps launch outflows with much higher velocities, factors 4

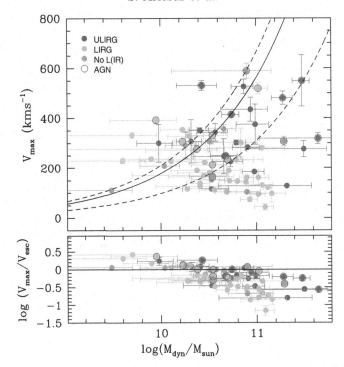

Figure 2. Top panel: Maximum outflow velocity as a function of the dynamical mass. Red and green dots distinguish ULIRGs and LIRGs systems. Orange dots are objects without L_{IR} determinations. Blue circles indicate objects with evidence of being affected by an AGN in the optical lines (see text). The solid line represents the mean escape velocity for r = 3kpc, considering a simple model of a truncated isothermal sphere of $r_{max}/r = 10$. The dotted lines correspond to values for $r_{max}/r = 1$ and 100 (see text). Bottom panel: The maximum outflow velocities are represented in units of the escape velocity for the case $r_{max}/r = 10$.

to 10 higher, than clumps in low-z U/LIRGs. For a given observed SF surface density, outflows in high-z clumps would be about one to two orders of magnitude more energetic than the outflows launched by clumps in U/LIRGs, i.e. the most extreme starburst galaxies in our low-z universe.

References

Arribas, S., Colina, L., Alonso-Herrero, A., *et al.* 2012, *A&A*, 541, A20
Arribas, S., Colina, L., Bellocchi, E., Maiolino, R., & Villar-Martín, M. 2014, *A&A*, 568, A14
Ceverino, D., Dekel, A., & Bournaud, F. 2010, *MNRAS*, 404, 2151
Di Matteo, T., Springel, V., & Hernquist, L. 2005, *Nature*, 433, 604
Genzel, R., Newman, S., Jones, T., *et al.* 2011, *ApJ*, 733, 101
Hopkins, P. F., Quataert, E., & Murray, N. 2012, *MNRAS*, 421, 3522
Lehnert, M. D., Nesvadba, N. P. H., Le Tiran, L., *et al.* 2009, *ApJ*, 699, 1660
Murray, N., Ménard, B., & Thompson, T. A. 2011, *ApJ*, 735, 66
Ostriker, E. C. & Shetty, R. 2011, *ApJ*, 731, 41
Silk, J. & Rees, M. J. 1998, *A&A*, 331, L1
Veilleux, S., Cecil, G., & Bland-Hawthorn, J. 2005, *ARA&A*, 43, 769

Galaxies in 3D across the Universe
Proceedings IAU Symposium No. 309, 2014
B. L. Ziegler, F. Combes, H. Dannerbauer, M. Verdugo, eds.

© International Astronomical Union 2015
doi:10.1017/S1743921314009351

Dissecting the 3D structure of elliptical galaxies with gravitational lensing and stellar kinematics

Matteo Barnabè[1,2], Chiara Spiniello[3] and Léon V. E. Koopmans[4]

[1]Dark Cosmology Centre, Niels Bohr Institute, University of Copenhagen,
Juliane Maries Vej 30, 2100 Copenhagen Ø, Denmark
email: **mbarnabe@dark-cosmology.dk**
[2]Niels Bohr International Academy, Niels Bohr Institute, University of Copenhagen,
Blegdamsvej 17, 2100 Copenhagen Ø, Denmark
[3]Max Planck Institute for Astrophysics, Karl-Schwarzschild-Strasse 1,
85740 Garching, Germany
[4]Kapteyn Astronomical Institute, University of Groningen, P. O. Box 800,
9700 AV Groningen, The Netherlands

Abstract. The combination of strong gravitational lensing and stellar kinematics provides a powerful and robust method to investigate the mass and dynamical structure of early-type galaxies. We demonstrate this approach by analysing two massive ellipticals from the XLENS Survey for which both high-resolution *HST* imaging and X-Shooter spectroscopic observations are available. We adopt a flexible axisymmetric two-component mass model for the lens galaxies, consisting of a generalised NFW dark halo and a realistic self-gravitating stellar mass distribution. For both systems, we put constraints on the dark halo inner structure and flattening, and we find that they are dominated by the luminous component within one effective radius. By comparing the tight inferences on the stellar mass from the combined lensing and dynamics analysis with the values obtained from stellar population studies, we conclude that both galaxies are characterised by a Salpeter-like stellar initial mass function.

Keywords. galaxies: elliptical and lenticular, cD - galaxies: structure - galaxies: kinematics and dynamics - gravitational lensing: strong

1. Introduction

Understanding the formation and evolution mechanisms of early-type galaxies (ETGs) remains a crucial challenge in present-day astrophysics. In recent years, high-resolution simulations of galaxy evolution, including both baryons and dark matter (e.g. Genel *et al.* 2014), have progressed enough that it is finally becoming possible to investigate in detail the physical properties of simulated systems and compare them to the corresponding properties of observed ellipticals in the local Universe. Whereas nearby ETGs have been the object of intense scrutiny, with stellar dynamics being the most often employed diagnostic tool (see e.g. Thomas *et al.* 2011; Cappellari *et al.* 2013), little is known about the detailed structure of ellipticals beyond redshift $z \approx 0.1$, despite the obvious importance of providing observational constraints to simulations throughout cosmic history.

At higher redshift, however, massive galaxies occasionally act as strong gravitational lenses: for such systems it becomes possible to supplement stellar kinematics with the lensing constraints to investigate the mass structure of these distant galaxies to a level that is comparable with what can be obtained for nearby objects (see e.g. Czoske *et al.* 2008; Koopmans *et al.* 2009; van de Ven *et al.* 2010; Barnabè *et al.* 2010, 2011). High quality combined data-sets, including extended kinematics, make it possible to disentangle the contributions of dark and luminous matter to the total mass budget, which in turns allows one to draw inferences also on the stellar initial mass function (IMF) of the system, by comparing the results with the stellar masses obtained from independent stellar population synthesis (SPS) analysis assuming a variety of IMF profiles. We demonstrate the power of this approach by applying a self-consistent combined lensing and dynamics analysis to two massive ETGs from the XLENS Survey (Spiniello *et al.* 2011) for which high signal-to-noise X-Shooter spectroscopic observations are at hand.

2. Observations

Both massive ellipticals SDSS J0936+0913 (velocity dispersion $\sigma \simeq 245$ kms^{-1}; redshift $z = 0.164$) and SDSS J0912+0029 ($\sigma \simeq 325$ kms^{-1}; $z = 0.190$) were first observed as part of the SLACS Survey (Bolton *et al.* 2008). The high-resolution images needed for the lensing modelling were obtained with *HST* ACS through the *F814W* filter. Elliptical B-spline models of the lens galaxies were subtracted off the images to isolate the structure of the gravitationally lensed background sources.

High signal-to-noise (S/N > 50) *UVB–VIS* X-Shooter spectra are available for both systems, and were used both to carry out SPS analysis and derive long-slit spatially resolved stellar kinematics up to about one effective radius, $R_{\rm e}$.

3. Modelling

A joint self-consistent modelling of all the available lensing and kinematic constraints to derive inferences on the mass and dynamical structure of the two systems is conducted with the fully Bayesian CAULDRON code (Barnabè & Koopmans 2007; Barnabè *et al.* 2012). The lens galaxy is described by a flexible and realistic two-component axisymmetric mass model consisting of (i) a generalised Navarro-Frenk-White (gNFW) halo characterised by four free parameters (the inner density slope γ, the axial ratio $q_{\rm h}$, the halo concentration parameter c_{-2} and the virial velocity $v_{\rm vir}$) and (ii) a mass profile for the baryonic component obtained by de-projecting the multi-Gaussian expansion fit (MGE) to the observed surface brightness distribution of the galaxy, characterised by one free parameter (the total stellar mass $M_\star$ that sets the normalisation of the luminous distribution). The stellar kinematics is modelled with the Jeans anisotropic method (JAM) developed by Cappellari (2008), including the meridional plane orbital anisotropy $b = \sigma_R^2 / \sigma_z^2$ as a free parameter.

4. Results and discussion

The full inferences on the model parameters obtained from the combined CAULDRON analysis are expressed as multivariate posterior probability distribution functions (PDFs), encapsulating all the information on the uncertainties in a statistically rigorous way. The 1D and 2D marginalised posterior PDFs for the individual parameters can be visualised as a familiar corner plot, as shown in Fig. 1 for galaxy J0912.

Both systems are found to be markedly dominated by the baryonic component within the inner $R_{\rm e}$, with a dark matter fraction $f_{\rm DM} = 0.06^{+0.10}_{-0.05}$ for J0936 and $f_{\rm DM} = 0.18^{+0.08}_{-0.08}$ for the more massive system J0912. The dark halo structure of both galaxies ($c_{-2} = 6.3^{+18.4}_{-4.1}$ and $v_{\rm vir} = 124^{+160}_{-88}$ kms^{-1} for J0936; $c_{-2} = 7.2^{+3.8}_{-2.7}$ and $v_{\rm vir} = 470^{+160}_{-130}$ kms^{-1} for J0912) is consistent with the concentration–virial velocity relation predicted from pure N-body simulations (Macciò *et al.* 2008): thus, within the fairly broad uncertainties,

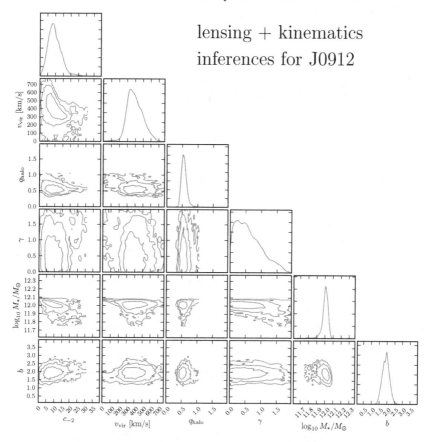

Figure 1. Marginalised 1D and 2D posterior PDFs for the six model parameters for galaxy J0912, using the combined constraints from lensing and kinematics. The three contours indicate the regions containing, respectively, 68%, 95% and 99.7% of the probability.

there is no obvious evidence for contraction or expansion of the dark halo in response to galaxy formation. The inner slope of the dark halo is basically unconstrained for J0936 ($\gamma = 0.92^{+0.72}_{-0.64}$), while for J0912 we can conclude that steep slopes are disfavoured ($\gamma = 0.46^{+0.41}_{-0.30}$). The two haloes differ in terms of flattening, with J0936 having a more spherical halo ($q_\mathrm{h} = 0.93^{+0.25}_{-0.18}$), while the halo of J0912 is clearly oblate ($q_\mathrm{h} = 0.54^{+0.10}_{-0.08}$).

The two lens galaxies have dissimilar dynamical structures within the probed region: J0936 ($b = 0.89^{+0.37}_{-0.31}$) is consistent with being a semi-isotropic system, while the velocity dispersion ellipsoid of J0912 is distinctly flattened ($b = 1.94^{+0.24}_{-0.21}$).

By comparing the stellar mass inferred from the combined lensing and dynamics modelling with the one derived from SPS analysis of X-Shooter spectra we can also put robust constraints on the normalisation of the stellar IMF, as illustrated in Fig. 2 for J0912. We determine that both galaxies are consistent with having a Salpeter IMF (i.e., slope $x = 2.35$ for a power-law IMF profile $dN/dm \propto m^{-x}$), in agreement with the findings for local massive ETGs from both spectroscopic and dynamical studies (e.g. van Dokkum & Conroy 2010; Cappellari *et al.* 2012). Moreover, we can rule out both a Chabrier and a "super-Salpeter" ($x \geqslant 3$) IMF with a high degree of confidence. The XLENS data-set makes it possible also to determine the IMF in a completely independent way from a spectroscopic simple stellar population analysis of the optical line-strength indices, using the approach outlined by Spiniello *et al.* 2014: the IMF slopes obtained with this method

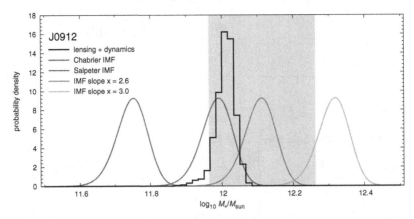

Figure 2. Comparison of the total stellar mass inferred for J0912 from the combined lensing and dynamics study (histogram) with $M_\star$ obtained from SPS analysis, for different values of the IMF slope (coloured curves). The green curve indicates $M_\star$ for the best-fit IMF slope determined from an independent method based on spectroscopic simple stellar population modelling of line-strength indices, with the grey band showing the 1-sigma uncertainties around that value.

($x = 2.10 \pm 0.15$ for J0936 and $x = 2.60 \pm 0.30$ for J0912) are fully consistent with the much tighter lensing and dynamics results. Finally, the constraints from these two complementary approaches can be combined to investigate the elusive low-mass cut-off of the IMF, as demonstrated by Barnabè et al. (2013). Work is ongoing to extend this analysis to the entire XLENS sample of 10 massive lens ETGs.

Acknowledgments

The Dark Cosmology Centre is funded by the Danish National Research Foundation.

References

Barnabè, M., Auger, M. W., Treu, T., et al. 2010, MNRAS, 406, 2339
Barnabè, M., Czoske, O., Koopmans, L. V. E., et al. 2011, MNRAS, 415, 2215
Barnabè, M. & Koopmans, L. V. E. 2007, ApJ, 666, 726
Barnabè, M., Spiniello, C., Koopmans, L. V. E., et al. 2013, MNRAS, 436, 253
Barnabè, M., Dutton, A. A., Marshall, P. J., et al. 2012, MNRAS, 423, 1073
Bolton, A. S., Burles, S., Koopmans, L. V. E., et al. 2008, ApJ, 682, 964
Cappellari, M. 2008, MNRAS, 390, 71
Cappellari, M., McDermid, R. M., Alatalo, K., et al. 2012, Nature, 484, 485
—. 2013, MNRAS, 432, 1862
Czoske, O., Barnabè, M., Koopmans, L. V. E., et al. 2008, MNRAS, 384, 987
Genel, S., Vogelsberger, M., Springel, V., et al. 2014, arXiv:1405.3749
Koopmans, L. V. E., Bolton, A., Treu, T., et al. 2009, ApJ, 703, L51
Macciò, A. V., Dutton, A. A., & van den Bosch, F. C. 2008, MNRAS, 391, 1940
Spiniello, C., Koopmans, L. V. E., Trager, S. C., et al. 2011, MNRAS, 417, 3000
Spiniello, C., Trager, S., Koopmans, L. V. E., & Conroy, C. 2014, MNRAS, 438, 1483
Thomas, J., Saglia, R. P., Bender, R., et al. 2011, MNRAS, 415, 545
van de Ven, G., Falcón-Barroso, J., McDermid, R. M., et al. 2010, ApJ, 719, 1481
van Dokkum, P. G. & Conroy, C. 2010, Nature, 468, 940

Galaxies in 3D across the Universe
Proceedings IAU Symposium No. 309, 2014
B. L. Ziegler, F. Combes, H. Dannerbauer, M. Verdugo, eds.

© International Astronomical Union 2015
doi:10.1017/S1743921314009363

Regrowth of stellar disks in mature galaxies: The two component nature of NGC 7217 revisited with VIRUS-W [†,◇]

Maximilian H. Fabricius,[1,2] Lodovico Coccato,[3] Ralf Bender,[1,2] Niv Drory,[4] Claus Gössl,[2] Martin Landriau,[4] Roberto P. Saglia,[1] Jens Thomas,[1] and Michael J. Williams[1,5]

[1] Max Planck Institute for Extraterrestrial Physics, Giessenbachstraße, 85748 Garching, Germany

[2] University Observatory Munich, Scheinerstraße 1, 81679 Munich, Germany

[3] European Southern Observatory, Karl-Schwarzschild-Straße 2, D-85748 Garching bei Muenchen, Germany

[4] McDonald Observatory, The University of Texas at Austin, 2515 Speedway, Stop C1402, Austin, Texas 78712-1206, USA

[5] Department of Astronomy, Columbia University, New York 10027, USA

Abstract. We have obtained high spectral resolution ($R \approx 9000$), integral field observations of the three spiral galaxies NGC 3521, NGC 7217 and NGC 7331 using the new fiber-based Integral Field Unit instrument VIRUS-W at the 2.7 m telescope of the McDonald Observatory in Texas. Our data allow us to revisit previous claims of counter rotation in these objects. A detailed kinematic decomposition of NGC 7217 shows that no counter rotating stellar component is present. We find that NGC 7217 hosts a low dispersion, rotating disk that is embedded in a high velocity dispersion stellar halo or bulge that is co-rotating with the disk. Due to the very different velocity dispersions (≈ 20 km s^{-1} vs. 150 km s^{-1}), we are further able to perform a Lick index analysis on both components separately which indicates that the two stellar populations are clearly separated in (Mgb,⟨Fe⟩) space. The velocities and dispersions of the faster component are very similar to those of the interstellar gas as measured from the [O III] emission. Morphological evidence of active star formation in this component further suggests that NGC 7217 may be in the process of (re)growing a disk inside a more massive and higher dispersion stellar halo.

Keywords. galaxies: bulges — galaxies: evolution — galaxies: formation — galaxies: structure

[†] This paper includes data taken at The McDonald Observatory of The University of Texas at Austin.

[◇] This paper contains data obtained at the Wendelstein Observatory of the Ludwig-Maximilians University Munich.

---

1. Introduction

The decomposition of galaxies into the structural components such as bulges, disks and bars is an essential tool to help the understanding of the formation of these systems. For a long time this has been limited to photometry alone, either through the decomposition of surface brightness profiles or the direct modelling of 2D images (e.g. Peng *et al.* 2002). More recently several approaches have been taken to decompose also spectra into multiple components (e.g. Vergani *et al.* 2007; Coccato *et al.* 2011; Johnston *et al.* 2012; Coccato *et al.* 2013, see also the parallel contribution by Coccato *et al.*). Some of these methods make explicitly use of the different kinematics of for instance two counter rotating components.

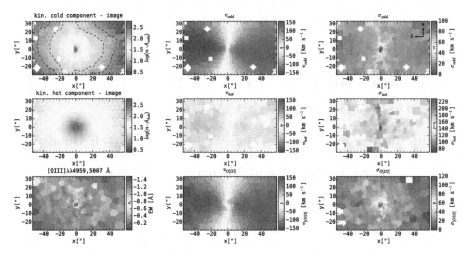

Figure 1. Two-dimensional maps of the of the stellar kinematics obtained from the double–Gaussian decomposition and the gas kinematics. We plot from left to right: the amplitude, the mean velocity and the velocity dispersion of the best fitting model. The upper row plots the moments for the cold component while the middle row plots the hot component. The amplitudes for the two components have been normalized such that their integrated sum equals the wavelength collapsed datacube. To highlight the flattening of the cold disk component, we added contours at two arbitrary values. The bottom row shows the summed equivalent width of the [O III]$\lambda\lambda$ 4959,5007 Å emission lines and the velocity and velocity dispersion obtained by the kinematics extraction routine from the brighter line at 5007 Å. In all maps positive y-values point to north and positive x-values to west. Reproduced from Fabricius *et al.* 2014.

While the spectral decomposition requires high signal to noise (SN) spectra, a large spectral resolution, and consequently large integration times, it has two significant advantages over the photometric decomposition: First, it is model free, it does not rely on the inwards extrapolation of an exponential profile for instance. Secondly, it allows us to study absorption line strengths of the superimposed components. Thus ages, metallicities and over-abundances of the stellar components can be studied separately.

NGC 7217 was previously claimed to host two counter rotating stellar disks (Merrifield & Kuijken 1994) based on double peaked line of sight velocity distributions (LOSVDs) and high velocity tails. This has been questioned by Buta *et al.* 1995 (hereafter B95) who claims that at least the tails can be explained by a massive but low surface brightness stellar halo. Using the new Integral Field Unit (IFU) spectrograph VIRUS-W we have obtained spectra that we analyse using a new algorithm to derive non-parametric LOSVDs but also an established algorithm to derive parametric two component models.

All results have been presented in Fabricius *et al.* (2014). Here we give a brief summary of our observations, the analysis techniques and our findings.

2. Data & Analysis

We observed NGC 3521, NGC 7217 and NGC 7331 using the fiber based Integral-Field Unit spectrograph (IFU) VIRUS-W at the 2.7 m (Fabricius *et al.* 2012) telescope of the McDonald Observatory in Texas. We used the stellar dynamics mode of the instrument that offers a spectral resolution of $R \approx 9000$ ($\sigma_{inst} = 15$ km s^{-1}) in the optical. The IFU has a total field of view of 105 arcsec×55 arcsec with a fill factor of 1/3 such that three dithered observations are needed to fill the field of view completely. We interleaved dithers with sky nods for background subtraction. The observations were carried out in

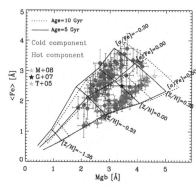

Figure 2. Mgb and $\langle$Fe$\rangle$ line strengths indices measured for the cold component (blue circles) and the hot component (red diamonds) in NGC 7217. Predictions from single stellar population models (Thomas *et al.* 2003) are also shown for comparison. The black, and green stars represent the mean values of the measurements by Gorgas *et al.* (2007) and Morelli *et al.* (2008) for a sample of bulges in spiral galaxies with the corresponding error bars representing the standard deviations of their measurements. The light blue star and its error bars represents the mean measurements and the standard deviation of a sample of early-type galaxies by Thomas *et al.* (2005). Reproduced from Fabricius *et al.* 2014.

dark time in August 2011, May 2011 and June 2012. Individual on-object integration times were limited to 1800 s to sample the sky regularly. The total on-object integration time per fiber amounts to 1.2 h, 1.5 h and 4.5 h for NGC 3521, NGC 7217 and NGC 7331 respectively.

For the primary kinematic extraction we employ a new version of the Maximum Penalized Likelihood which adds the treatment of nebular emission lines. This method allows us to derive non-parametric line-of-sight velocity distributions that show an obvious double component structure. We decompose the LOSVDs using two Gaussian models. In a second step we use the parametric spectral decomposition technique of Coccato *et al.* (2011) to confirm the result and to derive line strength indices for the two different components separately.

3. Results

In Fig. 1 we show the relative weights, the velocities and the velocity dispersions of the two different stellar components in NGC 7217. We also show the equivalent width, and the velocity and the velocity dispersion of the nebular emission as measured from the [O III] lines. We find that rather than two counter rotating stellar disks, NGC 7217 hosts a low dispersion (≈ 20 km s^{-1}), rotating ($v_{max} = 150$ km s^{-1}) disk that is embedded in a high velocity dispersion stellar halo (≈ 170 km s^{-1}) or bulge that is co-rotating ($v_{max} = 50$ km s^{-1}) with the disk. With the exception of the central arcseconds the low dispersion disk appears well aligned with the gas disk. Both the velocity fields and the dispersions are very similar, suggesting that if gas disk is a residual to the formation of the stellar disk, then the stellar disk is yet dynamically young. A structural analysis of the relative weights of the two gaussian components as function of position shows great similarity with the photometric decomposition by B95. The low dispersion component therefore is very likely identical to the exponential disk by B95 while the high dispersion halo represents his de Vaucouleurs' bulge (Fabricius *et al.* 2014).

The parametric spectral decomposition reveals that the lower dispersion component forms essentially a parallel sequence to the high dispersion component in the (Mgb,$\langle$Fe$\rangle$) absorption line strength diagram (see Fig. 3). Both sequences span a range of about 2 Å both in Fe and Mgb but the disk is offset by about an Å towards either larger Fe strengths, lower Mgb strengths or a combination of both.

4. Conclusions

Using high SN and large spectral resolution optical IFU data we have derived velocity and velocity dispersion maps for two separate, superimposed stellar components in NGC

7217. We were able to show that those correspond to a dominant, high velocity dispersion and slowly rotating stellar halo and an embedded, rotating stellar disk. The disk only contains about 30 % of the total light. Given its bluer color and the obvious sites of active star formation in imaging data it almost certainly has a lower mass to light ratio than the halo. Its mass fraction is therefore even lower that 30 %. The comparison of the line strengths with single stellar population models (Thomas *et al.* 2003) in the $(Mgb, \langle Fe \rangle)$ plane shows that the disk component is offset to lower $[\alpha/Fe]$ over-abundances but does not extend to higher metallicities than the high dispersion halo. This would be naturally explained by an accretion of relatively primordial material, that restarted the star formation at much lower metallicities but at the maximum over-abundance of the high dispersion component. The photometric and spectroscopic properties of the halo component are in several aspects more similar to an elliptical galaxy than to those of the bulges of spiral galaxies. If the halo is merger-built, than the disk would not have survived the merging event. We suggest that it must have formed after the merger, either from material that remained at large radii throughout the time of coalescence or from material that was accreted later through minor mergers or cold accretion from the intergalactic medium (Mazzuca *et al.* 2006; Eliche-Moral *et al.* 2010).

We are currently in the process of analysing the data for NGC 3521 and NGC 7331 that have similar previous claims of counter rotation Prada *et al.* (1996); Zeilinger *et al.* (2001).

Acknowledgements

We would like to thank the organizers for this wonderful conference. L. C. acknowledges financial support from the European Communitys Seventh Framework Program (/FP7/2007-2013/) under grant agreement No. 229517. This research has made use of the NASA/IPAC Extragalactic Database (NED) which is operated by the Jet Propulsion Laboratory, California Institute of Technology, under contract with the National Aeronautics and Space Administration.

References

Buta, R., van Driel, W., Braine, J., *et al.* 1995, *ApJ*, 450, 593
Coccato, L., Morelli, L., Corsini, E. M., *et al.* 2011, *MNRAS*, 412, L113
Coccato, L., Morelli, L., Pizzella, A., *et al.* 2013, *A&A*, 549, A3
Eliche-Moral, M. C., González-García, A. C., Balcells, M., *et al.* 2010, in *AIP Conference Series*, Vol. 1240, ed. V. P. Debattista & C. C. Popescu, 237–238
Fabricius, M. H., Coccato, L., & Bender, R. 2014, *MNRAS*, 441, 2212
Fabricius, M. H., Grupp, F., Bender, R., *et al.* 2012, in *SPIE Conference Series*, Vol. 8446
Gorgas, J., Jablonka, P., & Goudfrooij, P. 2007, *A&A*, 474, 1081
Johnston, E. J., Aragón-Salamanca, A., Merrifield, M. R., & Bedregal, A. G. 2012, *MNRAS*, 422, 2590
Mazzuca, L. M., Sarzi, M., Knapen, J. H., Veilleux, S., & Swaters, R. 2006, *ApJ*, 649, L79
Merrifield, M. R. & Kuijken, K. 1994, *ApJ*, 432, 575
Morelli, L., Pompei, E., Pizzella, A., *et al.* 2008, *MNRAS*, 389, 341
Peng, C. Y., Ho, L. C., Impey, C. D., & Rix, H.-W. 2002, *AJ*, 124, 266
Prada, F., Gutierrez, C. M., Peletier, R. F., & McKeith, C. D. 1996, *ApJ*, 463, L9+
Thomas, D., Maraston, C., & Bender, R. 2003, *MNRAS*, 343, 279
Thomas, D., Maraston, C., Bender, R., & Mendes de Oliveira, C. 2005, *ApJ*, 621, 673
Vergani, D., Pizzella, A., Corsini, E. M., *et al.* 2007, *A&A*, 463, 883
Zeilinger, W. W., Vega Beltrán, J. C., Rozas, M., *et al.* 2001, *Ap&SS*, 276, 643

Galaxies in 3D across the Universe
Proceedings IAU Symposium No. 309, 2014
B. L. Ziegler, F. Combes, H. Dannerbauer, M. Verdugo, eds.

© International Astronomical Union 2015
doi:10.1017/S1743921314009375

IFUs surveys, a panoramic view of galaxy evolution

Sebastián F. Sánchez and The CALIFA collaboration

Instituto de Astronomía,Universidad Nacional Autonóma de Mexico, A. P. 70-264, 04510, México,D. F. email: sfsanchez@astro.unam.mx

Abstract. We present here a brief summary of the currenly on-going IFU surveys of galaxies in teh Local Universe, describing their main characteristics, including their sample selections, instrumental setups, wavelength ranges, and area of the galaxies covered. Finally, we make an emphasis on the main characteristics of the CALIFA survey and the more recent results that has been recently published.

Keywords. galaxies: evolution, techniques: integral field spectroscopy

1. Introduction

Much of our recently acquired understanding of the architecture of the Universe and its constituents derives from large surveys (e.g., 2dFGRS, Folkes *et al.* 1999;SDSS, York *et al.* 2000; GEMS, Rix *et al.* 2004; VVDS, Le Fèvre *et al.* 2004; COSMOS, Scoville *et al.* 2007 ; GAMMA, Driver *et al.* 2009, to name but a few). Such surveys have not only constrained the evolution of global quantities such as the cosmic star formation rate, but also enabled us to link this with the properties of individual galaxies – morphological types, stellar masses, metallicities, etc.. Compared to previous approaches, the major advantages of this recent generation of surveys are: (1) the large number of objects sampled, allowing for meaningful statistical analysis to be performed on an unprecedented scale; (2) the possibility to construct large comparison/control samples for each subset of galaxies; (3) a broad coverage of galaxy subtypes and environmental conditions, allowing for the derivation of universal conclusions; and (4) the homogeneity of the data acquisition, reduction and (in some cases) analysis.

On the other hand, the cost of these surveys, in terms of telescope time, manpower, and involved time scales, is also unprecedented in astronomy. The user of such data products has not necessarily been involved in any step of designing or conducting the survey, but nevertheless takes advantage of the data by exploiting them according to her/his scientific interests. This new approch to observational astronomy is also changing our perception of the scientific rationale behind a new survey: While it is clear that certain planned scientific applications are key determinants to the design of the observations and 'drive' the survey, the survey data should at the same time allow for a broad range of scientific exploitation. This aspect is now often called *Legacy* value.

Current technology generally leads to surveys either in the imaging or in the spectroscopic domain. While imaging surveys provide two-dimensional coverage, they carry very little spectral information. This is also true for multiband photometric surveys such as COMBO-17 (Wolf *et al.* 2003), ALHAMBRA (Moles *et al.*. 2008) or the planned PAU project (Benítez *et al.*. 2009), which will still be unable to accurately capture individual spectral lines and measure, e.g., emission line ratios or internal radial velocity differences. Spectroscopic surveys such as SDSS or zCOSMOS, on the other hand, do provide more detailed astrophysical information, but they are generally limited to one spectrum per

Table 1. Comparison of IFU Surveys

Specification	MaNGA	SAMI	CALIFA	Atlas3D
Sample Size	10,000	3,400	600	260
Selection	$M > 10^9 M.$	$M > 10^{8.2} M.$	$45" < D_{25} < 80"$	$E/S0 M > 10^{9.8} M.$
Radial coverage	$1.5r_e$ (2/3), $2.5r_e$ (1/3)	$1\text{-}2r_e$	$>2.5r_e$	$<1r_e$
S/N at $1r_e$	15-30	10-30	~30	15
Wavelength range(Å)	3600-10300	3700-7350	3700-7500	4800-5380
Instrumental resolution	50-80 km/s	75/28 km/s	85/150 km/s	105 km/s
Input Spaxel Size	2.0″	1.6‴	2.7″	1″
Input Spaxels per object	$<3\times127^1$	3×61	3×331	1,431
Spatial FWHM	2″	2″	2.5″	1.5″
Telescope size	2.5m	3.5m	3.5m	4.2m

[1] This corresponds to the largest MaNGA bundle. Each MaNGA plate provides with 17 bundles of different amount of fibers: 2x19; 4x37; 4x61; 2x91; 5x127.

galaxy, often with aperture losses that are difficult to control. For example, the 3″ diameter of the fiber used in the SDSS corresponds to vastly different linear scales at different redshifts, without the possibility to correct for these aperture effects.

An observational technique combining the advantages of imaging and spectroscopy (albeit with usually quite small field of view) is Integral Field Spectroscopy (IFS). However, so far this technique has rarely been used in a 'survey mode' to investigate large samples. Among the few exceptions there is, most notably, the SAURON survey (de Zeeuw *et al.* 2002), focused on the study of the central regions of 72 nearby early-type galaxies and bulges of spirals. Others are (i) the PINGS project at the CAHA 3.5m of a dozen very nearby galaxies (Rosales-Ortega *et al.* 2010), (ii) the Disk Mass Survey (Bershady *et al.* 2010) that provided super high spectral resolution using PPAK and SparsePak of 46 face-on spirals; (iii) the VENGA project (Blanc *et al.* 2013), that is observing a sample of 32 face-on spirals with VIRUS-P; and (iv) and the SIRIUS project, currently studying 70 (U)LIRGS at $z < 0.26$ using different IFUs (Arribas *et al.* 2008), among others. However, despite the dramatic improvement over previous data provided by these 'surveys', they are all affected by non-trivial sample selection criteria and, most importantly, incomplete coverage of the full extent of the galaxies.

In the last few years this situation has changed dramatically, and it will change even more in the upcomming years, with the development of IFS surveys of galaxies, in particular Atlas3D (Cappellari *et al.* 2011), CALIFA (Sánchez *et al.* 2012), MaNGA (Law *et al.* 2014) and SAMI (Croom *et al.* 2012). We review here the current on-going or recently started surveys, making an emphasis on the result obtained by the CALIFA survey (Sánchez *et al.* 2012), the first of those surveys providing wih a panoramic view of galaxy properties.

2. Major recent and on-going surveys

Most of the listed on-going IFU surveys in the Local Universe have as a major goal to understand the nature of galaxies as a consequence of the evolution through cosmologocial times. Therefore, most of their differences are on the technical details of the selected IFU, the adopted setup and the sample selection. In general, all the surveys tries to select a sample representative of the population of galaxies at low redshifts. Therefore, most of them cover mostly all galaxy types, covering as much as possible the color-magnitude diagram. However the adopt different selection criteria. In the case of MaNGA and SAMI the main criteria was to cover a wide range of galaxy masses. In MaNGA it was forced that the distribution along masses was as flat as possible. SAMI adopted a different procedure, selecting a set of volume-limited sub-samples, together with a set of sub-samples centred in galaxy clusters. In the previous case the sample was derived from the SDSS catalogs, while in the former one it was extracted from the GAMMA survey. The CALIFA sample selection is based on their optical size, i.e., it is a diameter selection that guaranteed that all galaxies are observed in most of their optical extension, as we explain below. Atlas3D adopted a different scheme, their main science goal (to understand the building up of bulges and early-type galaxies). The selected early-type/red galaxies from a volume-limited sample up to ~42 Mpc (being the sample at lower redshiffs of the different surveys).

Figure 1 and Table 1 presents an schematic comparison between the different surveys. The differences between different surveys are clearly highlighted in there. While Atlas3D provides the highest spatial resolution and better spatial sampling for the individual galaxies, if offers a limited wavelength range and FoV compared to the size of the galaxies. In other cases, like MaNGA, the total number of objects is the main statistical advantage, since it would provide with the average characteristics properties of the main population of galaxies for many different galaxy types with better statistical significance than any of the other surveys. The penalty is that not all the galaxies will be sampled by the same number of fibers (ranging between 3x19 to 3x127), and therefore the physical sampling is different for different galaxies. In the case of SAMI the main adventage is the wider range of galaxy masses sampled and the higher spectral resolution in the red wavelength range. However the large redshift range implies a different physical sampling of the different galaxies. Finally, in the case of CALIFA, it provides the widest spatial coverage compared to the spatial size of the galaxies, and one of the best spatial physical resolutions. However, the sample is more limited than the one studiest by MaNGA or SAMI. In summary we consider that each of the IFUs surveys is very complementary.

3. CALIFA: A brief introduction

The Calar Alto Legacy Integral Field Area (CALIFA) survey (Snchez *et al.* 2012a) is an ongoing large project of the Centro Astronmico Hispano-Alemn at the Calar Alto observatory to obtain spatially resolved spectra for 600 local ($0.005< z <0.03$) galaxies by means of integral field spectroscopy (IFS). CALIFA observations started in June 2010 with the Potsdam Multi Aperture Spectrograph (PMAS, Roth *et al.* 2005), mounted to the 3.5m telescope, utilizing the large (74"×64") hexagonal field-of-view (FoV) offered by the PPak fiber bundle (Verheijen *et al.* 2004; Kelz *et al.* 2006). PPak was created for the Disk Mass Survey (Bershady *et al.* 2010). Each galaxy is observed using two different setups, an intermediate spectral resolution one (V1200, $R \sim 1650$), that cover the blue range of the optical wavelength range (3700-4700Å), and a low-resolution one (V500,$R \sim 850$, that covers the first octave of the optical wavelength range (3750-7500Å). A diameter-selected sample of 939 galaxies were drawn from the 7th data release of the Sloan Digital Sky Survey (SDSS, Abazajian *et al.* 2009) which is described in Walcher *et al.* (2014). From this mother sample the 600 target galaxies are randomly selected.

Combining the techniques of imaging and spectroscopy through optical IFS provides a more comprehensive view of individual galaxy properties than any traditional survey. CALIFA-like observations were collected during the feasibility studies (Mrmol-Queralt

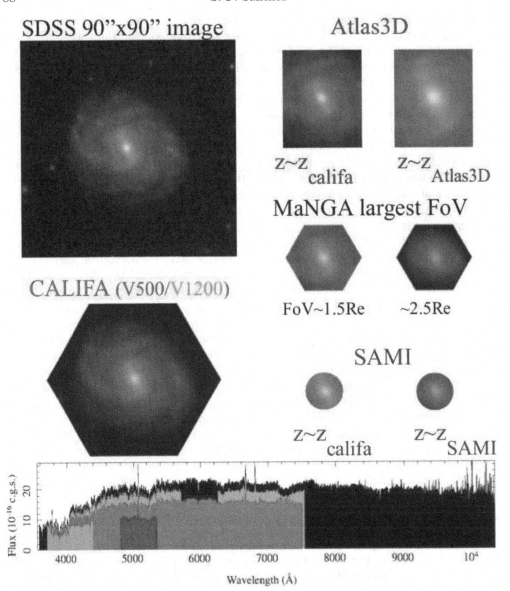

Figure 1. Schematic figure comparing the different on-going major IFU surveys. The top-left figure shows a 90"×90" post-stamp image of NGC5947 extracted from the SDSS imaging survey. The same image as it would be observed by the four different IFU surveys discussed in this article are presented in the different post-stamp figures. For MaNGA we have selected the bundle with the largest FoV, although we should remind the reader that only 1/5th of the survey will be observed using this bundle. Finally, the wavelength range covered by each survey is presented in the bottom panel. We should remind that the spectral resolution is different for each survey. For SAMI the largest resolution is achieved in the redder wavelength range covered by this survey.

et al. 2011; Viironen *et al.* 2012) and the PPak IFS Nearby Galaxy Survey (PINGS, Rosales-Ortega *et al.* 2010), a predecessor of this survey. First results based on those datasets already explored their information content (e.g. Snchez *et al.* 2011; Rosales-Ortega *et al.* 2011; Alonso-Herrero *et al.* 2012; Snchez *et al.* 2012b; Rosales-Ortega *et al.* 2012). CALIFA can therefore be expected to make a substantial contribution to our understanding of galaxy evolution in various aspects including, (i) the relative importance and consequences of merging and secular processes; (ii) the evolution of galaxies across the colormagnitude diagram; (iii) the effects of the environment on galaxies; (iv) the AGN-host galaxy connection; (v) the internal dynamical processes in galaxies; and (vi) the global and spatially resolved star formation history of various galaxy types.

Compared with other IFS surveys, CALIFA offers an unique combination of (i) a sample covering a wide range of morphological types in a wide range of masses, sampling the Color-Magnitude diagram for $M_g > -18$ mag; (ii) a large FoV, that guarantees to cover the entire optical extension of the galaxies up to $2.5r_e$ for an 80% of the sample; and (iii) an accurate spatial sampling, with a typical spatial resolution of $\sim$1 kpc for the entire sample, which allows to optical spatial resolved spectroscopic properties of most relevant structures in galaxies (spiral arms, bars, buges, Hii regions...). The penalty for a better spatial sampling of the galaxies is the somehow limited number of galaxies in the survey (e.g., MaNGA and SAMI). In terms of the spectral resolution, only the blue wavelength range is sampled with a similar spectral resolution than these two other surveys.

As a legacy survey, one of the main goals of the CALIFA collaboration is to grant public access of the fully reduced datacubes. In November 2012 we deliver our 1st Data Release (Husemann *et al.* 2013), comprising 200 datacubes corresponding to 100 objects †. After almost two years, and a major improvement in the data reduction, we present our 2nd Data Release (Garcia Benito *et al.*, in prep.), comprising 400 datacubes corresponding to 200 objects ‡, the 1st of October 2014.

4. CALIFA: Main Science Results

Figure 2 illustrate the dataproducts that can be derived from the IFU datasets obtained by the CALIFA survey, comprising a panoramic view of the spatial resolved spectroscopic prorperties of these galaxies, including information of the stellar populations, ionized gas, mass distribution and stellar and gas kinematics. Similar dataproducts are derived for any of the indicated projects: Atlas3D, MaNGA or SAMI.

Different science goals have been already addressed using this information: (i) New techniques has been developed to understand the spatially resolved star formation histories (SFH) of galaxies (Cid Fernandes *et al.*, 2013, 2014). We found the solid evidence that mass-assembly in the typical galaxies happens from inside-out (Pérez *et al.*, 2013). The SFH of bulges and early-type galaxies are fundamentally related to the total stellar mass, while for disk galaxies it is more related to the local stellar mass density (González Delgado *et al.*, 2013 & 2014); (ii) We developed new tools to detect and extract the spectroscopic information of HII regions (Sánchez *et al.*, 2012b), building the largest catalog currently available ($\sim$6,000 HII regions and aggregations). This catalog has been used to define a new oxygen abundance calibrator anchored with electron temperature measurements (Marino *et al.*, 2013). From these, we explored the proposed dependence of the Mass-Metallicity relation with SFR (Sánchez *et al.*, 2013), and the local Mass-Metallicity relation (Rosales-Ortega *et al.* 2012). We found that all galaxies in our sample present a common abundance radial gradient with a similar slope when normalized to the effective radius (Sánchez *et al.*, 2014), which agrees with the proposed inside-out scenario for galaxy growth. This characteristic slope is independent of the properties of the galaxies, and in particular of the presence or absence of a bar, contrary to previous results. More

---

† http://califa.caha.es/DR1/
‡ http://califa.caha.es/DR2/

Figure 2. Schematic representation of the different dataproducts that can be obtained for sample of galaxies with the state-of-the art IFU surveys. In this case it shows for a set of 100 galaxies extracted from the CALIFA survey six hexagons showing (following the clock arrows): (i) The true-color maps showing the light distribution; (ii) The stellar mass density distribution; (iii) The distribution of the luminosity weighted ages across the galaxies; (iv) a three-color image showing the ionized gas distribution for Hα, [NII]λ6584 and [OIII]λ5007; (v) the Hα intensity distribution and (vi) the ionized gas kinematics. Similar dataproducts are derived for Altas3D, MaNGA and SAMI datasets.

recently, this result has been confirmed by the analysis of the stellar abundance gradient in the same sample (Sánchez-Blázquez *et al.*, 2014). We also explore the progenitors of SuperNovae based on the gas phase metallicity and loca SFR at the location of the explosion (Gabany *et al.* 2014); (iii) We explore the origin of the low intensity, LINER-like, ionized gas in galaxies. These regions regions are clearly not related to star-formation activity, or to AGN activity. They are most probably relatd to post-AGB ionization in many cases (Kehrig *et al.*, 2012; Singh *et al.*, 2013; Papaderos *et al.* 2013); (iv) We explore the aperture and resolution effects on the data. CALIFA provides a unique tool to understand the aperture and resolution effects in larger single-fiber (like SDSS) and

IFS surveys (like MaNGA, SAMI). We explored the effects of the dilution of the signal in different gas and stellar population properties (Mast *et al.*, 2014), and proposed an new empirical aperture correction for the SDSS data (Iglesias-Páramo *et al.*, 2013); (v) CALIFA is the first IFU survey that allows gas and stellar kinematic studies for all morphologies with enough spectroscopic resolution to study (a) the kinematics of the ionized gas (García-Lorenzo *et al.*, 2014), (b) the effects of bars in the kinematics of galaxies (Barrera-Ballesteros *et al.* 2014a); (c) the effects of the intraction stage on the kinematic signatures (Barrera-Ballesteros *et al.*, submitted), (d) measure the Bar Pattern Speeds in late-type galaxies (Aguerri *et al.*, submitted), (iv) extend the measurements of the angular momentum of galaxies to previously unexplored ranges of morphology and ellipticity (Falcón-Barroso *et al.*, in prep.); and (v) finally we explore in detail the effects of galaxy interaction in the enhancement of star-formatio rate and the ignition of galactic outflows (Wild *et al.* 2014).

5. Conclusions

Along this review we have presented the main characteristics of IFU surveys in the Local Universe, describing their main advantage with respect to precedent techniques (like single fiber spectroscopic surveys or imaging surveys). We made a summary of some of the first attempts of performing a major IFU survey, and we describe the current on-going or recently finished ones. Finally, we present the main data-products that can be derived using these surveys and made a brief summary of the science goals achieved recently by the CALIFA survey, one of the most advanced on-going surveys that sample the full range of morphological types of galaxies.

Recent results already presented based on the pilot and even the main samples of the recently started MaNGA and SAMI surveys in this conference, and the quality shown by MUSE data (Bacon *et al.*,), illustrate how IFU surveys would fundamental tools in our understanding of the evolution of galaxies.

Acknowledgements

SFS thank the support of the Mexican CONACyt Grant 180125.

We thank K. Bundy, S. Croom, M. Bershady, M. Cappellari for providing us with fundamental information and useful comments that has helped in the redaction of this document.

We thank the full CALIFA collaboration for providing us with most of the information on the results from the survey, and in particular to R. Garcia Benito for providing us with Fig. 3.

References

Abazajian, K. N., *et al.* 2009, *ApJS*, 182, 543
Arribas, S., *et al.*, 2008, *A&A*, 479, 687
Barrera-Ballesteros, J. K., Falcón-Barroso, J., García-Lorenzo, B., *et al.* 2014, *A&A*, 568, A70
Benítez, N., *et al.* 2009, *ApJ*, 691, 241
Bershady, M. A., Verheijen, M. A. W., Swaters, R. A., *et al.* 2010, *ApJ*, 716, 198
Blanc, G. A., Weinzirl, T., Song, M., *et al.* 2013, *AJ*, 145, 138
Cappellari, M., Emsellem, E., Krajnović, D., *et al.* 2011, *MNRAS*, 413, 813
Cid Fernandes, R., Pérez, E., García Benito, R., *et al.* 2013, *A&A*, 557, A86
Cid Fernandes, R., González Delgado, R. M., García Benito, R., *et al.* 2014, *A&A*, 561, A130

Croom, S. M., Lawrence, J. S., Bland-Hawthorn, J., *et al.* 2012, *MNRAS*, 421, 872

de Zeeuw, P. T., Bureau, M., Emsellem, E., *et al.* 2002, *MNRAS*, 329, 513

Driver S. P., Norberg P., Baldry I. K., Bamford S. P., HopkinsA. M., Liske J., Loveday J., & Peacock J. A., 2009, *A&G*, 50, 5.12

Folkes, S., *et al.* 1999, *MNRAS*, 308, 459

Galbany, L., Stanishev, V., Mourão, A. M., *et al.* 2014, arXiv:1409.1623

Garcia-Lorenzo, B., Marquez, I., Barrera-Ballesteros, J. K., *et al.* 2014, arXiv:1408.5765

González Delgado, R. M., Pérez, E., Cid Fernandes, R., *et al.* 2014, *A&A*, 562, A47

González Delgado, R. M., Cid Fernandes, R., García-Benito, R., *et al.* 2014, *ApJ*, 791, L16

Husemann, B., Jahnke, K., Sánchez, S. F., *et al.* 2013, *A&A*, 549, A87

Iglesias-Páramo, J., Vílchez, J. M., Galbany, L., *et al.* 2013, *A&A*, 553, L7

Kelz, A., Verheijen, M. A. W., Roth, M. M., *et al.* 2006, *PASP*, 118, 129

Kehrig, C., Monreal-Ibero, A., Papaderos, P., *et al.* 2012, *A&A*, 540, A11

Law, D. R., MaNGA Team 2014, American Astronomical Society Meeting Abstracts #223, 223, #254.31

Le Fèvre, O., *et al.* 2004, *A&A*, 428, 1043

Marino, R. A., Rosales-Ortega, F. F., Sánchez, S. F., *et al.* 2013, *A&A*, 559, A114

Mármol-Queraltó, E., Sánchez, S. F., Marino, R. A., *et al.* 2011, *A&A*, 534, A8

Mast, D., Rosales-Ortega, F. F., Sánchez, S. F., *et al.* 2014, *A&A*, 561, A129

Moles, M., *et al.* 2008, AJ, 136, 1325

Papaderos, P., Gomes, J. M., Vílchez, J. M., *et al.* 2013, *A&A*, 555, L1

Pérez, E., Cid Fernandes, R., González Delgado, R. M., *et al.* 2013, *ApJ*, 764, L1

Rix, H.-W., *et al.* 2004, *ApJS*, 152, 163

Rosales-Ortega, F. F., Kennicutt, R. C., Sánchez, S. F., *et al.* 2010, *MNRAS*, 461

Rosales-Ortega, F. F., Sánchez, S. F., Iglesias-Páramo, J., *et al.* 2012, *ApJ*, 756, L31

Sánchez, S. F., Kennicutt, R. C., Gil de Paz, A., *et al.* 2012, *A&A*, 538, A8

Sánchez, S. F., Rosales-Ortega, F. F., Jungwiert, B., *et al.* 2013, *A&A*, 554, A58

Sánchez, S. F., Rosales-Ortega, F. F., Iglesias-Páramo, J., *et al.* 2014, *A&A*, 563, A49

Sanchez-Blazquez, P., Rosales-Ortega, F., Mendez-Abreu, J., *et al.* 2014, arXiv:1407.0002

Scoville, N., *et al.* 2007, *ApJS*, 172, 1

Singh, R., van de Ven, G., Jahnke, K., *et al.* 2013, *A&A*, 558, A43

Viironen, K., Sánchez, S. F., Marmol-Queraltó, E., *et al.* 2012, *A&A*, 538, A144

Walcher, C. J., Wisotzki, L., Bekeraité, S., *et al.* 2014, arXiv:1407.2939

Wild, V., Rosales-Ortega, F., Falcón-Barroso, J., *et al.* 2014, *A&A*, 567, A132

Wolf, C., Meisenheimer, K., Rix, H.-W., Borch, A., Dye, S., & Kleinheinrich, M. 2003, *A&A*, 401, 73

York, D. G., Adelman, J., Anderson, Jr., J. E., *et al.* 2000, *AJ*, 120, 1579

Galaxies in 3D across the Universe
Proceedings IAU Symposium No. 309, 2014
B. L. Ziegler, F. Combes, H. Dannerbauer, M. Verdugo, eds.

© International Astronomical Union 2015
doi:10.1017/S1743921314009387

Spectral synthesis of stellar populations in the 3D era: The CALIFA experience

R. Cid Fernandes[1], E. A. D. Lacerda[1], R. M. González Delgado[2], N. Vale Asari[1], R. García-Benito[2], E. Pérez[2], A. L. de Amorim[1], C. Cortijo-Ferrero[1], R. López Fernández[2], S. F. Sánchez[2,3] and the CALIFA collaboration

[1] Departamento de Física, Universidade Federal de Santa Catarina, Florianópolis, SC, Brazil
email: `cid@astro.ufsc,br`
[2] Instituto de Astrofísica de Andalucía (CSIC), Granada, Spain
[3] Instituto de Astronomía, Universidad Nacional Autonóma de Mexico, D. F., México

Abstract. Methods to recover the fossil record of galaxy evolution encoded in their optical spectra have been instrumental in processing the avalanche of data from mega-surveys along the last decade, effectively transforming observed spectra onto a long and rich list of physical properties: from stellar masses and mean ages to full star formation histories. This promoted progress in our understanding of galaxies as a whole. Yet, the lack of spatial resolution introduces undesirable aperture effects, and hampers advances on the internal physics of galaxies. This is now changing with 3D surveys. The mapping of stellar populations in data-cubes allows us to figure what comes from where, unscrambling information previously available only in integrated form. This contribution uses our STARLIGHT-based analysis of 300 CALIFA galaxies to illustrate the power of spectral synthesis applied to data-cubes. The selected results highlighted here include: (a) The evolution of the mass-metallicity and mass-density-metallicity relations, as traced by the mean stellar metallicity. (b) A comparison of star formation rates obtained from $H\alpha$ to those derived from full spectral fits. (c) The relation between star formation rate and dust optical depth within galaxies, which turns out to mimic the Schmidt-Kennicutt law. (d) PCA tomography experiments.

Keywords. galaxies: evolution – galaxies: formation – galaxies: fundamental parameters – galaxies: stellar content – galaxies: structure

1. Introduction

So much has changed so quickly in the way we study galaxies that an astronomer suddenly transported from a conference in the 1990s to a symposium like this one would be thoroughly stunned with things we now take for granted, like counting galaxies by the thousands, and unashamedly quoting their stellar masses, star formation rates and mean stellar ages as if these properties were trivially derived. This revolution happened due to the confluence of major advances in two fronts. First, mega-surveys like the SDSS came into existence. Secondly, models for the spectra of stellar populations with an appropriate quality and resolution to analyze actual galaxy spectra finally became available. This gave new life to old spectral synthesis methods to dig the fossil record of galaxy evolution out of their observed spectra, i.e., to transform F_λ into physical properties such as stellar mass, mean age, metallicity, extinction, and the whole star formation history (SFH).

Naturally, there are caveats. On the modeling side—skipping technical aspects and the semi-philosophical issue of whether to use index-based or full λ-by-λ spectral fitting methods—uncertainties in the evolutionary synthesis ingredients are currently an important limiting factor (e.g., Chen *et al.* 2010; Cid Fernandes *et al.* 2014). On the observational side, though much has been learned from the application of spectral synthesis to SDSS galaxies, the lack of spatial resolution severely limits interpretation of the results. Indeed, not knowing which photons come from where forces us to treat galaxies as point sources, mixing clearly distinct sub-components (bulge, disk, arms, bars) and their different stellar populations.

Here is where Integral Field Spectroscopy (IFS) surveys come to our rescue (see Sánchez contribution), combining the morphological information of images with the diagnostic power of spectroscopy. Besides producing maps of everything we previously had only for entire galaxies, these surveys will be instrumental in reverse engineering the meaning of global properties derived from spatially unresolved data like the SDSS and high-z targets.

We have been exploring the information on stellar populations within CALIFA datacubes by means of the spectral synthesis code STARLIGHT. In short, after some basic preprocessing steps the datacube is spatially binned into Voronoi zones whenever necessary to reach $S/N \geqslant 20$, and all zone spectra are extracted and fitted with STARLIGHT. The results are then packed into fits files and analyzed with PyCASSO†, a very handy tool to post-process and analyze STARLIGHT results in 3D. Pérez *et al.* (2013) and González Delgado *et al.* (2014a,b) present our first scientific results, while technical aspects are described in Cid Fernandes *et al.* (2013, 2014). Instead of reviewing this already published work, we dedicate these few pages to presenting some yet unpublished (and unpolished) results which further illustrate the richness of the manifold of physical properties derived from the application of stellar populations diagnostic tools to IFS surveys.

2. Chemical evolution

Recently, González Delgado *et al.* (2014b) analyzed the STARLIGHT-derived mean *stellar* metallicities ($Z_\star$) of 300 CALIFA data cubes. Metallicity was shown to correlate both with the stellar mass ($M_\star$) and the surface mass density ($\mu_\star$). $Z_\star$ is more closely related to $M_\star$ in spheroids, while in disks it is $\mu_\star$ who seems to govern the chemical evolution. Thus, as we had previously found for mean stellar ages, the balance between local ($\mu_\star$-driven) and global ($M_\star$-driven) processes affecting $Z_\star$ varies with the location within a galaxy (see also González Delgado's contribution in this same volume).

Another finding reported in that paper is that the $Z_\star$ values obtained considering only stars younger than 2 Gyr are higher than those for the whole population (1 Myr $< t <$ 14 Gyr). Although this is expected in terms of chemical evolution, $Z_\star$ is a non-trivial property to derive in composite stellar populations, and deriving its time dependence is even harder. In Fig. 1 we look in more detail at this remarkable result. Its left panel shows the $M_\star$-$Z_\star$ relation (MZR), breaking $Z_\star (< t)$ into six age ranges, from $t < 14$ Gyr (i.e., all stars) to $t < 1$ Gyr. Similarly, the right panel shows the evolution of the $\mu_\star$-$Z_\star$ relation (μZR). The curves are based on 300 galaxies (but 253418 spectra!), whose galaxy-wide average $Z_\star (< t)$ are binned along the horizontal axis and smoothed for cosmetic purposes.

Both plots show that $Z_\star$ increases steadily as galaxies age. The pace of chemical evolution varies with both $M_\star$ and $\mu_\star$, producing the systematic flattening of the MZR and μZR with time. In fact, stars of all ages have similar metallicities in the most massive (also the densest) galaxies, as these systems essentially completed their star formation and chemical evolution very long ago. Going down the $M_\star$ and $\mu_\star$ scales one finds slower and slower evolution. The SFHs of these galaxies are also slower, in the sense that their stellar mass was formed over a longer time-span.

† Python Califa Starlight Synthesis Organizer

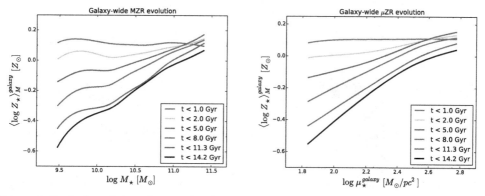

Figure 1. Evolution of the galaxy-wide $M_\star$ and $\mu_\star$ versus *stellar* metallicity relations. Each line corresponds to the average $Z_\star(< t)$ curve obtained from 300 galaxies ($\sim$ 250k spectra). $Z_\star(< t)$ values are computed including only stellar populations younger than $t = 1, 2, \ldots 14$ Gyr (from top to bottom), as derived from the STARLIGHT spectral decomposition.

The point to highlight here is that $Z_\star$ and its time evolution behave amazingly well, and for galaxies of all types. Studies of the MZR are usually based on nebular properties. Though useful, Z_{neb} can only be applied when the emission lines are powered by young stars, thus excluding regions where BPT-like diagnostic diagrams indicate the contribution of other ionizing agents (AGN, shocks, old stars). Furthermore, it is obviously impossible to trace the evolution of Z_{neb} for any single galaxy. The availability of reliable estimates of the stellar metallicity thus offers an independent (and in several ways more informative) way to address issues related to the chemical evolution of galaxies.

3. Star formation Rates: Nebular $\times$ stellar tracers

By construction, estimates of the recent star formation rate (SFR) involve the hypothesis that SFR = constant over some time interval T. In the case of the Hα luminosity, one has $L_{H\alpha} \propto \int_0^{``\infty"} \mathrm{SFR}(t) q_H(t) dt \sim \mathrm{SFR} \int_0^T q_H(t) dt$, where $q_H(t)$ is the rate of ionizing photons per unit initial mass of a burst of age t. Because $q_H(t)$ drops precipitously after $\sim 10^7$ yr, its integral converges on a time-span T of this same order.† Assuming $\mathrm{SFR}(t < T) \sim$ constant seems plausible when dealing with whole (or large chunks of) galaxies, since in this regime the data average over many regions formed in the last T years (e.g., a collection of HII regions of different ages and stellar masses). Doing the same for a single spaxel is at least dangerous, if not wrong, as one quickly realizes by imagining the case of a spaxel containing a single HII region, understood as an instantaneous burst where $L_{H\alpha}$ does not measure a rate (see also Calzetti's contribution).

In contrast, STARLIGHT assumes nothing about the SFH. Its more general and flexible non-parametric description of the SFH is thus in principle more suitable to derive SFRs, and these can be computed over any desired time scale. The caveat is that, unlike for ionizing photons, the optical continuum contains a non-trivial mixture of populations of all ages, and isolating the contribution due to a specific sub-population is an inevitably uncertain and degenerate process. Still, Asari *et al.* (2007) showed that the Hα and STARLIGHT-based SFRs agree remarkably well in SDSS star-forming galaxies, so we repeat this comparison for CALIFA data.

† Other tracers like the UV luminosity or the dust-reprocessed far-IR luminosity follow the same logic, but the corresponding time-integrals require much longer times to converge.

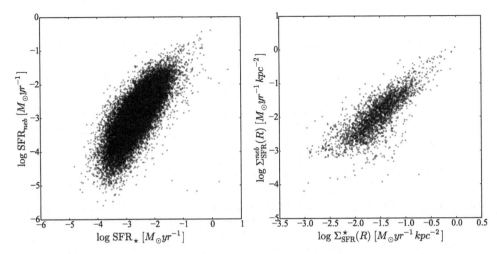

Figure 2. *Left:* Comparison of the SFRs estimated from Hα with those obtained from the spectral fits of the stellar continuum for ∼ 40k CALIFA spectra. *Right:* As in the left, but now comparing SFR surface densities (thus eliminating distance-induced correlations) for 20 radial bins per galaxy, hence averaging over HII regions, as implicitly required by the SFR(Hα) method. The STARLIGHT-based SFRs were computed for a time-scale of 25 Myr.

In Fig. 2 we compare the Hα and STARLIGHT-based SFRs for 40532 zones in 300 CAL-IFA galaxies, selected to belong to the star-forming wing in the [OIII]/Hβ vs. [NII]/Hα diagram. Despite all potential caveats pointed out above, the correlation is excellent. In the right panel we refine the analysis by comparing surface densities averaged over radial rings, thus mitigating stochastic effects and better complying with the SFR ∼ const. hypothesis implicit in the Hα method. The correlation is even better, and in fact cleaner.

One can look at this result as reinforcing the confidence on the spectral fits with STARLIGHT. One can also look at it in closer detail to map differences between nebular and stellar estimates of SFRs, and to investigate plausible causes such as differential extinction and IMF. Also, now that we have a SFR indicator independent of (but consistent with) emission lines, we can use it across the board, even when SFR(Hα) does not apply, as in AGN dominated regions. We leave this as a small demonstration of the power (and promise) of the combination of stellar and nebular diagnostics.

4. Schmidt-Kennicutt like relations

As any optical survey, CALIFA looks at the stellar, nebular, and dust content of galaxies. The missing piece of information is gas, the fuel for star formation and galaxy evolution. Molecular and atomic maps of CALIFA galaxies will tie in nicely with the information already at hand, allowing spatially resolved studies of the relation between gaseous content, SFR and chemical abundance for a varied and representative sample.

Meanwhile, let us explore dust as proxy for gas. Specifically, we correlate the V-band dust optical depth (τ_V) with the SFR surface density (Σ_{SFR}). Both quantities are derived entirely from the STARLIGHT spectral fits. This allows a more comprehensive view than one would obtain by restricting the analysis to spaxels whose emission lines are both (a) consistent with star-formation in BPT-like diagrams, and (b) strong enough to be reliably measured. Since τ_V ultimately represents the dust column density Σ_{dust}, which in turn presumably traces Σ_{gas}, to first order (and ignoring metallicity dependence in the

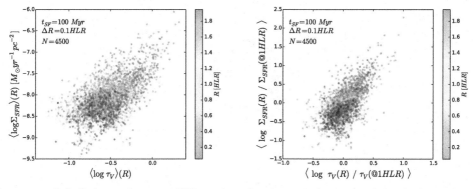

Figure 3. *Left:* Relation between SFR surface density and dust optical depth, both derived from the spectral synthesis of 300 CALIFA galaxies. Each point corresponds to radial averages of Σ_{SFR} and τ_V for an (inclination corrected) radial bin of width $\Delta R = 0.1$ Half Light Radius (HLR), color coded to reflect R. *Right:* As the left panel, but scaling the Σ_{SFR} and τ_V values of each galaxy by the corresponding values at $R = 1$ HLR. Barring a dust-to-gast conversion factor, these relations are analogous to the Schmidt-Kennicutt relation between Σ_{SFR} and Σ_{gas}.

dust-to-gas conversion factor) one expects $\Sigma_{SFR}(\tau_V)$ to behave as a Schmidt-Kennicutt $\Sigma_{SFR}(\Sigma_{gas})$ relation.

This expectation is borne out in Fig. 3, which shows the relation between Σ_{SFR} and τ_V we obtain with CALIFA. The plot contains 4500 points, each one with average measurements performed over radial bins of width = 0.1 Half Light Radius (HLR). Both Σ_{SFR} and τ_V increase towards the nucleus, as one infers from the color change from the upper-right to bottom-left of the plot. Despite the scatter, the relation is visibly super-linear, with a logarithmic slope in the neighborhood of 1.5, close to what is found with direct measurements of Σ_{gas} (Kennicutt & Evans 2012 and references therein). The right panel in Fig. 3 rescales the Σ_{SFR} and τ_V values of each galaxy to the corresponding values at $R = 1$ HLR. This rescaling trick, which ultimately expresses results for each galaxy in its own natural units, significantly reduces the scatter, probably due to the minimization of "second-parameter" effects.

Again, we show these plots as a teaser of what is nowadays doable thanks to the combination of spectral synthesis tools and 3D spectroscopic surveys. Empirical relations such as the ones in Fig. 3 can be worked from different angles. One can for instance use them to verify the existence of a SK-like relation within galaxies and investigate how the star-formation "efficiency" varies with, say, morphological type or mass-density. Alternatively, one may postulate a SK law and use Σ_{SFR} to derive Σ_{gas}, which in turn would be useful to study (i) $\Sigma_{dust}/\Sigma_{gas}$ dust-to-gas ratios, (ii) gas fractions and their relation to stellar and nebular metallicities, (iii) etc. Plenty of work ahead!

5. PCA tomography

Finally, we briefly mention an entirely different approach to explore data-cubes. The PCA tomography technique devised by Steiner *et al.* (2009) identifies correlations within data-cubes in a purely mathematical way, free of astrophysical pre-conceptions. Reverse engineering the meaning of the results is not always a simple task, although spectacular results achieved with Gemini data-cubes reveal the potential of this technique.

Fig. 4 gives a taste of what one finds for CALIFA data after masking emission lines and rescaling fluxes to account for the large dynamic range within the cube—PCA is highly sensitive to pre-processing steps. The left panel shows the image (tomogram) of

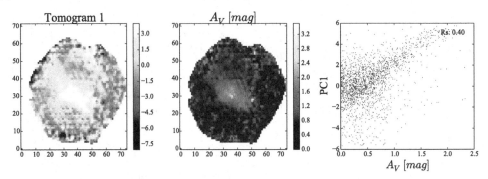

Figure 4. PCA tomography experiments with CALIFA data-cubes (see text for details).

the first PC in the data-cube of Arp 220. The central lane of high PC1 reflects the region of large extinction, as confirmed by its clean correspondence with the STARLIGHT-derived A_V map in the central panel and the PC1 vs. A_V plot on the right. Considering the few galaxies analyzed so far, we generally find that the first 2 or 3 PCs generally correlate well with some stellar population property. The velocity field, in particular, is invariably recognized in at least one PC. From our point of view this is actually a handicap of the method, as variance is waisted in mapping a property $(v_\star)$ which can easily be measured by more conventional means, so ways to circumvent this drawback are being implemented. We also note that relating PCs to previously known properties is an instructive exercise, although it kind of defeats the very purpose of processing the data through a physics-free algorithm, and letting the tomograms and eigenspectra hint at the underlying phenomena. In any case, there is still plenty more to be examined. Emission line and combined stellar plus nebular spectral cubes are still being PCA'ed, and one hopes that this technique helps unveiling the interplay between stars, gas and dust. Last but not least, PCA and other techniques may be used as noise filters and/or to identify subtle instrumental effects, so there is plenty to be done along this line too.

Acknowledgements

This contribution is based on data obtained by the CALIFA survey (http://califa.caha.es), funded by the Spanish MINECO grants ICTS-2009-10, AYA2010-15081, and the CAHA operated jointly by the Max-Planck IfA and the IAA (CSIC). The CALIFA Collaboration thanks the Calar Alto staff for the dedication to this project. Support from CNPq (Brazil) through grants 300881/2010-0 and 401452/2012-3 is duly acknowledged.

References

Asari, N. V., Cid Fernandes, R., Stasińska, G., *et al.* 2007, *MNRAS*, 381, 263
Chen, X. Y., Liang, Y. C., Hammer, F., *et al.* 2010, *A&A*, 515, A101
Cid Fernandes, R., Pérez, E., García Benito, R., *et al.* 2013, *A&A*, 557, 86
Cid Fernandes, R., González Delgado, R. M., Pérez, E., *et al.* 2014, *A&A*, 561, 130
González Delgado, R. M., Pérez, E., Cid Fernandes, R., *et al.* 2014a, *A&A*, 562, 47
González Delgado, R. M., Cid Fernandes, R., García-Benito, R., *et al.* 2014b, *ApJ*, 791, L16
Kennicutt, R. C. & Evans, N. J. 2012, *ARA&A*, 50, 531
Pérez, E., Cid Fernandes, R., González Delgado, R. M., *et al.* 2013, *ApJ*, 764, 1L
Steiner, J. E., Menezes, R. B., Ricci, T. V., & Oliveira, A. S. 2009, *MNRAS*, 395, 64

Galaxies in 3D across the Universe
Proceedings IAU Symposium No. 309, 2014
B. L. Ziegler, F. Combes, H. Dannerbauer, M. Verdugo, eds.

© International Astronomical Union 2015
doi:10.1017/S1743921314009399

The star formation history of galaxies in 3D: CALIFA perspective

R. M. González Delgado[1], R. Cid Fernandes[2], R. García-Benito[1], E. Pérez[1], A. L. de Amorim[2], C. Cortijo-Ferrero[1], E. A. D. Lacerda[2], R. López Fernández[1], S. F. Sánchez[1,3], N. Vale Asari[2] and CALIFA collaboration

[1]Instituto de Astrofísica de Andalucía (CSIC), Glorieta de la Astronomía s/n, E-18008 Granada, Spain
email: rosa@iaa.es
[2]Departamento de Física, Universidade Federal de Santa Catarina, P. O. Box 476, 88040-900, Florianópolis, SC, Brazil
[3]Instituto de Astronomía,Universidad Nacional Autonóma de Mexico, A. P. 70-264, 04510, México,D. F.

Abstract. We resolve spatially the star formation history of 300 nearby galaxies from the CALIFA integral field survey to investigate: a) the radial structure and gradients of the present stellar populations properties as a function of the Hubble type; and b) the role that plays the galaxy stellar mass and stellar mass surface density in governing the star formation history and metallicity enrichment of spheroids and the disks of galaxies. We apply the fossil record method based on spectral synthesis techniques to recover spatially and temporally resolved maps of stellar population properties of spheroids and spirals with galaxy mass from 10^9 to 7×10^{11} M$_\odot$. The individual radial profiles of the stellar mass surface density ($\mu_\star$), stellar extinction (A_V), luminosity weighted ages ($\langle \log age \rangle_L$), and mass weighted metallicity ($\langle \log Z/Z_\odot \rangle_M$) are stacked in seven bins of galaxy morphology (E, S0, Sa, Sb, Sbc, Sc and Sd). All these properties show negative gradients as a sight of the inside-out growth of massive galaxies. However, the gradients depend on the Hubble type in different ways. For the same galaxy mass, E and S0 galaxies show the largest inner gradients in $\mu_\star$; and Andromeda-like galaxies (Sb with log M$_\star$(M$_\odot$) $\sim$ 11) show the largest inner age and metallicity gradients. In average, spiral galaxies have a stellar metallicity gradient $\sim$ -0.1 dex per half-light radius, in agreement with the value estimated for the ionized gas oxygen abundance gradient by CALIFA. A global (M$_\star$-driven) and local ($\mu_\star$-driven) stellar metallicity relation are derived. We find that in disks, the stellar mass surface density regulates the stellar metallicity; in spheroids, the galaxy stellar mass dominates the physics of star formation and chemical enrichment.

Keywords. galaxies: evolution – galaxies: formation – galaxies: fundamental parameters – galaxies: stellar content – galaxies: structure

1. Introduction

Much of we know about galaxy properties has come from panoramic imaging or 1D spectroscopic surveys. While imaging provides useful 2D information of galaxy morphology and structural properties, spectroscopic surveys give information on the central or global stellar population and ionized gas properties and kinematics. However, galaxies are a complex mix of stars, interstellar gas, dust and dark matter, distributed in their disks and bulges, and resolved spatial information is needed to constrain the formation processes and evolution of the galaxy sub-components.

Integral Field Spectroscopy (IFS) observations can provide a unique 3D view of galaxies (two spacial plus one spectral dimensions), and allows to recover 2D maps of the stellar population and ionized gas properties, and kinematics. Until very recently it was not possible to obtain IFS data for sample larger than a few tens of galaxies. ALTAS3D (Cappellari *et al.* 2011), CALIFA (Sánchez *et al.* 2012), SAMI (Croom *et al.* 2012) and MaNGA (Bundy *et al.* 2014) surveys are the first in doing this steep forward, taking observations of several hundreds to several thousands of galaxies of the nearby Universe.

CALIFA is our currently-ongoing survey, observing 600 nearby galaxies with the PPaK IFU at the 3.5m telescope of Calar Alto observatory. The observations cover 3700-7000 Å with an intermediate spectral resolution (fwhm ~ 6 Å in the data presented in this contribution) and ~ 1arcmin^2 field of view with a final spatial sampling of 1 arcsec. Galaxies were selected from SDSS in the redshift range $0.005 \leqslant z \leqslant 0.03$, covering all the color magnitude diagram down to $M_r \leqslant$ -18, resulting in a sample containing all morphological types. An extended description of the survey, data reduction and sample can be found in Sánchez *et al.*(2012), Hussemann *et al.*(2012) and Walcher *et al.*(2014).

Previously, we derived the spatial resolved star formation history of the CALIFA galaxies on the first data release (DR1) using the fossil records that the stellar populations imprint in the galaxy spectra. This method dissects galaxies in space and time providing a 3D information that allows to retrieve when and where the mass and stellar metallicity were assembled as a function of look-back-time. We use the *STARLIGHT* code (Cid Fernandes *et al.*2005) to do a λ-by-λ spectral fit using different sets of single stellar population (SSP) models. These SSP are from a combination of Vazdekis *et al.* (2010) and Gonzáles Delgado *et al.*(2005) (labelled *GMe*), or from Charlot & Bruzual (2007) (labelled *CBe*).

Our scientific results from the first 100 CALIFA galaxies were presented in Pérez *et al.*(2013), Cid Fernandes *et al.*(2013, 2014) and Gonzáles Delgado *et al.*(2014a). One highlight result of these works is that the signal of downsizing is spatially preserved, with inner and outer regions growing faster for more massive galaxies, consequence of the inside-out growth of massive galaxies.

Here, based on the fossil records analysis of 300 CALIFA galaxies we present the results on: a) the radial structure and gradients of the stellar populations as a function of the Hubble type; and b) the role that plays the galaxy mass and stellar mass surface density in governing the star formation history and metallicity enrichment in ellipticals and in the bulge and disk components of galaxies.

Two complementary contributions to this one are presented in these proceedings: Sánchez presents CALIFA survey in the context of other contemporaneous IFS surveys such as SAMI and MaNGA; and Cid Fernandes *et al.* explains in details the methodology that we apply here, the uncertainties associated to the method, and results related with the growth of mass and metallicity in galaxies and their evolution.

2. Results

We present new results based on the radial structure of the present stellar population properties of 300 CALIFA galaxies that were observed with the V500 and V1200 setups and calibrated with the new pipeline 1.4 (see García-Benito *et al.* 2014 for details). 1D spatial radial profiles are obtained from the 2D maps, with an azimuthal averaging by an elliptical xy-to-R conversion. These 2D maps are created after collapsing the SFH in the time domain. The results are presented by stacking each galaxy individual radial profiles that has been previously normalized to a common metric that uses the half-light-radius (HLR) of each galaxy.

2.1. Hubble sequence: stellar population properties of galaxies in the tuning-fork diagram

One step to understand how galaxies form and evolve is classifying galaxies and studying their properties. Most of the massive galaxies in the near Universe are E, S0 and spirals (Blanton & Moustakas 2009), following well the Hubble tuning-fork diagram. The bulge fraction seems to be one of the main physical parameters that produce the Hubble sequence, increasing from late to early spirals. In this scheme, S0 galaxies are a transition class between the spiral classes and the elliptical one, with large bulges, but intermediate between Sa and E galaxies. On the other hand, galaxies properties such as color, mass, surface brightness, luminosity, and gas fraction are correlated with the Hubble type (Robert & Haynes 1994). This suggests that the Hubble sequence can illustrate possible paths for galaxy formation and evolution. If so, how is the spatial resolved stellar population properties of galaxies correlated with the Hubble type? Can the Hubble-tuning-fork scheme be useful to organize galaxies per galaxy mass and age or galaxy mass and metallicity?

CALIFA is a suitable benchmark to address these questions because it includes a significant amount of E, S0 and spirals. After a visual classification, the 300 galaxies were grouped in 41 E, 32 S0, 51 Sa, 53 Sb, 58 Sbc, 50 Sc, and 15 Sd. This sub-sample is a well representation of the morphological distribution of the whole CALIFA sample. Here we present the radial structure of the stellar mass surface density ($\mu_\star$), stellar extinction (A_V), luminosity weighted stellar age ($\langle \log age \rangle_L$), mass weighted stellar metallicity ($\langle \log Z/Z_\odot \rangle_M$), by stacking the galaxies by their Hubble type. First, we present how galaxies are distributed by stellar mass ($M_\star$) and their sizes in mass (a_{50}^M: radius that contains half of the mass) and in light (a_{50}^L: radius that contains half of the light, HLR). Most of the results discussed here are obtained with the GMe SSP models, but similar results are obtained with the CBe base (see Fig.1).

Galaxy stellar mass: We obtain the galaxy stellar mass ($M_\star$) after resolving spatially the SFH of each zone and hence, taking into account spatial variation of the stellar extinction and M/L ratio. Fig.1a shows the distribution of $M_\star$ as a function of Hubble type. The mass ranges from 10^9 to 7×10^{11} $M_\odot$ (GMe). We see a clear segregation in mass: galaxies with high bulge -to-disk ratio (E, S0, Sa) are the most massive ones ($\geqslant 10^{11}$ $M_\odot$), and galaxies with small bulges (Sc-Sd) have masses M $\leqslant 10^{10}$ $M_\odot$. The galaxy stellar mass distribution obtained with CBe models is similar to $M_\star$ with GMe base, but shifted by -0.25 dex due to the change of IMF (Chabrier with GMe and Salpeter with GMe SSP models)

Galaxy size: We take the advantage of our spatially resolved SFH and extinction maps to show that galaxies are more compact in mass than in light (González Delgado *et al.* 2014a), resulting in a ratio of the radius that contains half of mass with respecto to the ratio that contains half of the light (a_{50}^M/a_{50}^L) of 0.8. Galaxies are therefore typically 20% smaller in mass than how they appear in optical light. Fig.1b shows the distribution of a_{50}^M/a_{50}^L as a function of Hubble type, and a clear trend is observed. Sa-Sb-Sbc have the lowest a_{50}^M/a_{50}^L, due to the fact that they show a very prominent old bulge which have similar central properties to the spheroidal components of S0 and E, but a blue and extended disc which contributes to the light in the range Sa to Sbc.

Stellar mass surface density: The left panel in Fig.2 shows the radial profiles (in units of a_{50}^L) of log $\mu_\star$ obtained with GMe SSP base. Individual results are stacked in seven morphological bins. Error bars in the panel indicate the dispersion at one a_{50}^L distance in the galaxies of the Sa class, but it is similar for other Hubble type and radial distance. Negative gradients are detected in all galaxy types and increase from late type

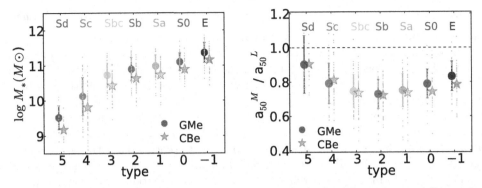

Figure 1. Left: Distribution of the galaxy stellar masses obtained from the spatially resolved spectral fits of each galaxy for each Hubble type of the galaxy of this work (grey small points). The coloured dots and stars are the mean galaxy stellar mass in each Hubble type obtained with the GMe and CBe SSP models. The bars show the dispersion in mass. Right: Relation between a_{50}^M/a_{50}^L and Hubble type. Symbols are as in the left panel of the figure.

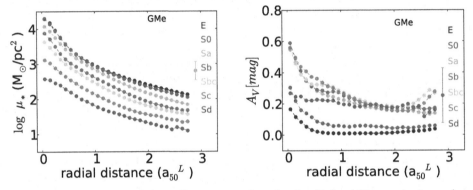

Figure 2. Radial profiles of the stellar mass surface density (in logarithm scale, $\log \mu_\star$) and the stellar extinction A_V as a function of Hubble type. The radial distance is in HLR units.

(Sd) to early type (S0, and E) galaxies. At a constant $M_\star$, spheroidal (S0 and E) are more compact than spirals, and S0 and E galaxies have similar compactness at all distances.

Stellar extinction: The right panel in Fig.2 shows the radial profiles A_V. All galaxy types show radial profiles that increase toward the center. Early type galaxies (E, and S0) are also extinguished toward the nucleus by 0.2-0.4 mag, while out of 1 HLR of the galaxies, the (old) stellar population are almost reddening-free. All spiral disks show ∼0.2-0.3 mag extinction. The bulges are significantly more extinguished up to 0.6 mag, except the bulges of late type spirals (Sd) that A_V is similar to the extinction in the disk.

Stellar ages: The left panel of Fig.3 shows the radial profiles of $\langle \log age \rangle_L$ (in yr). Symbols are as in Fig.2. Negative gradients are detected for all the Hubble types, suggesting that the quenching is progressing outwards, and the galaxies are growing inside-out, as we concluded with our mass assembly growth analysis (Pérez *et al.* 2012). Inner gradients are calculated between the galaxy nucleus and at 1 a_{50}^L, and the outer gradient between 1 and 2 a_{50}^L. The inner age gradient shows a clear behaviour with Hubble type, being maximum for spirals of intermediate type (Sb-Sbc). At $M_\star$ constant, Sb-Sbc galaxies have the largest age gradient. The age gradient in the outer disk (between 1 and 2 a_{50}^L) is smaller than the inner ones, but again it is larger for the spiral Sa-Sb-Sbc.

Stellar metallicity: The right panel in Fig.3 shows the radial profiles of mass weighted stellar metallicity obtained as explained in González Delgado *et al.*(2014b). Except for

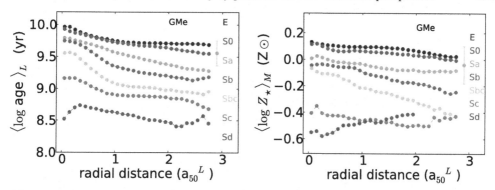

Figure 3. As in Fig.2 for the luminosity weighted ages and the mass weighted metallicity.

late spirals (Sc-Sd), spirals present negative gradients in average $\sim$ -0.1dex per HLR, similar to the value obtained for the nebular oxygen abundances obtained by CALIFA (Sánchez *et al.* 2013). For galaxies of equal stellar mass, the intermediate type spirals (Sbc) have the largest gradients. These negative gradients again are sights of the inside-out growth of the disks. However, the metallicity gradient for late spirals is positive indicating that these galaxies, in the low mass galaxy bins, may be formed from outside-in. This result is also in agreement with our results from the mass assembly evolution (Pérez *et al.* 2012).

2.2. *The role of $\mu_\star$ and $M_\star$ in the SFH and chemical enrichment of galaxies*

The galaxy stellar mass is also considered one of the fundamental properties as it also provides a measure of the galaxy formation and and evolution. The bimodal distribution of galaxies in the color-magnitude diagram that place them in the red sequence and blue cloud not only reflects the Hubble type, but also the dependence of the distribution on the galaxy stellar mass. Sorting galaxies by $M_\star$ we can study how their properties scale among the different classes. With CALIFA, thanks to the spatial information, we can check how important are the local ($\mu_\star$- driven) and global ($M_\star$-driven) processes in determining the star formation history and chemical enrichment in galaxies.

We find that there is a strong relation between the local values of $\mu_\star$ and the metallicity which is similar in amplitude to the global mass metallicity relation that exits between $\langle \log Z/Z_\odot \rangle_M$ and $M_\star$ over the whole 10^9 to 10^{12} $M_\odot$ range (González Delgado *et al.* 2014b). This means that local and global processes are important in the metallicity enrichment of the galaxies. However, the balance between local and global effects varies with the location within a galaxy. While in disks, $\mu_\star$ regulate the stellar metallicity, producing a correlation between log $\mu_\star$ and $\langle \log Z/Z_\odot \rangle_M$, in bulges and ellipticals is $M_\star$ who dominates the chemical enrichment (Fig.4). Furthermore, in spheroids the chemical enrichment happened much faster and earlier than in disks.

These results are in agreement with the analysis of the star formation history of galaxies (González Delgado *et al.* 2014a). We have shown that mean stellar ages (a first moment descriptor of the SFH) relate strongly to $\mu_\star$ in galactic disks, indicating that local properties dictate the pace of star-formation. The slower growth (hence younger ages) found at low $\mu_\star$ should lead to less metal enrichment, in agreement with the μZR realtion. Within bulges/spheroids, $M_\star$ is a much more relevant driver of the SFH. Most of the star formation activity in these regions was over long ago, leading to fast metal enrichment and little or no chemical evolution since those early days.

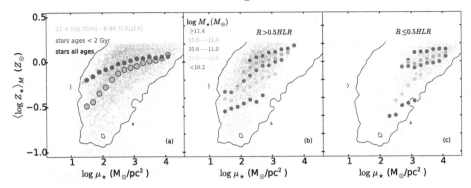

Figure 4. Left: Local stellar metallicity versus the local stellar mass surface density. The grey circles tracks the $\mu_\star$-binned stellar metallicity relation (μZR). Blue circles show the μZR obtained considering only star younger than 2 Gyr in the computation of the metallicity. Cyan stars show the CALIFA-based nebular μZR of Sánchez *et al.*(2013). Middle and right: The mean μZR obtained by breaking the 300 galaxies sample into five $M_\star$ intervals, restricting the analysis to spatial regions outwards and for the inner R= 0.5 HLR

Acknowledgements

This contribution is based on data obtained by the CALIFA survey (http://califa.caha.es), funded by the Spanish MINECO grants ICTS-2009-10, AYA2010-15081, and the CAHA operated jointly by the Max-Planck IfA and the IAA (CSIC). The CALIFA Collaboration thanks the CAHA staff for the dedication to this project. Support from CNPq (Brazil) through Programa Ciência sem Fronteiras (401452/2012-3) is duly acknowledged.

References

Blanton, M. R. & Moustakas, J. 2009, *ARAA*, 47, 159
Bundy, K., *et al.* 2014, https://www.sdss3.org/future/manga.php
Bruzual G., Charlot S. 2003, *MNRAS*, 344, 1000
Cappellari, M, Emsellem, E., *et al.*, 2011, *MNRAS*, 413, 813
Cid Fernandes, R., Mateus, A., Sodré, L., *et al.* 2005, *MNRAS*, 358, 363
Cid Fernandes, R., Pérez, E., García Benito, R., *et al.* 2013, *A&A*, 557, 86
Cid Fernandes, R., González Delgado, R. M., Pérez, E., *et al.* 2014, *A&A*, 561, 130
Croom, S., Lawrence, J. S., Bland-Hawthorn, J., *et al.*, 2012, *MNRAS*, 421, 872
García-Benito, R., *et al.*, 2014, *A&A*, submitted
González Delgado, R. M., Cerviño, M., Martins, *et al.* 2005, *MNRAS*, 357, 945
González Delgado, R. M., Pérez, E., Cid Fernandes, R., *et al.* 2014a, *A&A*, 562, 47
González Delgado, R. M., Cid Fernandes, R., García-Benito, R., *et al.* 2014b, *ApJ*, 791, L16
Husemann, B., Jahnke, K., Sánchez, S. F., *et al.* 2013, *A&A*, 549, A87
Pérez, E., Cid Fernandes, R., González Delgado, R. M., *et al.* 2013, *ApJ*, 764, 1L
Roberts, M. S. & Haynes, M. P., *ARAA*, 32, 115
Sánchez, S. F., Kennicutt, R. C., Gil de Paz, A., *et al.* 2012, *A&A*, 538, 8
Sánchez, S. F., Rosales-Ortega, F. F., Jungwiert, B., *et al.* 2013, *A&A*, 554, 58
Vazdekis, A., Sánchez-Blázquez, P., Falcón-Barroso, J., *et al.* 2010, *MNRAS*, 404, 1639
Walcher, C. J. , Wisotzki, L., Bekeraité, S., *et al.* 2014, *A&A*, arXiv1407.2939

Galaxies in 3D across the Universe
Proceedings IAU Symposium No. 309, 2014
B. L. Ziegler, F. Combes, H. Dannerbauer, M. Verdugo, eds.

© International Astronomical Union 2015
doi:10.1017/S1743921314009405

Extended nebular emission in CALIFA early-type galaxies

J. M. Gomes[1], P. Papaderos[1], C. Kehrig[2], J. M. Vílchez[2], M. D. Lehnert[3] and the CALIFA collaboration

[1] Instituto de Astrofísica e Ciências do Espaço, Universidade do Porto, CAUP, Rua das Estrelas, PT4150-762 Porto, Portugal
email: jean@astro.up.pt; papaderos@astro.up.pt
[2] Instituto de Astrofísica de Andalucía (CSIC), Glorieta de la Astronomía s/n Aptdo. 3004, E18080-Granada, Spain
[3] GEPI, Observatoire de Paris, UMR 8111, CNRS, Université Paris Diderot, 5 place Jules Janssen, 92190 Meudon, France

Abstract. The morphological, spectroscopic and kinematical properties of the warm interstellar medium (*wim*) in early-type galaxies (ETGs) hold key observational constraints to nuclear activity and the buildup history of these massive quiescent systems. High-quality integral field spectroscopy (IFS) data with a wide spectral and spatial coverage, such as those from the CALIFA survey, offer a precious opportunity for advancing our understanding in this respect. We use deep IFS data from CALIFA (califa.caha.es) to study the *wim* over the entire extent and optical spectral range of 32 nearby ETGs. We find that all ETGs in our sample show faint (Hα equivalent width EW(Hα)~0.5 ... 2 Å) extranuclear nebular emission extending out to $\geqslant$2 Petrosian$_{50}$ radii. Confirming and strengthening our conclusions in Papaderos *et al.* (2013, hereafter P13) we argue that ETGs span a broad *continuous* sequence with regard to the properties of their *wim*, and they can be roughly subdivided into two characteristic classes. The first one (type i) comprises ETGs with a nearly constant EW(Hα)~1–3 Å in their extranuclear component, in quantitative agreement with (even though, no proof for) the hypothesis of photoionization by the post-AGB stellar component being the main driver of extended *wim* emission. The second class (type ii) consists of virtually *wim*-evacuated ETGs with a large Lyman continuum (Ly_c) photon escape fraction and a very low ($\leqslant$0.5 Å) EW(Hα) in their nuclear zone. These two ETG classes appear indistinguishable from one another by their LINER-specific emission-line ratios. Additionally, here we extend the classification by P13 by the class i+ which stands for a subset of type i ETGs with low-level star-forming activity in contiguous spiral-arm like features in their outermost periphery. These faint features, together with traces of localized star formation in several type i&i+ systems point to a non-negligible contribution from young massive stars to the global ionizing photon budget in ETGs.

Keywords. galaxies: elliptical and lenticular, cD - galaxies: evolution - galaxies: formation

1. Introduction

Even though the presence of faint nebular emission (*ne*) in the nuclei of many early-type galaxies (ETGs) has long been established observationally (e.g., Sarzi *et al.* 2010; Kehrig *et al.* 2012), the nature of the dominant excitation mechanism of the warm interstellar medium (*wim*) in these systems remains a subject of debate. The *low-ionization nuclear emission-line region* (LINER) emission-line ratios, as a typical property of ETG nuclei, have prompted various interpretations (see, e.g., Yan & Blanton 2012), including low-accretion rate active galactic nuclei (AGN; e.g., Ho 2008), fast shocks (e.g. Dopita & Sutherland 1995), and hot, evolved ($\geqslant 10^8$ yr) post-AGB (pAGB) stars (e.g., Binette *et al.* 1994; Stasińska *et al.* 2008). High-quality integral field spectroscopy (IFS) data with

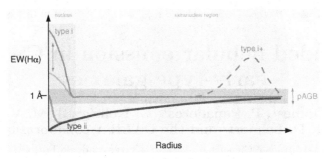

Figure 1. Schematic representation of the two main classes of ETGs, as defined in Papaderos *et al.* (2013) and Gomes *et al.* (2014). The EW(Hα) profiles of **type i** ETGs show in their extranuclear component nearly constant values within the narrow range between ∼0.5 Å and ∼2.4 Å, with a mean value of typically ∼1 Å. A few of these systems (labeled i+) additionally show an EW(Hα) excess in their periphery, which, as discussed in G14, is due to low-level star-forming activity. The defining property of **type ii** ETGs is a centrally very low (≤0.5 Å) mean EW(Hα), increasing then smoothly to ∼1 Å at their periphery. Regarding their nuclear properties, both ETG types show a large diversity, from systems with virtually *wim*-evacuated cores (≲0.1 Å) to galaxies with a compact nuclear EW(Hα) excess.

a wide spectral and spatial coverage, such as those from the *Calar Alto Legacy Integral Field Area* (CALIFA) survey (Sánchez *et al.* 2012; Sánchez 2014), offer an important opportunity to gain insight into the nature of nuclear and extranuclear gas excitation sources and advance our understanding on the evolutionary pathways of ETGs. Here we provide a brief summary of our results from an ongoing study of low-spectral-resolution ($R \sim 850$) CALIFA IFS cubes, observed with PMAS/PPAK (Roth *et al.* 2005; Kelz *et al.* 2006), for 20 E and 12 S0 nearby (<150 Mpc) galaxies. This sample was initially studied in Papaderos *et al.* (2013, hereafter P13) with main focus on the radial distribution of the EW(Hα) and Lyman continuum photon escape fraction (Ly_c) in ETGs, and has permitted a tentative subdivision of these systems into two main classes. A thorough 2D analysis of the same sample, including, e.g., EW(Hα) and stellar age maps, stellar and gas kinematics and gas excitation diagnostics will be presented in Gomes *et al.* (2014; hereafter G14). G14 also provide a detailed description of our IFS data processing and spectral modeling pipeline **Porto3D** and of the methods used to determine the radial distribution of various quantities of interest (e.g., EW(Hα)) both based on single-spaxel (**sisp**) determinations and a statistical analysis of **sisp** measurements within isophotal annuli (**isan**).

2. Results

Figure 2 illustrates on the example of the S0 galaxy NGC 1167 some of the quantities determined and discussed in G14. Panels **a&b** show, respectively, the Hα flux in 10^{-16} erg s^{-1} cm^{-2} (cf vertical bar to the right of the panel) and the EW(Hα) in Å, as determined after subtraction of the best-fitting synthetic stellar spectrum at each spaxel. The radial distribution of these two quantities as a function of the photometric radius $R^\star$ (arcsec) is shown in panels c) and d), respectively. **sisp** determinations obtained within the nuclear region, defined as twice the angular resolution of the IFS data (i.e. for $R^\star \leq 3.8$ arcsec) and in the extranuclear component are shown with red and blue circles, respectively. Open squares correspond to **isan** determinations within irregular isophotal annuli with the vertical bars illustrating the $\pm 1\sigma$ scatter of individual **sisp** measurements.

A subdivision of the EW(Hα) map into three intervals (panel e) is meant to help the reader distinguish between regions where the observed EW(Hα) is consistent, within the

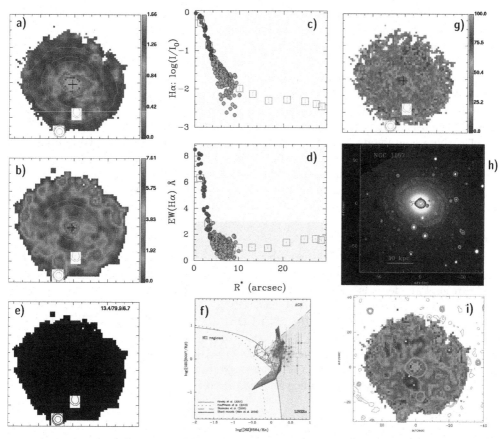

Figure 2. NGC 1167: a–d) Hα and EW(Hα) maps and radial profiles; e) Diagnostic subdivision of the EW(Hα) map (see discussion); f)) BPT diagram; g) Luminosity contribution of stars younger than 5 Gyr; Contours overlaid with a true-color SDSS image and the EW(Hα) map (h & i, respectively) depict faint spiral-like features in the periphery of the ETG, as revealed by image processing with an unsharp-masking technique.

uncertainties, with pure pAGB photoionization (0.5–2.4 Å; light blue), an additional gas excitation source is needed to account for the EW(Hα) ($\geq$2.4 Å; yellow), and where the EW(Hα) is by a factor $\geq$2 lower than that predicted from pAGB photoionization models ($\leq$0.5 Å; dark blue), hence Ly_c photon escape is important. The percentage of the spectroscopically studied area that is consistent with these three interpretations is indicated at the upper-right.

Panels f) shows sisp and isan determinations of the log([N II]6583/Hα) vs log([O III]5007/Hβ) diagnostic emission-line ratios after Baldwin *et al.* (1981, referred to in the following as BPT ratios). The meaning and color coding of the symbols is identical to that in panels c&d. The loci on the BPT diagrams that are characteristic of AGN and LINERs, and that corresponding to photoionization by young massive stars in HII regions are indicated by demarcation lines from Kauffmann *et al.* (2003, dotted curve), Kewley *et al.* (2001, solid curve) and Schawinski *et al.* (2007, dashed line). The grid of thin-gray lines roughly at the middle of each diagram depicts the parameter space that can be accounted for by pure shock excitation, as predicted by Allen *et al.* (2008) for a magnetic field of 1 μG, and a range of shock velocities between 100 and 1000 km s^{-1}, for gas densities between 0.1 and 100 cm^{-3}.

The luminosity contribution $\mathcal{L}_{5\mathrm{Gyr}}$ (%) of stars younger than 5 Gyr at the normalization wavelength (panel g) echoes the well established fact that ETGs are dominated by an evolved stellar component throughout their optical extent. Note the clear trend for an outwardly increasing $\mathcal{L}_{5\mathrm{Gyr}}$, which is consistent with inside-out galaxy growth, or even the presence of ongoing low-level star-forming activity in the galaxy periphery.

The overlaid contours on the true-color SDSS image (panel h) depict faint spiral-like features in the periphery of NGC 1167, as revealed by image processing with the flux-conserving unsharp-masking technique by Papaderos *et al.* (1998). As apparent from panel i, these contiguous low-surface brightness features are spatially correlated with moderately extended zones of enhanced EW(Hα). This indicates that they are not purely stellar relics from fading spiral arms that have long ceased forming stars, but sites of ongoing low-level star formation in the extreme periphery of NGC 1167 (see detailed discussion of this subject in G14).

Acknowledgements

Jean Michel Gomes acknowledges support from the Fundação para a Ciência e a Tecnologia (FCT) through the Fellowship SFRH/BPD/66958/2009 from FCT (Portugal). Polychronis Papaderos is supported by an FCT Investigador 2013 Contract, funded by FCT/MCTES (Portugal) and POPH/FSE (EC). JMG and PP acknowledge support by FCT under project FCOMP-01-0124-FEDER-029170 (Reference FCT PTDC/FIS-AST/3214/2012), funded by FCT-MEC (PIDDAC) and FEDER (COMPETE). This paper is based on data from the Calar Alto Legacy Integral Field Area Survey, CALIFA (http://califa.caha.es), funded by the Spanish Ministery of Science under grant ICTS-2009-10, and the Centro Astronómico Hispano-Alemán.

References

Allen, M. G., Groves, B. A., Dopita, M. A., *et al.* 2008, *ApJS*, 178, 20
Baldwin, J. A., Phillips, M. M., & Terlevich, R. 1981, *PASP*, 93, 5
Binette, L., Magris, C. G., Stasińska, G., & Bruzual, A. G. 1994, *A&A*, 292, 13
Dopita, M. A. & Sutherland, R. S. 1995, *ApJ*, 455, 468
Gomes, J. M., Papaderos, P., Kehrig, C., Vílchez, J. M., Lehnert, M. D., Sánchez, S., Ziegler, B., *et al.* 2014, in prep.
Ho, L. C. 2008, *ARA&A*, 46, 475
Kauffmann, G., Heckman, T. M., Tremonti, C., *et al.* 2003, *MNRAS*, 346, 1055
Kehrig, C., Monreal-Ibero, A., Papaderos, P., *et al.* 2012, *A&A*, 540, A11 (K12)
Kelz, A., Verheijen, M. A. W.., Roth, M. M., *et al.* 2006, *PASP*, 118, 129
Kewley, L. J., Dopita, M. A., Sutherland, R. S., Heisler, C. A., & Trevena, J. 2001, *ApJ*, 556, 121
Papaderos, P., Izotov, Y. I., Fricke, K. J., Thuan, T. X., & Guseva, N. G. 1998, *A&A*, 338, 43
Papaderos, P., Gomes, J. M., Vílchez, J. M., *et al.* 2013, *A&A*, 555, L1
Roth, M. M., Kelz, A., Fechner, T., *et al.* 2005, *PASP*, 117, 620
Sánchez, S. F., Kennicutt, R. C., Gil de Paz, A., *et al.* 2012, *A&A*, 538, A8
Sánchez, S., Advances in Astronomy, Issue: *Metals in 3 D: A Cosmic View from Integral Field Spectroscopy*, 2014a, in press
Sarzi, M., Shields, J. C., Schawinski, K., *et al.* 2010, *MNRAS*, 402, 2187
Schawinski, K., Thomas, D., Sarzi, M., *et al.* 2007, *MNRAS*, 382, 1415
Stasińska, G., Vale Asari, N., Cid Fernandes, R., *et al.* 2008, *MNRAS*, 391, L29
Trinchieri, G. & di Serego Alighieri, S. 1991, *AJ*, 101, 1647
Yan, R. & Blanton, M. R. 2012, *ApJ*, 747:61

Galaxies in 3D across the Universe
Proceedings IAU Symposium No. 309, 2014
B. L. Ziegler, F. Combes, H. Dannerbauer, M. Verdugo, eds.
© International Astronomical Union 2015
doi:10.1017/S1743921314009417

The SAMI Galaxy Survey: first 1000 galaxies

J. T. Allen[1,2] and the SAMI Galaxy Survey Team[3]

[1]Sydney Institute for Astronomy (SIfA), School of Physics, The University of Sydney, NSW
2006, Australia
[2]ARC Centre of Excellence for All-sky Astrophysics (CAASTRO)
email: j.allen@physics.usyd.edu.au
[3]Full list of team members is available at http://sami-survey.org/members

Abstract. The Sydney–AAO Multi-object Integral field spectrograph (SAMI) Galaxy Survey is an ongoing project to obtain integral field spectroscopic observations of ∼3400 galaxies by mid-2016. Including the pilot survey, a total of ∼1000 galaxies have been observed to date, making the SAMI Galaxy Survey the largest of its kind in existence. This unique dataset allows a wide range of investigations into different aspects of galaxy evolution.

The first public data from the SAMI Galaxy Survey, consisting of 107 galaxies drawn from the full sample, has now been released. By giving early access to SAMI data for the entire research community, we aim to stimulate research across a broad range of topics in galaxy evolution. As the sample continues to grow, the survey will open up a new and unique parameter space for galaxy evolution studies.

Keywords. galaxies: evolution – galaxies: kinematics and dynamics – galaxies: structure – techniques: imaging spectroscopy

1. Introduction

Large-scale spectroscopic galaxies surveys, such as the Sloan Digital Sky Survey (SDSS; York *et al.* 2000) and Galaxy And Mass Assembly (GAMA; Driver *et al.* 2009, 2011) survey, are well established and powerful tools for investigations of galaxy evolution. However, they are limited by only observing a single aperture in the centre of each galaxy, so are inherently unable to investigate the spatially distributed properties of galaxies.

Integral field spectrographs (IFSs) provide the information missed by traditional fibre-based surveys. The spatially resolved spectroscopic data produced by an IFS can be used to measure diverse quantities such as rotation curves, spatial distributions of star formation, and radial variation in stellar populations. Until recently, technical limitations made IFS observations difficult and time-consuming, limiting sample sizes to a few hundred at most, e.g. 260 in ATLAS-3D (Cappellari *et al.* 2011) and ∼600 in CALIFA (Sánchez *et al.* 2012).

A massive step forward is now being taken by instruments that can make IFS observations of multiple objects at once. The Sydney–AAO Multi-object Integral field spectrograph (SAMI) is one of the first of such instruments, with 13 integral field units (IFUs), each with a field of view of 15″, that can be deployed across a 1-degree patrol field (Croom *et al.* 2012). Each IFU consists of a bundle of 61 optical fibres lightly fused to have a high (∼75%) filling factor (Bryant *et al.* 2014a). SAMI is installed on the 3.9-m Anglo-Australian Telescope (AAT), feeding the existing AAOmega spectrograph (Sharp *et al.* 2006). For the first time, SAMI allows the rapid generation of large samples of IFS observations.

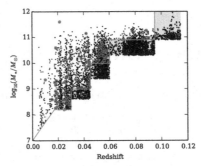

Figure 1. Stellar mass and redshift for all galaxies in the field (GAMA) regions of the SAMI Galaxy Survey (black points) and those in the EDR sample (red circles). The blue boundaries indicate the primary selection criteria, while the shaded regions indicate lower-priority targets. Large-scale structure within the GAMA regions is seen in the overdensities of galaxies at particular redshifts.

2. The SAMI Galaxy Survey

The SAMI Galaxy Survey is an ongoing project using SAMI to obtain integral field spectroscopic data for ~3400 galaxies, with an expected completion date of mid-2016. At the time of writing, ~1000 galaxies have been observed, including the pilot survey of ~100 galaxies. Early science investigations have included studies of galactic winds (Fogarty *et al.* 2012; Ho *et al.* 2014), the kinematic morphology–density relation (Fogarty *et al.* 2014), and star formation in dwarf galaxies (Richards *et al.*, MNRAS accepted); see also the related contributions by J. Bland-Hawthorn, L. M. R. Fogarty, I.-T. Ho, and N. Scott in this volume.

2.1. *Target selection*

The target galaxies for the SAMI Galaxy Survey are split into two samples: a field and group sample from the Galaxy And Mass Assembly (GAMA) survey G09, G12 and G15 fields, and a cluster sample from a set of eight galaxy clusters. The inclusion of a dedicated cluster sample extends the range of environmental densities to higher values than a simple mass- or luminosity-selected sample of galaxies would provide. Full details of the target selection are provided in Bryant *et al.* (2014b).

The target selection for the GAMA regions, consisting of a tiered set of volume-limited samples, is illustrated in Fig. 1. The figure also marks the galaxies included in the Early Data Release (EDR), discussed in Section 3.

2.2. *Data reduction*

The initial steps in the SAMI data reduction, up to the production of row-stacked spectra (RSS), are carried out using version 5.62 of 2DFDR†. The dedicated SAMI pipeline (Allen *et al.* 2014a) then performs flux calibration using separate observations of spectrophotometric standard stars, and corrects for telluric absorption based on simultaneous observations of a secondary standard star in the galaxy field. The set of ~7 dithered frames in each field are then resampled onto a regular grid and combined to produce a pair of datacubes (representing the blue and red arms of AAOmega) for each galaxy.

Fig. 2 illustrates the finished product for a single galaxy, showing the nature and quality of the data obtained. The complete data reduction process is described in Sharp *et al.* (2014).

† http://www.aao.gov.au/science/software/2dfdr

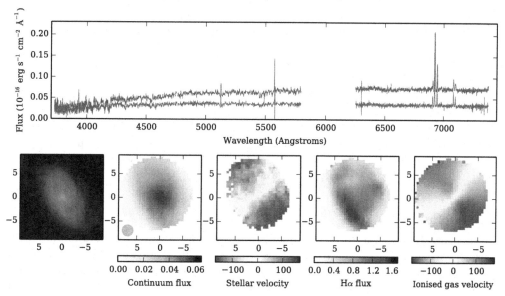

Figure 2. Example SAMI data for the galaxy 511867, with $z = 0.05523$ and $M_* = 10^{10.68} M_\odot$. Upper panel: flux for a central spaxel (blue) and one $3\rlap{.}''75$ to the North (red). Lower panels, from left to right: SDSS *gri* image; continuum flux map (10^{-16} erg s^{-1} cm^{-2} Å^{-1}); stellar velocity field (km s^{-1}); Hα flux map (10^{-16} erg s^{-1} cm^{-2}); Hα velocity field (km s^{-1}). The two velocity fields are each scaled individually. For the stellar velocity map, only spaxels with per-pixel signal-to-noise ratio >5 in the continuum are included. Each panel is $18''$ square, with North up and East to the left. The grey circle in the second panel shows the FWHM of the PSF.

2.3. *Data quality*

A comprehensive assessment of the quality of the data produced by SAMI is presented in Allen *et al.* (2014b); here we provide some key results.

Through the use of dome flat fields we achieve a flat field accuracy of 0.5–1% in most of the blue arm and 0.3–0.4% in the red arm. We use a combination of CuAr arc lamps and sky lines to achieve a wavelength calibration accuracy of $\simeq 0.1$ pixels or better. The continuum sky emission is typically subtracted with an accuracy of $\sim 1\%$.

Comparing synthetic $g - r$ colours to established photometry, we find a mean offset of 0.043 mag (SAMI data redder) with standard deviation 0.040 mag. The observed g-band magnitudes have a mean offset of -0.047 mag relative to SDSS photometry (SAMI data brighter), with a scatter of 0.27 mag, although a significant contribution to this scatter comes from difficulties in defining magnitudes for extended, irregular sources. The median full width at half maximum (FWHM) of the final point spread function in the datacubes is $2\rlap{.}''1$, with a typical range from $1\rlap{.}''5$ to $3\rlap{.}''0$.

2.4. *Ancillary data*

The SAMI Galaxy Survey is backed up by a wide range of ancillary datasets. By targetting the GAMA fields, we have available deep spectroscopic data as well as imaging from current and future surveys ranging from the UV to radio, including GALEX MIS, VST KiDS, VISTA VIKING, WISE, Herschel-ATLAS, GMRT and ASKAP (Driver *et al.* 2011). The spectroscopic information is crucial for robustly defining the environments of the SAMI Galaxy Survey targets (Brough *et al.* 2013), while the multi-wavelength imaging data provides a wealth of information on processes (e.g. dust-obscured star formation, radio AGN) that may not be apparent in the optical data. The cluster sample also has a variety of ancillary data including archival X-ray observations from *XMM-Newton* and

Chandra, and we have carried out a dedicated redshift survey to characterise the clusters and give robust classifications of cluster membership (Owers *et al.*, in prep.).

3. Early Data Release

We have recently made available a set of fully calibrated datacubes for 107 galaxies, forming the SAMI Galaxy Survey Early Data Release (EDR; Allen *et al.* 2014b). The galaxies were selected from the GAMA regions of the survey. Also available are the datacubes of the corresponding secondary standard stars and a table of ancillary data including quantities such as stellar mass, effective radius and surface brightness. All data can be downloaded from the SAMI Galaxy Survey EDR website†.

4. Conclusions

The SAMI Galaxy Survey now includes ∼1000 galaxies with IFS observations, and will grow to ∼3400 galaxies by mid-2016. This unique dataset is providing new insight into a broad range of topics in galaxy evolution, from star formation in dwarf galaxies to stellar kinematics in galaxy clusters. As the sample grows, further results will come into reach based on robust statistical analyses of the galaxy population, something that has rarely been possible with IFS data in the past. As a starting point, we have publicly released the datacubes for 107 galaxies, allowing the research community access for scientific investigations and preparation for the full dataset.

Acknowledgements

JTA acknowledges the award of an ARC Super Science Fellowship and a SIEF John Stocker Fellowship.

References

Allen J. T., *et al.*, 2014a, *Astrophysics Source Code Library*, ascl:1407.006
Allen J. T., *et al.*, 2014b, *MNRAS* submitted, arXiv:1407.6068
Brough S., *et al.*, 2013, *MNRAS*, 435, 2903
Bryant J. J., Bland-Hawthorn J., Fogarty L. M. R., Lawrence J. S., Croom S. M., 2014a, *MNRAS*, 438, 869
Bryant J. J., *et al.*, 2014b, *MNRAS* submitted, arXiv:1407.7335
Cappellari M., *et al.*, 2011, *MNRAS*, 416, 1680
Croom S. M., *et al.*, 2012, *MNRAS*, 421, 872
Driver S. P., *et al.*, 2009, *Astronomy and Geophysics*, 50, 12
Driver S. P., *et al.*, 2011, *MNRAS*, 413, 971
Fogarty L. M. R., *et al.*, 2012, *ApJ*, 761, 169
Fogarty L. M. R., *et al.*, 2014, *MNRAS*, 443, 485
Ho I.-T., *et al.*, 2014, *MNRAS* accepted, arXiv:1407.2411
Sánchez S. F., *et al.*, 2012, *A&A*, 538, A8
Sharp R., *et al.*, 2006, in SPIE Conference Series Vol. 6269
Sharp R., *et al.*, 2014, *MNRAS* submitted, arXiv:1407.5237
York D. G., *et al.*, 2000, *AJ*, 120, 1579

† http://sami-survey.org/edr

Galaxies in 3D across the Universe
Proceedings IAU Symposium No. 309, 2014
B. L. Ziegler, F. Combes, H. Dannerbauer, M. Verdugo, eds.

© International Astronomical Union 2015
doi:10.1017/S1743921314009429

The Kinematic Morphology-Density Relation from the SAMI Pilot Survey.

L. M. R. Fogarty[1,2] and the SAMI Galaxy Survey Team[3]

[1]Sydney Institute for Astronomy (SIfA), School of Physics, The University of Sydney, NSW 2006, Australia
[2]ARC Centre of Excellence for All-sky Astrophysics (CAASTRO)
email: l.fogarty@physics.usyd.edu.au
[3]Full list of team members is available at http://sami-survey.org/members

Abstract. We present the kinematic morphology-density relation in three galaxy clusters, Abell 85, 168 and 2399, using data from the SAMI Pilot Survey. We classify the early-type galaxies in our sample as fast or slow rotators (FRs/SRs) according to a measured proxy for their projected specific stellar angular momentum. We find each cluster contains both fast and slow rotators with and average fraction of SRs in the sample of $f_{SR} = 0.15 \pm 0.04$. We investigate this fraction within each cluster as a function of local projected galaxy density. For Abell 85 we find that f_{SR} increases at high local density but for Abell 168 and 2399 this trend is not seen. We find SRs not just at the centres of our clusters but also on the outskirts and hypothesise that these SRs may have formed in group environments eventually accreted to the larger cluster.

Keywords. galaxies: elliptical and lenticular, cD - galaxies: evolution - galaxies: formation

1. Introduction

The morphology-density relation (Dressler 1980) shows that the number density of early-type galaxies (ETGs, typically classified morphologically as lenticular (S0) and elliptical (E) galaxies) increases as one proceeds from low density to high density environments, at the expense of spiral galaxies whose number density declines comparatively. This robust result holds for galaxy clusters with diverse properties, including relaxed and unrelaxed systems, high and low mass clusters and some clusters with strong X-ray emission.

In recent decades the traditional definition of ETGs has been refined to include information gathered using 3D data from integral field spectroscopy (IFS). Building on results from the SAURON Survey (de Zeeuw et al. 2002), the ATLAS^{3D} Survey (Cappellari et al. 2011a) studied a volume limited sample of 260 ETGs out to 42 Mpc distance. The unprecedented detail of these 3D observations led to a re-classification of ETGs from a morphological basis to a kinematic one. The new framework (Emsellem et al. 2007, 2011) uses the kinematics of ETGs to classify them as fast rotators (FRs), disk-supported systems with high specific stellar angular momentum, or slow rotators (SRs), dispersion-supported systems with low specific stellar angular momentum. Analogous to the morphology-density relation Cappellari et al. (2011b) found that the fraction of SRs in the galaxy population increases in the highest density environments. This is the kinematic morphology-density relation. This relation can be used to study the formation and evolution of ETGs but requires large samples of integral field spectroscopy (IFS) data.

The Sydney-AAO Multi-Object Integral Field Spectrograph (SAMI, Croom et al. 2012) is a fibre-based multi-object spectrograph on the AAT capable of observing 13 galaxies at one time, thus increasing IFS survey efficiency by an order of magnitude. With SAMI

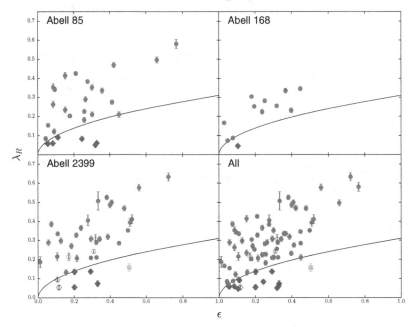

Figure 1. The $\lambda_R - \epsilon$ diagram for Abell 85 (top left), 168 (top right), 2399 (bottom left) and the entire SAMI Pilot Survey sample (bottom right). The blue circles are FRs and the red diamonds SRs. The turquoise square point indicates a double sigma galaxy - a disk supported system interloping into SR space. As such it is excluded from our statistics.

we have completed the SAMI Pilot Survey, presented here. We observed 106 galaxies across three galaxy clusters (Abell 85, 168 and 2399) with IFS data for each.

2. Galaxy Data and Kinematic Classification

The SAMI Pilot Survey was carried out over 10 nights at the AAT in September and October 2012. Full details of the observations are given in Fogarty *et al.* (2014) and Fogarty *et al.* in prep.. A total of 106 galaxies of all morphological types were successfully observed. The majority of these were cluster member galaxies, with a handful of foreground and background objects. Once our observations were complete we morphologically classified our sample by eye, in order to isolate the ETGs. In total 79 ETGs were observed.

SAMI feeds the double-beamed AAOmega spectrograph (Smith *et al.* 2004) which is fully configurable. For all SAMI Pilot Survey observations we used the 5700 Ådichroic with a mid-resolution 580V grating in the blue arm (covering the wavelength range $3700 - 5700$ Åwith R = 1700) and a high-resolution grating in the red arm (covering the wavelength range $6200 - 7300$ Åwith R = 4500). Each observation produces two fully-calibrated data cubes with variance and covariance information fully propagated through the reduction process. The data reduction procedure for SAMI is described fully in Allen *et al.* (2014) and Sharp *et al.* (2014).

The blue SAMI data cubes cover many important stellar absorption features and are used to derive stellar velocity and velocity dispersion maps for each galaxy. We use the penalised pixel fitting technique of Cappellari & Emsellem (2004) and the MILES empirical spectral templates (Sánchez-Blázquez *et al.* 2006). The kinematic maps are then used to calculate λ_R (Emsellem *et al.* 2007, 2011), a proxy for the projected specific stellar angular momentum for each galaxy. The classification system of Emsellem *et al.*

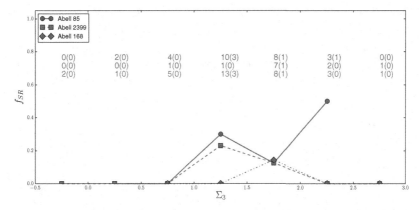

Figure 2. The kinematic morphology-density relation for Abell 85 (blue circles, solid line), 168 (red diamonds, dash-dot line) and 2399 (green squares, dashed line). The number of galaxies (SRs) is given for each bin over the bin.

(2011) is used whereby a galaxy is considered a SR if $\lambda_R < k\epsilon$ where the value of k is determined by the fiducial radius within which one measures λ_R and ϵ (galaxy ellipticity). Since our galaxies span a range of sizes we use three fiducial radii, Re/2, Re and 2Re for large, intermediate and small objects respectively. This does not bias our classification, as shown in Fogarty *et al.* (2014).

Figure 1 shows the $\lambda_R - \epsilon$ diagram for all three of our clusters and the entire sample combined. We find SRs in all three clusters studied and the fraction of SRs in the ETG population is 0.21 ± 0.08, 0.08 ± 0.08 and 0.12 ± 0.06 for Abell 85, 168 and 2399 respectively. The overall fraction for the entire sample is $f_{SR} = 0.15 \pm 0.04$. This is consistent with previous studies, which find an overall value of $f_{SR} \sim 0.15$ across a range of average global environments (D'Eugenio *et al.* 2013; Houghton *et al.* 2013).

3. The Kinematic Morphology-Density Relation

We examine the value of f_{SR} as a function of local projected galaxy density. This is parametrised by the third nearest neighbour surface density, Σ_3, and the method by which Σ_3 is calculated is described in Fogarty *et al.* (2014). Figure 2 shows the kinematic morphology-density relation for Abell 85 (blue circles), 168 (red diamonds) and 2399 (green squares). For Abell 168 our sampling with the SAMI Pilot sample is poor so it is not possible to draw any robust conclusions for this cluster. For Abell 85 we see the expected trend such that f_{SR} increases in the densest part of the cluster. However, this trend is not seen in Abell 2399 where the two central BCG candidate galaxies are in fact FRs. In both Abell 85 and 2399 we also see a high f_{SR} at low to intermediate densities within each cluster, indicating that SRs are not only found in the centres of clusters but also on their outskirts.

The roughly constant value of f_{SR} we find (consistent with previous studies, e.g. Houghton *et al.* (2013)) in all three clusters suggests that the formation mechanism for SRs is equally efficient in environments as diverse as the field, galaxy groups and low-mass to high-mass clusters. However, SRs are preferentially found at cluster centres in many previous studies (Cappellari *et al.* 2011b; D'Eugenio *et al.* 2013; Houghton *et al.* 2013). Here we see a more complex picture with one cluster obeying the expected kinematic morphology-density relation but with evidence as well for in-falling SRs. One possible explanation is that SRs could form preferentially as central galaxies in a halo, such as a galaxy group. These halos then grow into large clusters dominated by SRs or are eventually accreted to already existing clusters. The large SRs then make their way to the centres of their host clusters by dynamical friction. We hypothesise that the SRs on the outskirts of Abell 85 and 2399 were formed in groups and are joining their host clusters through merging processes. This is supported by the fact that all three of the clusters studied for the SAMI Pilot Survey are known cluster-cluster mergers.

4. Conclusions

We have kinematically classified the 79 ETGs of the SAMI Pilot Survey of three galaxy clusters, Abell 85, 168 and 2399. We calculated a SR fraction of 0.21 ± 0.08, 0.08 ± 0.08 and 0.12 ± 0.06 for each cluster respectively with an overall fraction of $f_{SR} = 0.15 \pm 0.04$ for the entire sample. This is consistent with previous work in this area and the hypothesis that the SR formation mechanism is equally efficient across many different global environments from the field to high-mass clusters.

In addition we examined the kinematic morphology-density relation for all three clusters. We find that although in Abell 85 we see the expected trend such that f_{SR} increases in the densest part of the cluster we also see significant evidence for the existence of SRs in low and intermediate density regions of both Abell 85 and 2399. We hypothesise that these SRs could have formed as centrals in groups and later fallen into the cluster as SRs.

Acknowledgements

The SAMI Galaxy Survey is based on observations made at the Anglo-Australian Telescope. The Sydney-AAO Multi-object Integral field spectrograph (SAMI) was developed jointly by the University of Sydney and the Australian Astronomical Observatory. The SAMI input catalogue is based on data taken from the Sloan Digital Sky Survey, the GAMA Survey and the VST ATLAS Survey. The SAMI Galaxy Survey is funded by the Australian Research Council Centre of Excellence for All-sky Astrophysics (CAASTRO), through project number CE110001020, and other participating institutions. The SAMI Galaxy Survey website is http://sami-survey.org/.

References

Allen, J. T., Croom, S. M., Konstantopoulos, I. S., *et al.* 2014, *MNRAS*, submitted
Cappellari, M. & Emsellem, E. 2004, *MNRAS*, 116, 138
Cappellari, M., Emsellem, E., Krajnović, D., *et al.* 2011, *MNRAS*, 413, 813
Cappellari, M., Emsellem, E., Krajnović, D., *et al.* 2011, *MNRAS*, 416, 1680
Croom, S. M., Lawrence, J. S., Bland-Hawthorn, J., *et al.* 2012, *MNRAS*, 421, 872
D'Eugenio, F., Houghton, R. C. W., Davies, R. L., *et al.* 2013, *MNRAS*, 429, 1258
de Zeeuw, P. T., Bureau, M., Emsellem, E., *et al.* 2002, *MNRAS*, 329, 513
Dressler A. 1980, *ApJ*, 236, 351
Emsellem, E., Cappellari, M., Krajnović, D., *et al.* 2007, *MNRAS*, 379, 401
Emsellem, E., Cappellari, M., Krajnović, D., *et al.* 2011, *MNRAS*, 414, 888
Fogarty, L. M. R., Scott, N., Owers, M. S., *et al.* 2014, *MNRAS*, 443 485
Houghton, R. C. W., Davies, R. L., D'Eugenio, F., *et al.* 2013, *MNRAS*, 436, 19
Sánchez-Blázquez, P., Peletier, R. F., Jiménez-Vicente, J., *et al.* 2006, *MNRAS*, 371, 703
Sharp, R., Allen, J. T., Fogarty, L. M. R., *et al.* 2014, *MNRAS*, submitted
Smith, G. A., Saunders, W., Bridges, T., *et al.* 2004, *SPIE*, 5492, 410

Galaxies in 3D across the Universe
Proceedings IAU Symposium No. 309, 2014
B. L. Ziegler, F. Combes, H. Dannerbauer, M. Verdugo, eds.
© International Astronomical Union 2015
doi:10.1017/S1743921314009430

The origin of metallicity gradients in massive galaxies at large radii

Michaela Hirschmann[1] and Thorsten Naab[2]

[1] UPMC-CNRS, UMR7095, Institut dAstrophysique de Paris, F-75014 Paris, France
email: hirschma@iap.fr
[2] Max-Plank-Institute für Astrophysik, Karl-Schwarzschild Strasse 1, D-85740 Garching,
Germany
email: naab@mpa-garching.mpg.de

Abstract. We investigate the origin of stellar metallicity gradients in massive galaxies at large radii ($r > 2R_{\mathrm{eff}}$) using ten cosmological zoom simulations of halos with $6 \times 10^{12} M_\odot < M_{\mathrm{halo}} < 2 \times 10^{13} M_\odot$. The simulations follow metal cooling and enrichment from SNII, SNIa and AGB winds. We explore the differential impact of an empirical model for galactic winds that reproduces the evolution of the mass-metallicity relation. At larger radii, the galaxies become more dominated by stars accreted from satellite galaxies in major and minor mergers. In the wind model, fewer stars are accreted, but they are significantly more metal poor resulting in steep global metallicity ($\langle \nabla Z_{\mathrm{stars}} \rangle = -0.35$ dex/dex) gradients in agreement with observations. Metallicity gradients of models without winds are inconsistent with observations. For the wind model, stellar accretion is steepening existing in-situ metallicity gradients by about 0.2 dex by the present day and is required to match observed gradients. Metallicity gradients are significantly steeper for systems, which have accreted stars in minor mergers. In contrast, galaxies with major mergers have relatively flat gradients, confirming previous results. We highlight the importance of stellar accretion for stellar population properties of massive galaxies at large radii, which provide important constraints for formation models.

Keywords. galaxies: evolution - galaxies: formation - methods: numerical

1. Introduction

It is a natural prediction of modern hierarchical cosmological models that the assembly of massive galaxies involves major and minor mergers although most stars in most galaxies have been made in-situ from accreted or recycled gas. Nonetheless, these mergers are expected to play a significant role for the structural and morphological evolution of the massive early-type galaxy population. One important structural galaxy property, which is thought to be strongly influenced by mergers, are the (in general negative) metallicity gradients observed early-on in massive, present-day elliptical (e.g. McClure & Racine 1969), but also in disk galaxies (e.g. Wyse & Silk 1989), typically within $1R_{\mathrm{eff}}$. Thanks to improved and more elaborated observational techniques, present-day metallicity gradients can nowadays be measured out to much larger radii, partly out to even $8R_{\mathrm{eff}}$, (e.g. La Barbera *et al.* 2012).

Previous studies (e.g. Kobayashi 2004) only discuss the emergence of inner gradients (up to 3 R_{eff}) at comparably poor spatial resolution. Here, we focus on the stellar accretion origin of metallicity gradients in high-resolution re-simulated massive galaxies *at large radii* ($2R_{\mathrm{eff}} < r < 6R_{\mathrm{eff}}$) in a full cosmological context. We particularly intend to explore the combined effect of strong galactic winds and of the individual merger and accretion histories on the in-situ formed and accreted stellar fractions and on the steepening/flattening of the metallicity gradients at these large radii.

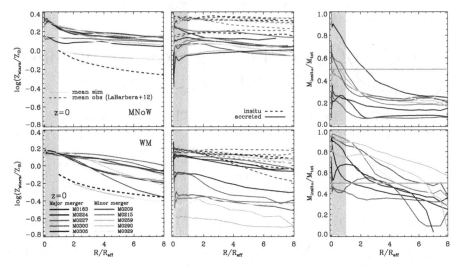

Figure 1. *Left column:* Total stellar metallicity gradients (mass weighted) at $z = 0$ for the ten main galaxies (different colors) simulated with the MNoW and WM model. The green solid lines indicate the average gradient at $2 < R/R_{\rm eff} < 6$. *Middle panels:* Metallicity gradients at $z = 0$ separated into stars formed in-situ (dashed lines) and accreted stars (solid lines). *Right panels:* Fraction of in-situ to total stellar mass as a function of radius.

2. High-resolution simulations of individual galaxy halos

We consider the 10 most massive high-resolution, cosmological zoom simulations (covering a mass range of $6 \times 10^{12} < M_{\rm halo} < 2 \times 10^{13} M_{\odot}$) presented in (Hirschmann *et al.* 2013) including a treatment for metal enrichment (SNII, SNIa and AGB stars) and a phenomenological feedback scheme for galactic winds (Oppenheimer & Dave 2006, 2008). The dark matter/gas particles have a mass resolution of $m_{\rm dm} = 2.5 \cdot 10^7 M_{\odot} h^{-1}$ and $m_{\rm gas} = m_{\rm star} = 4.2 \cdot 10^6 M_{\odot} h^{-1}$, respectively with a co-moving gravitational softening length for the gas and star particles of $400~h^{-1}{\rm pc}$ and for the dark matter particles of $890~h^{-1}{\rm pc}$ (Oser *et al.* 2010).

These cosmological zoom simulations were shown to be successful in suppressing early star formation at $z > 1$, in predicting reasonable star formation histories for galaxies in present day halos of $\sim 10^{12} M_{\odot}$, in producing galaxies with high cold gas fractions (30 - 60 per cent) at high redshift, and in significantly reducing the baryon conversion efficiencies for halos ($M_{\rm halo} < 10^{12} M_{\odot}$) at all redshifts in overall good agreement with observational constraints. Due to the delayed onset of star formation in the wind models, the metal enrichment of gas and stars is delayed and is also found to agree well with observational constraints.

3. Metallicity gradients of present-day massive galaxies

In Fig. 1, we show the total (mass-weighted) stellar metallicity gradients out to 8 $R_{\rm eff}$ for the main galaxies at $z = 0$ in the MNoW and the WM model (as indicated in the legend). For the MNoW galaxies, the (most) central metallicity has values of $Z/Z_{\odot} \sim 0.4$ and drops to $Z/Z_{\odot} \sim 0.1$ at large radii, the slopes reach a minimum value of -0.25 dex/dex. This gradient is mainly driven by the accreted stars (the in-situ distributions are almost all nearly flat), which have on average solar metallicity (top middle panel of Fig. 1) and dominate most systems outside $1R_{\rm eff}$ (top right panel of Fig. 1).

The WM galaxies (bottom panels of Fig. 1) have lower central metallicities ($Z/Z_\odot \sim$ 0.2) with much steeper outer gradients down to -0.76 dex/dex with a mean of -0.35 dex/dex. The reason for the steeper gradients in the WM compared to the MNoW model is twofold: on the one hand, the steeper gradients originate from the accretion of metal-poorer stellar populations. On the other hand, also the in-situ components show metallicity gradients contributing to the overall gradients. The latter is most likely due to infall of (particularly re-infall of previously ejected) metal-poor gas onto the galaxy which can be then turned into metal-poor stars as a consequence of an inside-out growth, the same process causing the metallicity gradients in disk galaxies. Late re-accretion of previously ejected gas occurs typically in the WM model due to the strong galactic winds, but not in the MNoW model, where the in-situ gradients are, therefore, relatively flat (see top middle panel of Fig. 1). However, despite of the partly negative in-situ gradients, we find that on average, the in-situ gradients are reduced by ~ 0.2 dex due to accretion of metal-poor stellar populations.

3.1. *Galaxy mergers and gradients*

The black-blue lines in Fig. 1 indicate galaxies which experienced at least one major galaxy merger since $z = 1$, while the red-yellow lines illustrate those having undergone only minor galaxy mergers. Present-day galaxies having experienced a recent major merger have typically flatter gradients than those with a more quiet merger history.

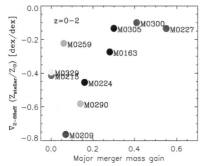

Fig. 2 quantifies the connection between the galaxy merger history and the steepness of the metallicity gradients: we show the fitted metallicity gradients at $z = 0$ for WM galaxies versus the mass gain by major mergers. The mass gain by major mergers considers the entire stellar mass which was brought into the main galaxy by major mergers since $z = 2$ normalised to the present-day stellar mass.

Figure 2. Stellar metallicity gradients (between $2 - 6R_{\mathrm{eff}}$) versus the mass gain through major mergers for the ten re-simulated galaxies in the WM simulations. The higher the major merger mass gain, the smaller is the slope of the metallicity gradient.

The fitted metallicity gradients correlate with the past merger history: for a major merger mass gain above 20 %, the metallicity gradients are flatter than -0.3 dex/dex. Instead, for lower x-values, the metallicity gradients are mostly more negative than -0.4 dex/dex. This is due to the fact that the accreted stars show a huge variety of metallicities from $Z/Z_\odot \sim -0.6$ to $Z/Z_\odot \sim +0.1$ (see bottom middle panel of Fig. 1) depending on the exact merger history: in case of a recent major merger, the accreted metallicity is significantly larger than without a major merger (as more massive galaxies have higher metallicity). The higher metallicity of the accreted stars together with the different mixing behaviour (violent relaxation) in case of major mergers flattens the total metallicity gradients.

This result bears an important implication for observations, as it can help to reconstruct the past assembly history for observed present-day metallicity gradients: The relation between the steepness of the gradients and the individual merger history implies that observed massive galaxies having steep outer gradients most likely have not experienced any major merger event after $z = 1$, but instead several minor mergers.

3.2. *Comparison with observations*

In a recent study of Pastorello *et al.* (2014), using the SLUGGS survey, they investigate metallicity gradients at least out to $2.5R_{\rm eff}$. For comparable stellar masses, their galaxies reveal slopes between $-1.15 - +0.18$ dex/dex. In addition, in a study by La Barbera *et al.* (2012), they investigate metallicity gradients of early-type galaxies ($M_{\rm stellar} > 3 \times 10^{10} M_\odot$) even out to $8 \times R_{\rm eff}$ using the SDSS-based Spider survey. For massive galaxies with $10^{11} < M_{\rm stellar} < 7 \times 10^{11} M_\odot$ they find outer metallicity gradients ($1-8 \times R_{\rm eff}$) in the range of $-0.29 - -0.74$ dex/dex depending on the stellar population model (illustrated by the black dashed lines in Fig 1).

Overall, our WM galaxies are able to cover such a broad range of slopes ($-0.8 - -0.1$ dex/dex) much better than the MNoW galaxies whose slopes are on average too flat (within the range of $-0.25 - +0.03$ dex/dex). The average metallicity gradient of the WM galaxies (-0.35 dex/dex) is in excellent agreement with the one of La Barbera *et al.* (2012). *As expected a priori, this implies that a strong stellar feedback is a key mechanism to be consistent with observed steep metallicity gradients in massive galaxies in the local Universe.*

4. Conclusions

Analysing zoom simulations of 10 massive galaxies suggests that the outer negative metallicity gradients (at radii $r > 2\ R_{\rm eff}$) of present-day massive galaxies are mainly determined by the accretion of stars with lower metallicity. Towards low redshift, stars accreted in low metallicity galaxies become more and more dominant at large radii and the metallicity gradients of in-situ formed stars in the wind model are enhanced by ~ 0.2 dex/dex by accretion of metal-poor systems. The model with galactic winds predicts steeper total metallicity gradients (on average -0.35 dex/dex) as the accreted stellar systems are significantly more metal-poor and despite of the fact that much less stellar mass in total is accreted compared to the no-wind model. We can conclude that stellar accretion (in minor mergers) of low mass satellites results in steep outer metallicity gradients as predicted from idealized experiments (Villumsen 1983) successfully matching the broad range of observed metallicity profiles of local galaxies at large radii.

Acknowledgements

MH acknowledges financial support from the European Research Council via an Advanced Grant under grant agreement no. 321323NEOGAL. TN acknowledges support from the DFG excellence cluster "Origin of the Universe".

References

Hirschmann, M., *et al.* 2013, *MNRAS*, 436, 2929
Kobayashi, C., 2004, *MNRAS*, 347, 740
La Barbera, F., *et al.* 2012, *MNRAS*, 426, 2300
McClure, R. & Racine R. 1969, *AJ*, 74, 1000
Oppenheimer, B. & Dave, R. 2006, *MNRAS*, 373, 1265
Oppenheimer, B. & Dave, R. 2008, *MNRAS*, 387, 577
Oser, L., *et al.* 2010, *ApJ*, 725, 2312
Pastorello, N., *et al.* 2014, *MNRAS*, 442, 1003
Villumsen, J., 1983, *MNRAS*, 204, 219
Wyse, R. & Silk, J. 1989, *ApJ*, 339, 700

Galaxies in 3D across the Universe
Proceedings IAU Symposium No. 309, 2014
B. L. Ziegler, F. Combes, H. Dannerbauer, M. Verdugo, eds.
© International Astronomical Union 2015
doi:10.1017/S1743921314009442

The Scaling of Star Formation: from Molecular Clouds to Galaxies

Daniela Calzetti

Department of Astronomy
University of Massachusetts
710 North Pleasant Street
Amhest, MA 01003, USA
email: calzetti@astro.umass.edu

Abstract. This is a review of the extant literature and recent work on the scaling relation(s) that link the gas content of galaxies to the measured star formation rates. A diverse array of observing techniques and underlying physical assumptions characterize the determination of these relations at different scales, that range from the tens of parsec sizes of molecular clouds to the tens of kpc sizes of whole galaxies. Different techniques and measurements, and a variety of strategies, have been used by many authors to compare the scaling relations, both within and across galaxies. Although the picture is far from final, the past decade has seen tremendous progress in this field, and more progress is expected over the next several years.

Keywords. galaxies: star formation – galaxies: ISM – ISM:clouds

1. Introduction

The studies that link star formation to its fuel, the gas, are at the foundation of all investigations of the evolution of galaxies across cosmic times. The early hypothesis by Schmidt (1959) of a correlation between star formation rate (SFR) and gas volume densities has evolved over the years to use the surface densities of both the SFR and the gas: $\Sigma_{SFR}/(M_\odot \; yr^{-1} \; kpc^{-2}) \propto [\Sigma_{gas}/(M_\odot \; pc^{-2})]^\gamma$, in order to employ only observable quantities and remove the (uncertain) measurement of the thickness of the emitting regions (e.g., Kennicutt 1989, 1998). Alternate, non–exclusive, formulations introduce dynamical timescales τ_{dyn}: $\Sigma_{SFR} = \epsilon_{ff}[\Sigma_{gas}/\tau_{dyn}]$ (e.g. Kennicutt 1998; Kennicutt, & Evans 2012), where ϵ_{ff} is the efficiency of star formation, i.e., the rate of conversion of gas to stars over one dynamical timescale. The gas/SFR ratio expresses the timescale over which the gas will be exhausted is star formation proceed at the same rate, thus defining a depletion timescale: $\tau_{dep} = \Sigma_{gas}/\Sigma_{SFR} = \tau_{dyn}/\epsilon_{ff}$. Following common practice, I will term the relation between SFR and gas surface densities the 'Schmidt–Kennicutt Law', or SK Law for short.

Advances in both space and ground facilities and instrumentation have opened a new era for the investigation of the scaling laws of star formation, pushing the frontier both in time, towards higher redshifts, and in space, towards smaller and smaller scales within galaxies. High angular resolution and/or deep measurements of both the dust–obscured and unobscured SFRs have been secured by space facilities like the Galaxy Evolution Explorer in the ultraviolet (UV), the Hubble Space Telescope in the UV–to–nearIR, and the Spitzer Space Telescope and the Herschel Space Observatory in the infrared, in concert with ground–based optical and nearIR observations of the nebular line emission from galaxies. Ground–based radio, millimeter, and sub–mm facilities and the Herschel Space Observatory have secured the HI, CO and dust continuum observations needed to trace the gas (and dust) emission.

In this review, I briefly summarize the basic results of the past ∼decade of investigations, with specific emphasis on the intermediate scales within nearby galaxies, i.e., the scales that are in–between the whole galaxy and the giant molecular clouds (GMCs), and range from a few hundred pc to several kpc. These scales are here referred to as 'sub–galactic', while measurements of whole galaxies are termed 'global'. In the nearby Universe, the study of sub–galactic scales has most benefitted from the facilities briefly described above. Because of the extensive literature on the subject, this review is inevitably incomplete. Omissions are unintentional.

2. The Global Relation

The general picture that has emerged since the first systematic analysis of local galaxies (Kennicutt 1989) is that more actively star forming galaxies convert gas into stars at a higher rate than less active galaxies. This continues to be supported by more recent data, both at low (Kennicutt 1998; Kennicutt, & Evans 2012) and high redshift (Daddi, et al. 2010a; Genzel, et al. 2010; Santini, et al. 2014). The fundamental relation between SFR density and gas density of galaxies has been variously interpreted as a non–linearity between Σ_{SFR} and Σ_{gas} ($\gamma > 1$, e.g., Kennicutt, & Evans 2012; Cormier, et al. 2014) or a decrease in the depletion timescale τ_{dep} for increasing SFR density or specific SFR, sSFR=SFR/Mass$_{stars}$ (e.g., Saintonge, et al. 2011; Tacconi, et al. 2013; Genzel, et al. 2014, see, also, Linda Tacconi's contribution to this Conference).

One of the recurring questions when measuring the Global SK Law is whether Σ_{gas} should be calculated including only the molecular (star–forming) component or the sum of all gas, the main contributors typically being neutral atomic and molecular (including CO–dark molecular gas, e.g., Wolfire, Hollenbach, & McKee 2010; Langer, et al. 2014). There is not necessarily a unique answer to this question, and it may depend on the actual details of the question. To the extent that the main question is: what is the relation between SFR and the gas reservoir?, the search for an answer should likely include all gas components. Although the atomic gas traced by HI does not participate directly in the formation of stars, the existence of a density boundary for HI, $\Sigma_{HI} \sim 10$ M$_\odot$ pc^{-2} (e.g., Kennicutt, et al. 2007), suggests that HI converts into H$_2$ above this density value (Krumholz, McKee, & Tumlinson 2009). If HI replenishes the molecular gas reservoir depleted by star formation and to the extent that a galaxy can be considered an isolated system on the timescales relevant for star formation ($\leqslant$100 Myr), HI ought to be included in calculations of the gas reservoir of a galaxy when deriving the Global SK Law (Boissier, et al. 2003). This is likely less of an issue at high redshift, where current evidence suggests that the molecular fraction in galaxies is higher than in local galaxies (Daddi, et al. 2010b; Tacconi, et al. 2013).

The higher rate of gas–to–stars conversion in more active galaxies can be re-interpreted as a change in the dynamical timescale relevant for the conversion of gas to stars, at fixed efficiency ϵ_{ff}; since the dynamical timescale decreases as the star formation rate increases, the relation between Σ_{SFR} and Σ_{gas}/τ_{dyn} becomes linear (e.g., Kennicutt 1998; Daddi, et al. 2010a; Genzel, et al. 2010; Krumholz, Dekel, & McKee 2012). However, there is currently no agreement on what timescale to use for the SK Law, whether the orbital, free–fall, and/or crossing time, and why (Kennicutt, & Evans 2012). The value of the efficiency of star formation, ϵ_{ff}, which is around a few percent, is also another parameter that has not been explained physically, so far. Finally, there is some evidence for the existence of two regimes for the Global SK Law, one for 'normal star–forming' galaxies and one for 'starburst' galaxies (Daddi, et al. 2010a; Genzel, et al. 2010), but the dichotomy represented by the two regimes may be driven, at least in part, by the choice of the CO–to–H_2 factor (Narayanan, *et al.* 2011). Thus, despite much progress in this area, many questions still remain open.

3. Down to the Scales of GMCs

Zooming into galaxies can offer better understanding of the physical underpinning of the SK Law, since we can approach the region sizes where star formation occurs. Star formation is more closely related to GMCs than to the more diffuse gas (as traced by HI), and it has become customary to represent the Sub–Galactic SK Law via a relation between Σ_{SFR} and Σ_{H2}, rather than Σ_{HI+H2} (e.g., Kennicutt, *et al.* 2007; Bigiel, *et al.* 2008, , and many others). Furthermore, recent investigations suggest that, within GMCs, star formation is closely associated with the dense molecular gas (A_V >7 mag), with a linear relation between SFR and $M_{gas,dense}$ (e.g., Lada, *et al.* 2012, 2013, see, also, the review by Charlie Lada at this Conference).

Independently of the gas phase that we decide to relate to star formation, the fundamental question remains: how are the different gas phases (loosely separated according to increasing gas density) interconnected, and how does this interconnection lead to the observed SFRs? This question is still unanswered.

The results on the SK Law of GMCs are difficult to relate to those of whole galaxies, because moving up in scale does not translate into a simple rescaling of quantities. The locus of GMCs in the Σ_{SFR}–Σ_{H2} plane is one–to–two orders of magnitude higher than the locus occupied by galaxies and sub–galactic regions (e.g., Heiderman, *et al.* 2010; Gutermuth, *et al.* 2011). Most of the area of galaxies is not occupied by GMCs and their associated star formation: the GMCs covering factor hovers around 5%–15% in local star–forming galaxies, with large variations from system to system (Leroy, *et al.* 2009). However, the covering factor alone does not account for the discrepancy between the GMC–scale SK Law and the Global SK Law. Large fractions of CO–emitting diffuse molecular gas need to be invoked, in order to move the GMC–scale SK Law to higher gas surface densities, while keeping the SFR surface density at low values (Heiderman, *et al.* 2010). Alternatively, the two Laws (GMCs and Global) may originate from different physical mechanisms, with the Global SK Law determined by stellar and supernova feedback, and the GMC–scale SK Law dominated by local processes, such as turbulence–regulated gravitational collapse (Faucher–Giguère, Quataert & Hopkins 2013, see, also, the review by Phil Hopkins at this Conference). Thus, the relation between star formation in GMCs and in galaxies as a whole is far from trivial. The SK Law at intermediate ($\approx$kpc) scales may provide insights on the link between the two.

4. The Sub–Galactic Relations

Extensive literature exists on the Sub–Galactic SK Law, where investigators have used a variety of approaches for the treatment of both the data and the statistics, with analyses that include both radial profiles and galaxy region bins, and covering the full spatial range between $\sim$200 pc and $\sim$2 kpc (a highly incomplete list includes: Heyer, *et al.* 2004; Kennicutt, *et al.* 2007; Bigiel, *et al.* 2008; Blanc, *et al.* 2009; Verley, *et al.* 2010; Liu, *et al.* 2011; Momose, *et al.* 2013; Shetty, *et al.* 2014). One common characteristic to these studies is the general lack of agreement on what the shape of the Sub–Galactic SK Law is, specifically in regard to the value of γ, that ranges from $\sim$ 0.6 to $\sim$ 1.8. For $\gamma \geqslant 1$, the value of γ carries information about the physical process(es) responsible for the conversion of gas into stars. Thus, determining whether a single value exists among galaxies and what this value is for sub–galactic regions has become an important quest.

Conversely, studies that find a sub–linear ($\gamma < 1$) Sub–Galactic SK Law require the introduction of components unrelated to star–formation, such as CO–emitting diffuse molecular gas (e.g. Shetty, Clark, & Klessen 2014).

At this point in time, it may still be difficult to establish whether a universal Sub–Galactic SK Law exists, and whether it can be used as a conduit for relating the GMC–scale to the Global SK Laws. The reason is the many biases and spurious effects that can be introduced in the measurement of either the SFR and/or the gas. The following list attempts at conveying the variety of such effects:

- kpc–size 'bins' within galaxies are not isolated systems, not even on the timescales of star formation. This affects measurements of Σ_{SFR}. Ionizing photons leak out of HII regions, at the 30%–60% level (e.g. Thilker, *et al.* 2002), and ionize gas up to a kpc from the point of origin; measurements of faint Hα cannot be readily converted to a SFR because of this effect. While the UV emission from evolved (>100 Myr) stellar populations and the mid–IR emission from the dust they heat only accounts for ≈20%–30% of the total emission from a galaxy (Johnson, *et al.* 2013; Leroy, *et al.* 2012), these numbers do not capture the effect of the clustering of star formation on Sub–Galactic measurements. Clustered star formation disperses in the field, e.g., from the spiral arms into the inter arm regions, in ∼100–300 Myr, or less (Crocker, *et al.* 2014), which produces the redder UV colors of the field relative to those of star–forming regions. This affects popular SFR indicators based on UV, IR, and other stellar continuum or dust emission bands. For reference, a galactic region with stellar populations in the age range 100 Myr–10 Gyr produces enough UV+IR emission to mimic (in unresolved conditions) star formation with values $\Sigma_{SFR}(UV + IR)$ ∼1–6×10^{-3} M$_\odot$ yr^{-1} kpc^{-2}, for star formation with an e–folding time of 5 Gyr and for constant star formation, respectively, using typical mean stellar densities (Barden, *et al.* 2005). Yet, this region contains no current star formation. Its main effect will be to bias measurements of faint star forming regions by providing a spurious 'pedestal' or 'excess' to their SFRs (Liu, *et al.* 2011), together with an overall scatter due to stellar population content variations among bins.

- Measurements of gas surface density are non–trivial. Although there is agreement in the literature that on Sub–Galactic scales star formation is better correlated with molecular gas, the tools to determine Σ_{H2} have both strengths and weaknesses. The most common approach, to measure H$_2$ via CO line emission can be biased by: (1) the dependency of the CO–to–H$_2$ conversion factor on metallicity, gas temperature, and turbulence (Bolatto, Wolfire, & Leroy 2013); (2) the variation of the ratio among different CO transitions within a galaxy (e.g., Koda, *et al.* 2012) and the fact that higher transitions trace smaller fractions of the molecular gas (e.g, Rigopoulou, *et al.* 2013); (3) presence of CO-dark, star–forming H$_2$; and (4) presence of CO-emitting, non–star–forming, diffuse H$_2$. The other popular approach, that uses a combination of dust and HI maps to derive the H$_2$ distribution can be affected by: (1) variations, at the level of 2X, in the dust surface densities, due to choice of fit parameters in the dust emission modeling (e.g., Kirkpatrick, *et al.* 2014); (2) presence of sub–mm excess in the dust emission, which affects estimates of the largest, by mass, dust component (Galametz, *et al.* 2014); and (3) the potential super–linear dependency of the dust–to–gas ratio on metallicity (Rémy–Ruyer, *et al.* 2014). While most of these problems affect both the Sub–Galactic and the Global SK Laws, local variations of conditions within a galaxy can exacerbate the impact on the Sub-Galactic SK Law (Koda, *et al.* 2012; Sandstrom, *et al.* 2013).

- The timescale of the association between star formation and gas clouds is likely short. As young stellar structures evolve, the gas clouds from which they formed dissolves, on scales as short as ∼20–30 Myr (Kawamura, *et al.* 2009). When coupled with the dispersal of star formation and the low covering factor of molecular clouds (see

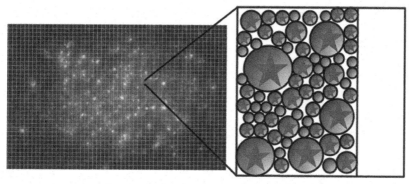

Figure 1. Schematic representation of the simulations by Calzetti, Liu & Koda (2012). **(Left)**: The nearby galaxy NGC7793 is shown in a three–color representation with a grid overlaid. The grid represents the 'blind selection' of galactic regions, discussed in section 5. **(Right)**: Molecular clouds (grey circles) populate each region (bin), with a cloud mass distribution that follows observed trends and limits. A maximum cloud mass of 3×10^7 M$_\odot$ (R$_{cloud,max}$ $\sim$300 pc) is adopted in the simulations. The empty region to the right of the bin is a schematic for a covering factor <1. Clouds are populated with star formation (blue stars) according to a pre–selected star formation law. Monte Carlo simulations are used to model a range of physical and geometric conditions in the regions.

above), this suggests that measurements of the Sub–Galactic SK Law are subject to significant effects of stochastic (random) sampling when the level of star formation is low, Σ_{SFR} <a few$\times 10^{-3}$ M$_\odot$ yr^{-1} kpc^{-2}, or when the bins are small, roughly <300 pc.

5. Measuring the Sub-Galactic Relations

There are basically two approaches that have been used to derive the Sub-Galactic SK Law in external galaxies: (1) measure it in regions where star formation is present, using a number of tracers; and (2) divide a galaxy into spatial bins and use all bins for the analysis. The first approach introduces an a–priori selection on the regions to be used, but avoids most of the problems linked to biased measures, which have been detailed in the previous section. The second approach, far more popular than the first, uses a blind selection method for the galactic regions, which avoids a–priori selections, but is subject to biases in the measurements of SFR and gas.

The impact of some of the biases listed in the previous section can be modeled using Monte Carlo simulations of the expected conditions in blindly selected galaxy regions (Calzetti, Liu & Koda 2012). Figure 1 shows a cartoon schematic of a simulated region, which we will call a bin. The simulations include a range of input star formation laws (parametrized by the power law $\beta = 1, 1.5, 2$), cloud mass functions, cloud covering distributions, and bin sizes, and produce, as outputs, the slope γ_{H2} and the scatter σ_{H2} about the best fit line of the Sub–Galactic SK Law as a function of bin size. This enables us to explore the effects of adding/subtracting a background to the stellar light distribution (to mimic the contribution of the light from evolved stars to measurements Σ_{SFR}), of gas thresholds to star formation, of different statistical estimators (bi–sector fit, bi–linear regression fit), and of adding noise, scatter, and data censoring.

Figure 2 shows the recovered γ_{H2} and σ_{H2} for a default model that includes a simple relation between SFR and H$_2$, with no additions of thresholds or backgrounds. The general trend is that both γ_{H2} and σ_{H2} decrease for increasing bin size, and they tend to converge to unity and to zero, respectively, independently of the input star formation power law, as parametrized by β. This is an effect of how the cloud mass distribution populates the

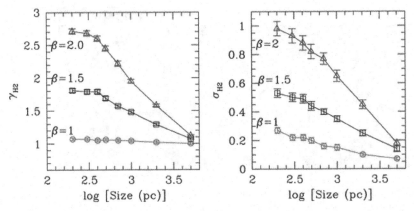

Figure 2. The recovered slope γ_{H2} of the SK Law (left), and scatter σ_{H2} about the best fit line (right) as a function of bin size, for the baseline Monte Carlo model in Calzetti, Liu & Koda (2012). The input star formation law has slope parametrized by β (($\beta=1$, 1.5, 2). The recovered slope and the scatter are always dependent on the bin size, except for the case of the slope when $\beta=1$. For $\beta > 1$, the transition between $\gamma_{H2} > \beta$ and $\gamma_{H2} < \beta$ occurs at the bin size corresponding to the transition between a stochastically sampled and a fully sampled cloud mass function. The latter situation converges towards a pure 'cloud counting' regime, where the input star formation law does not or only minimally impact the recovered slope.

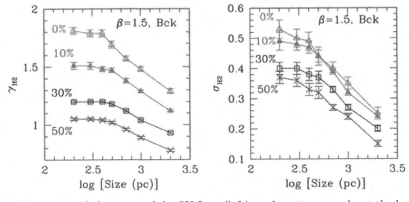

Figure 3. The recovered slope γ_{H2} of the SK Law (left), and scatter σ_{H2} about the best fit line (right) as a function of bin size, for the input star formation law with $\beta = 1.5$, and a range of background levels added to the SFRs. This simulates a situation in which the light used in the measurement of the SFR receives a contribution from evolved (unrelated to the star formation) stellar populations. The indicated fractions are the fraction of the background relative to the total. Sub–linear slopes are recovered for a 50% background contribution.

bin, changing from being stochastically sampled at small bin sizes to be fully sampled at large bin sizes, effectively 'counting clouds' in the latter regime. The convergence of γ_{H2} and σ_{H2} to 1 and 0, respectively, occurs at smaller bin sizes than those shown in Figure 2 for smaller values of the maximum cloud size, i.e., for $R_{cloud,max}$ <300 pc. The trends observed in Figure 2 are similar to the observational ones obtained by Liu, *et al.* (2011), who analyzed the Sub–Galactic SK Law of two nearby galaxies as a function of region size.

   Presence of a background added to the stellar light tracing the SFR also affects the recovered γ_{H2} (Figure 3), by lowering its values and pushing them towards unity or sub–unity for increasing background values. At 1 kpc, $\gamma_{H2} < 1$ for a 50% contamination by background light. A similar trend is observed for the scatter σ_{H2} about the best fit line.

The trends become more marked, with the decrease being steeper for both quantities, if $R_{cloud,max}$ <300 pc.

6. Conclusions

The connection between the scaling laws of star formation of whole galaxies and of GMCs remains elusive, despite much progress on the measurements of the SK Law at all scales (and at many redshifts) over the past ~decade. The intermediate scales represented by the ~kpc regions in galaxies may offer an opportunity to link the two SK Laws (or establish that they cannot be linked), but many potential biases affect the measurements of SFR and gas densities in regions that cannot be considered 'isolated'.

'Blind' region selections, like those employed by many studies, should be avoided, in favor of identifying areas of current star formation, which will mitigate the biases on SFR density derivations. If blind region selection cannot be avoided, careful attention should be devoted to regions with gas surface densities with Σ_{gas} <100-200 $M_\odot$ pc^{-2}, i.e., the surface densities typical of normal star–forming galaxies in the local Universe. These are divided between the 'stochastic' regime (Σ_{gas} <5-10 $M_\odot$ pc^{-2}) and the 'cloud counting' regime. One avenue that can be explored to overcome some of the biases and limitations is to perform measurements at a range of scales, which can be used, together with simulations, to disentangle the many effects contributing to the measured values of the slope γ_{H2} of the SK Law and the scatter σ_{H2} of the data about the best fit. These effects include, in addition to the 'true' law of star formation, contamination of SFRs by background light, non-star–forming gas, etc.. Combining high–angular–resolution data on both SFR and gas, by using the synergy between HST (and, in the future, JWST) and ALMA, will enable covering the full range of scales from kpc down to those of GMCs, and finally link the scaling laws of star formation at all scales.

Acknowledgements

I would like to thank Mark Heyer (University of Massachusetts) for many enlightening discussions. I would also like to thank the Conference Organizers for their invitation to a very inspiring and instructive meeting.

References

Barden M., Rix H. W., Somerville R. S., et al. 2005, *ApJ*, 635, 959
Blanc G., Heiderman A., Gebhardt K., et al. 2009, *ApJ*, 704, 842
Bigiel F., Leroy A., Fabian F., et al. 2008, *AJ*, 136, 2846
Boissier S., Prantzos N., Boselli A., et al. 2003, *MNRAS*, 346, 1215
Bolatto A. D., Wolfire M., & Leroy A. K. 2013, *ARA&A*, 51, 207
Calzetti D., Liu G., & Koda J. 2012, *ApJ*, 752, 98
Cormier D., Madden S. C., Lebouteiller V., et al. 2014, *A&A*, 564A, 121
Crocker A. F., Chandar R., Calzetti D., et al. 2014, submitted
Daddi E., Elbaz D., Walter F., et al. 2010a, *ApJ*, 714, L118
Daddi E., Bournaud F., Walter F., et al. 2010b, *ApJ*, 713, 686
Faucher–Giguère C.–A., Quataert E., & Hopkins P. F. 2013, *MNRAS*, 433, 1970
Galametz M., Albrecht M., Kennicutt R., et al. 2014, *MNRAS*, 439, 2542
Genzel R., Tacconi L. J., Gracia-Carpio J., et al. 2010, *MNRAS*, 407, 2091
Genzel R., Tacconi L. J., Lutz D., et al. 2014, (astroph/1409.1171)
Gutermuth R. A., Pipher J. L., Megeath S. T., et al. 2011, *ApJ*, 739, 84
Heiderman A., Evans N. J., Allen L. E., et al. 2010, *ApJ*, 723, 1019

Heyer M. H., Corbelli E., Schneider S. E., & Young J. S. 2004, *ApJ*, 602, 723

Johnson B. D., Weisz D. R., Dalcanton J. J., *et al.* 2013, *ApJ*, 772, 8

Kawamura A., Mizuno Y., Minamidani T., *et al.* 2009, *ApJS*, 184, 1

Kennicutt R. C. Jr., 1989, *ApJ*, 344, 685

Kennicutt R. C. Jr., 1998, *ApJ*, 498, 541

Kennicutt R. C., Calzetti D., Walter F., *et al.* 2007, *ApJ*, 671, 333

Kennicutt R. C., Evans N. J. 2012, *ARA&A*, 50, 531

Kirkpatrick A., Calzetti D., Kennicutt R., *et al.* 2014, *ApJ*, 789, 130

Koda J., Scoville N., Hasegawa T., *et al.* 2012, *ApJ*, 761, 41

Krumholz M. R., McKee C. F., & Tumlinson J. 2009, *ApJ*, 693, 216

Krumholz M. R., Dekel A., & McKee C. F. 2012, *ApJ*, 745, 69

Lada C. J., Forbrich J., Lombardi M., & Alves J. F. 2012, *ApJ*, 745, 19

Lada C. J., Forbrich J., Roman–Zuniga C., *et al.* 2013, *ApJ*, 778, 133

Langer W. D., Velusamy T., Pineda J. L., *et al.* 2014, *A&A*, 561A, 122

Leroy A. K., Walter F., Bigiel F., *et al.* 2009, *AJ*, 137, 4670

Leroy A. K., Bigiel F., de Blok W. J. G.., *et al.* 2012, *AJ*, 144, 3

Liu G., Koda J., Calzetti D., *et al.* 2011, *ApJ*, 735, 63

Momose R., Koda J., Kennicutt R. C., *et al.* 2013, *ApJ*, 772, L13

Narayanan D.., Krumholz M., Ostriker E. C., & Hernquist L. 2011, *MNRAS*, 418, 664

Rémy–Ruyer A., Madden S. C., Galliano F., *et al.* 2014, *A&A*, 536A, 31

Rigopoulou D., Hurley P. D., Swinyard B. M., *et al.* 2013, *MNRAS*, 434, 2051

Sandstrom K. M., Leroy A. K., Walter F., *et al.* 2013, *ApJ*, 777, 5

Santini D., Maiolino R., Magnelli B., *et al.* 2014, *A&A*, 562A, 30

Saintonge A., Kauffmann G., Wang J., *et al.* 2011, *MNRAS*, 415, 61

Shetty R., Kelly B. C., Rahman N., *et al.* 2014a, *MNRAS*, 437, L61

Shetty R., Clark P. C., & Klessen R. S. 2014b, *MNRAS*, 442, 2208

Schmidt M., 1959, *ApJ*, 129, 243

Tacconi L. J., Neri R., Genzel R., *et al.* 2013, *ApJ*, 768, 74

Thilker D. A., Walterbos R. A. M.., Braun R., & Hoopes C. G. 2002, *AJ*, 124, 3118

Verley S.., Corbelli E., Giovanardi C., & Hunt L. 2010, *A&A*, 510A, 64

Wolfire M. G., Hollenbach D., & McKee C. F. 2010, *ApJ*, 716, 1191

Galaxies in 3D across the Universe
Proceedings IAU Symposium No. 309, 2014
B. L. Ziegler, F. Combes, H. Dannerbauer, M. Verdugo, eds.

© International Astronomical Union 2015
doi:10.1017/S1743921314009454

DYNAMO Survey: An Upclose View of Clumpy Galaxies

David Fisher[1] and the DYNAMO[2] Team

Centre for Astrophysics and Supercomputing, Swinburne University of Technology, P. O. Box 218, Hawthorn, VIC 3122, Australia
email: dfisher@swin.edu.au
[2] http://adsabs.harvard.edu/abs/2014MNRAS.437.1070G

Abstract. We highlight recent results on the DYNAMO survey of turbulent, clumpy disks galaxies found at z=0.1. Bright star forming DYNAMO galaxies are found to be very similar in properties to star forming galaxies in the high redshift Universe. Typical star formation rates of turbulent DYNAMO galaxies range 10-80 $M_\odot$ yr^{-1}. Roughly 2/3 of DYNAMO galaxies have Hα kinematics that are consistent with rotation. The typical gas velocity dispersion of DYNAMO galaxies is $\sigma_{H\alpha} \sim 20-60$ km s^{-1}. We show that, when convolved to the same resolution, maps of Hα emission in DYNAMO galaxies have essentially identical morphology as that of $z \sim 1-3$ galaxies. Finally, DYNAMO galaxies have high molecular gas fractions $f_{mol} \sim 20-35\%$. We note that DYNAMO galaxies are not dwarfs, typical masses are $M_{star} \sim 0.8-8 \times 10^{10}$ $M_\odot$. These data are all consistent with a scenario in which despite being at relatively low redshift the DYNAMO galaxies are forming stars similarly to that observed in the high-redshift Universe, that is to say star formation is occurring in very massive ($M_{clump} \sim 10^9$ $M_\odot$), very large ($r_{clump} \sim 300$ pc) clumps of gas.

Keywords. galaxies: elliptical and lenticular, cD - galaxies: evolution - galaxies: formation

1. Introduction

Over 80% of the stars in our universe formed during an epoch at redshifts $z = 1-3$ (Hopkins & Beacom 2006). Star forming galaxies during this epoch have irregular, *clumpy* morphologies (Abraham *et al.* 1996), are very gas rich (Tacconi *et al.* 2013), and that gas is very turbulent (Genzel *et al.* 2011). The clumps in these galaxies are super-giant (300-1000pc) star-forming complexes, which are an order of magnitude larger than similar regions in local galaxies (Wisnioski *et al.* 2012). Clump masses are often interpreted as a response to the turbulent ISM containing them, high random motions lead to large Jeans' masses, roughly consistent with observed clump masses at $z \sim 1-3$ (Elmegreen *et al.* 2008).

Because they are typically observed at $z > 1$, clumps are mostly unresolved (Jones *et al.* 2010), even with HST or AO observations. The picture of how star formation and hence galaxy evolution proceeds during the period when most of the stars were forming is therefore incomplete. A detailed description of the size, density, and kinematics resolving individual clumps in turbulent disks is therefore needed.

2. DYNAMO Sample

The **DYNAMO** sample is a set of nearby galaxies that have strikingly similar kinematics, star formation properties and gas content as high redshift clumpy disk galaxies. The survey includes AAT and 2.5 m integral field observations of Hα of 95 spiral galaxies with the highest Hα luminosity from SDSS (excluding AGNs).

DYNAMO Hα maps convolved to $z \sim 2$ resolution

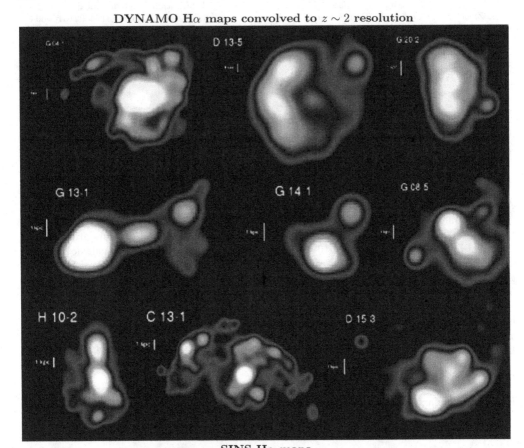

SINS Hα maps

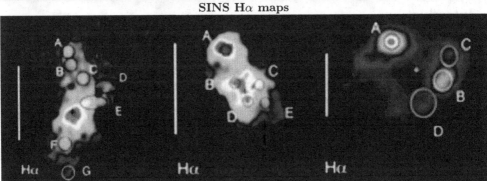

Figure 1. The top 9 panels show HST ACS/WFC maps of Hα emission from the DYNAMO survey of galaxies. The maps have been convolved with a Gaussian that sets the physical resolution to ~ 800 pc. The galaxies are therefore at a similar resolution to how they would appear if they were located at $z \sim 2$. In the bottom panel we show maps from the SINS survey of galaxies (Genzel *et al.* 2014). When viewed at the same resolution, DYNAMO galaxies appear indistinguishable in morphology to the clumpy, turbulent galaxies of the distant Universe.

DYNAMO galaxies are indeed very good analogues of clumpy $z \sim 2$ galaxies. In Fig. 1 we show Hα maps of DYNAMO galaxies taken with the *Hubble Space Telescope* ACS/WFC narrow band filters. The maps have been convolved with a Gaussian whose full width at half maximum corresponds to 850 pc. These maps, therefore have similar

Figure 2. Left panel: HST Hα map of, target galaxy DYNAMO G13-1, at the full resolution of HST (∼ 250 pc resolution). The white bar shows 0.2 arcsec. Right panel: Simulating the same object as if it were located at $z = 2$. Arrows in both maps indicate the location of 3 clumps, identified in the unblurred map. In the $z = 2$ map the 3 clumps would appear as one large clump, possibly leading to specious conclusions.

resolution to Hα maps targeting galaxies in $z \sim 1 - 3$ Universe. We also show three Hα maps taken from Genzel *et al.* (2011) for comparison. When convolved to the same resolution, similarity between DYNAMO galaxies and high-z is quite remarkable. For example, star forming rings are common in high-z clumpy galaxies (e.g. Genzel *et al.* 2014). Indeed we detect ring-like distributions of Hα emission in a number of DYNAMO galaxies, most prominently in G04-1, D13-5 and G20-2 (top row in Fig. 1). We also detect chains of star forming clumps (G13-1, H10-2). 'Chain galaxies' have been known about in the high redshift Universe for many years, and have been shown in Hα maps of $z \sim 1-3$ galaxies by a number of surveys (e.g. Genzel *et al.* 2011; Wisnioski *et al.* 2012; Swinbank *et al.* 2012).

In Fig 2 we compare the morphology of clumps at resolution set to match a $z = 2$ galaxy, to the full resolution of a DYNAMO galaxy. It is clear that the factor of ∼ 3× finer spatial resolution of a galaxy at $z = 0.1$ significantly affects the identification of clumps. In the low-resolution Hα maps, set to match observations of a $z \sim 2$ galaxy, the ionized gas is arranged into 3 clumps that are fairly large. At the full HST resolution of a $z = 0.14$ DYNAMO galaxy (shown left in Fig 2) the clumps of ionized gas emission are smaller, and those clumps in the low resolution map. The typical diameter of clumps sizes in DYNAMO galaxies, when measured in the $z \sim 0.1$ resolution maps is $< d_{clump} > \sim$ 0.6 kpc. When making the same measurement of the Hα maps that have been blurred to match $z \sim 2$ the typical size increases to $< d_{clump} > \sim$ 2.2 kpc. This is just over a factor of 2× increase in clump sizes. In the $z \sim 2$ blurred maps only ∼ 44% of the clumps that are identified at $z \sim 0.1$ resolution are detected. The other clumps are either undetectable due to lower surface brightness or blend into a single clump are lower resolution.

Perhaps the most critical similarity of DYNAMO galaxies to high redshift star forming galaxies is that both are quite gas rich and very actively forming stars. Using star formation rates derived from integral field spectroscopic maps Green *et al.* (2014) estimates the star formation rates of DYNAMO galaxies range from ∼ $1 - 80$ $M_{\odot}$ yr^{-1}, and from Sloan Digital Sky photometry we determine stellar masses of ∼ $0.1 - 10 \times 10^{10}$ $M_{\odot}$ for the same galaxies. Fisher *et al.* (2014) finds that DYNAMO galaxies have molecular gas fractions, $f_{mol} = M_{mol}/(M_{mol} + M_{star}) = 20 - 35\%$. In Fig. 3 (left) we compare the gas fractions of 3 DYNAMO galaxies to that of 3 surveys of galaxies. The gas fractions of DYNAMO galaxies are found to be extreme outliers compared to local galaxies of the COLD GASS survey (Saintonge *et al.* 2011). The gas fractions of DYNAMO galaxies are very similar to what is observed in high-z samples (eg. PHIBSS Tacconi *et al.* 2013).

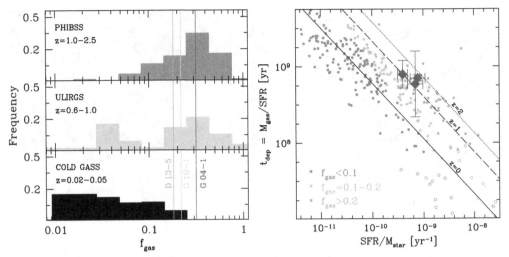

Figure 3. Both panels are replotted from Fisher *et al.* (2014). **Left** shows molecular gas fractions of three galaxies from the DYNAMO survey are compared to the distribution of gas fractions from 3 surveys. From bottom to top, those surveys of molecular gas are the z=0 COLD GASS survey (Saintonge *et al.* 2011), the survey of intermediate redshift ULIRGS (Combes *et al.* 2013), and the PHIBSS survey of $z \sim 1-3$ galaxies (Tacconi *et al.* 2013). The gas fractions of DYNAMO galaxies (vertical lines) are more consistent with high redshift galaxies, than those of the low-z Universe. **Right** shows the relationship between gas depletion time ($t_{dep} = M_{mol}/SFR$) and specific star formation rate ($SFR/M_{\star}$). The data is taken from the same samples as in left panel. The points are colored by gas fractions, as indicated by the key. The DYNAMO galaxies are forming stars in a manner consistent with their gas fractions.

Furthermore, DYNAMO galaxies are forming stars appropriately for their gas fraction, not their redshift. Galaxies with large specific star formation rates are known to have more efficient star formation (ie. shorter gas depletion timescales, Saintonge *et al.* 2011). However, galaxies at higher redshift, with larger gas fractions, have longer gas depletion times than a galaxy with the same specific star formation rate in the nearby Universe. Fisher *et al.* (2014) finds that DYNAMO galaxies have $t_{dep} \sim 0.5\ Gyr$. Again they are found to be more similar to galaxies of the high-z Universe than nearby galaxies.

References

Abraham, *et al.* 1996, *ApJS*, 107, 1
Bassett, *et al.* 2014, *MNRAS*, 442, 3206
Combes, *et al.* 2013, *A&A*, 550, A41
Elmegreen, B. G., Bournaud, F., & Elmegreen, D. M. 2008, *ApJ*, 688, 67
Fisher, *et al.* 2014, *ApJ*, 790, L30
Genzel, *et al.* 2014, *ApJ*, 785, 75
Genzel, *et al.* 2011, *ApJ*, 733, 101
Green, *et al.* 2010, *Nature*, 467, 684
Green, *et al.*, P. J. 2014, *MNRAS*, 437, 1070
Jones, *et al.* 2010, *MNRAS*, 404, 1247
Saintonge, *et al.* 2011, *MNRAS*, 415, 61
Swinbank, *et al.* 2012, *ApJ*, 760, 130
Tacconi, *et al.*, 2013, *ApJ*, 768, 74
Wisnioski, *et al.*, C. 2012, *MNRAS*, 422, 3339

Galaxies in 3D across the Universe
Proceedings IAU Symposium No. 309, 2014
B. L. Ziegler, F. Combes, H. Dannerbauer, M. Verdugo, eds.

© International Astronomical Union 2015
doi:10.1017/S1743921314009466

Counter-rotating disks in galaxies: dissecting kinematics and stellar populations with 3D spectroscopy

Lodovico Coccato[1], Lorenzo Morelli[2,3], Alessandro Pizzella[2,3], Enrico Maria Corsini[2,3], Elena Dalla Bontà[2,3] and Maximilian Fabricius[4,5]

[1]European Southern Observatory, Karl-Schwarzschild-Straße 2, D-85748, Garching bei München, Germany.
[2]Dipartimento di Fisica ed Astronomia "Galileo Galilei", Università di Padova, vicolo dell'Osservatorio 3, 35122 Padova, Italy.
[3]INAF-Osservatorio Astronomico di Padova, vicolo dell'Osservatorio 5, 35122 Padova, Italy.
[4]Max Planck Institute for Extraterrestrial Physics, Giessenbachstraße, D-85748 Garching, Germany.
[5]University Observatory Munich, Scheinerstraße 1, 81679 Munich, Germany.
email: lcoccato@eso.org

Abstract. We present a spectral decomposition technique that separates the contribution of different kinematic components in galaxies from the observed spectrum. This allows to study the kinematics and properties of the stellar populations of the individual components (e.g., bulge, disk, counter-rotating cores, orthogonal structures). Here, we discuss the results of this technique for galaxies that host counter-rotating stellar disks of comparable size. In all the studied cases, the counter-rotating stellar disk is the less massive, the youngest and has different chemical content (metallicity and α-elements abundance ratio) than the main galaxy disk. Further applications of the spectral decomposition technique are also discussed.

Keywords. galaxies: elliptical and lenticular - galaxies: formation

1. Introduction

Galaxies can have a complex kinematic structure, due to the superimposition of multiple components that have different kinematics and stellar populations, such as bulge and disk (in lenticulars and spirals), host spheroid and polar ring (in polar ring galaxies), and counter-rotating stellar disks or kinematically decoupled cored (mainly - but not only - in early-type galaxies).

The interesting case of large-scale counter-rotating galaxies, i.e., those that host two stellar components of comparable sizes that rotate along opposite directions, is the topic of this work. The prototype of this class of objects is the E7/S0 galaxy NGC 4550, whose counter-rotating nature was first discovered by Rubin *et al.* (1992).

2. Spectral decomposition

The spectral decomposition technique by Coccato *et al.* (2011) is an implementation of the penalized pixel fitting code by Cappellari & Emsellem (2004). For a given spectrum, the code builds two synthetic stellar templates (one for each stellar component) as a linear combination of stellar spectra from an input stellar library. It then convolves these two stellar templates with two Gaussian line-of-sight velocity distributions (LOSVDs), which are characterized by different velocity and velocity dispersion. The input galaxy

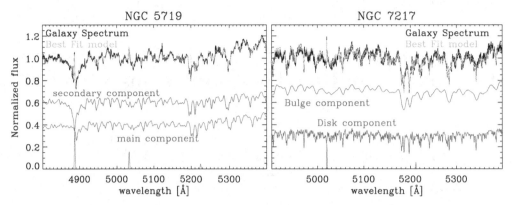

Figure 1. Example of the fit results of the spectral decomposition technique, applied to disentangle the contribution of two counter-rotating stellar disks (left panel, from Coccato *et al.* 2011), and the contribution of the stars in the bulge and in the disk (right panel, from Fabricius *et al.* 2014). Black: observed galaxy spectrum; green: best fit model; blue and red: spectra of the two stellar components; purple: gas emission lines.

and stellar spectra are normalized to their continuum level, therefore the contribution of each component to the observed galaxy spectrum is measured in terms of light. Gaussian functions are added to the convolved synthetic templates to account for ionized-gas emission lines. Multiplicative Legendre polynomials are included in the fit to match the shape of the galaxy continuum, and are set to be the same for the two synthetic templates. This accounts for the effects of dust extinction and variations in the instrument transmission. These steps are embedded in a iterative χ^2 minimization loop.

The spectral decomposition code returns the spectra of two best-fit synthetic stellar templates and ionized-gas emissions, along with the best-fitting parameters of light contribution, velocity, and velocity dispersion. The line strength of the Lick indices of the two counter-rotating components are then extracted from the two best-fit synthetic templates, and used to infer the properties of the stellar populations of the two components.

Figure 1 shows some examples of results of the spectral decomposition.

3. Applications and results

The technique has been successfully applied to disentangle (i) counter-rotating stellar components in galaxies (Coccato *et al.* 2011, 2013; Pizzella *et al.* 2014), (ii) the kinematics of stars in bulge and disk (Fabricius *et al.* 2014), and (iii) the kinematics of the host galaxy and orthogonal disk in polar disk galaxies (Coccato *et al.* 2014).

Here, we describe the results obtained for a sample of galaxies that host counter-rotating stellar disks of comparable size: NGC 3959, NGC 4183, NGC 4550, and NGC 5719. Observations were obtained with the VIMOS integral field unit at the VLT (Chile), except for NGC 4138, which was observed with the B&C spectrograph at the 1.22-m telescope at Padova University (Italy). In all the studied galaxies, we detect the presence of an extended secondary stellar component, which is counter-rotating with respect the more luminous and massive stellar component. In addition, the secondary component is rotating along the same direction as the ionized gas. The measurements of the equivalent width of the absorption line features (Hβ, Mgb, Fe5270, and Fe5335) reveal that the secondary stellar component is always younger than the main stellar component, and it is more metal poor. Such a difference in stellar population is particularly pronounced in NGC 5719 (Figure 2).

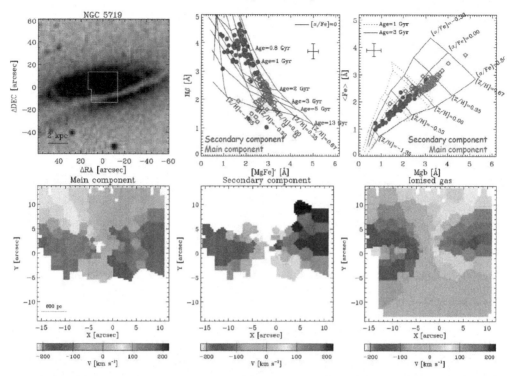

Figure 2. Top left panel: image of the Sa galaxy NGC 5719, the yellow region indicates the footprint of the VIMOS/IFU field of view. Top central and right panels: line strengths of the Lick indices measured on the main (red) and secondary (blue) stellar components: Hβ, combined magnesium-iron index [MgFe]' (Gorgas *et al.* 1990), Mgb, and average iron index <Fe> (Thomas *et al.* 2003). The black lines show the prediction of single stellar population models (Thomas *et al.* 2011). The bottom panels show the measured two-dimensional velocity fields of the main (left) and secondary (central) stellar components, and the ionized gas component (right). Adapted from Coccato *et al.* (2011).

In the case of IFU observations, the spectral decomposition allows also to investigate the morphology of the decoupled structures by comparing their light contribution to the reconstructed image of the galaxy. It is therefore possible to unveil the real extension and geometrical properties such as orientation, ellipticity, and scale length of the decoupled disks. In Figure 3 we compare the reconstructed images and the surface brightness radial profiles of the counter-rotating disks observed in NGC 3593. The secondary components overshines the main stellar disk within ∼ 10″. A single-component fit would reveal the presence of the secondary component only in the inner regions, by revealing the kinematic signature of a kinematically decoupled core (KDC, e.g. Bender 1988).

The combination of the luminosity profile and stellar mass-to-light ratio, as inferred from the stellar population analysis, shows that the secondary component has a lower mass density radial profile than the main component, also in the inner regions where it dominates in light.

4. Conclusions

The stellar population properties and kinematics of the counter-rotating stellar disks are consistent with the formation scenario where a pre-existing galaxy acquired external gas on retrograde orbits from a companion galaxy or the external medium (e.g. Thakar & Ryden 1996; Pizzella *et al.* 2004). Depending on the amount of acquired material, the

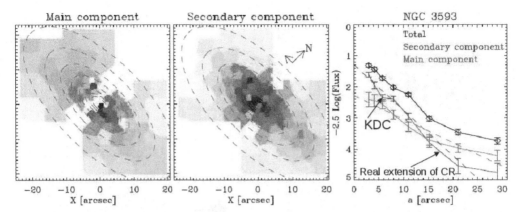

Figure 3. Maps (left and central panels) and radial profiles (right panel) of the surface brightness of the main and the secondary stellar components of NGC 3593. The black, red, and blue solid lines correspond to the total surface brightness and to the surface brightness of the main and the secondary stellar components, respectively. The dashed red and dashed blue lines correspond to the best-fitting exponential disks to the surface brightness of the main and the secondary stellar components, respectively. Dashed ellipses represent the boundaries of the elliptical annuli where the median surface brightnesses were computed. Adapted from Coccato *et al.* (2013).

new gas removes the pre-existing gas and sets onto a disk in retrograde rotation. The new gas forms thevv stars of the counter-rotating secondary component. In the case of NGC 5719 we have a direct of the on-going acquisition process (Vergani *et al.* 2007).

Despite of their discovery more than 30 years ago, counter-rotating galaxies still represent a challenging subject (see Bertola & Corsini 1999; Corsini 2014 for reviews). We have showed that the spectral decomposition allows to measure the properties of the stellar populations of both components, and to probe the predictions of the different formation scenarios. Forthcoming large integral field spectroscopic surveys such as MANGA (Bundy *et al.* 2014), will allow to increase the statistics on these objects, providing fundamental clues to constrain their formation scenarios.

References

Bender, R. 1988, *A&A*, 202, L5

Bertola, F. & Corsini, E. M. 1999, in *IAU Symposium*, Vol. 186, Galaxy Interactions at Low and High Redshift, ed. J. E. Barnes & D. B. Sanders, 149

Bundy, K., Bershady, M. A., Law, D., *et al.* 2014, *ApJ* submitted

Cappellari, M. & Emsellem, E. 2004, *PASP*, 116, 138

Coccato, L., Iodice, E., & Arnaboldi, M. 2014, *A&A* in press (ArXiv:1409.1235)

Coccato, L., Morelli, L., Corsini, E. M., *et al.* 2011, *MNRAS*, 412, L113

Coccato, L., Morelli, L., Pizzella, A., *et al.* 2013, *A&A*, 549, A3

Corsini, E. M. 2014, in *Astronomical Society of the Pacific Conference Series*, Vol. 486, Astronomical Society of the Pacific Conference Series, ed. E. Iodice & E. M. Corsini, 51

Fabricius, M. H., Coccato, L., Bender, R., *et al.* 2014, *MNRAS*, 441, 2212

Gorgas, J., Efstathiou, G., & Aragon Salamanca, A. 1990, *MNRAS*, 245, 217

Pizzella, A., Corsini, E. M., Vega Beltrán, J. C., & Bertola, F. 2004, *A&A*, 424, 447

Pizzella, A., Morelli, L., Corsini, E. M., *et al.* 2014, *A&A* in press (ArXiv:1409.3086)

Rubin, V. C., Graham, J. A., & Kenney, J. D. P. 1992, *ApJ*, 394, L9

Thakar, A. R. & Ryden, B. S. 1996, *ApJ*, 461, 55

Thomas, D., Maraston, C., & Bender, R. 2003, *MNRAS*, 339, 897

Thomas, D., Maraston, C., & Johansson, J. 2011, *MNRAS*, 412, 2183

Vergani, D., Pizzella, A., Corsini, *et al.* 2007, *A&A*, 463, 883

Galaxies in 3D across the Universe
Proceedings IAU Symposium No. 309, 2014
B. L. Ziegler, F. Combes, H. Dannerbauer, M. Verdugo, eds.

ⓒ International Astronomical Union 2015
doi:10.1017/S1743921314009478

LOFAR and Radio-Loud AGN

Wendy L. Williams[1,2] and Huub J. A. Röttgering[1] on behalf of the LOFAR collaboration

[1]Leiden Observatory, Leiden University, P.O. Box 9513, 2300 RA Leiden, The Netherlands
[2]Netherlands Institute for Radio Astronomy (ASTRON), PO Box 2, 7990AA Dwingeloo, The Netherlands

emails: `wwilliams@strw.leidenuniv.nl`, `rottgering@strw.leidenuniv.nl`

Abstract. At very low frequencies, the new pan-European radio telescope, LOFAR, is opening the last unexplored window of the electromagnetic spectrum for astrophysical studies. Operating at frequencies from 15 to 240 MHz, its superb sensitivity, high angular resolution, large field of view and flexible spectroscopic capabilities represent a dramatic improvement over previous facilities at these wavelengths. LOFAR will carry out a broad range of fundamental astrophysical studies in a number of key science topics including the formation and evolution of clusters, galaxies and black holes. In this contribution we describe some of the capabilities of LOFAR and present some recent results from the ongoing imaging efforts. We also discuss the impact of LOFAR on our studies of radio-loud AGN. Our recent study of the evolution of radio-loud AGN as a function of host stellar mass shows a clear increase in the fraction of lower mass galaxies which host radio-loud AGN at $1 < z < 2$ while the fraction for higher mass galaxies remains the same. This shows that the upturn in the radio luminosity function is driven by increasing AGN activity among low mass galaxies at higher redshifts. New LOFAR observations will allow us to build statistically large samples at high redshifts to constrain this evolution for the different accretion modes of AGN.

Keywords. Galaxies: active, surveys, telescopes, Instrumentation: interferometers

1. LOFAR Surveys and Recent Results

The Low Frequency Radio Array (LOFAR) is a new low freqency radio telescope spanning several European countries. LOFAR has been described in detail elsewhere (see van Haarlem *et al.* 2013), but we provide here a few highlights: The array comprises several 'stations' each containing two groups of simple dipole antennas. The Low Band Array (LBA) operates at $10 - 80$ MHz and the High Band Array (HBA) at $110 - 240$ MHz. There are 37 stations in the Netherlands, with baselines up to ~ 80 km, and a further eight stations in other European countries: five in Germany, and one each in the UK, Sweden, and France† which extends the array to ~1200 km. The electronically phased 'beams' enable multiple pointings to be observed simultaneously, making LOFAR extremely efficient for surveying large areas. They also make LOFAR highly flexible in terms of frequency choices.

The primary scientific drivers for LOFAR are encapsulated in the six Key Science Projects: (i) Epoch of Reionisation (EoR); (ii) Surveys; (iii) Transients; (iv) Cosmic Rays; (v) Cosmic Magnetism; and (vi) Solar Science and Space Weather. The Surveys project is most relevant here, and together with projects like Cosmic Magnetism, will provide images of nearby galaxies, AGN, clusters and large deep extragalactic fields. Three fundamental areas of astrophysics that have driven the design of the planned LOFAR

---

† up-to-date information on the current status of LOFAR is available at `http://www.astron.nl/~heald/lofarStatusMap.html`, powered by Google Maps.

surveys are: (i) forming massive galaxies at the epoch of reionisation, (ii) magnetic fields and shocked hot gas associated with the first bound clusters of galaxies, and (iii) star formation processes in distant galaxies. To achieve the diverse and broad goals of the LOFAR surveys, a tiered approach will be used (for details see Röttgering *et al.* 2010): Tier-1 will be an all sky survey at 15, 30, 60 and 120 MHz; Tier-2 will be a medium deep survey over 1000 square degrees at 30, 60, 120 and 200 MHz; and Tier-3 will cover about 100 square degrees down to an extreme depth of 6 μJy rms at 150 MHz. The depth versus frequency of the proposed surveys is given in Fig. 1. Another important motivation of LOFAR is to provide the entire international astronomical community with unique surveys of the radio sky that have a long-lasting legacy value for a broad range of astrophysical research.

At present the measured noise in LOFAR images is dominated by calibration and imaging errors and some commissioning and technical research is still needed to obtain maps with the theoretical noise levels. However, the maps that are already being produced are the deepest ever at these low frequencies and high resolutions. Early-science observations are producing some excellent imaging results; we highlight here a few recent results. For nearby galaxies, (de Gasperin *et al.* 2012) have studied Virgo A (M87) in detail. The EoR group has produced images 15″ resolution and noise levels of 100 μJy/beam across a wide band of 120 − 180 MHz (see Yatawatta *et al.* 2013, for some recent results) and their more recent work continues to push the noise levels down proportionally with increasing integration time. Extragalactic fields that have been imaged with LOFAR in the LBA include the Groth Strip and Boötes (see van Weeren *et al.* submitted) with images at 55 − 70 MHz (noise 4 − 5 mJy, resolution 20″), 40 − 50 MHz (6 − 7 mJy, 30″), and 30 − 40 MHz (11 mJy, 40″). Current work includes that of directional dependent ionospheric calibration which results in noise levels of ∼ 100μJy at 150 MHz (in the Toothbrush cluster field, van Weeren in prep.) The first release of the Multi-frequency Sky Snapshot survey (MSSS; Heald *et al.* submitted) is expected soon. To date we have HBA observations to the Tier-1 depth of all the targeted famous extragalactic deep fields (including Boötes, Groth Strip, Lockman Hole, ELAIS-N1, COSMOS, GOODS-N, H-ATLAS, XMM-LSS, NEP) and several cluster fields (including the Toothbrush and Sausage clusters), and

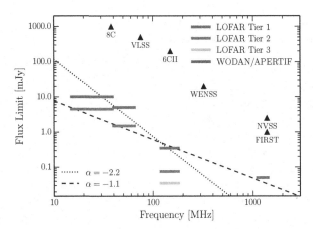

Figure 1. Flux limits (5σ) of the proposed LOFAR surveys compared to other existing radio surveys. The triangle represent existing surveys: NVSS, FIRST, WENSS, 6C, VLSS and 8C. The lines represent different power-laws ($S \sim \nu^{\alpha}$ with $\alpha = -1.1$ and -2.2) to illustrate how, depending on the spectral indices of the sources, the LOFAR surveys will compare to other current surveys. The proposed APERTIF/WODAN survey is also shown, which will use APERTIF on the WSRT.

work is underway to produce the ionospheric-calibrated deep images. In Cycle 2 of LO-FAR observations we initiated the Tier-1 HBA survey, by observing 42 pointings to the required Tier-1 depth. These first observations of this northern sky survey were chosen to lie within an area that is covered by Sloan Digital Sky Survey (SDSS) imaging and spectroscopy and is due to be covered by the Hobby-Eberly Telescope Dark Energy Experiment (HETDEX) in 2014 – 2017.

2. LOFAR and the evolution of radio loud AGN

Radio galaxies have been shown to comprise two different populations: High and low excitation (Laing *et al.* 1994; Tadhunter *et al.* 1998), each which may have a separate and different feedback effect. The first mode, often referred to as 'cold-mode', 'radiative-mode', or 'high-excitation', is characterised by strong optical emission lines, MIR emission and/or absorption associated with a dusty torus, along with possible X-ray emission. Here material is accreted on to the black hole through a radiatively efficient accretion disc. The second mode of AGN activity occurs with little radiated energy, but can still lead to the production of highly energetic radio jets. Here, the strong emission lines are absent and they lack other evidence for accretion disks. These 'radiatively inefficient', 'hot-mode', or 'low-excitation' AGN are thought to be fueled by radiatively inefficient advection-dominated accretion flows (ADAFs; e.g. Narayan & Yi 1995).

The radio-loud AGN population is quite well-understood in the local universe within the context of these two accretion modes (see e.g. Best & Heckman 2012; Janssen *et al.* 2012), however, this is not the case at higher redshifts. In particular, it remains unclear how the radio-loud AGN population evolves for different host stellar masses. The number density of powerful radio sources is two to three orders of magnitude greater at a redshift of two to three compared to the local Universe (e.g. Schmidt 1968; Rigby *et al.* 2011, and references therein). Locally, the fraction of galaxies which host a radio source, the radio-loud fraction, is a very steep function of host galaxy stellar mass (Best *et al.* 2005), increasing to > 30 per cent at stellar masses above 5×10^{11} M$_\odot$ for radio luminosities > 10^{23} W Hz^{-1}. These two facts suggest that there must be an increase in the prevalence of radio activity among galaxies of lower mass.

Using the SDSS value-added spectroscopic sample of radio loud galaxies (Best & Heckman 2012) and the VLA-COSMOS radio sample (Schinnerer *et al.* 2004, 2007) matched

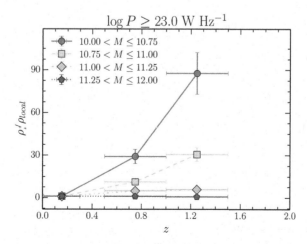

Figure 2. Relative space density of radio-loud sources with respect to the local density as a function of redshift for all sources with radio powers greater than the cutoff.

to a K_s-selected catalog of the COSMOS/UltraVISTA field (Muzzin *et al.* 2013) to compile two samples of radio loud AGN going out to $z = 2$. We have constructed radio luminosity functions binned by host stellar mass in three redshift ranges, $0.01 \leqslant z < 0.3$, $0.5 < z \leqslant 1.0$ and $1.0 < z \leqslant 2.0$. Considering the relative comoving space density of radio-loud sources with respect to the local comoving space density, we find that at $1 < z < 2$ the space density of the least massive galaxies hosting radio-loud AGN is *90* times greater relative to the local space density, while that of higher mass galaxies hosting radio-loud AGN remains the same (see Fig. 2). The radio-loud fraction as a function of stellar mass is observed to be clearly increasing with redshift and becoming shallower with mass. Equivalently, the fraction of lower mass galaxies hosting radio sources increases faster with redshift than the fraction for high mass galaxies. This is likely driven by increasing cold-mode accretion driven by mergers at higher redshifts. This increase can thus explain the upturn in the Radio Luminosity Function (RLF) at high redshift which is important for understanding the impact of AGN feedback in galaxy evolution

3. Prospects

LOFAR has been producing excellent quality interferometric data for a number of years now. The commissioning period is over and Cycle 1 observations started in November 2013. LOFAR Surveys have already started imaging a handful of deep fields towards the surveys Tier-1 depth, and the target noise levels for the three tiers of the HBA Surveys are within reach. The combination of the LOFAR surveys data and other ancillary multi-wavelength data available in many of the extra-galactic deep fields will make LOFAR a great tool for studies of AGN, distant clusters and galaxies.

Acknowledgements

LOFAR, the Low Frequency Array designed and constructed by ASTRON, has facilities in several countries, that are owned by various parties (each with their own funding sources), and that are collectively operated by the International LOFAR Telescope (ILT) foundation under a joint scientific policy.

References

Best, P. N., Kauffmann, G., Heckman, T. M., Brinchmann, J., *et al.* 2005, *MNRAS*, 362, 25
Best, P. N. & Heckman, T. M. 2012 *MNRAS*, 421, 1569
de Gasperin, F., Orrú, E., Murgia, M., *et al.* 2012, *A&A*, 547, A56
Hickox, R. C., Jones, C., Forman, W. R., *et al.* 2009, *ApJ*, 696, 891
Janssen, R. M. J., Röttgering, H. J. A., Best, P. N., & Brinchmann, J., 2012, *A&A*, 541, A62
Laing, R. A., Jenkins, C. R., Wall, J. V., *et al.* 1994, *The Physics of Active Galaxies*, 54, 201
Muzzin, A., Marchesini, D., Stefanon, M., *et al.* 2013, *ApJS*, 206, 8
Narayan, R. & Yi, I. 1995, *ApJ*, 452, 710
Rigby, E. E., Best, P. N., Brookes, M. H., *et al.* 2011, *MNRAS*, 416, 1900
Röttgering, H. J. A. 2010, *ISKAF2010 Science Meeting*
Schinnerer, E., Carilli, C. L., Scoville, N. Z., *et al.* 2004, *AJ*, 128, 1974
Schinnerer, E., Smolčić, V., Carilli, C. L., *et al.* 2007, *ApJS*, 172, 46
Schmidt, M., 1968, *ApJ*, 151, 393
Tadhunter, C. N., Morganti, R., Robinson, A., *et al.* 1998, *MNRAS*, 298, 1035
van Haarlem, M. P., Wise, M. W., Gunst, A. W., *et al.* 2013, *A&A*, 556, A2
Yatawatta, S., de Bruyn, A. G., Brentjens, M. A., *et al.* 2013, *A&A*, 550, A136

Galaxies in 3D across the Universe
Proceedings IAU Symposium No. 309, 2014
B. L. Ziegler, F. Combes, H. Dannerbauer, M. Verdugo, eds.
© International Astronomical Union 2015
doi:10.1017/S174392131400948X

Discovery of Carbon Radio Recombination Lines in M82

Leah K. Morabito[1], J. B. R. Oonk[1,2], Francisco Salgado[1], M. Carmen Toribio[2], Xander Tielens[1] and Huub Röttgering[1]

[1]Leiden Observatory, Leiden University,
P. O. Box 9513, 2300 RA Leiden, The Netherlands
email: morabito@strw.leidenuniv.nl
[2]Netherlands Institute for Radio Astronomy (ASTRON),
Postbus 2, 7990 AA Dwingeloo, The Netherlands

Abstract. Cold, diffuse HI clouds are a key component of the interstellar medium (ISM), and play an important role in the evolution of galaxies. Carbon radio recombination lines (CRRLs) trace this ISM stage, and with the enormous sensitivity of LOFAR we have already begun to map and constrain the physical properties of this gas in our own Galaxy. Using LOFAR's low band antenna, we have observed M82 and present the first ever extragalactic detection of CRRLs. We stack 22 lines to find a 8.5-sigma detection. The line peak to continuum ratio is ~ 0.003, with a FWHM of 31km s^{-1}. The CRRL feature is consistent with an origin in the cold, neutral medium in the direction of the nucleus of M82.

Keywords. galaxies: ISM

1. Introduction

M82 is a nearby (3.52 ± 0.02 Mpc; Jacobs *et al.* 2009) nuclear starburst galaxy that is part of a larger group of galaxies, and is thought to be interacting with M81 (Yun *et al.* 1993). M82 is host to many phases of the interstellar medium (ISM). A concentration of atomic hydrogen in the nuclear region is observed (Yun *et al.* 1993), where high density gas tracers associated with star formation trace a rotating ring of molecular gas (e.g., Kepley *et al.* 2014). The nuclear region also contains numerous HII regions (e.g., McDonald *et al.* 2002; Gandhi *et al.* 2011) and supernova remnants (SNRs; e.g., Muxlow *et al.* 1994; Fenech *et al.* 2010). The SNRs show turnovers in their spectra at low frequencies that imply the presence of ionized gas (Wills *et al.* 1997). The detailed interaction of these phases of the ISM is not yet understood. Carbon radio recombination lines can help by probing the cold, diffuse medium in M82.

When free electrons recombine with ions, recombination lines are produced. When the quantum numbers of the transitions are high enough ($n \gtrsim 50$), the recombination lines appear in the radio regime and are called radio recombination lines (RRLs). These RRLs come in two different "flavours": classical and diffuse. Classical RRLs are those observed above 1 GHz, where they trace densities of about 100cm^{-3} and temperature $T_e \sim 10^4$K gas, such as that associated with HII regions. These RRLs have been in our Galaxy as well as a handful of bright nearby starbursting galaxies, including M82 (e.g., Shaver *et al.* 1977; Rodriguez-Rico *et al.* 2004).

Below 1 GHz, we observe diffuse RRLs. These trace a different phase of the ISM, with a lower density ($n_e \sim 0.01 - 1.0$cm^{-3}) and temperature ($T_e \sim 100$K). These lines are seen only from carbon, which has an ionization potential of 11.3 eV, lower than that of hydrogen. Not much is yet known about diffuse RRLs, but LOFAR is beginning a survey that over the course of the next several years will map out the Galactic CRRL absorption/emission on large (degrees) to small (3-10 arcsec) scales.

2. Modelling CRRLs

The observed properties of CRRLs, line width and integrated optical depth, depend sensitively on the temperature and density of the gas. Figure 1 shows the dependence of integrated optical depth on quantum number (frequency) for two different gas phases. It is clear that measurements over a range of frequencies will place tighter constraints on the gas properties.

Low frequency RRLs are not in local thermodynamic equilibrium (LTE) and we use detailed non-LTE models (Salgado *et al.*, in prep.; also e.g., Walmsley & Watson 1982; Shaver 1975) to calculate the departure coefficients (b_n), which describe the departure from LTE. These departure coefficients depend on the input electron temperature, T_e, and density, n_e. When combined with an emission measure, $\mathrm{EM_{CII}} \sim n_e^2 L\mathrm{pc\ cm^{-6}}$, the departure coefficients can be translated directly into integrated optical depths. Singly ionised carbon has an additional dielectronic-like recombination process that must be included in the models, and results in large integrated optical depths at low frequencies. This drives the need for sensitive low frequency instruments with enough spectral and spatial resolution to resolve the lines and localize them to a particular source.

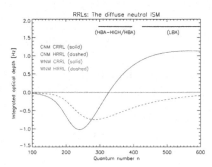

Figure 1. Integrated optical depth vs. quantum number (frequency). Cold neutral medium (CNM) and warm neutral medium (WNM) models (Salgado *et al.*, in prep.). The frequency ranges for LOFAR's low band antenna (LBA; 10-90 MHz) and high band antenna (HBA; 110-250) are marked. Models for both carbon RRLs and hydrogen RRLs are shown.

3. The Low Frequency Array

The Low Frequency Array (LOFAR) consists of 24 core and 16 remote stations in the Netherlands, and 8 international stations spread across Europe. The low band antenna (LBA) covers frequencies of 10-80 MHz, while the high band antenna (HBA) covers frequencies of 120-240 MHz. The wide instantaneous bandwidth of 96 MHz can be flexibly split to form multiple beams using LOFAR's unique phased array capabilities. Within a single observation LOFAR can therefore observe tens to hundreds of CRRL transitions. We can thus measure the dependence of the CRRL line properties as a function of frequency, to derive the physical properties of the associated gas by matching these with our models. Where necessary it also possible to stack individual line transitions in order to achieve a detection (e.g., Oonk *et al.* 2014).

4. Extragalactic CRRLs

Due to the stimulated nature of CRRLs (Shaver 1978), the line intensities of these absorption features are independent of redshift, and we will be able to probe the cold neutral medium (CNM) out to redshifts $\gtrsim 5$. From our RRL models (Salgado *et al.*, in prep.) we expect RRL peak optical depths of about 10^{-3} for typical HI column densities of $10^{21}\mathrm{cm^{-2}}$. LOFAR is currently surveying the Northern hemisphere down to depths of $\sim 100\mu\mathrm{Jy}$ at $\sim 150\mathrm{MHz}$, which will yield a few hundred radio galaxies bright enough for CRRL searches. Although this sample will be dominated by FRI and FRII galaxies, there will still be many of the nearby well-studied starforming galaxies like M82.

The sensitivity, large field of view, frequency resolution and coverage of the Square Kilometre Array (SKA) will revolutionize the study of low-frequency recombination lines. The increased sensitivity compared to LOFAR will expand the number of extragalactic sources searchable for RRLs by almost 3 orders of magnitude, including a few hundred nearby starforming galaxies. The sensitivity of the SKA1-LOW will be good enough that we will be able to spatially resolve the brightest 25 percent of these nearby galaxies.

LOFAR is already capable of detecting CRRLs in extragalactic sources (Morabito *et al.* 2014, submitted), and we have begun a pilot survey for extragalactic CRRLs which will set the framework on which to base our future SKA studies.

5. Observations and Data Reduction

We used LOFARs LBA to observe M82 in the frequency range of 30-78 MHz (observation details given in the table to the right). The calibrator was processed using the radio observatory's calibrator pipeline, and the amplitude solutions were transferred to the target. We imaged each channel separately for a total of 69 subbands, deliberately chosing low-resolution imaging parameters to speed up the processing time. We used only

Project ID	LC0_043
RA (J2000)	09h55m52.7s
DEC (J2000)	+69d40m46s
Observing date	2013 Feb 21
On-source time	5.0 h
Frequency range	30-78 MHz
No. SB	244
SB width	0.1953 MHz
Channels per SB	128
Velocity res.	6-16 km s^{-1}
Simultaneous Cal.	3C196

the 24 core stations to make the images. This provided resolutions of $\sim 200 \times 300$ arcsec. The images were smoothed to 400×400 arcsec. We rejected five subbands that were excessively noisy, and extracted the spectra from elliptical regions in each subband.

There were 22 subbands with α-transition ($\Delta n = 1$) CRRLs within 48-64MHz. We trimmed noisy edge channels and fit the continuum in each subband with a low (first or second) order polynomial, after blanking the expected line locations. The subbands were each divided by their continuum fit, and we subtracted unity to place the continuum at zero.

The CRRLs are not detected individually, so we stack the 22 transitions together. Before doing this, we must know what radial velocity to use to predict the rest-frame line centres for stacking. To determine the radial velocity we cross-correlated a template spectrum of CRRLs (sampled it in the same way as the observed spectrum) and the observed spectrum. We repeat this process for a range of redshifts, and find a peak at $v = 219$km s^{-1} ($z = 0.00073$).

Using this radial velocity, we stacked the continuum-subtracted subbands containing CRRLs at the rest-frame of M82. All subbands were first converted to radial velocity, relative to the local standard of rest (LSR). The individual points make up a composite spectrum that has irregular sampling, with more dense, evenly spaced points towards the central region, roughly ± 200km s^{-1} around the CRRL feature. After testing various smoothing methods to reduce the noise, we used a first order Savitzky-Golay filter (Savitzky & Golay 1964) with a width of 31 data points to smooth the final spectrum (see Figure 7)

To test the validity of the detection we performed a series of checks: we used a 'jackknife' procedure to rule out that the stacked spectrum could be dominated by one subband; we extracted the spectrum from the background sky in the images and stacked the same subbands in the same manner; we introduced random velocity shifts to each subband and stacked in the same manner. From these tests, we conclude that the feature is likely to be real.

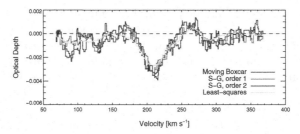

Figure 2. The stacked spectrum, smoothed in a variety of ways that each produce a similar absorption feature.

6. Results

The final spectrum is an average CRRL line profile for quantum levels $478 \leqslant n \leqslant 508$. We fit the CRRL feature with a Gaussian of depth 3×10^{-3}, a FWHM of 31 km s^{-1}, a centre of 211 km s^{-1}, and an additive offset of -2×10^{-4}. The centroid velocity places the line within the nuclear region of M82, as inferred from other gas tracers (Kepley *et al.* 2014; Wills *et al.* 1997). A full analysis of the detected CRRL feature and its association with other gas tracers is presented in Morabito *et al.*, 2014 (submitted).

7. Conclusions

We have detected, for the first time, CRRLs in an extragalactic source, M82. The CRRL line centroid indicates that gas that it traces is located in the nuclear region of M82. This single detection is not sufficient to constrain the gas temperature, density, and emission measure. Thus we are working towards detecting CRRLs at higher frequencies.

The implications of this detection are far-reaching. The intensity, and therefore observability, of these CRRL absorption lines scales only with flux, and is redshift-independent. The number of sources in which we could potentially observe CRRL absorption at low frequencies with LOFAR is many hundreds. This is the driver for a pilot survey to gather a census of extragalactic sources of CRRLs and start to constrain the evolution of the cold neutral medium over cosmological times.

Acknowledgements

LKM acknowledges financial support from NWO Top LOFAR project, project n. 614.001.006.

References

Fenech, D., Beswick, R., Muxlow, T. W. B., Pedlar, A., & Argo, M. K. 2010, *MNRAS*, 408, 607
Gandhi, P., Isobe, N., Birkinshaw, M., *et al.* 2011, *PASJ*, 63, 505
Jacobs, B. A., Rizzi, L., Tully, R. B., *et al.* 2009, *AJ*, 138, 332
Kepley, A. A., Leroy, A. K., Frayer, D., *et al.* 2014, *ApJ*, 780, L13
McDonald, A. R., Muxlow, T. W. B., Wills, K. A., Pedlar, A., & Beswick, R. J. 2002, *MNRAS*, 334, 912
Muxlow, T. W. B., Pedlar, A., Wilkinson, P. N., *et al.* 1994, *MNRAS*, 266, 455
Oonk, J. B. R., van Weeren, R. J., Salgado, F., *et al.* 2014, *MNRAS*, 437, 3506
Rodriguez-Rico, C. A., Viallefond, F., Zhao, J.-H., Goss, W. M., & Anantharamaiah, K. R. 2004, *ApJ*, 616, 783
Savitzky, A. & Golay, M. J. E. 1964, *Analytical Chemistry*, 36, 1627
Shaver, P. A. 1975, *Pramana*, 5, 1
Shaver, P. A., Churchwell, E., & Rots, A. H. 1977, *A&A*, 55, 435
Shaver, P. A. 1978, *A&A*, 68, 97
Walmsley, C. M. & Watson, W. D. 1982, *ApJ*, 260, 317
Wills, K. A., Pedlar, A., Muxlow, T. W. B., & Wilkinson, P. N. 1997, *MNRAS*, 291, 517
Yun, M. S., Ho, P. T. P., & Lo, K. Y. 1993, *ApJ*, 411, L17

Galaxies in 3D across the Universe
Proceedings IAU Symposium No. 309, 2014
B. L. Ziegler, F. Combes, H. Dannerbauer, M. Verdugo, eds.
© International Astronomical Union 2015
doi:10.1017/S1743921314009491

Disk Galaxies in the Magneticum Pathfinder Simulations

Rhea-Silvia Remus[1], Klaus Dolag[1,2], Lisa K. Bachmann[1], Alexander M. Beck[1,2], Andreas Burkert[1,3], Michaela Hirschmann[4] and Adelheid Teklu[1]

[1] Universitäts-Sternwarte München, Scheinerstr. 1, D-81679 München, Germany
[2] MPI for Astrophysics, Karl-Schwarzschild Strasse 1, D-85748 Garching, Germany
[3] MPI for Extraterrestrial Physics, Giessenbachstrasse 1, D-85748 Garching, Germany
[4] Institut d'Astrophysique de Paris, UPMC-CNRS, UMR7095, Boulevard Aragon, F-75014 Paris, France
email: rhea@usm.lmu.de

Abstract. We present *Magneticum Pathfinder*, a new set of hydrodynamical cosmological simulations covering a large range of cosmological scales. Among the important physical processes included in the simulations are the chemical and thermodynamical evolution of the diffuse gas as well as the evolution of stars and black holes and the corresponding feedback channels. In the high resolution boxes aimed at studies of galaxy formation and evolution, populations of both disk and spheroidal galaxies are self-consistently reproduced. These galaxy populations match the observed stellar mass function and show the same trends for disks and spheroids in the mass–size relation as observations from the SDSS. Additionally, we demonstrate that the simulated galaxies successfully reproduce the observed specific angular-momentum–mass relations for the two different morphological types of galaxies. In summary, the *Magneticum Pathfinder* simulations are a valuable tool for studying the assembly of cosmic and galactic structures in the universe.

1. Introduction

The observational properties of galaxies have been studied in a multitude of surveys, but until recently corresponding simulations including baryonic physics were limited to isolated or merging systems and cosmological zoom-in simulations of individual galaxies selected from larger cosmological boxes of low resolution. With increasing computational power and improved models of the physics that drive the evolution of baryonic structures it has now become possible to simulate large cosmological volumes with full baryonic treatment, providing sufficiently large samples of galaxies formed within the cosmological ΛCDM model to study the statistical properties of galaxies and compare them with observations. These new simulations will significantly improve our understanding of the formation and evolution of all types of galaxies, their dynamical properties, and the changes of morphology across cosmic time. Especially, they will shed light on the two dominant processes of galaxy formation, namely mergers and secular evolution. We present one such set of new cosmological simulations, called *Magneticum Pathfinder*, and show first results on the simulated galaxy properties in comparison to observations.

2. The Magneticum Pathfinder Simulations

The Magneticum Pathfinder simulations (Dolag *et al.*, in prep.) are a set of hydrodynamical cosmological boxes with volumes ranging from $(896 \text{ Mpc}/h)^3$ to $(18 \text{ Mpc}/h)^3$ and resolutions of $m_{\mathrm{Gas}} = 2.6 \cdot 10^9 M_{\odot}/h$ up to $m_{\mathrm{Gas}} = 3.9 \cdot 10^5 M_{\odot}/h$, performed with a modified version of GADGET (Springel 2005). They include radiative cooling, star formation

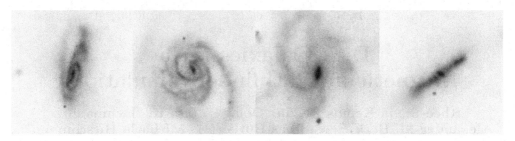

Figure 1. Examples of disk galaxies from the Magneticum Pathfinder Simulations.

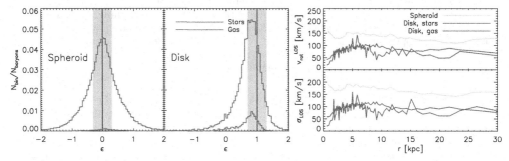

Figure 2. *Left:* Distribution of the circularity ϵ of a spheroidal (left) and a disk galaxy (right) for stars (solid red) and gas (dash-dotted red) within $0.1R_{\rm vir}$, normalized to the total stellar and gas mass of the galaxy. Gray shaded areas mark the selection criterion for spheroid (left) and disk (right) galaxies used in this work. *Right:* Line-of-sight (LOS) velocity (upper) and LOS velocity dispersion (lower panel) versus radius for the stellar component of the spheroid (cyan) and the disk (blue) galaxy and the gas component of the disk galaxy (dashed blue). The spheroidal clearly has a higher dispersion relative to its rotation velocity, while for the disk both the stellar and gas component are in very good agreement, as expected from observations.

and kinetic winds following Springel & Hernquist (2003), stellar evolution and metal enrichment according to Tornatore *et al.* (2007) and Wiersma *et al.* (2009), and black hole evolution and feedback of Springel *et al.* (2005), Fabjan *et al.* (2010) and Hirschmann *et al.* (2014). Furthermore, we implemented several improvements for smoothed particle hydrodynamics that more accurately treat turbulence and viscosity and result in a better modeling of galaxies (Dolag *et al.* 2005, Beck *et al.*, in prep., for details). Additionally, we include thermal conduction according to Dolag *et al.* (2004). We adopt a WMAP7 (Komatsu *et al.* 2011) ΛCDM cosmology with $h = 0.704$, $\Omega_m = 0.272$ and $\Lambda = 0.728$, and use SUBFIND (Springel *et al.* 2001, Dolag *et al.* 2009) to identify and extract structures from galaxy clusters down to galaxies.

3. Galaxies in the Magneticum Pathfinder Simulations

To study galactic dynamics in detail, high resolution is required. Therefore, we use the largest box with sufficient resolution (box length of 48 Mpc/h and resolution of $m_{\rm DM} = 3.6 \cdot 10^7 M_\odot/h$, $m_{\rm Gas} \approx 7.3 \cdot 10^6 M_\odot/h$, and slightly smaller stellar mass) for reproducing the observed diversity of morphologies and kinematic properties with adequate statistics. We follow $(2 \cdot 576)^3$ particles, employing a softening length of $\epsilon_{\rm DM,Gas} = 1.4$ kpc/h for dark matter and gas and $\epsilon_* = 0.7$ kpc/h for stars. The box contains 621 halos down to a mass of $5 \cdot 10^{11} M_\odot/h$, of which five are small clusters and approximately 40 are groups above $10^{13} M_\odot/h$ at redshift $z = 0$. Thus, we can study the dynamics and properties of galaxies down to normal Milky Way type galaxies; however, our resolution is insufficient

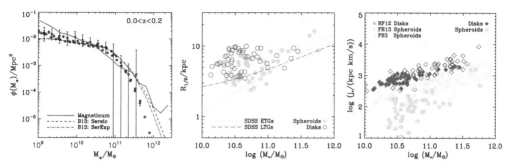

Figure 3. *Left:* Stellar mass functions for $z < 0.2$ in our simulation (solid line) compared to observations (diamonds, data points from Cole *et al.* 2001, Bell *et al.* 2003, Panter *et al.* 2004 and Pérez-González *et al.* 2008). For high stellar masses observational estimations are uncertain as discussed in Bernardi *et al.* 2013 (B13, using a Sérsic model (dashed) and a Sérsic-bulge-plus-exponential-disk model (dash-dotted line)). *Middle:* Mass–size relation of the simulated disks (blue open circles) and spheroids (cyan filled circles). The dashed blue and solid cyan lines represent the mean size as function of mass from Shen *et al.* 2003 for SDSS early-type and late-type galaxies. *Right:* Specific angular momentum of the stellar component of the simulated disks (blue filled diamonds) and spheroids (cyan filled circles) as function of the stellar mass. Open symbols represent the observations of disk galaxies from Romanowsky & Fall 2012 (RF12, blue diamonds) and for ellipticals from Fall & Romanowsky 2013 (FR13, cyan plussed circles) and Fall 1983 (F83, cyan open circles).

to properly resolve dwarf galaxies. Fig. 1 shows some typical disk galaxies formed in this simulation.

3.1. *The Classification of Galaxies*

We successfully reproduce a population of disk as well as spheroidal galaxies due to the improvements in the numerical methods as well as the included prescriptions of physical processes. To distinguish the populations and classify the galaxies, we use a procedure similar to Scannapieco *et al.* (2008): we align each galaxy along its principal axis of inertia of the stars within $0.1R_{\rm vir}$ and subsequently classify it according to the circularity parameter $\epsilon = j_z/j_{\rm circ}$, where j_z is the specific angular momentum of each particle with respect to the z-axis and $j_{\rm circ}$ is the specific angular momentum expected for a circular orbit. We classify a galaxy as spheroidal if more than 40% of both stellar and gas particles lie within $-0.3 \leqslant \epsilon \leqslant 0.3$, and as disk if more than 40% of its gas and more than 30% of its stellar particles are within $0.7 \leqslant \epsilon \leqslant 1.3$. Fig. 2 shows the distribution of ϵ for a spheroid and a disk (left panel). We identify 51 clearly rotationally supported disks and 51 clearly dispersion-dominated spheroids from our sample of 621 galaxies at $z = 0$. Our criterion is fairly strict, excluding systems like S0 galaxies, spheroids with gas disks, merging systems and disk galaxies with bar-like structures from the classification.

3.2. *Population Characteristics*

The stellar mass function for all central galaxies (independent of their classification type) is shown in Fig. 3 (left panel). A similar analysis for other boxes of the Magneticum simulations has been shown by Hirschmann *et al.* (2014) and Bachmann *et al.* (2014). Our simulation is in good agreement with the observations, apart from an overestimate at the low-mass end, where the simulation does not resolve the feedback associated with black holes. Another important characteristic of galaxy properties is the mass–size relation: for the same mass, spheroidal galaxies are more compact than their disky counterparts. As shown in the middle panel of Fig. 3, our disks and spheroids show the same behaviour as the observations, even though our calculation of the stellar half-mass radius is only a crude

approximation to the observed half-light radius determined from Sérsic fits (Shen *et al.*
2003). More details on the dynamical properties of our sample of galaxies are presented by
Remus *et al.* (in prep.). The right panel of Fig. 3 shows the specific angular-momentum–
mass relation of the stars in our disk and spheroidal galaxies compared to observations
from Fall (1983), Romanowsky & Fall (2012) and Fall & Romanowsky (2013). Again, our
simulated galaxies are in excellent agreement with the observations, showing the clear
distinction between the specific stellar angular momentum of disks and spheroids, which
is discussed in more detail by Teklu *et al.* (in prep).

4. Conclusions

The Magneticum simulations can successfully reproduce disk galaxies as well as
spheroids, with enough resolution to study dynamical properties of those galaxies. We
demonstrated that Magneticum galaxies match the observed properties of present-day
disk and spheroid galaxies, such as the stellar mass function and the mass–size and
angular-momentum–mass relations. This renders possible a statistical approach to the
different evolution mechanisms of both disks and spheroidals, which will be done in several forthcoming studies.

Acknowledgements

The Magneticum team acknowledges support from the DFG Excellence Cluster Universe,
the SFB-Transregio TR33, the Computational Center C2PAP, and the Leibniz-Rechen-
zentrum via projects pr86re and pr83li.

References

Bachmann, L. K., *et al.* 2014, arXiv:1409.3221
Bell, E. F., McIntosh, D. H., Katz, N., & Weinberg, M. D. 2003, *ApJS*, 149, 289
Bernardi, M., *et al.* 2013, *MNRAS*, 436, 697
Cole, S., *et al.* 2001, *MNRAS*, 326, 255
Dolag, K., Jubelgas, M., Springel, V., Borgani, S., & Rasia, E. 2004, *ApJ*, 606, L97
Dolag, K., Vazza, F., Brunetti, G., & Tormen, G. 2005, *MNRAS*, 364, 753
Dolag, K., Borgani, S., Murante, G., & Springel, V. 2009, *MNRAS*, 399, 497
Fabjan, D., *et al.* 2010, *MNRAS*, 401, 1670
Fall, S. M. 1983, in *IAU Symposium*, Vol. 100, Internal Kinematics and Dynamics of Galaxies,
     ed. E. Athanassoula, 391–398
Fall, S. M. & Romanowsky, A. J. 2013, *ApJ*, 769, L26
Hirschmann, M., *et al.* 2014, *MNRAS*, 442, 2304
Komatsu, E., *et al.* 2011, *ApJS*, 192, 18
Panter, B., Heavens, A. F., & Jimenez, R. 2004, *MNRAS*, 355, 764
Pérez-González, P. G., *et al.* 2008, *ApJ*, 675, 234
Romanowsky, A. J. & Fall, S. M. 2012, *ApJS*, 203, 17
Scannapieco, C., *et al.* 2008, *MNRAS*, 389, 1137
Shen, S., *et al.* 2003, *MNRAS*, 343, 978
Springel, V. 2005, *MNRAS*, 364, 1105
Springel, V., Di Matteo, T., & Hernquist, L. 2005, *MNRAS*, 361, 776
Springel, V. & Hernquist, L. 2003, *MNRAS*, 339, 289
Springel, V., White, S. D. M., Tormen, G., & Kauffmann, G. 2001, *MNRAS*, 328, 726
Tornatore, L., Borgani, S., Dolag, K., & Matteucci, F. 2007, *MNRAS*, 382, 1050
Wiersma, R. P. C., Schaye, J., & Smith, B. D. 2009, *MNRAS*, 393, 99

Galaxies in 3D across the Universe
Proceedings IAU Symposium No. 309, 2014   © International Astronomical Union 2015
B. L. Ziegler, F. Combes, H. Dannerbauer, M. Verdugo, eds.   doi:10.1017/S1743921314009508

Rings of star formation:
Imprints of a close galaxy encounter

Jorge Moreno[1,†,2]

[1]Department of Physics and Astronomy, University of Victoria, Finnerty Road, Victoria,
British Columbia, V8P 1A1, Canada
[2]Department of Physics and Astronomy, California State Polytechnic University, Pomona, CA
91768, USA
[†]CITA National Fellow
e-mail: jorgemoreno@csupomona.edu

Abstract. In this talk, I report results from galaxy merger simulations, which suggest the existence of a ring of star formation produced by close galaxy encounters. This is a generic feature of all galaxy interactions, provided that the disc spins are sufficiently aligned. This signature can be used to identify close galaxy pairs that have actually suffered a close interaction.

Keywords. galaxies: galaxies: formation – evolution – interactions

1. Introduction

Over thirty five years ago, Larson & Tinsley (1978) first recognised interactions as a promising avenue for triggering star formation in galaxies. This has been a subject of intense work by numerical simulators (Barnes & Hernquist 1991, 1996; Cox *et al.* 2006; Di Matteo *et al.* 2007, 2008; Cox *et al.* 2008). With the help of large surveys, such as the Sloan Digital Sky Survey, interactions are now established as a prime driver for igniting star-forming episodes (Ellison *et al.* 2008, 2010, 2011; Patton *et al.* 2011; Scudder *et al.* 2012; Patton *et al.* 2013), active galactic nuclei (Ellison *et al.* 2011, 2013b), and for the creation of tidal tails in galaxies with close companions (Casteels *et al.* 2013).

Indeed, it is common practice to use the presence of tidal tails as an indicator of past interaction (Kartaltepe *et al.* 2012; Hung *et al.* 2013, 2014). However, a problem with this approach is that for many orbits, tidal tails may dissipate long before the two galaxies merge together. Moreover, some orbital configurations may be more conductive to creating tidal tails than others. In other words, it could be the case that visual classification (via the presence of tails) might miss a fraction of truly interacting galaxy pairs.

In this presentation, I report a signature produced by galaxy interactions: a ring of star formation. This feature may serve as an alternative to identifying galaxies that have experienced a close encounter. A more exhaustive study of this ring-like structure is reserved for future work (Moreno *et al.*, in prep).

2. Methods, Results & Discussion

I employ a suite of 75 SPH (smoothed-particle hydrodynamics) merger simulations (Springel 2005), comprised of three disc-spin orientations selected from Robertson *et al.* (2006): the "e", "f", and "k" orientations – meant to represent aligned, perpendicular, and anti-aligned spins, respectively. See Torrey *et al.* (2012) and Moreno *et al.* (in prep) for details. For each of the three orientations considered, we focus on five eccentricities ($\epsilon = \{0.85, 0.90, 0.95, 1.0, 1.05\}$) and five impact parameters ($b = \{2, 4, 8, 12, 16\}$ kpc).

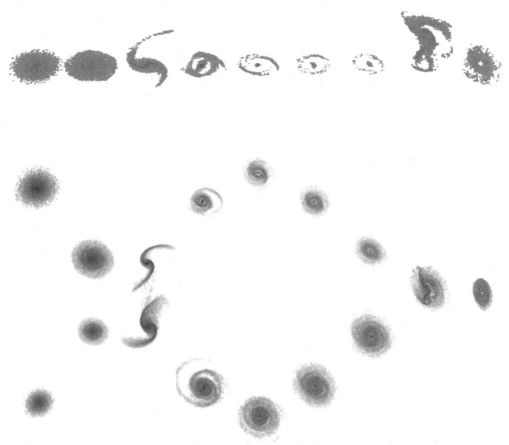

Figure 1. Interacting galaxies with strongly-aligned disc-spin orientation. *Left-to-right:* Various stages of interaction and merging: incoming, close pericentric passage, receding, re-approaching, and merging. *Top:* The morphology of star formation in the secondary (smaller) galaxy (in pink). *Bottom:* The gas-morphology of the two interacting galaxies (in blue – zoomed-out scale).

Figure 1 shows the results for one of our orbits ($\epsilon = 1.05$, $b = 16$ kpc, "e" orientation). In blue (bottom), we show the morphology of the gas component for both galaxies. From left-to-right, each stage of the merger sequence is presented: incoming (1-2), first pericentric passage (3), receding (4-5), re-approaching (6-7), merging (8-9). Tidal tails are only evident briefly (3). In pink (top), we show the morphology of star formation (zoomed-in scale). After first pericentric passage, tidal tails wrap around into a ring of off-nuclear dense gas. This accumulation of material triggers a stable star-forming ring-like structure, which survives for a prolonged period.

This feature takes ∼0.2-1 Gyr after first pericentric passage to appear. It is triggered for all orbits in the "e" and "f" orientations (aligned and perpendicular spins), regardless of ϵ and b. However, when the disc spins are anti-aligned ("k" orientation), this ring-like feature is absent (with the exception of a few anomalous orbits where the two disc interpenetrate one another).

Figure 2. Interacting galaxies with anti-aligned orientation. Same format as in Figure 1.

3. Conclusions

In this talk, I report the existence of ring-like off-nuclear star formation produced by close galaxy interactions. If the two discs are sufficiently aligned, a ring-like feature is triggered on the secondary galaxy soon after first passage.

In principle, this is a promising way of identifying those galaxy pairs that have actually experienced a close encounter in the past – which, in turn, would allow more refined calculations of the galaxy merger rate (Patton *et al.* 1997; Bluck *et al.* 2012; López-Sanjuan *et al.* 2013) – but see Moreno (2012) and Moreno *et al.* (2013) for caveats.

One could argue that, if observed, there might be other causes for the existence of this ring. For instance, a ring-like structure is detected in M31 (Gordon *et al.* 2006). However, based on its off-centre nature, simulations suggest that this was caused instead by a direct off-nuclear collision (Block *et al.* 2006; Dierickx *et al.* 2014).

It is my hope that these findings motivate observational investigations by integral-field spectroscopic surveys, such as CALIFA (Sánchez *et al.* 2014), SAMI (Croom *et al.* 2012), MaNGA (Bundy *et al.*, in prep), and the future HECTOR survey (Lawrence *et al.* 2012).

Acknowledgements

The computations in this paper were run on the Odyssey cluster supported by the FAS Division of Science, Research Computing Group at Harvard University. I acknowledge the Natural Science and Engineering Research and the Canadian Institute for Theoretical

Astrophysics for funding. This presentation would not have been possible without the support of my collaborators: Paul Torrey, Sara Ellison, David Patton, Asa Bluck, Gunjan Bansal and Lars Hernquist. Lastly, I wish to thank Volker Springel for kindly giving me permission to use his GADGET-3 code, Alexia Lewis for pointing out the M31-M32 encounter and references to it, and the organisers of *Galaxies in 3D across the universe* for putting together such an exciting, fruitful, and memorable meeting!

References

Barnes, J. E. & Hernquist, L. E. 1991, *ApJL*, 370, L65

Barnes, J. E. & Hernquist, L., 1996, *ApJ*, 471, 115

Block, D. L., Bournaud, F., Combes, F., *et al.* 2006, *Nat*, 443, 832

Bluck, A. F. L., Conselice, C. J., Buitrago, F., *et al.* 2012, *ApJ*, 747, 34

Casteels, K. R. V., Bamford, S. P., Skibba, R. A., *et al.* 2013, *MNRAS*, 429, 1051

Cox, T. J., Jonsson, P., Primack, J. R., & Somerville, R. S. 2006, *MNRAS*, 373, 1013

Cox, T. J., Jonsson, P., Somerville, R. S., Primack, J. R., & Dekel, A. 2008, *MNRAS*, 384, 386

Croom, S. M., Lawrence, J. S., Bland-Hawthorn, J., *et al.* 2012, *MNRAS*, 421, 872

Dierickx, M., Blecha, L., & Loeb, A. 2014, *ApJL*, 788, L38

Di Matteo, P., Combes, F., Melchior, A.-L., & Semelin, B. 2007, *AAP*, 468, 61

Di Matteo, P., Bournaud, F., Martig, M., *et al.* 2008, *A&A*, 492, 31

Ellison, S. L., Patton, D. R., Simard, L., & McConnachie, A. W. 2008, *AJ*, 135, 1877

Ellison, S. L., Patton, D. R., Simard, L., McConnachie, A. W., Baldry, I. K., & Mendel, J. T. 2010, *MNRAS*, 407, 151

Ellison, S. L., Patton, D. R., Mendel, J. T., & Scudder, J. M. 2011, *MNRAS*, 418, 2043

Ellison, S. L., Mendel, J. T., Scudder, J. M., Patton, D. R., & Palmer, M. J. D. 2013, *MNRAS*, 430, 3128

Ellison, S. L., Mendel, J. T., Patton, D. R., & Scudder, J. M. 2013, *MNRAS*, 435, 3627

Gordon, K. D., Bailin, J., Engelbracht, C. W., *et al.* 2006, *ApJL*, 638, L87

Kartaltepe, J. S., Dickinson, M., Alexander, D. M., *et al.* 2012, *ApJ*, 757, 23

Larson, R. B. & Tinsley, B. M. 1978, *ApJ*, 219, 46

Lawrence, J., Bland-Hawthorn, J., Bryant, J., *et al.* 2012, *PROCSPIE*, 8446,

López-Sanjuan, C., Le Fèvre, O., Tasca, L. A. M., *et al.* 2013, *A&A*, 553, A78

Hung, C.-L., Sanders, D. B., Casey, C. M., *et al.* 2013, *ApJ*, 778, 129

Hung, C.-L., Sanders, D. B., Casey, C. M., *et al.* 2014, *ApJ*, 791, 63

Moreno, J. 2012, *MNRAS*, 419, 411

Moreno, J., Bluck, A. F. L., Ellison, S. L., *et al.* 2013, *MNRAS*, 436, 1765

Patton, D. R., Pritchet, C. J., Yee, H. K. C., Ellingson, E., & Carlberg, R. G. 1997, *ApJ*, 475, 29

Patton, D. R., Ellison, S. L., Simard, L., McConnachie, A. W., & Mendel, J. T. 2011, *MNRAS*, 412, 591

Patton, D. R., Torrey, P., Ellison, S. L., Mendel, J. T., & Scudder, J. M. 2013, *MNRAS*, 433, L59

Robertson, B., Bullock, J. S., Cox, T. J., Di Matteo, T., Hernquist, L., Springel, V., & Yoshida, N. 2006, *ApJ*, 645, 986

Sánchez, S. F., Rosales-Ortega, F. F., Iglesias-Páramo, J., *et al.* 2014, *A&A*, 563, A49

Scudder, J. M., Ellison, S. L., Torrey, P., Patton, D. R., & Mendel, J. T. 2012, *MNRAS*, 426, 549

Springel, V. 2005, *MNRAS*, 364, 1105

Torrey, P., Cox, T. J., Kewley, L., & Hernquist, L. 2012, *ApJ*, 746, 108

Galaxies in 3D across the Universe
Proceedings IAU Symposium No. 309, 2014
B. L. Ziegler, F. Combes, H. Dannerbauer, M. Verdugo, eds.
© International Astronomical Union 2015
doi:10.1017/S174392131400951X

A kinematic analysis of the Giant star-forming Region of N11

Sergio Torres-Flores[1], Rodolfo Barbá[1], Jesús Maíz Apellániz[2], Mónica Rubio[3] and Guillermo Bosch[4]

[1]Departamento de Física, Universidad de La Serena,
Av. Cisternas 1200 Norte, La Serena, Chile
email: storres@dfuls.cl
[2]Centro de Astrobiología (INTA-CSIC), ESAC campus, P.O. Box 78,
28691 Villanueva de la Cañada, Madrid, Spain
[3]Departamento de Astronomía, Universidad de Chile, Casilla 36-D, Santiago, Chile
[4]Facultad de Ciencias Astronómicas y Geofísicas, Universidad Nacional de la La Plata,
Paseo del Bosque s/n, 1900 La Plata, Argentina

Abstract. In this work we present high resolution spectroscopic data of the giant star-forming region of N11, obtained with the GIRAFFE instrument at the Very Large Telescope. By using this data set, we find that most of the Hα emission lines profiles in this complex can be fitted by a single Gaussian, however, multiple emission line profiles can be observed in the central region of N11. By adding all the spectra, we derive the integrated Hα profile of this complex, which displays a width (σ) of about 12 km s^{-1} (corrected by instrumental and thermal width). We find that a single Gaussian fit on the integrated Hα profile leaves remaining wings, which can be fitted by a secondary broad Gaussian component. In addition, we find high velocity features, which spatially correlate with soft diffuse X-ray emission.

1. Introduction and Data

Giant HII regions (GHIIR) are characterized by displaying integrated emission lines which have supersonic widths (Smith & Weedman 1972, Melnick *et al.* 1999), however, the origin of these motions is still unclear. In addition, some authors have reported the existence of a low-intensity broad component in the integrated emission line profiles of GHIIR. Some authors suggested that this broad component is produced by a highly turbulent diffuse component Melnick *et al.* (1999), while others authors argue that this component arises as a superposition of multiple bubbles in expansion Chu & Kennicutt (1994). In order to study these phenomena, we present a kinematic analysis of the ionized gas in the star-forming region N11 (see previous works in Rosado *et al.* 1996, Nazé *et al.* 2004). We have used spectroscopic data taken with the GIRAFFE instrument (MEDUSA mode), at the Very Large Telescope to derive a data cube with a size of 19'×16', where the MEDUSA fibers were located in a quasi-regular grid over N11. In the following we describe part of our results, which are based on the kinematics of the Hα emission line.

2. Summary of the main results

In the left panel of Figure 1 we plot an example of the Hα data cube over an Hα image of N11B, which is a star-forming complex in N11 (Barbá *et al.* 2003). On this image (where the north is up and east is to the left), we can see the Hα emission line, which displays slightly asymmetric profiles, despite the fact that we are observing an actively star-forming system. At the south of N11B we can observe asymmetric Hα profiles, which are produced by high velocity components and/or small expanding structures. We note

154   S. Torres-Flores *et al.*

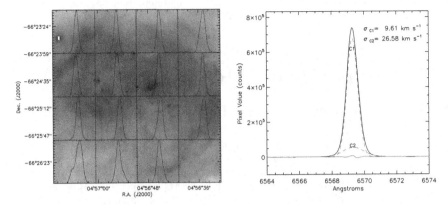

Figure 1. *Left panel:* An example of our spectroscopic Hα data cube, where the profiles are superimposed on a Hα image of N11B. *Right panel:* Integrated Hα profile of N11, derived from the whole Hα data cube (black line). Dashed red lines represent the Gaussian fits, and the solid red line shows the residual between the Gaussian fits and the observations.

that we have found high velocity features close to the central cavity in N11, where soft X-ray emission has been reported.

In the right panel of Figure 1 we show the integrated Hα profile of N11 (solid black line), which has been derived from the entire GIRAFFE/MEDUSA data cube (19'×16'). In order to determine the width of this profile, we have fitted a single Gaussian on it, obtaining a width of $\sigma \sim 12$ km s^{-1} (once corrected by instrumental and thermal widths). However, we detected the presence of remaining wings in the residual emission. For this reason, we have fitted a secondary Gaussian component to remove these wings. This exercise produce a narrow high-intensity component and a broad low-intensity component, which are plotted in Figure 1 with red dashed lines. On the same figure we plot the difference between the observed integrated profile and the Gaussian fits, which is negligible (red solid line). We found that the narrow component has a corrected width of $\sigma \sim 10$ km s^{-1}. This width, which may arises from the contribution of the stellar feedback, supernova events and gravity, is lower than the observed value for the 30 Dor nebula (Torres-Flores *et al.* 2013), suggesting that the evolutionary stage of the nebula may play an important role in the width of the integrated Hα profile.

Acknowledgements

We would like to thank the conference organizers. ST-F acknowledges the financial support of the project CONICYT PAI/ACADEMIA 7912010004.

References

Barbá, R. H., Rubio, M., Roth, M. R., & García, J. 2003, *AJ*, 125,1940
Chu, Y-H. & Kennicutt, R. C., Jr. 1994, *ApJ*, 425, 720
Melnick, J., Tenorio-Tagle, G., & Terlevich, R. 1999, *MNRAS*, 302, 677
Nazé, Y., Antokhin, I. I., Rauw, G., Chu, Y.-H., Gosset, E., & Vreux, J.-M. 2004, *A&A*, 418, 841
Rosado, M., Laval, A., Le Coarer, E., Georgelin, Y. P., Amram, P., *et al.* 1996, *A&A*, 308, 588
Torres-Flores, S., Barbá, R., Maíz Apellániz, J., Rubio, M., Bosch, G., Hénault-Brunet, V., & Evans, C. J. 2013, *A&A*, 555A, 60
Smith, M. G. & Weedman, D. W. 1972, *ApJ*, 172, 307

Galaxies in 3D across the Universe
Proceedings IAU Symposium No. 309, 2014
B. L. Ziegler, F. Combes, H. Dannerbauer, M. Verdugo, eds.

© International Astronomical Union 2015
doi:10.1017/S1743921314009521

3D spectroscopy of Wolf-Rayet HII galaxies

C. Kehrig[1], E. Pérez-Montero[1], J. M. Vílchez[1], J. Brinchmann[2], D. Kunth[3], F. Durret[3], J. Iglesias-Páramo[1] and J. Hernández-Fernández[4]

[1] Instituto de Astrofísica de Andalucía - Apartado de correos 3004, 18080 Granada, Spain
[2] Leiden University - PO Box 9513, 2300 RA Leiden, The Netherlands
[3] Institut d'Astrophysique de Paris - 98 bis boulevard Arago, 75014 Paris, France
[4] Universidade de São Paulo (IAG) - Rua do Matão 1226, 05508-090 SP, Brazil

Abstract. Wolf-Rayet HII galaxies are local metal-poor star-forming galaxies, observed when the most massive stars are evolving from O stars to WR stars, making them template systems to study distant starbursts. We have been performing a program to investigate the interplay between massive stars and gas in WR HII galaxies using IFS. Here, we highlight some results from the first 3D spectroscopic study of Mrk 178, *the closest metal-poor WR HII galaxy*, focusing on the origin of the nebular HeII emission and the aperture effects on the detection of WR features.

Keywords. galaxies: dwarf — ISM: abundances — ISM: HII regions — stars: Wolf-Rayet

1. Introduction

HII galaxies are local, dwarf starburst systems (e.g., Kehrig, Telles, & Cuisinier 2004; Westera *et al.* 2004), which show low metallicity $[1/50 \lesssim Z/Z_\odot \lesssim 1/3]$ (e.g., Kehrig *et al.* 2006; Pérez-Montero *et al.* 2009; Cairós *et al.* 2009, 2010). Wolf-Rayet (WR) signatures (commonly a broad feature at ~ 4680 Å or blue bump), indicating the presence of WR stars, have been found in the spectra of some HII galaxies (e.g., Kunth & Sargent 1981; Kehrig *et al.* 2013). This is an important observational fact since according to recent stellar evolution models for single rotating/non-rotating massive stars, hardly any WRs are expected in metal-poor environments (Leitherer *et al.* 2014). Studying the WR content in HII galaxies is crucial to test stellar evolutionary models at low metallicities. We have initiated a program to investigate HII galaxies with WR features using integral field spectroscopy (IFS; e.g., Kehrig *et al.* 2013; Pérez-Montero *et al.* 2013). So far, we have observed 15 WR galaxies with the optical IFUs: PMAS at the 3.5m telescope at CAHA and INTEGRAL at the 4.2m WHT in ORM. IFS has many benefits in a study of this kind, in comparison with long-slit spectroscopy. Using IFS one can locate and find WRs where they were not detected before, not only because it samples a larger area of the galaxy, but also because IFS can increase the contrast of the WR bump emission against the galaxy continuum, thus minimizing the WR bump dilution. Also, IFS is a powerful technique to probe issues related with aperture effects, and allows a more precise spatial correlation between massive stars and nebular properties (e.g., Kehrig *et al.* 2008, 2013).

2. Results and Conclusions

We summarize here some recent results on Mrk178, one of *the most metal-poor nearby WR galaxies* (see Kehrig *et al.* 2013, for more details):

1) The origin of high-ionization nebular lines (e.g. HeIIλ4686), apparently more frequent in high-z galaxies, is still an open question. One widely favored mechanism for He$^+$-ionization involves hot WRs, but it has been shown that nebular HeIIλ4686 is not

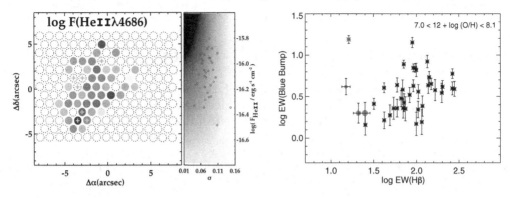

Figure 1. Mrk178 - *Left panel*: intensity map of the nebular HeIIλ4686 line; the spaxels where we detect WR features are marked with green crosses. *Right panel*: EW(WR blue bump) vs EW(Hβ). Asterisks show values obtained from SDSS DR7 for metal-poor WR galaxies; the red one represents Mrk178. The three blue circles, from the smallest to the biggest one, represent the 5, 7 and 10 arcsec-diameter apertures from our IFU data centered at the WR knot of Mrk178, at which the SDSS fiber was centered too (Kehrig *et al.* 2013).

always accompanied by WR signatures, thus WRs do not explain He^+-ionization at all times (Kehrig *et al.* 2008, 2011; Shirazi & Brinchmann 2012). In Mrk178, we find nebular HeIIλ4686 emission spatially extended reaching well beyond the location of the WR stars (Fig. 1, left-panel). The excitation source of He^+ in Mrk178 is still unknown.

2) From the SDSS spectra, we have found a too high EW(WR bump) value for Mrk178, which is the most deviant point among the metal-poor WR galaxies in Fig. 1, right-panel. Using our IFU data, we have demonstrated that this curious behaviour is caused by aperture effects, which actually affect, to some degree, the EW(WR bump) measurements for all galaxies in Fig.1. We have also shown that using too large an aperture, the chance of detecting WR features decreases, and that WR signatures can escape detection depending on the distance of the object and on the aperture size. Therefore, WR galaxy samples/catalogues constructed on single fiber/long-slit spectrum basis may be biased!

Acknowledgements

This work has been partially funded by research project AYA2010-21887-C04-01 from the Spanish PNAYA.

References

Cairós L. M., Caon N., Zurita C., *et al.* 2009, *A&A*, 507, 1291
Cairós L. M., Caon N., Zurita C., Kehrig C., Roth M., & Weilbacher P. 2010, *A&A*, 520, A90
Kehrig C., Telles E., & Cuisinier F. 2004, *AJ*, 128, 1141
Kehrig C., Vílchez J. M., Telles E., Cuisinier F., & Pérez-Montero E. 2006, *A&A*, 457, 477
Kehrig C., Vílchez J. M., Sánchez S. F., *et al.* 2008, *A&A*, 477, 813
Kehrig C., Oey M. S., Crowther P. A., *et al.* 2011, *A&A*, 526, A128
Kehrig C., Pérez-Montero E., Vílchez J. M., *et al.* 2013, *MNRAS*, 432, 2731
Kunth D. & Sargent W. L. W. 1981, *A&A*, 101, L5
Leitherer C., *et al.* 2014, *ApJS*, 212, id.14
Pérez-Montero E., García-Benito R., Díaz A. I., Pérez E., & Kehrig, C. 2009, *A&A*, 497, 53
Pérez-Montero E.,Kehrig C.,Brinchmann J., *et al.* 2013, Advances in Astronomy, 2013, id.837392
Shirazi M. & Brinchmann J. 2012, *MNRAS*, 421, 1043
Westera P., Cuisinier F., Telles E., & Kehrig C. 2004, *A&A*, 423, 133

Galaxies in 3D across the Universe
Proceedings IAU Symposium No. 309, 2014
B. L. Ziegler, F. Combes, H. Dannerbauer, M. Verdugo, eds.

© International Astronomical Union 2015
doi:10.1017/S1743921314009533

Simulations on the survivability of Tidal Dwarf Galaxies

Sylvia Ploeckinger[1,2], Simone Recchi[1,3], Gerhard Hensler[1,4] and Pavel Kroupa[5]

[1]Department for Astrophysics, University of Vienna,
Türkenschanzstr. 17, 1180 Vienna, Austria
[2]email: sylvia.ploeckinger@univie.ac.at
[3]email: simone.recchi@univie.ac.at
[4]email: gerhard.hensler@univie.ac.at
[5]Helmholtz-Institut fur Strahlen- und Kernphysik
Nussallee 1416, D-53115 Bonn, Germany
email: pavel@astro.uni-bonn.de

Abstract. We present detailed numerical simulations of the evolution of Tidal Dwarf Galaxies (TDGs) after they kinematically decouple from the rest of the tidal arm to investigate their survivability. Both the short-term (500 Myr) response of TDGs to the stellar feedback of different underlying stellar populations as well as the long-term evolution that is dominated by a time dependent tidal field is examined. All simulated TDGs survive until the end of the simulation time of up to 3 Gyr, despite their lack of a stabilising dark matter component.

Keywords. galaxies: dwarf - galaxies: evolution - galaxies: formation

1. Introduction

The formation of TDGs, as actively star-forming gaseous over-densities embedded in the tidal arms of interacting galaxies, is studied with observations in various wavelengths and their formation is modelled within large-scale simulations of galaxy interactions (see Duc & Renaud 2013, for a review). Whereas they are easily identifiable when they are still connected to the tidal arm, their typical lifetimes remain under discussion. The contribution of ancient TDGs to the population of low-mass galaxies depends on both their formation rate as well as on their survivability. The estimates of dwarf galaxies (DGs) with a tidal origin cover a wide range from 6% (Kaviraj *et al.* 2012) to 100% (Okazaki & Taniguchi 2000). Long-term chemo-dynamical simulations of TDGs do not only allow for a better estimate of their typical lifetimes but can also constrain the physical and chemical properties of ancient TDGs.

2. Results

Early response to stellar feedback: The initial conditions of the simulated TDGs represent a spherical, pressure supported gas cloud with cold over-densities that serve as the seeds for immediate star formation (SF). Within 20 Myr after the simulations start, the total SF rate of the TDGs reaches $5 \times 10^{-2} M_\odot \mathrm{yr}^{-1}$. We study the response of dark matter (DM)-free DGs to different stellar feedback scenarios. In the low feedback case, the initial mass function (IMF) of each star cluster is truncated at a maximal star mass up to which at least a single star can be formed and that is dependent on the mass of the star cluster ($m_{\max} - M_{\mathrm{ecl}}$ relation and IGIMF theory, for details see Kroupa & Weidner 2003; Weidner *et al.* 2013). For the high feedback case the IMF is assumed to be

158 S. Ploeckinger *et al.*

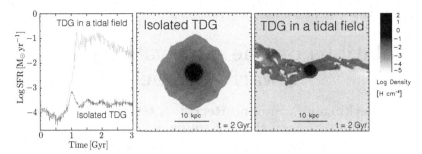

Figure 1. Comparison of the SFRs of initially identical TDGs with and without a tidal field (left panel). A face-on density slice is shown for the isolated TDG (middle panel) as well as the TDG in a tidal field (right spanel) at the peri-center passage of the orbiting TDG (t = 2 Gyr).

completely filled up to a maximal star mass of $120M_\odot$ for every star cluster. In the first 300 Myr, the stellar feedback regulates the SF significantly stronger in the high feedback case, resulting in a 6 times lower stellar mass after 500 Myr than in the low feedback case ($8.35 \times 10^5 M_\odot$ and $5.73 \times 10^6 M_\odot$, respectively). After 300 Myr, both simulated TDGs regulate their SF to a constant rate below $10^{-4} M_\odot \mathrm{yr}^{-1}$. In both cases, the initially high SF does not lead to a disruption of the TDG (Ploeckinger *et al.* 2014a).

Long term evolution: Starting from a smooth, warm, rotating gas cloud in the tidal field of a massive galaxy, we simulate the long-term evolution of TDGs. In order to carve out the effect of the tidal field, a comparison simulation of an identical but isolated TDG is performed. In the first Gyr the SF is comparable. After 1 Gyr, as the TDG in the tidal field is approaching the peri-center of its orbit, the TDG is compressed and subsequently the SF increases by more than 2 orders of magnitude, but both TDGs survive until the end of the simulation time at t = 3 Gyr (see Fig. 1 and Ploeckinger *et al.* 2014b, subm.).

3. Conclusions

We simulate pressure supported and rotating TDGs, with low and high stellar feedback, and exposed them to ram pressure and a tidal field. All simulated TDGs ($Z = 0.1-0.3Z_\odot$) survive until the end of the simulation (up to 3 Gyr). This serves as an additional sign that TDGs can turn into long-lived objects, complementary to indications from observations (Dabringhausen & Kroupa 2013; Duc *et al.* 2014) and simulations of galaxy interactions (e. g. Bournaud & Duc 2006; Hammer *et al.* 2010; Fouquet *et al.* 2012).

References

off

Bournaud F. & Duc P.-A. 2006, *A&A*, 456, 481
Dabringhausen J. & Kroupa P. 2013, *MNRAS*, 429, 1858
Duc P.-A., Paudel S., McDermid R. M., *et al.* 2014, *MNRAS*, 440, 1458
Duc P.-A. & Renaud F. 2013. Vol. 861 of Lecture Notes in Physics, Springer
Fouquet S., Hammer F., Yang Y., Puech M., & Flores H. 2012, *MNRAS*, 427, 1769
Hammer F., Yang Y. B., Wang J. L., Puech M., Flores H., & Fouquet S. 2010, *ApJ*, 725, 542
Kaviraj S., Darg D., Lintott C., Schawinski K., & Silk J. 2012, *MNRAS*, 419, 70
Kroupa P. & Weidner C. 2003, *ApJ*, 598, 1076
Okazaki T. & Taniguchi Y. 2000, *ApJ*, 543, 149
Ploeckinger S., Hensler G., Recchi S., Mitchell N., & Kroupa P. 2014a, *MNRAS*, 437, 3980
Ploeckinger S., Recchi S., Hensler G., & Kroupa P. 2014b, *MNRAS* subm.
Weidner C., Kroupa P., & Pflamm-Altenburg J. 2013, *MNRAS*, 434, 84

Galaxies in 3D across the Universe
Proceedings IAU Symposium No. 309, 2014
B. L. Ziegler, F. Combes, H. Dannerbauer, M. Verdugo, eds.

© International Astronomical Union 2015
doi:10.1017/S1743921314009545

News from the isolated ellipticals NGC 5812, NGC 7507, and NGC 7796

Tom Richtler[1], Ricardo Salinas[2], Richard Lane[1], Michael Hilker[3], Juan Pablo Caso[4] and Lilia P. Bassino[4]

[1]Departamento de Astronomía, Universidad de Concepción, Chile
[2]Department of Physics and Astronomy, Michigan State University, East Lansing, Michigan, USA
[3]European Southern Observatory, Garching, Germany
[4]Universidad Nacional de La Plata and IALP (CONICET-UNLP), La Plata, Argentina

Abstract. We report on ongoing photometric and spectroscopic work on a sample of isolated elliptical galaxies. We investigate their globular cluster systems, and use the kinematics of globular clusters and the integrated galaxy light to constrain their dark halos, which are not found in the cases of NGC 5812 and NGC 7507.

Keywords. galaxies: elliptical and lenticular, cD - galaxies: evolution - galaxies: formation - galaxies: halo

1. Introduction

Isolated galaxies are expected to evolve differently from galaxies in groups or clusters. This should be manifest particularly in the structures of their visible and dark halos, where globular clusters (GCs) can be tracers for both components. We collected photometric data, using VLT, Gemini South, and 4m Blanco telescope, of a dozen isolated ellipticals. Besides characterising their GC systems, we want to identify clusters suitable for the use as dynamical tracers for testing cosmological simulations (Niemi *et al.* 2010).

2. NGC 5812 - alone in the dark?

NGC 5812 (distance 28 Mpc) is a galaxy with a strong intermediate-age stellar component. The accompanying dwarf galaxy at a projected distance of about 20 kpc shows an extended tidal tail. Photometric data come from the 4m-Blanco telescope. Lane *et al.* (2013) describes its morphology, photometric properties and cluster system, which is rather poor. We obtained mask spectroscopy with GMOS/Gemini-South (GS-2013B-Q51; PI: Richtler) and got radial velocities for 25 GCs out to a projected radius of 20 kpc. The brightest object has about $M_R \approx -12$, confirming the existence of clusters of this brightness as statistically predicted in the CMD shown by Lane *et al.* One cluster has a strongly deviating velocity of 1440 km/s (systemic velocity 1970 km/s), while 24 objects show a tight distribution around the systemic velocity of NGC 5812 with a dispersion of only 95 km/s. This low value is puzzling. A sample of 24 GCs should already permit a reliable estimation of the total velocity dispersion of the cluster system. The central velocity dispersion of 200 km/s (HyperLeda) requires with our photometric model a stellar M/L_R-value of about 3.5 under isotropy. While we await a more complete dynamical analysis, we anticipate that it will be difficult to reconcile the low velocity dispersion with the existence of a massive dark halo. NGC 5812 thus seems to be another candidate among isolated ellipticals for having less dark matter than expected.

3. NGC 7507 - little dark matter

NGC 7507 (distance 27 Mpc) belongs to the galaxies with a Keplerian decline of the projected velocity dispersion (Salinas *et al.* 2012). It also has a quite poor GC system (Caso *et al.* 2013) and we could not use GCs as dynamical tracers. The kinematic data within 1 arcmin stem from long-slit observations. By deep mask spectroscopy, obtained with Gemini/GMOS (GS-2009B-Q84; PI: Salinas) we extended the radius with measured velocity dispersions out to 130 arcsec (Lane *et al.* 2014). The adjacent regions agree excellently, but at 70 arcsec, we observe a bump in the velocity dispersion profile, which according to Schauer *et al.* (2014) might be the relic of a former merging event. At larger radii, the velocity dispersion profile is again well described by the stellar mass alone. Radial anisotropy and the inclusion of rotation may help to accommodate some dark matter.

4. NGC 7796 - an isolated cluster elliptical

Deep imaging with VLT/VIMOS in B and R (89.B-457; PI: Salinas) builds the database for our investigation of the GC system of NGC 7796 (distance 50 Mpc), which is a massive old galaxy without striking substructure. However, a companion dwarf galaxy shows tidal tails and multiple nuclei/star cluste bluer than their parent galaxy. In contrast to many other isolated ellipticals, the GC system of NGC 7796 with estimated 2000 members rivals some cluster ellipticals. We derive a specific frequency of $S_N = 2.6 \pm 0.5$. We see the familiar bimodal distribution in B-R which characterises an old cluster system. The density profiles of blue and red GCs agree within the uncertainties. Therefore the growth of the halo probably did not happened through later accretion of dwarf galaxies, as expected for an isolated elliptical. More insight will come from comparing the kinematics of metal-poor and metal-rich GCs. The kinematical literature data for the galaxy centre result in M/L_R=6.5 for the stellar component under isotropy, consistent with an old metal-rich population, but do not permit solid statements regarding the dark matter content. Assuming MOND, M/L_R becomes 6.2, and the X-ray data of O'Sullivan *et al.* (2007) are consistent with the MONDian prediction. In comparison with NGC 7507 and NGC5812, NGC7796 may indicate that besides accretion, the epoch of star formation determines the richness of a GC system.

Acknowledgements

TR acknowledges support from FONDECYT project Nr. 1100620, the BASAL Centro de Astrofísica y Tecnologías Afines (CATA) PFB-06/2007, and a visitorship at ESO/Garching.

References

Caso J. P., Richtler T., Bassino, L. P., *et al.* 2013, *A&A*, 555, 56
Lane R., Salinas R., & Richtler, T. 2013, *A&A*, 549, 148
Lane R., Salinas R., & Richtler, T. 2014, *A&A*, accepted
Niemi S., Nurmi P., Heinämäcki, P., & Saar, E. 2010, *MNRAS*, 344, 1000
O'Sullivan E., Sanderson A. J. R., & Ponman, T. 2007, *MNRAS*, 380, 1409
Richtler T., Salinas R., Lane R., Hilker M., & Schirmer M. 2014, *A&A*, submitted
Salinas R., Richtler T., Bassino, L. P., *et al.* 2012, *A&A*, 538, A87
Schauer A. T. P.h., Remus R. S., Burkert, A., & Johansson P. H. 2014, *ApJ*, 783, 32

Galaxies in 3D across the Universe
Proceedings IAU Symposium No. 309, 2014
B. L. Ziegler, F. Combes, H. Dannerbauer, M. Verdugo, eds.

© International Astronomical Union 2015
doi:10.1017/S1743921314009557

Dwarf ellipticals in the eye of SAURON: dynamical & stellar population analysis in 3D

Agnieszka Ryś[1,2], Jesús Falcón-Barroso[1,2], Glenn van de Ven[3] and Mina Koleva[4]

[1]Instituto de Astrofísica de Canarias, 38200 La Laguna, Tenerife, Spain
email: arys@iac.es
[2]Departamento de Astrofísica, Universidad de La Laguna, 38205 La Laguna, Tenerife, Spain
[3]Max Planck Institute for Astronomy, Königstuhl 17, 69117 Heidelberg, Germany
[4]Sterrenkundig Observatorium, Ghent University, Krijgslaan 281, S9, B-9000 Gent, Belgium

Abstract. We present the dynamical and stellar population analysis of 12 dwarf elliptical galaxies (dEs) observed using the SAURON IFU (WHT, La Palma). We demonstrate that dEs have lower angular momenta than their presumed late-type progenitors and we show that dE circular velocity curves are steeper than the rotation curves of galaxies with equal and up to an order of magnitude higher luminosity. Transformation due to tidal harassment is able to explain all of the above, unless the dE progenitors were already compact and had lower angular momenta at higher redshifts. We then look at the star formation histories (SFHs) of our galaxies and find that for the majority of them star formation activity was either still strong at a few Gyr of age or they experienced a secondary burst of star formation roughly at that time. This latter possibility would be in agreement with the scenario where tidal harassment drives the remaining gas inwards and induces a secondary star formation episode. Finally, one of our galaxies appears to be composed exclusively of an old population ($\gtrsim$12 Gyr). Combining this with our earlier dynamical results, we conclude that it either was ram-pressure stripped early on in its evolution in a group environment and subsequently tidally heated (which lowered its angular momentum and increased compactness), or that it evolved *in situ* in the cluster's central parts, compact enough to avoid tidal disruption.

Keywords. galaxies: elliptical and lenticular, dwarf - galaxies: evolution - galaxies: formation

1. Introduction

dEs are the most common galaxy class in dense environments, outnumbering all other classes combined. Their low masses make them ideal test beds for studying different mechanisms that shape galaxies, since both external influences and internal feedback mechanisms are far more extreme in dwarfs than in massive galaxies. They also make them challenging to observe. dEs are a surprisingly inhomogeneous class, which has made it difficult to relate different dE subtypes to each other, as well as to place the whole class in the larger context of galaxy assembly and (trans)formation processes.

The idea behind the present project (Ryś *et al.* 2013) was to obtain large-scale two-dimensional *maps* of kinematic and stellar population properties for objects for which (with very few and small-scale exceptions) only one-dimensional profiles were available before. With integral-field data it is possible not only to reduce the uncertainties on radial properties, but also to characterize potential substructures far more accurately, and to provide detailed information on the dynamical structure of observed systems.

2. Results

We find that the specific angular momenta of dEs are significantly lower than those of the late-type galaxies of the CALIFA survey (Sánchez *et al.* 2012). This leads to the conclusion that when designing possible transformation paths we need to include processes that lower stellar angular momentum. We compare our dEs circular velocity profiles to those of massive (Catinella *et al.* 2006) and dwarf (Swaters *et al.* 2009) late-type galaxies and find that dE curves are steeper and their absolute values higher than those of late-type galaxies of comparable luminosity. Therefore, we need to account for this increase in concentration in our search for late-to-early type transformation mechanisms. Furthermore, we investigate whether any correlation exists between the dark matter fraction and the 3D (i.e. intrinsic) clustrocentric distance and we find that galaxies in the cluster outskirts tend to have higher dark-to-stellar matter ratios. All this implies that processes like ram-pressure stripping (Gunn & Gott 1972) alone are not able to explain the observed characteristics of the dE class. Tidal harassment (Moore *et al.* 1998) is currently the only available scenario with which we can explain the above findings, unless the dE progenitors were already compact and had lower angular momenta at high redshift (Ryś *et al.* 2014).

We look at resolved star formation histories of our galaxies and find that the majority of them either still exhibited strong star formation (SF) at a few Gyr ago or they experienced a secondary burst of SF that occured roughly at that time. We interpret this as a tentative piece of evidence in favor of the harassment scenario, where such episodes of SF are expected as the remaining gas is driven inwards as a result of tidal forces. We also find that the old populations of dEs are roughly coeval with those of giant ETGs, in principle making it possible for dEs to be the builing blocks of the outskirts of massive galaxies.

3. Conclusions and future work

We have presented the analysis of stellar kinematic and absorption line-strength maps for a sample of 12 Virgo Cluster and field dwarf elliptical galaxies. This is to date the largest sample of dEs for which integral-field data is available, and offers the largest spatial coverage so far. Nevertheless, we hope that once large samples for environments of varying density have been analyzed, the findings presented here will be strengthened/confirmed. It will also be of vital importance to be able to test our observational results against a larger suite of simulations of enough spatial resolution, going down to masses we require – and testing a range of initial conditions (orbits, masses, angular momenta, galaxy types and infall epochs) as well as interaction types. In this way observations could be interpreted against a more robust theoretical backbone.

References

Catinella B., Giovanelli R., & Haynes M. P. 2006, *ApJ*, 640, 751
Gunn J. E. & Gott III J. R. 1972, *ApJ*, 176, 1
Moore B., Lake G., & Katz N. 1998, *ApJ*, 495, 139
Ryś A., Falcón-Barroso J., & van de Ven G. 2013, *MNRAS*, 428, 2980
Ryś A., van de Ven G., & Falcón-Barroso J. 2014, *MNRAS*, 439, 284
Sánchez S. F., *et al.* 2012, *A&A*, 538, A8
Swaters R. A., Sancisi R., van Albada T. S., & van der Hulst J. M. 2009, *A&A*, 493, 871

Galaxies in 3D across the Universe
Proceedings IAU Symposium No. 309, 2014
B. L. Ziegler, F. Combes, H. Dannerbauer, M. Verdugo, eds.

© International Astronomical Union 2015
doi:10.1017/S1743921314009569

Galactic bulges: the importance of early formation scenarios vs. secular evolution

Marja K. Seidel[1,2], R. Cacho[3], T. Ruiz-Lara[4,5], J. Falcón-Barroso[1,2],
I. Pérez[4,5], P. Sánchez-Blázquez[6], F. P. A. Vogt[7], M. Ness[8],
K. Freeman[7] and S. Aniyan[7,9]

[1]Instituto de Astrofísica de Canarias, E-38200 La Laguna, Tenerife, Spain
[2]Departamento de Astrofísica, Universidad de La Laguna, E-38205 La Laguna, Tenerife, Spain
email: `mseidel@iac.es`
[3]Dpto. de Astrofísica y CC. de la Atmósfera, Universidad Complutense de Madrid, Spain
[4]Dpto. de Física Teórica y del Cosmos, Universidad de Granada, Facultad de Ciencias
(Edificio Mecenas), 18071 Granada, Spain
[5]Instituto Universitario Carlos I de Física Teórica y Computacional, Universidad de Granada,
18071 Granada, Spain
[6]Dpto. de Física Teórica, Universidad Autónoma de Madrid, E-28049 Cantoblanco, Spain
[7]Research School of Astronomy and Astrophysics, Australian National University, Canberra,
ACT 2611, Australia
[8]Max-Planck-Institut für Astronomie, Königstuhl 17, D-69117 Heidelberg, Germany
[9]European Southern Observatory, Karl-Schwarzschild-Str. 2, 85748 Garching, Germany

Abstract. We study the stellar content of three galactic bulges with the high resolution gratings
(R=7000) of the WiFeS integral field unit in order to better understand their formation and
evolution. In all cases we find that at least 50% of the stellar mass already existed 12 Gyrs ago,
more than currently predicted by simulations. A younger component (age between ∼1 to ∼8
Gyrs) is also prominent and its present day distribution seems to be much more affected by
morphological structures, especially bars, than the older one. This in-depth analysis supports
the notion of increasing complexity in bulges which cannot be achieved by mergers alone, but
requires a non-negligible contribution from secular evolution.

Keywords. galaxies: individual (NGC 5701, NGC 6753, NGC 7552), galaxies: bulges, galaxies:
evolution, galaxies: kinematics and dynamics, galaxies: stellar content, techniques: spectroscopic

1. Introduction

Galactic bulges, considered as a deviation from the exponential profile in the centres
of galaxies, hold crucial clues to refine our understanding of galaxy formation and evolution. Yet, the details of the interplay of collapse, mergers and secular processes are
still not fully understood. The review by Kormendy & Kennicutt (2004) (and references
therein) summarizes the two main bulge formation mechanisms: i) merger-driven and
ii) secularly evolved. The advances in instrumentation, spectral libraries (e.g., MILES,
Sánchez-Blázquez *et al.* 2006), and stellar population (SP) studies via full-spectral fitting
techniques (e.g. STECKMAP: Ocvirk *et al.* 2006) now allow us to characterize the stellar
content of galaxies on a spatially resolved basis, and test the different bulge formation
mechanisms. To that end, we conducted a pilot study using the WiFeS integral field unit
(and its high resolution mode; R=7000) to link the stellar kinematics and SP parameters
(and their radial distribution). The three bulges analyzed are NGC 5701, NGC 6753 and
NGC 7552 and their unsharp masks are shown in Fig. 1 along with their kinematics. The
high quality of our data allows us to characterize the 2D distribution of different stellar
populations (i.e. young, intermediate, and old, see Fig. 1, middle and bottom row).

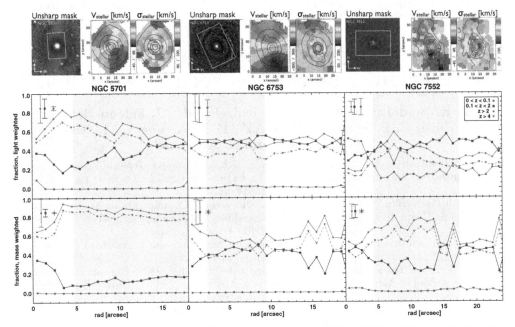

Figure 1. *Top row:* Unsharp masks (Spitzer image for NGC 5701, HST WFPC2 F814W filter for NGC 6753 and NGC 7552) with the WiFeS FoV (25×38 arcsec) in green, stellar velocity and stellar velocity dispersion map. *Below:* Fractions of young (blue dots), intermediate (green squares) and old SP (red rhombus) as a function of radius. Top left corner: corresponding uncertainties. *Middle row:* L-weighted and *Bottom row:* M-weighted results. Shaded regions show the central (< 0.3 kpc) and inner parts (0.5 kpc < r < 2 kpc).

2. Conclusions: Old bulge structures rejuvenated by bars

Our main conclusions are threefold (as illustrated in Fig. 1):

• The stellar velocity dispersion is higher in regions where the light of the old population is dominant

• The fraction of the mass of the old stellar population (existing already at z∼2) is 50-80% in the inner 2 kpc of all bulges, hinting towards high star formation (SF) during the collapse in the early cosmic web

• The barred galaxies (NGC 5701, NGC 7552) show a significant amount of light of the young SP and mass of the intermediate SP in the centres, likely indicating the influence of the bars driving necessary material towards the centre to fuel this second SF period.

These conclusions are based on a sample that cannot be regarded as representative. But as all our bulges show consistent features albeit their different nature (mass, morphology), it hints towards a common evolution. More details in Seidel *et al.* (2014, submitted).

Acknowledgements

This research is supported by the Spanish Ministry of Economy and Competitiveness (AYA2010-21322-C03-02 & AYA2009-11137) and of Science and Innovation (AYA2011-24728 & Consolider-Ingenio CSD2010-00064) and by the Junta de Andalucía (FQM-108).

References

Kormendy J. & Kennicutt, Jr., R. C. 2004, *Annual Review of Astronomy & Astrophysics*, 42, 603-683

Sánchez-Blázquez, P., Peletier, R. F., Jiménez-Vicente, J., Cardiel, N., Cenarro, A. J., Falcón-Barroso, J., Gorgas, J., Selam, S., & Vazdekis, A. 2006, *MNRAS*, 371, 703-718

Ocvirk, P. , Pichon, C., Lançon, A., & Thiébaut, E. 2006, 365, 74-84

Galaxies in 3D across the Universe
Proceedings IAU Symposium No. 309, 2014   © International Astronomical Union 2015
B. L. Ziegler, F. Combes, H. Dannerbauer, M. Verdugo, eds.   doi:10.1017/S1743921314009570

Radial gradients in the SLUGGS survey

Caroline Foster and the SLUGGS collaboration

email: cfoster@aao.gov.au

Abstract. The SAGES Legacy Unifying Galaxies and GlobularS (SLUGGS) survey probes the outskirts of early-type galaxies (ETGs). SLUGGS uses the DEIMOS spectrograph on Keck to simultaneously obtain globular cluster and spatially resolved galaxy spectra. The galaxy spectra allow for the reconstruction of kinematic maps out to ~ 3 effective radii (R_e), while globular clusters (GCs) push the galactocentric limit out to ~ 8-$10 R_e$. Stellar population parameters (abundances) can also be recovered. This provides a unique dataset for exploring the outer gradients (abundance and/or kinematic transition). We briefly present recent and salient results of the SLUGGS survey so far.

Keywords. galaxies: elliptical and lenticular - galaxies: kinematic - galaxies: abundances - galaxies: formation

1. Introduction

The two-phase galaxy formation scenario has recently been proposed in order to explain the fast growth in size of massive ETGs since redshift $z \sim 2$ (e.g. Daddi *et al.* 2005). Under this scenario, massive galaxies form in two phases: 1) early central bulge formation via multiple major mergers or dissipative processes at high redshift (e.g. Ceverino *et al.* 2014) and 2) subsequent series of multiple minor merger events increasing the size by preferably adding stars to the outskirts (e.g. Naab *et al.* 2009; van Dokkum *et al.* 2010). Hence, these in-situ bulge and the ex-situ halo components should be readily detectable observationally. Such transitions are also predicted in other formation scenarios (e.g. Hoffman *et al.* 2009).

The SAGES Legacy Unifying Galaxies and GlobularS (SLUGGS) survey (Brodie *et al.* 2014) uses a combination of the Stellar Kinematics using Multiple Slits (SKiMS) technique (Proctor *et al.* 2009) and globular cluster data to push the galactocentric boundary beyond what is typically allowed by integral field spectrographs (IFSs) in order to probe the faint outskirts of local massive ETGs. The SLUGGS sample contains 25 ETGs (ellipticals and lenticulars) from a representative range of environments. Data for 3 "bonus" galaxies are also being gathered.

In this short contribution, we briefly present our data (Section 2) and highlight some recent results (Section 3).

2. Data

The SLUGGS data and products are described in detail in Brodie *et al.* (2014) and more information can be found on our website (http://sluggs.swin.edu.au/). Here we give a brief overview.

The SLUGGS survey comprises an imaging and a follow-up spectroscopic component. The imaging data combine pre-existing archival data from various sources with wide-field multi-band imaging using Subaru/Suprime-cam. Spectra are obtained using the DEIMOS spectrograph on Keck in multi-slit mode. The instrument setup allows coverage of the near-infrared Calcium II Triplet around 8500 Å. The obtained spectral resolution is

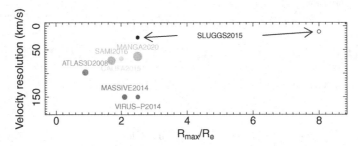

Figure 1. Comparison of the spatial extent and velocity resolution of SLUGGS with other past and current IFS surveys as labelled. The size of the symbol is proportional to the logarithm of the sample size. The GC and galaxy light components of SLUGGS are shown as hollow and filled black circles, respectively. The SLUGGS survey probes an area in this parameter space that was previously unexplored.

FWHM ~ 1.5 Å, affording a velocity resolution of ~ 24 and ~ 12 km s^{-1} for the galaxy and GC components of the survey, respectively. As a result, the SLUGGS survey probes a yet unexplored area of parameter space (see Fig. 1).

3. Radial gradients

Kinematic transitions have already been detected as part of SLUGGS (e.g. Foster *et al.* 2013; Arnold *et al.* 2014), thereby supporting the two-phase galaxy formation paradigm. In Pastorello *et al.* (2014), we find a remarkable trend in the outer stellar metallicity profiles such that massive galaxies have shallower outer profiles, indicating they have undergone a more active accretion history than low mass galaxies (e.g. White 1980). In parallel, Usher *et al.* (2012) and Pota *et al.* (2013) present the globular cluster abundances and kinematics, respectively, reaching out to several effective radii. The abundances and kinematics of SLUGGS galaxies are combined with that of their respective globular cluster systems in Pastorello *et al.* (in prep.) and Alabi *et al.* (in prep.), respectively.

References

Arnold J. A., *et al.* 2014, *ApJ*, 791, 80
Brodie J. P., *et al.* 2014, *ApJ* in press, arXiv:1405.2079
Ceverino D., Dekel A., Tweed D., & Primack J., 2014, ArXiv e-prints
Daddi E., *et al.* 2005, *ApJ*, 626, 680
Foster C., Arnold J. A., Forbes D. A., Pastorello N., Romanowsky A. J., Spitler L. R., Strader
    J., & Brodie J. P., 2013, *MNRAS*, 435, 3587
Hoffman L., Cox T. J., Dutta S., & Hernquist L., 2009, *ApJ*, 705, 920
Naab T., Johansson P. H., & Ostriker J. P., 2009, *ApJ*, 699, L178
Pastorello N., Forbes D. A., Foster C., Brodie J. P., Usher C., Romanowsky A. J., Strader J.,
    & Arnold J. A., 2014, *MNRAS*, 442, 1003
Pota V., *et al.* 2013, *MNRAS*, 428, 389
Proctor R. N., Forbes D. A., Romanowsky A. J., Brodie J. P., Strader J., Spolaor M., Mendel
    J. T., & Spitler L., 2009, *MNRAS*, 398, 91
Usher C., *et al.* 2012, *MNRAS*, 426, 1475
van Dokkum P. G., *et al.* 2010, *ApJ*, 709, 1018
White S. D. M., 1980, *MNRAS*, 191, 1P

Galaxies in 3D across the Universe
Proceedings IAU Symposium No. 309, 2014   © International Astronomical Union 2015
B. L. Ziegler, F. Combes, H. Dannerbauer, M. Verdugo, eds.   doi:10.1017/S1743921314009582

Continuous Mid-Infrared Star Formation Rate Indicators

A. J. Battisti[1], D. Calzetti[1], B. D. Johnson[2] and D. Elbaz[3]

[1]Department of Astronomy, University of Massachusetts, Amherst, MA 01003, USA
email: `abattist@astro.umass.edu`
[2]Institut d'Astrophysique de Paris, UMR 7095, 75014, Paris, France
[3]Laboratoire AIM-Paris-Saclay, CEA/DSM/Irfu, CNRS, Université Paris Diderot,
Saclay, pt courrier 131, 91191 Gif-sur-Yvette, France

Abstract. We present continuous, monochromatic star formation rate (SFR) indicators over the mid-infrared wavelength range of $6 - 70$ μm. We use a sample of 58 star forming galaxies (SFGs) in the *Spitzer*-SDSS-*GALEX* Spectroscopic Survey (SSGSS) at $z < 0.2$, for which there is a rich suite of multi-wavelength photometry and spectroscopy from the ultraviolet through to the infrared. The data from the *Spitzer* infrared spectrograph (IRS) of these galaxies, which spans $5 - 40$ μm, is anchored to their photometric counterparts. The spectral region between $40 - 70$ μm is interpolated using dust model fits to the IRS spectrum anchored by *Spitzer* 70 and 160 μm photometry. Since there are no sharp spectral features in this region, we expect these interpolations to be robust. This spectral range is calibrated as a SFR diagnostic using several reference SFR indicators to mitigate potential bias. Our band-specific continuous SFR indicators are found to be consistent with monochromatic calibrations in the local universe, as derived from *Spitzer*, WISE, and Herschel photometry. Additionally, in the era of the James Webb Space Telescope this will become a flexible tool, applicable to any SFG up to $z \sim 3$.

Keywords. galaxies: star formation - infrared: galaxies - stars: formation

1. Introduction

Star formation is a fundamental parameter of galaxies that gives key constraints on how galaxies are able to form and evolve (Somerville *et al.* 2012). For this reason, great efforts have been made to calibrate a wide range of the electromagnetic spectrum that can be linked to processes involved with recent star formation (see review by Kennicutt & Evans 2012). In particular, infrared (IR) wavelength calibrations are proving to be critical to understanding galaxies in the early Universe. However, difficulties arise in utilizing local calibrations because the regions of rest-frame wavelengths probed by a given band will vary with redshift. Therefore, in order to fully utilize current and future deep IR imaging surveys, continuous single-band SFR indicators will be imperative.

2. Results

After anchoring the spectroscopy and dust models of Draine & Li (2007) to the global IR photometric values, we normalize each spectrum by its reference SFR. This reference value is taken to be the average of the calibrations of Kennicutt *et al.* (2009), Rieke *et al.* (2009), Calzetti *et al.* (2010), and Hao *et al.* (2011), which incorporate the full suite of data available for this sample. The SFR-normalized IRS spectra are averaged together to create a composite spectrum for the group (see Figure 1). The SFR-normalized IRS spectra are also convolved with the filter response of specific bands as functions of redshift (see Figure 1). Two of the wavelength-continuous calibrations are shown in Figure 2,

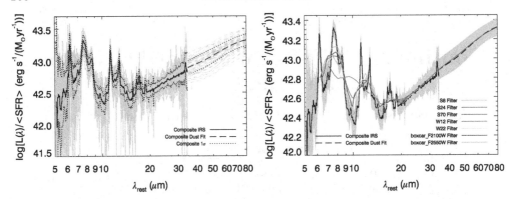

Figure 1. Left: Normalized IRS luminosity, $L(\lambda)/\langle SFR \rangle$, for our SFG galaxies (gray solid lines). The composite spectrum is the thick black line and the standard deviation from of this average is the black dotted line. The dust continuum fit for each galaxy (gray dashed lines) along with the average (black dashed line) and 1σ standard deviation is also shown. Right: The solid colored lines show the composite after each spectrum is convolved with the *Spitzer*, WISE, and JWST-MIRI filters as functions of redshift.

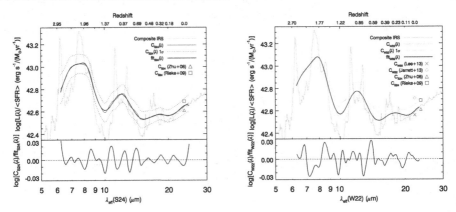

Figure 2. Example of the conversion factor for a *Spitzer* and WISE band using our SFG galaxies (solid colored lines), along with their uncertainty (dotted colored lines). The solid black line is the fit to this conversion, $fit_x(\lambda)$. The local conversion factors from the literature are also shown for comparison (colored symbols).

and they demonstrate that our calibrations are typically accurate to within 30%. It is important to emphasize that the accuracy of such an application is dependent on the shape of the SED of SFGs as a function of redshift. The extent to which this condition holds is being examined in detail.

References

Calzetti, D., Wu, S.-Y., Hong, S., *et al.* 2010, *ApJ*, 714, 1256

Draine, B. T. & Li, A. 2007, *ApJ*, 657, 810

Hao, C.-N., Kennicutt, R. C., Johnson, B. D., Calzetti, D., *et al.* 2011, *ApJ*, 741, 124

Kennicutt, R. C., Jr. & Evans, N. J., II 2012, ARA&A, 50, 531

Kennicutt, R. C., Hao, C.-N., Calzetti, D., *et al.* 2009, *ApJ*, 703, 1672

Rieke, G. H., Alonso-Herrero, A., Weiner, B. J., *et al.* 2009, *ApJ*, 692, 556

Somerville, R. S., Gilmore, R. C., Primack, J. R., & Domínguez, A. 2012, *MNRAS* 423, 1992

Galaxies in 3D across the Universe
Proceedings IAU Symposium No. 309, 2014
B. L. Ziegler, F. Combes, H. Dannerbauer, M. Verdugo, eds.

© International Astronomical Union 2015
doi:10.1017/S1743921314009594

The beauty of resolution: The SN Ib factory NGC 2770 spatially resolved

C. C. Thöne[1], L. Christensen[2], J. Gorosabel[1,3,4] and A. de Ugarte Postigo[1,2]

[1]Instituto de Astrofísica de Andalucía, Glorieta de la Astronomía s/n, 18008 Granada, Spain. email: cthoene@iaa.es
[2]Dark Cosmology Centre, Juliane Maries Vej 30, DK-2100 Copenhagen, Denmark
[3]Unidad Asociada Grupo Ciencias Planetarias UPV/EHU-IAA/CSIC, Departamento de Física Aplicada I, E.T.S., Ingeniería, Universidad del País Vasco UPV/EHU, E-48013 Bilbao, Spain
[4]Ikerbasque, Basque Foundation for Science, E-48008 Bilbao, Spain

Abstract. The late-type spiral NGC 2770 hosted 3 Type Ib supernovae (SNe) in or next to star-forming regions in its outer spiral arms. We study the properties of the SN sites and the galaxy at different spatial resolutions to infer propeties of the SN progenitors and the SF history of the galaxy. Several 3D techniques are used and, for the first time, we present images of metallicity, shocks and stellar population ages from OSIRIS/GTC imaging with tunable narrowband filters.

Keywords. supernovae: individual: SN 2008D, SN 2007uy, SN 1999eh; galaxies: abundances; techniques: high angular resolution

1. Introduction and Observations

NGC 2770 is a late-type spiral (.SAS5*) at DL 30 Mpc with log M*=10.3 $M_\odot$ and SFR=1.1 $M_\odot$/yr, largely neglected until the serendipitous discovery of an X-ray outburst (XRO 080109) marking the shock-breakout of SN 2008D (Soderberg *et al.* 2008). The galaxy had hosted 2 other Type Ib SNe in the past, SN 2007uy and SN 1999eh. Our goal is to infer properties of the SN progenitors by studying their environments and investigate the SF history of the galaxy.

A first study used longslit spectra at 4 slit positions including the SN sites and the center of the galaxy (Thöne *et al.* 2009). We refine this study using the VIMOS IFU with 0.66" sampling ($\sim$ 100pc) at 4 pointings covering Hβ to [SII] and tunable narrowband filters at OSIRIS/GTC with 0.25" sampling ($\sim$30pc) imaging [NII], Hα in steps of 12Å allowing to disentangle the two lines and [SII] in steps of 20Å.

2. Resolved gas properties

TF imaging in Hα reveals that NGC 2770 has to be reclassified as a barred spiral galaxy and the bar is also obvious in IR imaging from UKIDSS or 2MASS. Otherwise, the galaxy shows a regular velocity field with some small disturbances or a warp in the S-E part of the disk. Some turbulence occurs inside larger SF regions and the ones building the bar.

NGC 2770 shows a smooth negative metallicity gradient. The high resolution TF data reveal gradients within the SF regions whose outskirts are more metal poor. Some projection effects might play a role if the SF regions are not transparent. Shocked material traced by [SII]/Hα is visible at the edge of SF regions, especially evident in the TF imaging due to the higher resolution. The bar SF regions have rather low [SII]/Hα values.

Comparing Hα and an offband filter we can infer the Hα EW from the TF data, which is related to the age of the SF regions. As expected for inside-out SF there is an age

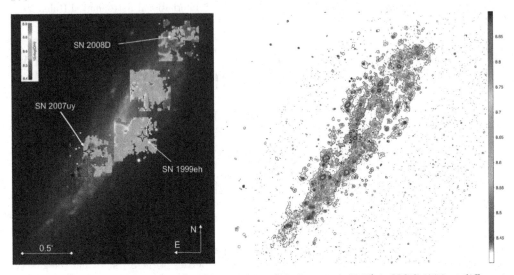

Figure 1. Gas metallicity using the N2 parameter (Marino *et al.* 2013) in NGC 2770 at different resolutions: Left: VIMOS/IFU, Right: Narrowband TF imaging. The TFs show a large amount of details even within single SF regions averaged out in the lower resolution VIMOS data.

gradient with younger ages in the outskirts. Infact, the youngest regions are found in the third and outermost spiral arm that only stretches across the northern half of the galaxy.

3. Conclusions on the SN progenitors

SNe Ib are suggested to come from either single WR stars of M > 35 $M_\odot$, corresponding to a lifetime of ~6 Myr, or binary systems where the companion strips the star of its H envelope. Possible differences in metallicity to other SN types are still debated (see e.g. Leloudas *et al.* 2011; Modjaz *et al.* 2011). A binary origin is suggested by a recent IFU study (Kuncarayakti *et al.* 2013), infact, a possible binary companion to a Ib SN has recently been detected in HST imaging (Folatelli *et al.* 2014; Bersten *et al.* 2014).

The SN sites in NGC 2770 have Hα EWs of 30–100Å implying single star progenitors of < 25 $M_\odot$. SN 2008D lies on a bridge connecting two SF regions while SN 2007uy and 1999eh are associated with SF regions, although not the brightest ones. SN 2007uy and 1999eh have metallicities of 12+log(O/H)~8.5, close to the expected value at their distance from the center, 2008D (12+log(O/H)~8.6) lies in the upper end of the metallicity at its distance. We conclude that a binary system seems to be a likely progenitor for SNe 2008D and 2007uy while SN 1999eh could be consistent with a single star origin.

References

Soderberg A. M., *et al.* 2008, *Nature* 453, 469
Thöne, C. C., *et al.* 2009, *ApJ*, 698, 1307
Marino, R. A., *et al.* 2013, *A&A*, 559, A114
Leloudas, G., *et al.* 2011, *A&A,*, 530, A95
Modjaz, M., *et al.* 2011, *ApJ*, 731, L4
Kuncarayakti *et al.* 2013, *AJ*, 146, id. 30
Folatelli, G., *et al.* 2014, *ATEL*, 6375
Bersten, M., *et al.* 2014, astro-ph/1403.7288

Galaxies in 3D across the Universe
Proceedings IAU Symposium No. 309, 2014
B. L. Ziegler, F. Combes, H. Dannerbauer, M. Verdugo, eds.

© International Astronomical Union 2015
doi:10.1017/S1743921314009600

OMEGA: OSIRIS Mapping of Emission-Line Galaxies in A901/2

Bruno Rodríguez del Pino, Ana L. Chies-Santos, Alfonso Aragón-Salamanca, Steven P. Bamford and Meghan E. Gray

School of Physics and Astronomy, The University of Nottingham,
University Park, Nottingham, NG7 2RD, UK
email: `ppxbr@nottingham.ac.uk`

Abstract. This work presents the first results from an ESO Large Programme carried out using the OSIRIS instrument on the 10m GTC telescope (La Palma). We have observed a large sample of galaxies in the region of the Abell 901/902 system (z~0.165) which has been extensively studied as part of the STAGES project. We have obtained spectrally and spatially resolved H-alpha and [NII] emission maps for a very large sample of galaxies covering a broad range of environments. The new data are combined with extensive multi-wavelength observations which include HST, COMBO-17, Spitzer, Galex and XMM imaging to study star formation and AGN activity as a function of environment and galaxy properties such as luminosity, mass and morphology. The ultimate goal is to understand, in detail, the effect of the environment on star formation and AGN activity.

Keywords. galaxies: clusters - galaxies: AGN - galaxies: star formation

1. Introduction

The properties of the galaxies change as a function of environment. Galaxies in low-density regions tend to be blue, star-forming objects with disk morphology, whereas those living in denser regions tend to be red, passive objects with more spheroidal morphologies (Dressler 1980; Bamford *et al.* 2009). The environment also seems to affect the probability of a galaxy hosting an Active Galactic Nucleus (AGN), at least in the high-luminosity end (Kauffmann *et al.* 2004; Popesso & Baviano 2006). By looking at a multi-cluster system at z ~ 0.165, our main goal is to understand better the role played by the environment in the transformation of galaxies from actively star-forming to passive, with the subsequent morphological change, and the relation between AGN activity and environment.

2. OMEGA survey

The OSIRIS Mapping of Emission-Line Galaxies in A901/2 (OMEGA) survey has been designed to study AGN and star formation activity as a function of environment in the A901/2 multi-cluster system at z~0.165. This system, which spans~0.5 × 0.5 deg^2 (~5x5 Mpc^2) contains a wide variety of environments and has been extensively studied by the STAGES collaboration (Gray *et al.* 2009). Besides that, there is already data from this system in different wavelengths, including *XMM-Newton*, *GALEX*, *Spitzer*, 2dF, GMRT, Magellan, and the 17-band COMBO-17 photometric redshift survey.

Despite the wealth of data aleready available, it is crucial to have optical diagnostics such as emission line spectra to really understand the roles of obscured and unobscured star formation, and the fraction of low-luminosity AGN with no X-Ray emission. To obtain these missing optical diagnostics we have used the OSIRIS instrument in Narrow-band Tunable Filter mode at the GTC to map the Hα and [N II] emission lines in the

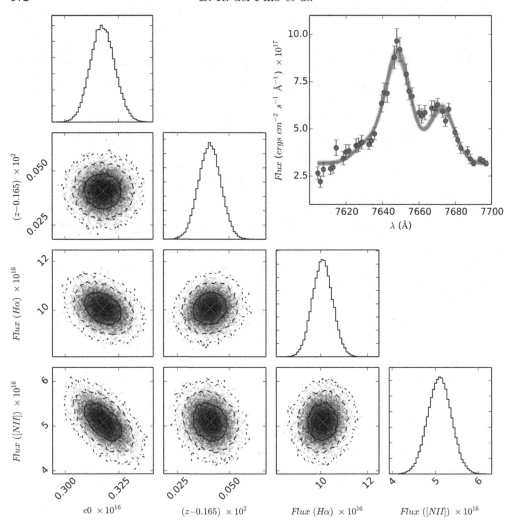

Figure 1. Parameter-parameter likelihood distributions with one- (solid), two- (dashed) and three- (dash-dotted) sigma contours obtained for the four-parameter model of the spectrum of galaxy 42713. On the top right we show the spectrum together with the model using the median values of the parameters' probability distribution, which are also plotted as a red cross on the two-dimensional distributions.

A901/2 multi-cluster system. These two lines contain a lot of information, since the Hα line is a very good star formation indicator (Kennicutt 1998), and in combination with the [N II] line can provide diagnostics about the AGN activity in the galaxies, using the WHAN diagram introduced by Cid Fernandes *et al.* 2010.

To map these two emission lines in the whole system, we divided the ∼0.5 × 0.5 deg² field into 20 circular fields of 8 arcmin diameter each (OSIRIS field of view). For each field a series of 'monochromatic' images was taken with a fixed wavelength step between them. For an optimal deblending of the two lines this step was set to 7Å. The wavelength across the field of view also changes as a function of distance to the optical center, following expression (29) in Gonzalez *et al.* 2014. In this way, Hα and [N II] are mapped by taking sufficient steps in wavelength based on the mean redshift of each field. In this work we present the results from the analysis of the two densest fields.

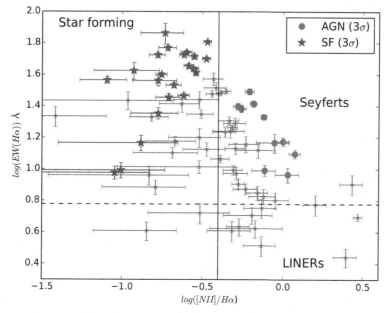

Figure 2. A diagnostic plot of [N II]/Hα vs. EW (Hα), the so-called WHAN diagram (Cid Fernandes *et al.* 2010). We only show detections, using R_{PSF} aperture measurements, where both Hα and [N II] fall in the OMEGA probed wavelength range: a total of 57 ELGs. From these we find 9 3σ-detected AGN hosts and 17 SF galaxies.

3. Analysis and first results

To build up the low-resolution spectra out of these images we performed aperture photometry using two different apertures: one matching the seeing of the images and another one encompassing the full extent of the galaxies. An example of one of these spectra is shown in Figure 1. As our main goal is to measure the fluxes in the two lines we need to model our data. Our model is a composite of three gaussians (because there is another weaker [N II] line at λ) and the continuum.

To exploit all the information contained in our data we use an algorigthm based on Markov Chain Monte Carlo (MCMC) techniques, which not only provides the maximum likelihood of the model parameters but also generates their full probability densitiy distributions. In our case we employ the software emcee (Foreman-Mackey *et al.* 2013). An example of the performance of the algorithm is shown in figure 1. From the output of the MCMC runs we selected a sample based on the probability of a detection in Hα, and a redshift consistent with one or the two lines being within the wavelength range.

From the analysis of the two densest fields we use the WHAN diagnostic diagram (Cid Fernandes *et al.* 2010) in Figure 2 to evaluate the probability of the galaxies hosting an AGN, star formation or both. By looking at the spatial distribution of these sources we find that both AGN and star-forming galaxies tend to avoid the densest regions. The Hα Luminosity Function, shown in Figure 3, when compared with other studies, shows fewer high star-forming objects than the field, but more than in dense clusters. Using the SFR vs Stellar Mass relation, there is also evidence of a widespread suppression of the star formation in the cluster galaxies compared with the field ones. Besides that, our SFR estimates based on the Hα flux agree well with other SFR indicators.

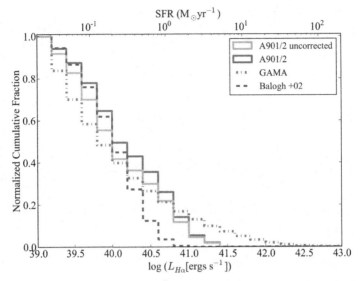

Figure 3. The cumulative Hα LF for the highest density regions of A901/2 (both corrected and uncorrected for contamination and completeness). For comparison we show the Hα LF from the cluster A1689 at $z = 0.18$ (Balogh *et al.* 2002) and the field from the GAMA survey at $z = 0.1$–0.2 (Gunawardhana *et al.* 2013), according to the legend. The studied regions of OMEGA have SFRs that fall in between those of the cluster A1689 and the field.

4. Conclusions and future work

The analysis of only one tenth (2 fields out of 20) of the data of the OMEGA survey has shown that Tunable Filter observations are a very powerful tool to obtain emission line fluxes in a large number of galaxies in single redshift windows. The results found in this work make the analysis of the whole data set very promising as we will be able to show the environmental effects on AGN and star formation activity in ~ 1000 objects in the multi-cluster system at z=0.165.

A complete version of the data reduction, analysis and first results of this survey can be found in Chies-Santos *et al.* in prep.

References

Balogh, M. L., Couch, W. J., Smail, I., Bower, R. G., & Glazebrook, K. 2002, *MNRAS*, 335, 10
Bamford, S. P., Nichol, R. C., Baldry, I. K., *et al.* 2009, *MNRAS*, 393, 1324
Cid Fernandes, R., Stasińska, G., Schlickmann, M. S., *et al.* 2010, *MNRAS*, 403, 1036
Dressler, A. 1980, *ApJ*, 236, 351
Foreman-Mackey, D., Hogg, D. W., Lang, D., & Goodman, J. 2013, *PASP*, 125, 306
Gonzalez, J. J., Cepa, J., Gonzalez-Serrano, J. I., & Sanchez-Portal, M. 2014, arXiv:1407.0159
Gunawardhana, M. L. P., Hopkins, A. M., Bland-Hawthorn, J., *et al.* 2013, *MNRAS*, 433, 2764
Gray, M. E., Wolf, C., Barden, M., *et al.* 2009, *MNRAS*, 393, 1275
Kauffmann, G., White, S. D. M., Heckman, T. M., *et al.* 2004, *MNRAS*, 353, 713
Kennicutt, R. C., Jr. 1998, *ARA&A*, 36, 189
Popesso, P. & Biviano, A. 2006, *A&A*, 460, 23

Galaxies in 3D across the Universe
Proceedings IAU Symposium No. 309, 2014
B. L. Ziegler, F. Combes, H. Dannerbauer, M. Verdugo, eds.

© International Astronomical Union 2015
doi:10.1017/S1743921314009612

3D Studies of Galaxies in Compact Groups

Claudia Mendes de Oliveira[1], Sergio Torres-Flores[2], Philippe Amram[3], Henri Plana[4] and Benoit Epinat[3]

[1]Instituto de Astronomia, Geofísica e Ciências Atmosféricas da Universidade de São Paulo,
Cidade Universitária, CEP:05508-900, São Paulo, SP, Brazil
email: claudia.oliveira@iag.usp.br

[2]Departamento de Física, Universidad de La Serena,
Av. Cisternas 1200 Norte, La Serena, Chile

[3]Aix Marseille Université, CNRS, LAM (Laboratoire d'Astrophysique de Marseille)
UMR 7326, 13388, Marseille, France

[4]Laboratório de Astrofísica Teórica e Observacional, Universidade Estadual de Santa Cruz,
Rodovia Ilhéus-Itabuna Km 16 45650-000 Ilhéus BA, Brazil

Abstract. Fabry-Perot data of compact group galaxies have been used to show that the Tully-Fisher relation in any photometric band, for galaxies with $v_{max} > 100$ km/s, is very similar to that for galaxies in other less dense environments. In the low-luminosity end, however, a few compact group galaxies fall above the relation apparently because they are too bright for their mass. Here we show that if the mass is properly computed from spectral energy distribution fitting or mass modelling, for the low-luminosity galaxies, their positions in the *stellar-mass* or *baryonic* Tully-Fisher relation are what is expected for a normal Tully-Fisher relation and the outlying positions observed in the B and K Tully-Fisher relation could be explained by brightening of the low-luminosity interacting galaxies due to strong star formation or AGN activity.

Keywords. galaxies: spiral, elliptical and lenticular - galaxies: kinematics and dynamics, galaxies: scaling relations

1. Introduction

There has been increased interest in the past years in the use of the Tully-Fisher relation (TF) as a means of quantifying galaxy evolution as a function of redshift. However, before using the TF as such, one must understand how its slope and intercept changes with the density of the environment.

Galaxies in Hickson compact groups are often interacting galaxies, which show peculiarities in either their morphologies or kinematics. These are good laboratories where to study the effect of the dense environment on the Tully-Fisher relation. It is then of importance to find out if galaxies in the dense environments of compact groups follow well the TF relation of galaxies in less dense environments.

As shown in Mendes de Oliveira *et al.* (2003) and Torres-Flores *et al.* (2010), after a careful determination of the position angles, inclinations and kinematic centers of the galaxies, using Fabry-Perot velocity maps, and taking into account the presence of non-circular motions, most of the HCG galaxies lie on the B-band TF relation, with a few low-mass outliers. The low-mass galaxies tend to present a low rotational velocity for their observed luminosity or a high luminosity for their observed rotational velocity. If included in the fit, these low-mass galaxies affect strongly the determination of the slope and zero-point of the TF relation, biasing studies of the evolution of the TF with redshift. Studying the K-band, stellar and baryonic Tully-Fisher relation we can investigate if the mass of the galaxies is truncated due to interactions or if a burst of star formation could be strong enough to increase the luminosity of these objects, pushing them off the TF-relations.

2. Results

In order to compare the Tully-Fisher relation of compact group galaxies with that of galaxies in less dense environments we used the GHASP sample as a control sample. The kinematic properties of the GHASP sample was studied by using Fabry-Perot data and their Tully-Fisher relation was published in Torres-Flores *et al.* (2011). However, this sample is composed of mostly massive galaxies. Given that we are also interested in the location of the low-mass HCG galaxies on the Tully-Fisher relation defined by galaxies in less dense environments, we have complemented the GHASP sample with the gas-rich galaxies studied by McGaugh (2012). As McGaugh (2012) lists the stellar and gaseous mass of the gas-rich sample, these galaxies can be used as a comparison of the stellar and baryonic Tully-Fisher relations.

In the case of the HCG galaxies, B-band absolute magnitudes were estimated by using the apparent magnitudes listed in the Hickson catalogue, considering the different distances for each galaxy and correcting for Galactic extinction. A similar analysis was performed for the K-band absolute values, where the apparent magnitudes were taken from the 2MASS database. The stellar masses were derived from the K-band luminosities and the mass-to-light ratio obtained through the use of optical colors (see Bell *et al.* 2003). Rotational velocities for each galaxy were taken from Mendes de Oliveira *et al.* (2003) and Torres-Flores *et al.* (2010).

The results for the B-band, K-band and stellar Tully-Fisher relations are plotted in Figure 1. The main results of this analysis are: (1) the bulk of the galaxies in compact groups (those with $v_{rot} > 100$ km/s) follow the Tully-Fisher relation; (2) there are a few outliers at the low mass end for the B and K Tully-Fisher relations and if these are included in the fits they will affect both the slope and the intercept of the relation; (3) for the lowest mass galaxies in the compact groups, brightening in the B or K band due to interactions or AGN activity will affect the TF; (4) there are no outliers in the stellar and baryonic TF, once the masses of the low-mass galaxies are computed using spectral energy distribution fitting (SED) or mass modelling analysis.

3. Conclusions

3D kinematic studies of galaxies in compact groups have revealed the importance of measuring correctly the kinematic parameters, center, position angle and inclination of these interacting galaxies, as well as taking into account the contribution to its kinematics of non-circular motions when studying their TF relation. This is relevant not only for compact group galaxies but for interacting galaxies in general. 3D velocity maps are then a requirement when studying the evolution of galaxy properties in different environments and lookback times, given the increasing numbers of interacting galaxies with redshift.

Here we show that, in addition to the above, in order to correctly compute the slope and intercept of the TF relation for interacting galaxies one must be aware that specially for galaxies with maximum rotational velocities smaller than 100 km s^{-1}, masses have to be obtained from SED fitting or mass modelling, given that the B or K magnitudes of these galaxies could be brightened by star formation or nuclear activity, biasing the results for interacting galaxies.

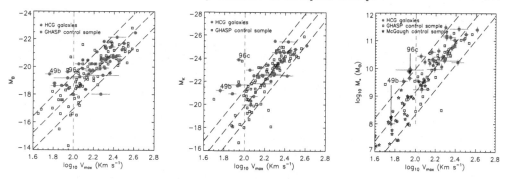

Figure 1. B-band (left-hand panel), K-band (middle-panel) and stellar (right-hand panel) TF relations for the HCG sample (red filled dots). In the three panels, we have divided the GHASP sample for galaxies that display symmetric rotation curves (filled black squares) and asymmetric rotation curves (empty squares), following Torres-Flores *et al.* (2011). The black dashed lines represent the linear fit and a dispersion of 1 on the GHASP data. Gas-rich galaxies (McGaugh 2012) are shown by empty stars. The red vertical lines indicate a rotational velocity of 100 km s^{-1}. Note that the outliers of the relations in the first two panels follow better on the relation for the stellar TF (third pannel). The baryonic TF is shown in Torres-Flores *et al.* (2013).

Acknowledgements

We acknowledge support FAPESP and CNPq. ST-F also acknowledges the support of the project CONICYT PAI/ACADEMIA 7912010004

References

Bell, E. F., McIntosh, D. H., Katz, N., & Weinberg, M. D. 2003 *ApJS*, 149, 289.

Mendes de Oliveira, C., Amram, P., Plana, H., & Balkowski, C. 2003, *AJ*, 126, 2635.

McGaugh, S. 2012, *AJ*, 143, 40.

Torres-Flores S., Mendes de Oliveira, C., Amram, P., Plana, H., Epinat, B., Carignan, C., & Balkowski, C. 2010, *A&A*, 521A, 59.

Torres-Flores, S., Epinat, B., Amram, P., Plana, H., & Mendes de Oliveira, C. 2011, *MNRAS*, 416, 1936.

Torres-Flores, S., Mendes de Oliveira, C., Plana, H., Amram, P., & Epinat, B. 2013, *MNRAS*, 432, 3085.

Galaxies in 3D across the Universe
Proceedings IAU Symposium No. 309, 2014
B. L. Ziegler, F. Combes, H. Dannerbauer, M. Verdugo, eds.

© International Astronomical Union 2015
doi:10.1017/S1743921314009624

A Herschel and CARMA view of CO and [C II] in Hickson Compact groups

Katherine Alatalo[1]*, Philip N. Appleton[1] and Ute Lisenfeld[2]

[1] Infrared Processing and Analysis Center
California Institute of Technology, Pasadena, California 91125, USA
*email: kalatalo@caltech.edu
[2] Departamento de Física Teórica y del Cosmos, Universidad de Granada, Granada, Spain

Abstract. Understanding the evolution of galaxies from the starforming blue cloud to the quiescent red sequence has been revolutionized by observations taken with *Herschel* Space Observatory, and the onset of the era of sensitive millimeter interferometers, allowing astronomers to probe both cold dust as well as the cool interstellar medium in a large set of galaxies with unprecedented sensitivity. Recent *Herschel* observations of of H_2-bright Hickson Compact Groups of galaxies (HCGs) has shown that [C II] may be boosted in diffuse shocked gas. CARMA CO(1–0) observations of these [C II]-bright HCGs has shown that these turbulent systems also can show suppression of SF. Here we present preliminary results from observations of HCGs with *Herschel* and CARMA, and their [C II] and CO(1–0) properties to discuss how shocks influence galaxy transitions and star formation.

Keywords. galaxies: elliptical and lenticular, cD - galaxies: evolution - galaxies: formation

1. Introduction

Compact groups are defined as "small, relatively isolated systems of typically four or five galaxies in close proximity to one another" (Hickson 1997). They tend to have a high fraction of early-type galaxies (E/S0), evidence of tidal interactions, are high density with low velocity dispersion (Hickson 1997) and are generally deficient in H I (Verdes-Montenegro *et al.* 2001; Borthakur, Yun & Verdes-Montenegro 2010). Compact groups appear to go through an evolution (Fig. 1; Verdes-Montenegro *et al.* 2001) that can be traced based on its neutral gas depletion (Verdes-Montenegro *et al.* 2001), with galaxies the most advanced group stages no longer containing their own interstellar medium, and instead being surrounded by a common envelope (Rasmussen *et al.* 2008).

Johnson *et al.* (2007) documented a marked "infrared gap" in compact group galaxies, with very few galaxies observed between the star-forming cloud and the quiescent cloud. Cluver *et al.* (2013) showed that the "gap" galaxies in compact groups tended to have warm hydrogen emission (traced by the *Spitzer* Infrared Spectrograph) that was enhanced beyond the point that simple radiative processes could explained, deemed Molecular Hydrogen Emission Galaxies (MOHEGs; Ogle *et al.* 2006). Cluver *et al.* (2013) were able to show that the enhanced H_2 emission was most easily explained via shocks. With the onset of the Wide-field Infrared Survey Explorer (WISE), this infrared transition zone was shown to be more universal, with the bifurcation between early-type and late-type galaxies present in a large sample of galaxies (Alatalo *et al.* 2014a). Understanding this transition takes both a broad survey approach, as well as carefully selected case studies.

Using the Photodetector Array Camera and Spectrometer (PACS; Poglitsch *et al.* 2010) on *Herschel* (Pilbratt *et al.* 2010), we observed ISM cooling lines, primarily [C II] and [O I] of HCGs, and directly compare the ISM cooling to the molecular gas, traced

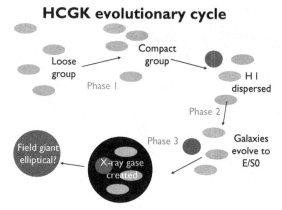

Figure 1. A schematic of the proposed HCG evolutionary picture. The H I observations taken by Verdes-Montenegro *et al.* (2001) and Borthakur, Yun & Verdes-Montenegro (2010) support the Hickson (1997) picture in which galaxies start in a loose group, move into the compavt group phase (during which the H I in the individual galaxies are dispersed) until the gas is located within a common envelope. During the evolution of the ISM, the galaxies within the group are also transitioning (Johnson *et al.* 2007; Walker *et al.* 2010, 2013; Cluver *et al.* 2013).

via the Carbon Monoxide (CO) line, observed using the Combined Array for Research in Millimeter Astronomy (CARMA).

2. Observations and Reduction

11 HCGs were observed with CARMA over the course of three semesters between 12 Mar 2013 – 16 Jun 2014, and one, HCG 96 was taken from the archive, totaling 12 HCGs and 14 individual galaxies. Raw data were reduced and calibrated identically to the ATLAS^{3D} galaxies (Alatalo *et al.* 2013). Figure 2 shows the integrated intensity and mean velocity map of HCG 96, as an example of the data quality that has been attained by the CARMA observations. CO in many of the HCGs have also been observed with single dish telescopes (Martinez-Badenes *et al.* 2012; Lisenfeld *et al.* 2014). For the most part, the fluxes from CARMA agree with the single dish fluxes, although in the cases of larger, more extended sources, CARMA recovers more flux than the single dish in cases where only a single pointing was observed.

Herschel PACS [C II] and [O I] observations were taken of 11 HCGs, and the data were reduced using Herschel Interactive Processing Environment (HIPE) software package CIB13-3069, and data were analyzed identically to HCG 57, described in Alatalo *et al.* (2014b). [C II] was detected in all 11 groups, and [O I] was detected in 9. Of the 11 PACS-imaged HCGs, 9 were detected in CO (Lisenfeld *et al.* 2014) and 8 were imaged using CARMA.

3. Discussion

In normal galaxies, [C II] can be a reliable tracer of SF (de Looze *et al.* 2011), but this relation begins to break down as soon as one begins investigating the outliers. Observations have shown that in local galaxies, $L_{[C\ II]}/L_{FIR}$ decreases in intense starburst environments (e.g. local ULIRGs; Malhotra *et al.* 2001; Luhman *et al.* 2003) and high-z quasars, but the "deficit" may relate to the intensity of the UV radiation field. Díaz-Santos *et al.* (2013) observed that $L_{[C\ II]}/L_{FIR}$ decreases as a function of infrared luminosity in LIRGs, and thus in the most prolific star-formers, [C II] underestimates the true star formation rate. Figure 3a shows the inverse of the Díaz-Santos *et al.* (2013) result when one observes a selection of shocked galaxies (including Stephan's Quintet, Taffy and HCG 57). In most "normal" galaxies, $L_{[C\ II]}/L_{FIR}$ rarely exceeds 1%. In these

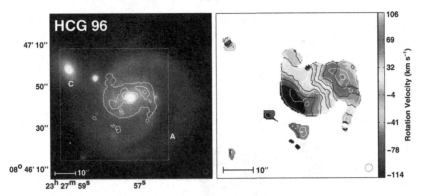

Figure 2. An example of the CO data from CARMA. **(Left):** The Sloan Digital Sky Survey *g, r, i* image underlies the moment0 map of HCG 96a and 96c (labeled on the diagram), which were both detected in the CARMA pointing. **(Right):** The mean velocity map of HCG 96a and 96c. In HCG 96a, the velocity structure traces the spirals, and in HCG 96c, the molecular gas is centralized, and regularly rotating.

shock-dominated objects Peterson *et al.* 2012; Cluver *et al.* 2013; Appleton *et al.* 2013; Alatalo *et al.* 2014b, there are many more regions that exceed this 1% "ceiling". These examples show that one must exercise caution when using $L_{[C\ II]}$ to determine the SF in an object, but it also shows that [C II] is capable of pinpointing turbulence in systems.

Compact group galaxies whose ISM is turbulent, likely due to direct collisions that cause shocks, are also the galaxies that appear to contain the least efficient molecular gas. This observation is supported by the fact that both Stephan's Quintet and HCG 57 show enhanced $[C\ II]/L_{FIR}$ and signs of suppressed SF (Guillard *et al.* 2012; Appleton *et al.* 2013; Alatalo *et al.* 2014b). Lisenfeld *et al.* (2014) studied the CO properties of MOHEGs and non-MOHEGs in HCGS, and, although several objects in their sample showed a suppressed SF, they did not find a statistical difference between the star formation efficiency of MOHEGs compared to non-MOHEG sources. This might mean that excited H2 emission does not influence the star formation of an entire galaxy.

Figure 3b compares the H_2 mass to the SF rate. The colorization is based on the distance off of the Kennicutt-Schmidt relation ($\Sigma_{SFR} = 2.5 \times 10^{-4} \Sigma_{gas}^{1.4}$; Kennicutt 1998). Interestingly, this figure shows that many of the HCG galaxies studied for this survey lay below the K-S relation, with several showing considerable (> 10) suppression. Two MOHEGs (including HCG 57a) represent the most suppressed objects. Given that galaxies in HCGs are much more likely to experience frequent interactions than field galaxies, they are ideal test beds for how turbulence can impact star formation. The fact that these systems are more likely to lie below the K-S relation than above is intriguing. However, a larger number of objects and a more detailed study is neceesary to confirm this trend. The [C II] imaging from Herschel will allow for a much more in-depth look at galaxies, determining areas of high turbulence within galaxies, and investigating whether those regions show signs of suppression. The recent study on HCG 57a, which used this method, seems to confirm the utility of this method.

We are following up on these galaxies to determine where they lie in context, including whether they are found within the optical green valley or the infrared transition zone, to see if the suppression of star formation is a cause, or a symptom of this transition. Further [C II] and CO(1–0) are needed of a larger sample of transitioning galaxies to fully understand this population, and in particular, how turbulence and shocks impact the supply of the cold molecular gas, as well as the how existing molecular gas forms stars.

K. A. is supported by funding through Herschel, a European Space Agency Cornerstone Mission with significant participation by NASA, through an award issued by JPL/Caltech. U. L. acknowledges support by the research projects AYA2011-24728 from the Spanish Ministerio de

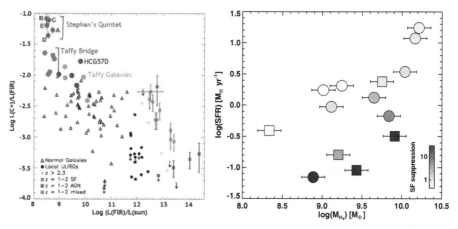

Figure 3. (Left): The L([C II])/L$_{FIR}$ vs L$_{FIR}$ of starforming galaixes and ULIRGs (Malhotra *et al.* 2001) and high$-z$ sources (Stacey *et al.* 2010). There is an approximate L([C II])/L$_{FIR}$ upper limit of 1%, until we include sources known to have shocks, including the Taffy Galaxies (Peterson *et al.* 2012), Stephan's Quintet (Appleton *et al.* 2013) and HCG 57 (Alatalo *et al.* 2014b). **(Right):** The total star formation rate calculated from Bitsakis *et al.* (2014), normalized to a Salpeter initial mass function (Salpeter 1955) compared to the molecular gas mass derived from the CO(1–0) data from CARMA (using the L$_{CO}$–to–M$_{H_2}$ conversion of $2. \times 10^{20}$ cm^{-2} (K km s^{-1})$^{-1}$; Bolatto *et al.* 2013). The colorization is based on the distance off of the relation derived in Kennicutt (1998).

Ciencia y Educación and the Junta de Andalucía (Spain) grants FQM108. Support for CARMA construction was derived from the Gordon and Betty Moore Foundation, the Kenneth T. and Eileen L. Norris Foundation, the James S. McDonnell Foundation, the Associates of the California Institute of Technology, the University of Chicago, the states of California, Illinois, and Maryland, and the National Science Foundation. Ongoing CARMA development and operations are supported by the National Science Foundation under a cooperative agreement, and by the CARMA partner universities. *Herschel* is an ESA space observatory with science instruments provided by European-led Principal Investigator consortia and with important participation from NASA. This work is based [in part] on observations made with the *Spitzer* Space Telescope, which is operated by the Jet Propulsion Laboratory, California Institute of Technology under a contract with NASA.

References

Alatalo, K., *et al.* 2013, *MNRAS*, 432, 1796

Alatalo, K., *et al.* 2014, *ApJL*, in press, (arXiv:1409.2489)

Alatalo, K., *et al.* 2014, *ApJ*, in press, (arXiv:1409.5482)

Appleton, P. N., *et al.* 2013, *ApJ*, 777, 66

Bitsakis, T., *et al.* 2010, *A&A*, 517, 75

Bitsakis, T., *et al.* 2011, *A&A*, 533, 142

Bitsakis, T., *et al.* 2014, *A&A*, 565, 25

Bolatto, A. D., Wolfire, M., & Leroy, A. K. 2013, *ARA&A*, 51, 207

Borthakur, S., Yun, M. S., & Verdes-Montenegro, L. 2010, *ApJ*, 710, 385

Cluver, M. E., *et al.* 2013, *ApJ*, 765, 93

de Looze, I., *et al.* 2011, *MNRAS*, 416, 2712

Díaz-Santos, T., *et al.* 2013, *ApJ*, 774, 68

Guillard, P., *et al.* 2012, *ApJ*, 749, 158

Hickson, P. 1997, *ARA&A*, 35, 357

Johnson, K. E., *et al.* 2007, *AJ*, 134, 1522

Kennicutt, R. C. 1998, *ApJ*, 498, 541

Lisenfeld, U., *et al.* 2014, *A&A*, in press, (arXiv:1407.4731)

Luhman, M. L., *et al.* 2003, *ApJ*, 594, 758

Malhotra, S., *et al.* 2001, *ApJ*, 561, 766

Martinez-Badenes, V., Lisenfeld, U., Espada, D., *et al.* 2012, *A&A*, 540, A96

Ogle, P., Whysong, D., & Antonucci, R. 2006, *ApJ*, 647, 161

Peterson, B. W., *et al.* 2012, *ApJ*, 751, 11

Pilbratt, G. L., *et al.* 2010, *A&A*, 518, L1

Poglitsch, A., *et al.* 2010, *A&A*, 518, L2

Rasmussen, J., *et al.* 2008, *MNRAS*, 388, 1245

Salpeter, E. E. 1955, *ApJ*, 121, 161

Stacey, G. J., *et al.* 2010, *ApJ*, 724, 957

Verdes-Montenegro, L., *et al.* 2001, *A&A*, 377, 812

Walker, L. M., *et al.* 2010, *AJ*, 140, 1254

Walker, L. M., *et al.* 2012, *AJ*, 143, 69

Walker, L. M., *et al.* 2013, *ApJ*, 775, 129

Galaxies in 3D across the Universe
Proceedings IAU Symposium No. 309, 2014
B. L. Ziegler, F. Combes, H. Dannerbauer, M. Verdugo, eds.

© International Astronomical Union 2015
doi:10.1017/S1743921314009636

Models of AGN feedback

Françoise Combes

Observatoire de Paris
61 Av. de l'Observatoire F-75 014 Paris, France
email: francoise.combes@obspm.fr

Abstract. The physical processes responsible of sweeping up the surrounding gas in the host galaxy of an AGN, and able in some circumstances to expel it from the galaxy, are not yet well known. The various mechanisms are briefly reviewed: quasar or radio modes, either momentum-conserving outflows, energy-conserving outflows, or intermediate. They are confronted to observations, to know whether they can explain the M-sigma relation, quench the star formation or whether they can also provide some positive feedback and how the black hole accretion history is related to that of star formation.

Keywords. galaxies: elliptical and lenticular, cD - galaxies: evolution - galaxies: formation

1. Introduction

AGN feedback is invoked to prevent massive galaxies to form in too high abundance, and to understand why the fraction of baryons in galaxies is very low, and decreasing with the galaxy mass (e.g. Behroozi *et al.* 2013).

It is clear that the growth of massive black holes in the center of galaxies releases enough energy to have a large impact on the galaxy host, if this energy is efficiently coupled to the matter. Assuming the famous relation between black hole and bulge mass (e.g. Gültekin *et al.* 2009), M_{BH}=1-2 10^{-3} M_{bul}, and the radiative efficiency of accretion onto a black hole of 10%, E_{BH} =0.1$M_{BH}c^2$, the binding energy of the bulge is $E_{bul} \sim M_{bul}\sigma^2$, with σ its velocity dispersion. Typically, the ratio of these two energies is $E_{BH}/E_{bul} > 100$. There is therefore no problem in global energy, however, this energy could be radiated away through an elongated cavity perpendicular to the galaxy plane, and the true impact on the galaxy is unknown.

One can distinguish two main modes for AGN feedback (see e.g. the recent review by Fabian, 2012):

– the **quasar mode**, through radiative processes or winds, when the AGN luminosity is high, close to the Eddington limit. This is the case for young QSOs, at high redshift. The Eddington limit is $L_{Edd} = 4\pi GM_{BH}m_pc/\sigma_T$, with σ_T being the Thomson scattering cross-section between charged particles and photons. When writing the equilibrium between the gravity and radiative pressure on the infalling ionized gas, combined with the Virial relation, this gives $M_{BH} \sim f \sigma_T\sigma^4$, with f the gas fraction, which is quite close to the $M_{BH} - \sigma$ relation. When applying the same considerations at larger radii, this time for neutral gas, submitted to radiation pressure on dust, with the cross-section σ_d, this leads a limitation on the mass than can be accumulated on the bulge and the corresponding masses are in the ratio $M_{bul}/M_{BH}=\sigma_d/\sigma_T \sim 1000$, which is surprisingly close to the observed ratio.

– the **radio mode**, or kinetic mode with radio jets, when the AGN luminosity is low, $L < 0.01$ L_{Edd}, corresponding typically to low z, massive galaxies, like the local radio ellipticals. The process is then not destructive, and keeps a balance between cooling and heating. This is the mode occuring in cool core clusters, from the AGN of the central bright galaxy. It is also associated to low-luminosity AGN in Seyferts. It might be combined with a radiatively inefficient flow ADAF (Advection dominated Accretion Flow).

The best evidence of AGN feedback is found in cool core clusters of galaxies, where the active nucleus in the central galaxy moderates the cooling flow, through its radio jet, creating bubbles in the hot gas. This can have an impact up to radii larger than the central galaxy, at 100kpc scales (McNamara *et al.* 2009). About 70 percent and possibly 90 percent of clusters have a cool core (Edge *et al.* 1992). The clusters that are not cooling flows are unrelaxed and most probably recent mergers of sub-clusters.

In the following, we will detail the AGN feedback moderating cooling flows, describe recent results on molecular outflows, driven by starbursts and AGN, and the nature of their mechanisms and energetics. We finish by considerations on the modes of quenching.

2. AGN feedback moderating gas flow in cool core clusters

It is frequent in cool core clusters to see the extended impact of the central radio jet, under bubbles and cavities imprinted on the hot X-ray gas, and cold filaments and streaks observed with $H\alpha$ and CO emission. In the prototypical Perseus cluster, a large network of cold molecular gas filaments has been observed (Salomé *et al.* 2006). Recently Canning *et al.* (2014) have identified a population of very young (a few Myr), compact star clusters, over kiloparsecs scales, related to the network of filaments. These stars could then enrich the stellar halo of the central galaxy NGC 1275.

Observations suggest that the filament network is a consequence of both inflow and outflow. Some of the hot gas cools and fuels the central AGN, which jets entrain the molecular gas in the central galaxy coming from previous cooling episodes. The gas uplifted has been metal enriched in the galaxy before mixing with the cooling gas, and therefore can radiate in CO. The bubbles create inhomogeneities and further cooling. The latter is not only occuring in the center of the cluster, but far away, up to 50 kpc, due to the inhomogeneous density and compressions there. The observed velocity of the gas in filaments is much lower than free-fall (Salomé *et al.* 2008). Numerical simulations have reproduced this process, with buoyant bubbles triggering compression and gas cooling at the surface of cavities, mixing with the cold gas dragged upwards (Revaz *et al.* 2007).

A large variety of simulations of AGN feedback in clusters or massive elliptical galaxies have been performed, with different degrees of approximaions and sophistication (Brueggen *et al.* 2007, Cattaneo & Teyssier 2007, Dubois *et al.* 2010, Gaspari *et al.* 2011, 2012). They vary in the cooling rate, or the assumption for accretion (boosted Bondi rate, but cold gas accretion gives better results). None is self-consistent and can predict the accretion rate, which depends on subgrid physics. The radiation pressure mechanism appears insufficient to impact a significant part of the cluster. Mechanical feedback with jets or winds is required. The most recent simulations (Gaspari *et al.* 2012, see Figure 1), succeed in moderating the cooling, and keep the cool core structure, while previous ones were highly destructive. The efficiency ϵ, ratio between kinetic energy in the jet and accretion energy, can be scaled to the structure, $\epsilon = 3 \ 10^{-4}$ for ellipticals and $5 \ 10^{-3}$ for clusters.

Are the starbursts related to AGN activity in simulations? Thacker *et al.* (2014) have performed a comparison between the various models in diagrams of star formation rates (SFR) versus BH accretion rates (BHAR). During a merger, accreted gas fuels both the SFR and BHAR. A BHAR delay is expected since SN feedback is too strong at the beginning (Wild *et al.* 2010).

184 F. Combes

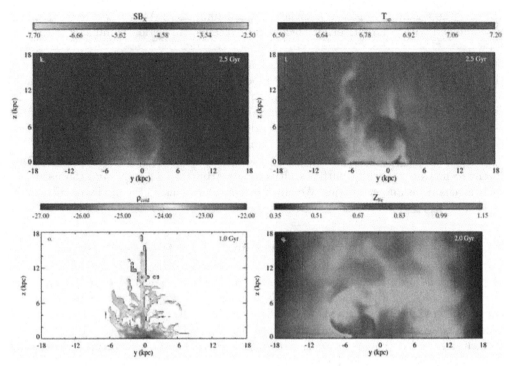

Figure 1. Simulation of AGN feedback in a cool core cluster, with the X-ray surface brightness (top left), the temperature (top right), the cold gas filaments (bottom left) and metallicity (bottom right), from Gaspari *et al.* (2012).

3. Molecular Outflows

In recent years, the discovery of many massive molecular outflows has given support to AGN feedback. In the prototypical Mrk 231 both a nuclear starburst and the AGN contribute to the gas outflow, of ~ 700 M$_\odot$/yr (Feruglio *et al.* 2010). Some outflows have been resolved with a size of a few kpc, significant to impact the galaxy, and provide some quenching. The kinematic power is of the order of 2 10^{44} erg/s, and the energy of the AGN is required. High density is traced in the outflows through HCN, HCO$^+$ (Aalto *et al.* 2012). Even more powerful outflows are observed at high redshift (Maiolino *et al.* 2012).

Cicone *et al.* (2014) reported interesting relations between outflow rates and AGN properties, for the molecular outflows known. For AGN-hosts, the outflow rate correlates with the AGN power (see Fig 2). The momentum of the outflow is proportional to the photon momentum and is about 20 L$_{AGN}$/c. This is interesting to constrain the physical mechanisms of the flow. This boosted momentum can be explained by energy-driven outflows (Zubovas & King 2012).

The question of the very existence of fast molecular outflows is still unsolved. Outflowing gas is accelerated by a shock, and heated to 10^6-10^7K. Molecules should be dissociated at such temperatures. Even if cold clumps are carried out in the flow, there should exist some shock signature. One solution proposed is that radiative cooling is quick enough to reform molecules in a large fraction of the outflowing material (Zubovas & King 2014). With V$\sim$1000 km/s, and dM/dt $\sim$1000 M$_\odot$/yr, efficient cooling produces multi-phase media, with triggered star formation.

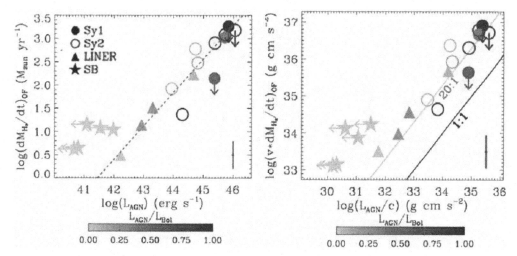

Figure 2. Relation between outflow rate and AGN luminosity (left), and outflow momentum with L_{AGN}/c, from Cicone *et al.* (2014). Colors indicate the L_{AGN}/L_{bol} ratio. The various symbols distinguish the AGN types, and the stellar symbols are starbursts.

4. Mechanisms

A very schematic view of the shocks generated by an AGN wind is displayed in Fig 3 left. The shocked ISM after the discontinuity region (region c) cools efficiently (free-free emission, metal cooling), and the flow becomes unstable; if the ratio of the radiation pressure to gas pressure is lower than 0.5, then the gas collapse in clumps, detaching from the hot flow (Krolik *et al.* 1981). The region (c) becomes multiphase, with Rayleigh-Taylor instabilities. The time-scale for cooling is much lower than 1 Myr, and star formation is induced. This may explain the tight correlation between starbursts and AGN. Also it might then be difficult to disentagnle supernovae or AGN outflows. All could have been triggered by the AGN (Zubovas & King 2014).

4.1. *Energy or momentum conserving outflows*

As shown in Fig 3, if the region (b) of the shocked wind is cooling efficiently, then energy is not conserved, but only the momentum, this is the case of a momentum-conserving outflow.

But for very fast winds $v_{in} > 10\,000$ km/s and up to 50 000 km/s, radiative losses are slow, and almost all the energy can be conserved in this case, called energy-conserving flow (Faucher-Giguère & Quataert 2012). Then $dM_{in}/dt\,v_{in}^2 \sim dM_s/dt\,v_s^2$, and there can be a boost in the outflow momentum, as large as $v_{in}/(2\,v_s) \sim 50$! This explains why the molecular outflows are observed with a momentum flux $>> L_{AGN}/c$. It is the push by the hot post-shock gas which boosts the momentum and the velocity v_s of the swept-up material. In some cases, even slow winds $v_{in} \sim 1000$ km/s driven by radiation pressure on dust, could be energy-conserving. The phenomenon is analogous to the adiabatic phase, or Sedov-Taylor phase in supernovae remnants. This momentum boost increases the efficiency of supernovae feedback in galaxies.

Within this framework, it was possible to build a realistic model and outflow solution for the typical case of Mrk231, A momentum flux of 15 L_{AGN}/c has been obtained (Faucher-Giguère & Quataert 2012).

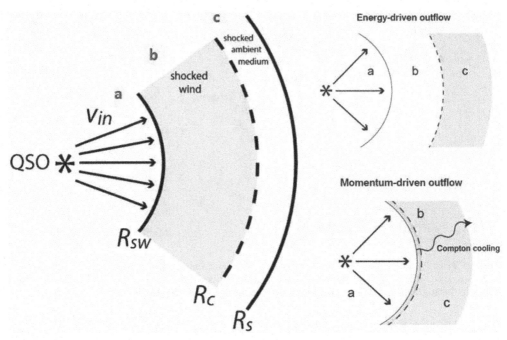

Figure 3. Schematic radial representation of the AGN wind and shock induced in the surrounding ISM. (a) is the wind region; delimited by a reverse shock (R_{SW}) (b) is the shocked wind, ending at the discontinuity surface (R_C, dash line), (c) is the shocked ISM, bounded by the forward shock (R_S), expanding into the unperturbed ISM. The two panels at right indicate the cases of energy-driven outflow (top) where the region (b) is not cooling and energy is conserved there, and momentum-driven outflow (bottom), where the region (b) cools rapidly, loses its pressure support, and becomes thin (from Costa *et al.* 2014).

4.2. *Winds launch*

Towards the very center, ultra-fast outflows (UFO) or relativistic winds have been observed in UV absorption lines (BAL quasars), or from X-rays coronae (Tombesi *et al.* 2010, 2014).

These winds can be launched from accretion disks, through several physical mechanisms. Thermal heating (Compton) makes the gas reach the escape velocity. The radiation pressure on electrons (super Eddington regime), or even radiation pressure on dust, or magnetic driving could be the source (Proga 2003, 2005). More realistically, all driving mechanisms may act together.

Simulations of the **quasar mode** have been performed taking into account the multi-phase medium (e.g. Nayakshin 2014). Most of the outflow kinetic energy escapes through the voids. Cold gas is pushed by ram-pressure, but there is more feedback on low-density gas. Both positive and negative feedbacks are observed. The simulations account for the M-σ relation.

Simulations of the **radio mode** have also been performed with a fractal structure of the gas (Wagner & Bicknell 2011). Relativistic jets produce a very efficient feedback, and impact on the galaxy, in spite of the porosity of the ISM.

4.3. *Positive AGN feedback*

Many simulations reveal signs of positive feedback (e.g. Silk 2005, Dubois *et al.* 2013). The phenomenon is more difficult to observe, but some systems do show evidence of jet-induces star formation, like the Minkowki object, Centaurus A, or for instance the young,

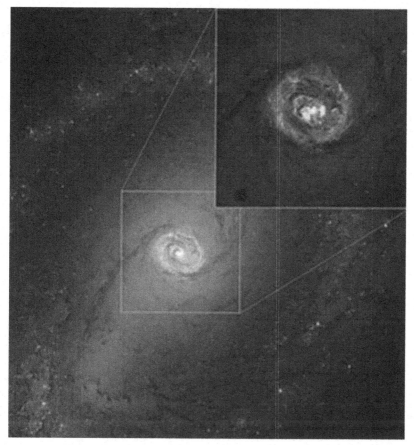

Figure 4. The NGC 1433 barred spiral is a Seyfert 2 of low-luminosity. The CO(3-2) ALMA map of the nuclear disk shown here superposed on the HST image has revealed an outflow on the minor axis, the smallest molecular flow detected up to now (Combes *et al.* 2013).

restarted radio loud AGN 4C12.50. The outflow is located 100 pc from the nucleus where the radio jet interacts with the ISM (Morganti *et al.* 2013, Dasyra & Combes 2012).

4.4. *Feedback in low-luminosity AGN*

Feedback can also play a role in low-luminosity AGN, in moderating star formation. The modest impact here is compensated by the large numbers of such objects. Recently the ALMA observations at 25 pc resolution of the CO(3-2) line in the barred spiral NGC 1433 has revealed a molecular ouflow on the minor axis (Combes *et al.* 2013, see Fig 4.). About 7% of the molecular mass detected belongs to the outflow, with a velocity $\sim$ 100 km/s, uncertain because of ill-known inclination. This is the smallest flow detected up to now. The kinetic luminosity is $\sim 2.3 \ 10^{40}$ erg/s, and the flow momentum is > 10 L_{AGN}/c.

The molecular gas is still fueling the AGN, as shown by the computed gravity torques of the bar (Smajic *et al.* 2014).

5. Modes of quenching

There are two main modes to consider, a rapid one, with time-scales of 100 Myr, in galaxy mergers, with feedback from supernovae and from AGN, and a slow one, with

2-4 Gyr time-scales, either through morphological quenching after bulge formation, or through strangulation, where the replenishment of the disk in gas is stopped (Schawinski *et al.* 2014). Mild feedback could also delay star formation, in a secular way. It is however observed that the co-evolution of galaxies and black holes is no longer tight for low-mass galaxies, governed by secular evolution (e.g. Kormendy & Ho 2013).

It is not sure whether a feedback loop is required to explain the M-σ relation. Simulations of torque limited growth of the central BH have retrieved the relation, without any feedback (Angles-Alcazar *et al.* 2013).

The galaxy and BH grow mostly through gas accretion, sometimes mergers (but they are not essential). The numerical evolution has been computed with different M-σ relations for seeds. However, the effects of initial conditions are quickly erased. The growth of the BH, $dM_{BH}/dt \sim$ SFR with scatter, and no feedback loop is required.

AGB feedback in mergers have been invoked for a long time for theoretical reasons, and simulated with the usual sub-grid physics (e.g. Springel *et al.* 2005, Hopkins *et al.* 2006). However, the amount of feedback, and how efficiently it couples with the galaxy disks is unknown. Recently, Gabor & Bournaud (2014) claim that the AGN quenching effect on star formation is negligible, although a significant amount of hot gas is expelled in the intergalactic medium.

Several modes have been simulated, with more sophisticated details: **quasar mode**, when $dM_{BH}/dt > 0.01$ Eddington rate, and the energy is released spherically; **radio-jets** otherwise, with a velocity V= 10^4 km/s imposed in a cylinder perpendicular to the disk.

The efficiency to form stars can be reduced by a factor 7, and simulations do show a decrease of the baryon concentration (Dubois *et al.* 2013).

Costa *et al.* (2014) have reproduced the two cases energy and momentum driven winds, and find the energy-driven mode much more efficient, with outflow momenta > 10 L_{Edd}/c. The entrained cold gas has masses $> 10^9$ $M_\odot$ if the after-shock gas cools with metals.

6. Conclusions

AGN feedback is very efficient in cool core clusters to moderate the cooling. The mechanism here is mechanical with radio jets and cold gas accretion.

Molecular outflows are now observed frequently around AGN, with velocities v=200-1200 km/s, outflowing masses 10^7-10^9 $M_\odot$, and load factors >3.

The mechanisms range from quasar modes (winds) with powerful AGN, to radio modes (jets), for more massive galaxies with lower Eddington ratios (and lower redshift). The fueling could be either cosmic mass accretion or mergers.

Molecular outflows in AGN-hosts reveal boosted momentum flux of $\sim$20 L_{AGN}/c This can only be obtained through energy-conserving flows.

Acknowledgements

Thanks for the organisers, and in particular the chair Bodo Ziegler for such an exciting and well-organised conference. I acknowledge the European Research Council for the Advanced Grant Program Number 267399-Momentum.

References

Aalto, S., Garcia-Burillo, S., Muller, S. *et al.*, 2012, *A&A* 537, A44
Angles-Alcazar, D., Ozel, F., & Davé, R., 2013, *ApJ* 770, 5
Behroozi, P. S., Wechsler, R. H., & Conroy, C., 2013, *ApJ* 770, 57

Brueggen, M., Heinz, S., Roediger, E., *et al.*, 2007, *MNRAS* 380, L67

Canning, R. E. A., Ryon, J. E., Gallagher, J. S., *et al.*, 2014, *MNRAS* in press, arXiv1406.4800

Cattaneo, A. & Teyssier, R., 2007, *MNRAS* 376, 1547

Cicone, C., Maiolino, R., Sturm, E., *et al.*, 2014, *A&A* 562, A21

Combes, F., Garcia-Burillo, S., Casasola, V., *et al.*, 2013, *A&A* 558, A124

Costa, T., Sijacki, D., & Haehnelt, M. G., 2014, *MNRAS* in press arXiv1406.2691

Dasyra, K. & Combes F., 2012, *A&A* 541, L7

Dubois, Y., Devriendt, J., Slyz, A., & Teyssier, R., 2010, *MNRAS* 409, 985

Dubois, Y., Pichon, C., Devriendt, J., *et al.*, 2013, *MNRAS* 428, 2885

Edge, A. C., Stewart, G. C., & Fabian, A. C., 1992, *MNRAS* 258, 177

Fabian, A. C., 2012, *ARAA* 50, 455

Faucher-Giguère, C-A. & Quataert, E., 2012, *MNRAS* 425, 605

Feruglio, C., Maiolino, R., Piconcelli, E., *et al.*, 2010, *A&A* 518, L155

Gabor, J. & Bournaud, F., 2014, *MNRAS* 441, 1615

Gaspari, M., Melioli, C., Brighenti, F., & D'Ercole, A., 2011, *MNRAS* 411, 349

Gaspari, M., Brighenti, F., & Temi, P., 2012, *MNRAS* 424, 190

Gültekin, K., Richstone, D. O., Gebhardt, K., *et al.*, 2009, *ApJ* 698, 198

Hopkins, P., Hernquist, L., Cox, T. J., *et al.*, 2006, *ApJS* 163, 1

Kormendy, J. & Ho, L. C., 2013, *ARAA* 51, 511

Krolik, J. H., McKee, C. F., & Tarter, C. B., 1981, *ApJ* 249, 422

Maiolino, R., Gallerani, S., Neri, R., *et al.*, 2012, *MNRAS* 425, L66

McNamara, B. R., Kazemzadeh, F., Rafferty, D. A., *et al.*, 2009, *ApJ* 698, 594

Morganti, R., Fogasy, J., Paragi, Z., *et al.*, 2013, *Science* 341, 1082

Nayakshin, S., 2014, *MNRAS* 437, 2404

Proga, D. 2003, *ApJ* 585, 406

Proga, D. 2005, *ApJ* 630, L9

Revaz, Y., Combes, F., & Salomé P., 2007, *A&A* 477, L33

Salomé, P., Combes, F., Edge, A., *et al.*, 2006, *A&A* 454, 437

Salomé, P., Combes, F., Revaz, Y., *et al.*, 2008, *A&A* 484, 317

Schawinski, K., Urry, C. M., Simmons, B. D., *et al.*, 2014, *MNRAS* 440, 889

Silk, J., 2005, *MNRAS* 364, 1337

Smajic, S., Moser, L., Eckart, A., *et al.*, 2014, *A&A* in press, arXiv1404.6562

Springel, V., Di Matteo, T., & Hernquist, L., 2005, *MNRAS* 361, 776

Thacker, R. J., MacMackin, C., Wurster, J., & Hobbs, A., 2014, *MNRAS* in press, arXiv1407.0685

Tombesi, F., Cappi, M., Reeves, J. N., *et al.*, 2010, *A&A* 521, A57

Tombesi, F., Tazaki, F., Mushotzky, R. F., *et al.*, 2014, *MNRAS* in press arXiv1406.7252

Wagner, A. Y. & Bicknell, G. V., 2011, *ApJ* 728, 29

Wild, V., Heckman, T., & Charlot, S., 2010, *MNRAS* 405, 933

Zubovas, K. & King, A., 2012, *ApJ* 745, L34

Zubovas, K. & King A., 2014, *MNRAS* 439, 400

Galaxies in 3D across the Universe
Proceedings IAU Symposium No. 309, 2014
B. L. Ziegler, F. Combes, H. Dannerbauer, M. Verdugo, eds.

© International Astronomical Union 2015
doi:10.1017/S1743921314009648

The Narrow Line Region in 3D: mapping AGN feeding and feedback

Thaisa Storchi Bergmann[1,2]

[1]Instituto de Física,UFRGS, Campus do Vale, Porto Alegre, RS, Brazil
[2]Visitor Professor at the Harvard-Smithsonian Center for Astrophysics
*email: thaisa@ufrgs.br

Abstract. Early studies of nearby Seyfert galaxies have led to the picture that the Narrow Line Region is a cone-shaped region of gas ionized by radiation from a nuclear source collimated by a dusty torus, where the gas is in outflow. In this contribution, I discuss a 3D view of the NLR obtained via Integral Field Spectroscopy, showing that: (1) although the region of highest emission is elongated (and in some cases cone-shaped), there is also lower level emission beyond the "ionization cone", indicating that the AGN radiation leaks through the torus; (2) besides outflows, the gas kinematics include also rotation in the galaxy plane and inflows; (3) in many cases the outflows are compact and restricted to the inner few 100pc; we argue that these may be early stages of an outflow that will evolve to an open-ended, cone-like one. Inflows are observed in ionized gas in LINERs, and in warm molecular gas in more luminous AGN, being usually found on hundred of pc scales. Mass outflow rates in ionized gas are of the order of a few $M_\odot$ yr^{-1}, while the mass inflow rates are of the order of tenths of $M_\odot$ yr^{-1}. Mass inflow rates in warm molecular gas are $\approx$ 4–5 orders of magnitude lower, but these inflows seem to be only tracers of more massive inflows in cold molecular gas that should be observable at mm wavelengths.

Keywords. galaxies - active, galaxies - nuclei, galaxies - kinematics, galaxies - ISM

1. Introduction

The physical processes that couple the growth of supermassive black holes (SMBH) to their host galaxies occur in the inner few kpc (Hopkins & Quataert 2010) when the nucleus becomes active due to mass accretion to the SMBH (Ferrarese & Ford 2005; Kormendy & Ho 2013). The radiation emitted by Active Galactic Nuclei (AGN) ionizes the gas in the vicinity of the nucleus, forming the Narrow Line Region (NLR). AGN-driven winds (Ciotti & Ostriker 2010; Elvis 2000) interact with the gas and produce outflows that are observed in the NLR reaching velocities of hundreds of km s^{-1} (Storchi-Bergmann *et al.* 2010; Fisher *et al.* 2013). Relativistic jets emanating from the AGN also interact with the gas of the NLR (Wang & Elvis 2009). Both types of outflow produce feedback, which is a necessary ingredient in galaxy evolution models to avoid producing over-massive galaxies (Fabian 2012). The importance of the NLR stems from the fact that it is spatially resolved, extending from hundreds to thousands of parsecs from the nucleus and exhibits strong line emission. These properties allow the observation of the interaction between the AGN and the circumnuclear gas in the galaxy: (1) via the observation of the NLR geometry and excitation properties that constrain the AGN structure and ionizing source; (2) via the gas kinematics that maps the processes of the AGN feeding and feedback.

In this contribution, I show results from a "3D" study of the NLR of nearby AGNs by our research group AGNIFS (AGN Integral Field Spectroscopy). I present maps of the gas flux distribution and kinematics, as well as estimates of gas mass inflow and outflow

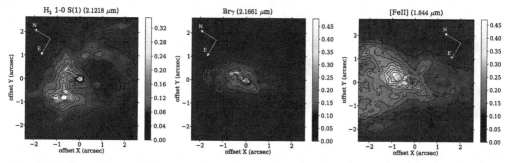

Figure 1. Gas flux distributions in the emission lines $H_2\lambda2.12\mu m$, $Br\gamma$ and [Fe II]$1.644\mu m$ within the inner $\approx$200 pc radius of NGC 1068 from Riffel *et al.* (2014a). The green contours are from the flux distribution in the optical [O III]$\lambda5007$ emission line.

rates, from observations of both the ionized and hot molecular gas.

We have used two Gemini instruments in these studies: the Near-IR Integral Field Spectrograph NIFS (McGregor *et al.* 2003), with the adaptive optics module ALTAIR and the Gemini Multi-Object Spectrograph Integral Field Unit (GMOS-IFU). With NIFS, we have performed observations in the Z, J, H and K bands of the near-IR, at an angular resolution of about 0.′′1 and spectral resolution R$\approx$5300. With the GMOS-IFU, we have mostly observed the spectral region covering the emission lines [O I]$\lambda6300$, $H\alpha$+[N II] and [S II]$\lambda\lambda6717,31$, at a spectral resolution R$\approx$2000. The NIFS field-of-view (hereafter FOV) is $3'' \times 3''$, while the GMOS-IFU FOV can be either $3.''5\times5''$ in one-slit mode, or $7'' \times 5''$, in two-slit mode. In a few cases we have performed mosaiced observations to cover a larger FOV.

2. An hourglass-shaped/conical outflow

I begin with the NLR of the prototypical Seyfert 2 galaxy NGC 1068, that we have observed with NIFS+ALTAIR in the near-IR. At the adopted distance of 14.4 Mpc (Riffel *et al.* 2014a), the spatial resolution of our observations is $\approx$7 pc. We illustrate in Fig.1 the flux distributions in the emission lines of $H_2\lambda2.12\mu m$ of the warm (T$\approx$2000K, Riffel *et al.* (2014a)) molecular gas, compared with those of $Br\gamma$ and [Fe II]$\lambda1.644\mu m$. While the warm molecular gas is distributed in a $\approx$100 pc ring, the $Br\gamma$ flux distribution is very similar to that in the optical [O III]$\lambda5007$ emission line (green contours). The rightmost panel shows that the ionized flux distribution in [Fe II] is distinct from the two previous ones, showing an hourglass-shaped structure, resembling that seen in some planetary nebulae. We attribute these differences to: (1) the fact that the molecular gas is in the galaxy disc, while both the gas emitting $Br\gamma$ and [Fe II] extends to high galactic latitudes, where they are in outflow; (2) the $Br\gamma$ and [O III] emission are restricted to the region of highest ionization, where the collimated radiation from the AGN incides; (3) the [Fe II] emission comes from the larger partially ionized region, that extends beyond the fully ionized region, and is thus a better tracer of the whole outflow from the AGN.

In Barbosa *et al.* (2014) we show a comparison between channel maps obtained along the [Fe II] and H_2 emission line profiles. While the [Fe II] emission is observed in outflow (following the hourglass shape that extends to high galactic latitudes) up to blueshifts and redshifts of $\approx$ 800 km s^{-1}, the H_2 emission is only observed at low velocities ($\approx$100 km s^{-1}), consistent with a kinematics dominated by rotation (and expansion in this case, as pointed out in the paper) in the galaxy disk.

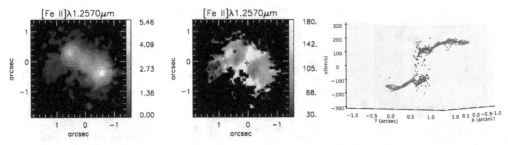

Figure 2. Left panel: gas flux distribution in the line [Fe II]1.644μm of the inner ≈300 pc of NGC 5929 (Riffel *et al.* 2014b). The green contours are from a radio image. Central panel: [Fe II] velocity dispersion, showing a "stripe" of high velocity dispersion. Right panel: a rendering of the velocity field showing that the highest velocity dispersions correspond to two components (one in blueshift and the other in redshift), interpreted as due to an equatorial outflow.

The NLR of NGC 1068 thus has the expected "conical" shape, where the gas is in outflow. But these characteristics seem to be more the exception than the rule. In fact, Fisher *et al.* (2013), in a long-slit study of the NLR of a sample of ≈60 active galaxies, using the instrument STIS aboard the Hubble Space Telescope, have concluded that in only about 1/3 of their sample, the kinematics of the NLR is consistent with a conical outflow. Although our 3D studies have so far covered a smaller sample, our results support also varied morphologies and velocity fields for the observed outflows.

3. Other types of outflow

Let's consider the case of NGC 5929 (Riffel *et al.* 2014b). Our NIFS observations of the NLR of this Seyfert 2 galaxy allow us to resolve 0.″1=18 pc. The left panel of Fig.2 shows the flux distribution in the [Fe II]λ1.64μm emission line over the inner ≈ 300 pc, that is most extended along the direction connecting two radio hot spots shown in green in the figure. The central panel shows a "stripe" of high velocity dispersion values running perpendicularly to the radio jet. Enhanced velocity dispersion is observed also at the loci of the radio hot spots, where it can be attributed to interaction between the radio jet and the ambient gas.

A closer look at the emission-line profiles along the high velocity dispersion stripe of Fig. 2 shows that they are better fitted by two velocity components, one blueshifted and the other redshifted, as illustrated in the rightmost panel of the figure. This 3D rendering of the velocity field shows that most of the gas is rotating with a major axis approximately coincident with the orientation of the radio axis, but there is also an outflow along the equatorial plane of the AGN. In Riffel *et al.* (2014b) we have interpreted this as originating in an accretion disk wind running almost parallel to the surface of the disk (Proga & Kallman 2004) and/or in a wind emanating from the AGN torus (Elitzur 2012).

We have found outflows perpendicular to the ionization axis in three other cases. In Arp 102B (Couto *et al.* 2013), observed with GMOS-IFU, the Hα emission-line kinematics reveal gas in rotation within the inner kiloparsec, but showing also an outflow perpendicular to the ionization axis and the radio jet within the inner ≈ 200 pc.

In the Seyfert 2 galaxy NGC 2110, also using GMOS-IFU, we (Schnorr Müller *et al.* 2014a) found elongated line emission over the whole FOV, that extends to ≈ 800 pc from the nucleus, intepreted as due to ionization by a collimated nuclear source. The gas kinematics reveal that most of this gas is in rotation in the galaxy disk, while within the inner ≈ 300 pc there is a high velocity dispersion component that we have attributed to a nuclear outflow.

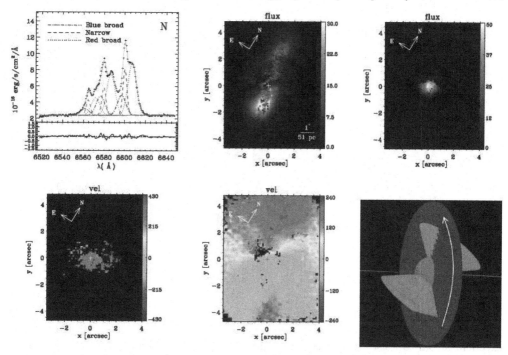

Figure 3. Top left: fit of the Hα+[N II] emission line profiles of the inner 50 pc of NGC 1386, showing a central narrow plus two broad components (nuclear outflow); top center: [NII]λ6584 flux distribution of the narrow component, showing the region of strongest emission; top right: flux distribution of the broad components; bottom left: velocity of the red broad component; bottom center: velocity fied of the narrow component; bottom right: scenario of a nuclear outflow (pink) and ionization cones (yellow) that intercept the disk (blue) (Lena *et al.* 2014).

Yet another case of a compact nuclear outflow is observed in the Seyfert 2 galaxy NGC 1386 (Lena *et al.* 2014), where our GMOS-IFU observations covered the inner ≈ 500 pc. We found a similar result to that found in NGC 2110: while the gas emission is elongated, extending up to the borders of the FOV, and can thus be attributed to ionization by a collimated nuclear source, the kinematics shows again that most of the gas is in rotation in the galaxy disk. Only within the inner ≈ 50 pc there is a region of high velocity dispersion and double components, interpreted as due to a nuclear outflow. We show this in Fig. 3, where the cartoon illustrates our preferred explanation for this case, but the general idea should also apply to the other galaxies with compact outflows discussed above.

4. Inflows

The observations of the NLR discussed above have shown that in many cases the gas emission is elongated or cone-shaped but the kinematics is dominated by rotation in the galaxy disk. Only in the innermost region we observe outflows. The AGN thus works as a flashlight that illuminates the ambient gas that is usually rotating in the galaxy disk. This fact has allowed us to see also inflows in the gas, which are usually associated to nuclear spiral arms or dusty filaments seen in structure maps (Simões Lopes *et al.* 2007). We have observed inflows in ionized gas around LINER nuclei, where outflows are very weak or absent, making it easier to see the inflows. Examples of galaxies where we have observed inflows in ionized gas are: NGC 1097 (Fathi *et al.* 2006), NGC 6951

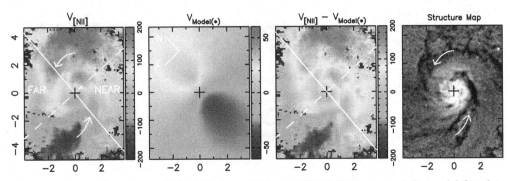

Figure 4. Left: gas velocity field of the inner 500 pc of NGC 7213; center left: model fitted to the stellar velocity field; center right: residuals between the gas and model stellar velocity field; right: structure map.

(Storchi-Bergmann *et al.* 2007), M 81 (Schnorr Müller *et al.* 2011) and NGC 7213 (Schnorr Müller *et al.* 2014b).

The case of NGC 7213 is illustrated in Fig. 4. The nucleus of this galaxy is classified as LINER/Seyfert 1, and we have obtained GMOS-IFU data of the inner $\approx$ 500 pc. The excellent signal-to-noise ratio of the data has allowed us to derive the stellar kinematics, revealing high velocity dispersion ($\approx$ 200 km s^{-1}) and rotation (amplitude of $\approx$50 km s^{-1}). Gas emission is observed over the whole FOV, presenting also rotation, but with an amplitude much larger than that of the stars. The leftmost panel of Fig. 4 shows the gas velocity field, that can be compared with the model fitted to the stellar velocity field, shown in the second panel. The third panel (to the right) shows the difference between the velocity field of the gas and that of the stars. Large residuals are found in redshift in the near side of the galaxy and in blueshift in the far side. We note that these residuals follow a spiral pattern that is correlated with a dusty spiral seen in a structure map built from an HST F606W image of the galaxy, shown in the rightmost panel of the figure. Under the assumption that the gas is in the plane of the galaxy (this is justified in the paper), we can only conclude that the excess blueshifts and redshifts are due to gas inflows along the spirals. The estimated mass inflow rate along these spirals are 0.4 M$_\odot$ yr^{-1}at $\approx$ 400 pc and 0.1 M$_\odot$ yr^{-1}at $\approx$ 100 pc of the nucleus.

We have found signatures of inflows also in warm molecular gas via 3D observations of the H$_2$ emission line at 2.122μm in the near-IR using NIFS. This line allows us to observe inflows even in luminous Seyfert galaxies. Our studies indicate that this line is usually excited by X-rays from the AGN, that penetrate in the galaxy disk. The disk origin is supported by the H$_2$ kinematics that is usually quite distinct from that of the ionized gas also observed in the near-IR. While the ionized gas (e.g. H$^+$ and [Fe II]) shows higher velocity dispersion, rotation and outflows, the H$_2$ kinematics shows lower velocity dispersion, rotation and inflows, supporting its location in the galaxy disk. H$_2$ inflows have been found in NGC 4051 (Riffel *et al.* 2008), Mrk 1066 (Riffel & Storchi-Bergmann 2011), Mrk 79 (Riffel *et al.* 2013). In the case of Mrk 766 (Schonell *et al.* 2014) we have found rotation in a compact disk, supporting also that this gas is feeding the AGN. Mass inflow rates in warm molecular gas are 4-5 orders of magnitudes lower than that observed in ionized gas. Nevertheless, the warm molecular gas may only be the heated skin of a much larger cold molecular gas reservoir (and inflow) that should be observable at mm wavelenghts (e.g. Combes *et al.* 2014).

5. Conclusions

I have reported 3D observations of the NLR of nearby active galaxies using integral field spectroscopy at spatial resolutions of 10–100 pc. These observations reveal that the AGN radiation works as a flashlight illuminating, heating and ionizing the gas in the circumnuclear regions. Ionized gas is seen not only along the AGN collimation axis, but over all directions (although at a lower intensity), indicating escape of radiation through the walls of the AGN torus. The illuminated, heated or ionized gas may be in outflow, inflow or rotating in the galaxy disk. The outfows are observed following a hollow cone or hourglass geometry or in a compact structure in the inner 50-300 pc; we suggest that these compact outflows may be the initial phase of an outflow that will evolve to an cone-like outflow. Beyond this region, rotation and inflows in the galaxy disk are observed. In a few LINERs studied, there is no outflow, and only gas rotating and inflowing in the galaxy disk are observed. Outflow velocities are in the range 200 – 800 km s^{-1} and the mass outflow rates are of the order of a few M$_\odot$ yr^{-1}. Inflows are observed along nuclear spirals, usually on $\approx$ 100 pc scales, at typical mass inflow rates in the range 0.1–1M$_\odot$ yr^{-1}. These inflow rates are much larger than the typical AGN accretion rates ($\approx$ 10^{-3}M$_\odot$ yr^{-1}), and will probably lead to the formation of new stars in the circumnuclear region (e.g. Storchi-Bergmann *et al.* 2012).

References

Allington-Smith, J., *et al.* 2002, *PASP*, 114, 892
Barbosa, F. K. B., *et al.* 2014, *MNRAS*, in press
Combes, F., *et al.* 2014, *A&A*, 565, 97
Couto, G., *et al.* 2013, *MNRAS*,435, 2892
Ciotti L, Ostriker JP, & Proga D. 2010, *ApJ* 717, 707
Elitzur M., 2012, *ApJL*, 747, L33.
Elvis, M. 2000, *ApJ*, 545, 63
Fabian, A. C., 2012, *ARAA*, 50, 455
Fathi, K., *et al.* 2006, *ApJ*, 641, L25
Ferrarese, L. & Ford, H., 2005, *SSRv*, 116, 523
Fisher, T., *et al.* 2013, *ApJS*, 209, 1
Hopkins P. & Quataert E., 2010, *MNRAS*, 407, 1529
Kormendy, J., & Ho, L. C, 2013, *ARA&A*, 51, 511
Lena, D., *et al.* 2014, submitted
McGregor, P. J., *et al.*, 2003, *Proceedings of the SPIE*, 4841, 1581
Müller Sánchez, F., Davies, R. I., Genzel, R., *et al.* 2009, *ApJ*, 691, 749
Proga, D. & Kallman, T. R. 2004, *ApJ*, 616, 688
Riffel, R., *et al.* 2008, *MNRAS*, 385, 1129
Riffel, R. A. & Storchi-Bergmann, T., 2011, *MNRAS*, 411, 469.
Riffel, R. A.; Storchi-Bergmann, T., & Winge, C. 2013, *MNRAS*, 430, 2249
Riffel, R. A., Vale, T. B., Storchi-Bergmann, T., & McGregor, P. J. 2014, *MNRAS*, 442, 656
Riffel, R. A., Storchi-Bergmann, T., & Riffel, R. 2014, *ApJ*, 780, L24
Schnorr-Müller, A., *et al.* 2011, *MNRAS*, 413, 149
Schnorr Müller, A., *et al.* 2014, *MNRAS*, 437, 1708
Schnorr Müller, A., *et al.* 2014, *MNRAS*, 438, 3322
Schonell, A. Jr., *et al.* 2014, *MNRAS*, in presss
Simões Lopes R. D., Storchi-Bergmann T., Saraiva M. F., & Martini P., 2007, *ApJ*, 655, 718.
Storchi-Bergmann, T., *et al.* 2007, *ApJ*, 670, 959.
Storchi-Bergmann, T., *et al.* 2010, *MNRAS*, 402, 819
Storchi-Bergmann, T., *et al.* 2012, *ApJ*, 755, 87
Wang, J., Elvis, M., *et al.* 2009, *ApJ*, 704, 1195

Galaxies in 3D across the Universe
Proceedings IAU Symposium No. 309, 2014   © International Astronomical Union 2015
B. L. Ziegler, F. Combes, H. Dannerbauer, M. Verdugo, eds.   doi:10.1017/S174392131400965X

Investigating AGN/Starburst activities through ALMA multi-line observations in the mid-stage IR-bright merger VV 114

Toshiki Saito[1,2], Daisuke Iono[2,3], Min S. Yun[4], Junko Ueda[2], Daniel Espada[2,5], Yoshiaki Hagiwara[2], Masatoshi Imanishi[2,3,6], Kentaro Motohara[7], Kouichiro Nakanishi[2,3,5], Hajime Sugai[8], Ken Tateuchi[7], Minju Lee[1,2] and Ryohei Kawabe[1,2,3]

[1]Department of Astronomy, The University of Tokyo,
7-3-1 Hongo, Bunkyo-ku, Tokyo 133-0033, Japan
[2]National Astronomical Observatory of Japan,
2-21-1 Osawa, Mitaka, Tokyo, 181-8588, Japan
[3]The Graduate University for Advanced Studies (SOKENDAI),
2-21-1 Osawa, Mitaka, Tokyo 181-0015, Japan
[4]Department of Astronomy, University of Massachusetts, Amherst, MA 01003, USA
[5]Joint ALMA Observatory,
Alonso de Córdova 3107, Vitacura, Casilla 19001, Santiago 19, Chile
[6]Subaru Telescope, 650 North A'ohoku Place, Hilo, HI 96720, USA
[7]Institute of Astronomy, The University of Tokyo,
7-3-1 Hongo, Bunkyo-ku, Tokyo 133-0033, Japan
[8]Kavli Institute for the Physics and Mathematics of the Universe (WPI),
The University of Tokyo, 5-1-5 Kashiwanoha, Kashiwa, Chiba 277-8583, Japan email:
`toshiki.saito@nao.ac.jp`

Abstract. We present ALMA cycle 0 observations of the luminous merger VV 114. One of the main goals is to investigate mechanisms of molecular line ratio enhancement. Regions with the high ^{12}CO (1–0)/^{13}CO (1–0) and ^{12}CO (3–2)/^{12}CO (1–0) is located at a central filamentary structure ($\sim$ 6 kpc) in VV 114. The filament consists of the eastern nucleus and the overlap region, where the galaxy disks are colliding. We also investigate these molecular line ratios on the Kennicutt-Schmidt law. VV 114 fills a gap between the "starburst" sequence and the "normal disk" sequence, and regions with the high ratios show the high $\Sigma_{\rm SFR}$ and $\Sigma_{\rm H_2}$. We suggest that the high ratios in VV 114 are due to star-forming activities in the both progenitor's nuclei and the merger-induced overlap region.

Keywords. galaxies: interaction - galaxies: starburst - ISM: molecules

1. Introduction

Molecular line ratios are important diagnostic tools of physical and chemical properties of extragalactic objects (Aalto 2007; Imanishi & Nakanishi 2014). ^{12}CO (3–2)/^{12}CO (1–0) line ratio ($R_{3-2/1-0}$) can trace conditions of molecular gas excitation (temperature and density) directly (Papadopoulos *et al.* 2012), while ^{12}CO (1–0)/^{13}CO (1–0) line ratio ($R_{12/13}$) is observationally known as a tracer of starburst activities (Casoli *et al.* 1992). Although many observations were carried out to understand the enhancement of these ratios in luminous galaxies (e.g., Casoli *et al.* 1992; Aalto *et al.* 1995; Glenn & Hunter 2001; Papadopoulos *et al.* 2012), the limited angular resolution and sample size (e.g., Aalto *et al.* 1997; Sliwa *et al.* 2012) prevent us from obtaining a detailed understanding of the physics behind the line ratio enhancement. We observed these molecular line ratios with high resolution cycle 0 ALMA (0".5 – 2".0) to investigate gas conditions in

VV 114 and the connection between gas and star-forming activities on the sky and the Kennicutt-Schmidt law (Kennicutt 1998).

VV 114 is a nearby ($D_L = 82$ Mpc) gas-rich ($M_{H_2} = 5.1 \times 10^{10}$ M$_\odot$, Yun *et al.* 1994) interacting LIRG ($L_{IR} = 4.7 \times 10^{10}$ L$_\odot$, Armus *et al.* 2009). An obscured AGN in the eastern galaxy is revealed by mid-IR spectroscopy, X-ray, and submm molecular line observations (Alonso-Herrero *et al.* 2002; Grimes *et al.* 2006; Iono *et al.* 2013), suggesting that both starburst and AGN activities might have been triggered by the ongoing merger.

2. Observations and Results

Observations toward VV 114 were carried out as the ALMA cycle 0 program (ID = 2011.0.00467.S; PI = D. Iono) using fourteen – twenty 12 m antennas. The synthesized beam size of band 3 ($\nu_{obs} \simeq 110$ GHz) and band 7 ($\nu_{obs} \simeq 345$ GHz) are 1" – 2" and 0".5, respectively. Although we made images of 10 molecular lines and continuum emission (Iono *et al.* 2013; Saito *et al.* 2013), we present the ^{12}CO (1–0) and ^{13}CO (1–0) images here.

Integrated intensity maps of the ^{12}CO (1–0) and the ^{13}CO (1–0) are shown in Fig. 1a and 1b, respectively. The ^{12}CO (1–0) emission traces northern and southern arm-like features ($\sim 12 \times 10$ kpc) and the strongest peak does not coincide with the both nuclei, but is located at the overlap region. The two arms coincide with dust lanes (Fig. 1e). On the other hand, the ^{13}CO (1–0) emission shows a filamentary structure (~ 6 kpc) at the CO centroid of VV 114. The ^{13}CO filament coincides with the Paschen alpha emission (Fig. 1f, Tateuchi *et al.* 2012). This means that the ^{13}CO line is a better tracer of star-forming activities than the ^{12}CO (1–0) line.

2.1. *Line Ratios on the sky*

Spatial distributions of the $R_{3-2/1-0}$ and $R_{12/13}$ in VV 114 are shown in Fig. 1c and 1d, respectively. The northern and southern arms show the low $R_{3-2/1-0}$ and $R_{12/13}$ (0.2 – 0.5 and < 10), while the both nuclei and the overlap region show the high ratios (0.5 – 0.8 and 20 – 50), respectively. The eastern nucleus shows the highest $R_{3-2/1-0}$ (~ 0.8), while the lower $R_{3-2/1-0}$ is measured near the western nucleus (~ 0.4). The difference may be due to a presence of strong heating sources. The $R_{3-2/1-0}$ of VV 114 is consistent with that of the early-stage merger, the Antennae (Ueda *et al.* 2012). Iono *et al.* (2013) found an obscured AGN and nuclear starbursts in the eastern galaxy using dense gas tracers (HCN and HCO$^+$), while the western galaxy only shows extended star formation (Grimes *et al.* 2006; Tateuchi *et al.* 2012).

2.2. *Line Ratios on the Kennicutt-Schmidt law*

The Kennicutt-Schmidt law with the line ratios is shown in Fig. 2a and 2b. Generally, we can see that star-forming regions in VV 114 fill a gap between two sequences ("starburst" and "normal disk", Daddi *et al.* 2010). VV 114 may be in a transition phase from spirals to a luminous merger. Highly excited regions in VV 114 ($R_{3-2/1-0} > 0.6$) already show gas concentrations and intense star-forming activities, while diffuse gas along with the arms is inactive. Similarly, the high $R_{12/13}$ regions also show active star formation.

3. Conclusions

We investigate molecular line ratios ($R_{3-2/1-0}$ and $R_{12/13}$) in the luminous merging galaxy VV 114 with cycle 0 ALMA. Both ratios are elevated (0.5 – 0.8 and 20 – 50,

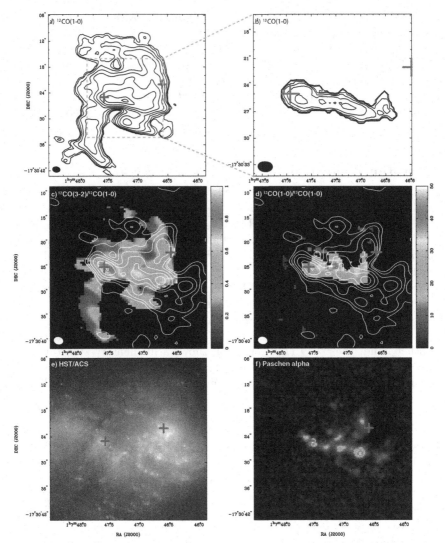

Figure 1. (a) ^{12}CO (1–0) integrated intensity image of VV 114. The red crosses show the positions of the nuclei defined by the peak positions of the Ks-band observation (Tateuchi *et al.* 2012). The black ellipse is the synthesized beam. (b) ^{13}CO (1–0) integrated intensity image of VV 114. (c) The $R_{3-2/1-0}$ image. The contour is the Paschen alpha image. The ratio in color scale ranges from 0 to 1. (d) The $R_{12/13}$ image. The ratio in color scale ranges from 0 to 40. (e) HST/ACS image of VV 114 (Credit: NASA, ESA, the Hubble Heritage (STScI/AURA)-ESA/Hubble Collaboration, and A. Evans (University of Virginia, Charlottesville/NRAO/Stony Brook University). (f) Paschen alpha image of VV 114 obtained with miniTAO/ANIR (Tateuchi *et al.* 2012).

respectively) in the overlap region as well as the progenitor's nuclei. Moreover, intense star-forming regions (with high gas surface density) show the elevated ratios. We suggest from the distribution of the ratios on the sky and the Kennicutt-Schmidt law that the elevated $R_{3-2/1-0}$ and $R_{12/13}$ are due to star-forming activities. Details will be provided in an upcoming paper (Saito *et al.* 2014 in prep.).

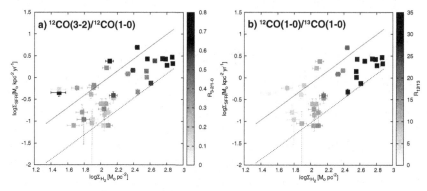

Figure 2. (a) The Keniccutt-Schmidt law with the $R_{3-2/1-0}$. The ratio in grey scale ranges from 0 to 0.8. The dashed and dotted lines are the "starburst" sequence and "normal disk" sequence, respectively (Daddi *et al.* 2010). (b) The Kennicutt-Schmidt law with the $R_{12/13}$. The ratio in grey scale ranges from 0 to 40.

Acknowledgements

This paper makes use of the following ALMA data: ADS/JAO.ALMA#2011.0.00467.S. ALMA is a partnership of ESO (representing its member states), NSF (USA) and NINS (Japan), together with NRC (Canada) and NCS and ASIAA (Taiwan), in cooperation with the Republic of Chile. The Joint ALMA Observatory is operated by ESO, AUI/NRAO, and NAOJ. We used a script developed by Y. Tamura for this calculation (http://www.ioa.s.u-tokyo.ac.jp/~ytamura/Wiki/?Science%2FUsingRADEX). TS, J. Ueda, and K. Tateuchi are financially supported by a Research Fellowship from the Japan Society for the Promotion of Science for Young Scientists. D. Iono was supported by the ALMA Japan Research Grant of NAOJ Chile Observaory, NAOJ-ALMA-0011 and JSPS KAKENHI Grant Number 2580016.

References

Aalto, S., Booth, R. S., Black, J. H., & Johansson, L. E. B. 1995, *A&A*, 300, 369
Aalto, S., Radford, S. J. E., Scoville, N. Z., & Sargent, A. I. 1997, *ApJ*, 475, L107
Aalto, S. 2007, *New Astr.*, 51, 52
Alonso-Herrero, A., Rieke, G. H., Rieke, M. J., & Scoville, N. Z. 2002, *AJ*, 124, 166
Armus, L., Mazzarella, J. M., Evans, A. S., *et al.* 2009, *PASP*, 121, 559
Bournaud, F., Powell, L. C., Chapon, D., & Teyssier, R. 2011, *IAU Symposium*, 271, 160
Casoli, F., Dupraz, C., & Combes, F. 1992, *A&A*, 264, 55
Daddi, E., Elbaz, D., Walter, F., *et al.* 2010, *ApJ*, 714, L118
Glenn, J. & Hunter, T. R. 2001, *ApJS*, 135, 177
Grimes, J. P., Heckman, T., Hoopes, C., *et al.* 2006, *ApJ*, 648, 310
Imanishi, M. & Nakanishi, K. 2014, *AJ*, 148, 9
Iono, D., Ho, P. T. P., Yun, M. S., *et al.* 2004, *ApJ*, 616, L63
Iono, D., Saito, T., Yun, M. S., *et al.* 2013, *PASJ*, 65, L7
Kennicutt, R. C., Jr. 1998, *ApJ*, 498, 541
Papadopoulos, P. P., van der Werf, P. P., Xilouris, E. M., *et al.* 2012, *MNRAS*, 426, 2601
Saito, T., Iono, D., Yun, M., *et al.* 2013, *Astronomical Society of the Pacific Conference Series*, 476, 287
Sliwa, K., Wilson, C. D., Petitpas, G. R., *et al.* 2012, *ApJ*, 753, 46
Tateuchi, K., Motohara, K., Konishi, M., *et al.* 2012, *Publication of Korean Astronomical Society*, 27, 297
Ueda, J., Iono, D., Petitpas, G., *et al.* 2012, *ApJ*, 745, 65
Yun, M. S., Scoville, N. Z., & Knop, R. A. 1994, *ApJ*, 430, L109

Galaxies in 3D across the Universe
Proceedings IAU Symposium No. 309, 2014
B. L. Ziegler, F. Combes, H. Dannerbauer, M. Verdugo, eds.

© International Astronomical Union 2015
doi:10.1017/S1743921314009661

S7 : Probing the physics of Seyfert Galaxies through their ENLR & HII Regions

Michael A. Dopita[1,2], Prajval Shastri[3], Julia Scharwächter[4], Lisa J. Kewley[1,5], Rebecca Davies[1], Ralph Sutherland[1], Preeti Kharb[3], Jessy Jose[3], Harish Bhatt[3], S. Ramya[3], Elise Hampton[1], Chichuan Jin[6], Julie Banfield[7], Ingyin Zaw[8], Shweta Srivastava[9] and Bethan James[10]

[1]RSAA, Australian National University, Cotter Road, Weston Creek, ACT 2611, Australia
[2]Astronomy Department, King Abdulaziz University, P. O. Box 80203, Jeddah, Saudi Arabia
email: Michael.Dopita@anu.edu.au

[3]Indian Institute of Astrophysics, Koramangala 2B Block, Bangalore 560034, India
[4]LERMA, Observatoire de Paris, 61 Avenue de l'Observatoire, 75014 Paris, France
[5]Institute for Astronomy, University of Hawaii, 2680 Woodlawn Drive, Honolulu, HI, USA
[6]Department of Physics, University of Durham, South Road, Durham DH1 3LE, UK
[7]CSIRO Astronomy & Space Science, P. O. Box 76, Epping NSW, 1710 Australia
[8]New York University (Abu Dhabi) , 70 Washington Sq. S, New York, NY 10012, USA
[9]Gorakhpur University, Gorakhpur, Uttar Pradesh, India
[10] Institute of Astronomy, Cambridge University, Madingley Road, Cambridge CB3 0HA, UK

Abstract. Here we present the first results from the *Siding Spring Southern Seyfert Spectroscopic Snapshot Survey* (S7) which aims to investigate the physics of $\sim$ 140 radio-detected southern active Galaxies with $z < 0.02$ through Integral Field Spectroscopy using the Wide Field Spectrograph (WiFeS). This instrument provides data cubes of the central 38×25 arc sec. of the target galaxies in the waveband $340 - 710$nm with the unusually high resolution of $R = 7000$ in the red ($530 - 710$nm), and $R = 3000$ in the blue ($340 - 560$nm). These data provide the morphology, kinematics and the excitation structure of the extended narrow-line region, probe relationships with the black hole characteristics and the host galaxy, measures host galaxy abundance gradients and the determination of nuclear abundances from the HII regions. From photoionisation modelling, we may determine the shape of the ionising spectrum of the AGN, discover whether AGN metallicities differ from nuclear abundances determined from HII regions, and probe grain destruction in the vicinity of the AGN. Here we present some preliminary results and modelling of both Seyfert galaxies observed as part of the survey.

Keywords. galaxies: elliptical and lenticular, cD - galaxies: evolution - galaxies: formation

1. Introduction

The study of nearby Seyfert galaxies offers potential insights not only into the physics of active galaxies themselves, but also into the galactic environments which feed the nuclear activity. According to the "standard" unified model of AGN (Antonucci & Barvainis 1990; Antonucci 1993) and its extensions (Dopita 1997), the Seyfert 1 galaxies are seen pole-on relative to the accretion disk, and these display very broad permitted lines originating in rapidly moving gas close to the central engine. In the Seyfert 2 galaxies, the thick, dusty and toroidal accretion disk obscures the central engine, and an Extended Narrow Line Region (ENLR) often confined within an "ionisation cone" is observed.

The nature of this ENLR can provide vital clues about the nature of the central black hole, and the mechanisms which produce the extreme UV (EUV) continuum. Seyfert galaxies are known to occupy a very restricted range of line ratios when plotted on the well-known BPT diagram (Baldwin *et al.* 1981) which plots [N II] $\lambda6584$/Hα vs. [O III] $\lambda5007$/Hβ or on the other diagrams introduced by Veilleux & Osterbrock (1987) involving the [S II] $\lambda6717,31$/Hα ratio or the [O I] $\lambda6300$/Hα ratio in the place of the [N II] $\lambda6584$/Hα ratio. It now seems clear that this is because the ENLR is (in perhaps all cases) radiation pressure dominated (Dopita & Groves 2002; Groves, Dopita & Sutherland 2004b; Groves *et al.* 2004). In this model, radiation pressure (acting upon both the gas and the dust) compresses the gas close to the ionisation front so that at high enough radiation pressure, the density close to the ionisation front scales as the radiation pressure, and the local ionisation parameter in the optically-emitting ENLR becomes constant. This results in an ENLR spectrum which is virtually independent of the input ionisation parameter. For dusty ENLR the radiation pressure comes to dominate the gas pressure for $\log U \gtrsim -2.5$, and the optical emission spectrum becomes invariant with the input ionisation parameter for $\log U \gtrsim 0.0$. In this condition, the observed density in the ENLR should drop off in radius in lockstep with the local intensity of the radiation field; $n_e \propto r^{-2}$, and the EUV luminosity can be inferred directly from a knowledge of the density and the radial distance.

Despite this constancy of the emission spectrum at high ionisation parameter, the emission line ratios remain sensitive to the form of the input EUV spectrum. Although the spectral energy distribution of Seyfert galaxies has now been studied from the far-infrared all the way up to hard X-ray and even γ−ray energies (*e.g.* Done *et al.* (2012); Jin *et al.* (2012a); Jin, Ward & Done (2012a)), the spectral region between 13.6 and ~ 150eV remains inacessible to either ground- or space-based observation. However, the ENLR spectrum is most sensitive to the form of this EUV spectrum, and its gross features can be inferred by a method reminiscent of the energy balance or Stoy technique used to estimate the effective temperature of stars in planetary nebulae (Stoy 1933; Kaler 1976; Preite-Martinez & Pottasch 1983). As the radiation field becomes harder, the heating per photoionisation increases, and the sum of the fluxes of the forbidden lines becomes greater relative to the recombination lines. Furthermore, individual line ratios are sensitive in different ways to the form of the EUV spectrum, and this can be exploited to infer the form of the EUV spectrum. However, for this method to work, we need to have a well-constrained knowledge of the chemical abundances in the ENLR.

A fair number of high-resolution imaging surveys of Seyferts in the [O III] $\lambda5007$ line have been undertaken, notably by Pogge (1988a,b, 1989); Haniff, Wilson & Ward (1988); Mulchaey, Wilson & Tsvetanov (1996a,d); Falke *et al.* (1998) and Schmitt *et al.* (2003). However, up to the present there have been relatively few systematic spectroscopic studies of the ENLR of nearby Seyferts at optical wavelengths (Cracco *et al.* 2011). Notable amongst these is the multi-slit work by Allen *et al.* (1999). However, the advent of integral field spectroscopy has made such studies much more efficient, and it is now possible to gather as much data in a night as Allen *et al.* (1999) did in three years! Motivated by these ideas, we have undertaken the *Siding Spring Southern Seyfert Spectroscopic Snapshot Survey* (S7) and here describe the methodology and first results of this survey.

2. The S7 Survey

The S7 survey is an integral field survey in the optical of ~ 140 southern Seyfert and LINER galaxies. It uses the Wide Field Spectrograph (WiFeS) mounted on the Nasmyth focus of the ANU 2.3m telescope (Dopita *et al.* 2010). This instrument provides data cubes of the central 38×25 arc sec. at a spatial resolution of 1.0 arc sec. It covers the waveband $340 - 710$nm with the unusually high resolution of $R = 7000$ in the red ($530 - 710$nm), and $R = 3000$ in the blue ($340 - 560$nm). The typical throughput of the instrument (top of the atmosphere to back of the detector) is $20 - 35\%$ (Dopita *et al.* 2010), which provides an excellent sensitivity to faint low surface brightness ENLR features.

The sample objects for S7 were selected from the Véron-Cetty & Véron (2006) catalogue of active galaxies, which is the most comprehensive compilation of active galaxies in the literature. Since we wish to investigate the interaction of the bipolar plasma jets with the Narrow Line Region and the ISM of the host galaxy, we limited the sample to galaxies with radio flux densities high enough to permit radio aperture synthesis observations. We adopted the following selection criteria:

- Declination < 10 degrees to avoid WiFeS observations at too great a zenith distance,
- Galactic latitude < |20| degrees (with a few exceptions) to avoid excessive galactic extinction.
- Radio flux density at 20cm $\lesssim 20mJy$ for those targets with declination N of -40 deg, which have NVSS measurements, and
- Redshift < 0.02. This criterion ($D < 80$ Mpc) ensures that the spatial resolution of the data is better than 400 pc arcsec^{-1}, sufficient to resolve the ENLR, and to ensure that the important diagnostic [S II] lines are still within the spectral range of the WiFeS high-resolution red grating.

All the data are reduced using the using the `PyWiFeS` pipeline written for the instrument (Childress *et al.* 2014). In brief, this produces a data cube which has been wavelength calibrated, sensitivity corrected (including telluric corrections), photometrically calibrated, and from which the cosmic ray events have been removed. With this software, a full night's worth of data can be reduced on a laptop in about 2 Hr. The data cube is then passed through the integral field spectrograph toolkit `LZIFU` (Ho *et al.* 2014 in prep.) to extract gas and stellar kinematics and other parameters from the WiFeS data. `LZIFU` performs simple stellar population (SSP) synthesis fitting to model the continuum, and fits the emission lines as 3-component Gaussians.

The standard data products of the S7 survey include images and BPT diagrams such as shown for one galaxy, NGC 6890 in Figure 1, and nuclear spectra – examples of which are given in Figure 2. In addition we have (for the gaseous component) emission line flux maps, line ratio maps, gas velocity maps, and gas velocity dispersion maps. From these we can derive abundance, star formation and dust extinction maps. For the stellar components we have continuum maps, extinction, stellar velocity and stellar velocity dipsersion maps. From these we can derive age distribution, metallicity etc.

3. First Results

In the first data S7 data release of about 50 Seyferts and LINERS (due before the end of 2014), a wide range of properties will be explored. Here we have only space to point out a couple of highlights of the research accomplished so far.

3.1. *Form of the EUV spectrum*

As described above, the ENLR spectrum provides strong constraints on the EUV spectrum of the Seyfert nucleus, provided that the ionisation parameter and chemical abundances in the ENLR can be inferred. As a first test example of the technique Dopita *et al.* (2014) examined NGC 5427 in some detail. Using off-nuclear integrations as well as the on-nucleus field, we were able to establish the abundance gradient and determine nuclear abundances from strong emission lines of individual HII regions using the `Pyqz` photoionisation grid interpolation routine described in Dopita *et al.* (2013). With the inferred nuclear abundances, a tight constraint on both $\log U$ and the shape of the EUV

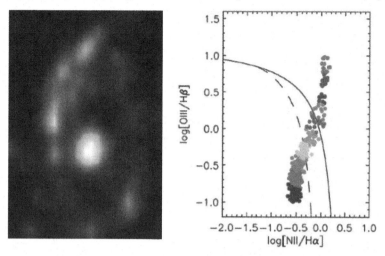

Figure 1. An example of data products produced in the S7 survey. The Left-hand panel is an image of NGC 6890 in Hα (red), [N II] λ6584 (green) and [O III] λ5007 (blue), boxcar smoothed over one arc sec. The H II regions appear reddish or gold depending on their excitation, and trace out the spiral arms, while the nucleus and the ENLR appears blue. On the right-hand panel we show the BPT diagram for the individual spaniels, colour coded according to the angular distance from the nucleus. This forms a tight mixing sequence between pure AGN line ratios (magenta) and pure HII region emission (brown). Compare *e.g.* Dopita *et al.* (2014); Davies *et al.* (2014a).

spectraum was obtained using the same **Mappings** IV code that was employed for the HII region analysis. This demonstrated that the hot accretion disk + hard Compton model of Done *et al.* (2012); Jin *et al.* (2012a); Jin, Ward & Done (2012a) worked well. The intermediate Compton component introduced by these authors in order to fit the X-ray data of luminous Seyferts seems to be absent. However, our models seem to indicate that such a component arises naturally within the inner ENLR itself, provided that the ionisation parameter is sufficient to Compton heat the plasma up to $\sim 10^6$K.

3.2. *Coronal Emission Regions*

Nuclear spectra of very high quality have been extracted for all the observed S7 galaxies using a 2.0 arc sec. aperture. Although the analysis of these is not yet complete, it is already clear that not only are there a number of mis-classified galaxies in the Véron-Cetty & Véron (2006) catalogue, but also the that classification scheme used for Seyferts (S3b, S1.8, S3h etc.) is too fine, and perhaps undermines a sense of unity and continuity between the different classes of object. It is clear from our data that important systematic trends exist object to object, and here we seek to illustrate this within the restricted range of objects showing clear coronal emission from species such as [Fe V], [Fe VII], [Fe X] and [Fe XIV], see Figure 2. The objects shown here are not all the objects for which we have detected coronal emission – strong coronal emission can also be seen in the Type I objects, Fairall 51 and NGC 7469. It is evident from Figure 2 that the relative strength and excitation of the coronal emission is correlated with the Hα line width, and with the electron density in the [O III] - emitting region, as evidenced by the [O III] λλ4363/5007 ratio. The density in the low-excitation gas as revealed by the [S II] λλ6717/6731 ratio remains around $n_e \sim 10^4$cm^{-3}, and the strength of these [S II] lines and the [N II] λ6584 line relative to the broad component of Hα also changes systematically with Hα line width. This strongly suggests that these low excitation lines arise in a region which is physically distinct from the region emitting the coronal species.

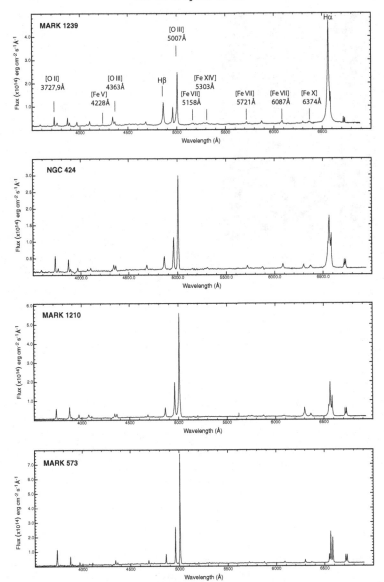

Figure 2. A selection of S7 galaxy nuclear spectra showing strong coronal line emission. The main coronal species and a number of the other lines are identified on the first panel. The panels are ordered in terms of the electron density as indicated by the [O III] $\lambda\lambda4363/5007\text{Å}$ ratio (MARK 1239 being the densest). Note that this order matches the order of the $H\alpha$ line width and the [N II] $\lambda6584/H\alpha$ ratio. This suggests strong correlations between the region of the narrow line emission, the electron density, and the strength of the coronal features.

These properties are consistent with the model advocated by Mullaney *et al.* (2009). In this, the coronal lines arise in a dense gas launched from the dusty inner torus ($10^{17} < R/\text{cm} < 10^{18}$) at very high local ionisation parameter $\log U \sim -0.4$, and is accelerated by radiation pressure to a terminal velocity of a few hundred km s^{-1}. At this ionisation parameter the gas is Compton heated to $\sim 10^6\,\text{K}$ or greater and the dust in the coronal emission region is destroyed, allowing the forbidden iron lines to reach such high intensity relative to the hydrogen lines. Systematic changes in width, excitation and and density are consistent with different inner torus radii for the objects shown in Figure 2 - MARK 1239 having a small inner torus, and MARK 573 having a larger one.

Acknowledgements

M. D. and L. K. acknowledge the support of the Australian Research Council (ARC) through Discovery project DP130103925. M. D. also acknowledges support under the King Abdulaziz University *HiCi* program. This research has made use of the NASA/IPAC Extragalactic Database (NED), which is operated by the Jet Propulsion Laboratory, California Institute of Technology, under contract with the National Aeronautics and Space Administration, the NASA Astrophysics Data System (ADS), and SAOImage DS9 (Joye & Mandel 2003), developed by the Smithsonian Astrophysical Observatory.

References

Allen, M. G., Dopita, M. A., Tsvetanov, Z. I., & Sutherland, R. S. 1999, *ApJ*, 511, 686
Antonucci, R. & Barvainis, R. 1990, *ApJ*, 363, L17
Antonucci, R. 1993, *ARA&A*, 31, 473
Baldwin, J. A., Phillips, M. M., & Terlevich, R. 1981, *PASP*, 93, 5
Childress, M. J. , Vogt, F. P. A. , Nielsen, J., & Sharp, R. G., 2014, *Ap&SS*, 349, 617
Cracco, V, *et al.* 2011, *MNRAS*, 418, 2630
Davies, R. L., Rich, J. A., Kewley, L. J., & Dopita, M. A. 2014, *MNRAS*, 439, 3835
Dopita, M. 1997, *PASA*, 14, 230
Dopita, M. A., Groves, B. A., Sutherland, R. S., Binette, L., & Cecil, G. 2002, *ApJ*, 572, 753
Dopita, M., Hart, J., McGregor, P., Oates, P., Bloxham, G., & Jones, D. 2007, *Ap&SS*, 310, 255
Dopita, M., Rhee, J., Farage, C., *et al.* 2010, *Ap&SS*, 327, 245
Dopita, M. A, Sutherland, R. S., Nicholls, D. C., Kewley, L. J., & Vogt, F. P. A. 2013, *ApJS*, 208, 10
Dopita, M. A. *et al.* 2014, *A&A*, 566, 41
Haniff, C. A.,Wilson, A. S., & Ward, M. J. 1988, *ApJ*, 334, 104
Done, C., Davis, S. W., Jin, C., Blaes, O., & Ward, M. 2012, *MNRAS*, 420, 1848
Falcke, H., Wilson, A., & Simpson, C. 1998, *ApJ*502, 199
Groves, B. A., Dopita, M. A., & Sutherland, R. S. 2004, *ApJS*, 153, 9
Groves, B. A., Dopita, M. A., & Sutherland, R. S. 2004, *ApJS*, 153, 75
Groves, B. A., Cecil, G., Ferruit, P., & Dopita, M. A. 2004, *ApJ*, 611, 786
Jin, C., Ward, M., Done, C., & Gelbord, J. 2012, *MNRAS*, 420, 1825
Jin, C., Ward, M. & Done, C. 2012, *MNRAS*, 422, 3268
Kaler, J. B. 1976, *ApJ*, 210, 843
Mulchaey, J. S., Wilson, A. S., & Tsvetanov, Z. I., 1996a, *ApJS*102, 309
Mulchaey, J. S., Wilson, A. S., & Tsvetanov, Z. I., 1996b, *ApJ*, 467, 197
Mullaney, J. R., Ward, M. J., Done, C., Ferland, G. J., & Schurch, N. 2009, *MNRAS*, 394, L16
Pogge, R. W., 1998a, *ApJ*, 328, 519
Pogge, R. W., 1988b, *ApJ*, 332, 702
Pogge, R. W., 1989, *ApJ*, 345, 730
Preite-Martinez, A. & Pottasch, S. R. 1983, *A&A*, 126, 31
Schmitt, H. R., Donley, J. L., Antonucci, R. R. J., Hutchings, J. B., & Kinney, A. L., 2003, *ApJS*, 148, 327
Stoy, R. H. 1933, *MNRAS*, 93, 588
Veilleux, S. & Osterbrock, D. E., 1987, *ApJS*, 63, 295
Véron-Cetty, M-P. & Véron, P. *A&A*455, 773

Galaxies in 3D across the Universe
Proceedings IAU Symposium No. 309, 2014
B. L. Ziegler, F. Combes, H. Dannerbauer, M. Verdugo, eds.

© International Astronomical Union 2015
doi:10.1017/S1743921314009673

Why galaxies care about Type Ia supernovae?

Noelia Jiménez,[1] Patricia B. Tissera[2] and Francesca Matteucci[3]

[1] Institut d'Estudis Espacials de Catalunya (ICE, IEEC/CSIC), E-08193 Bellaterra, Barcelona
[2] Physics Department, Universidad Andres Bello, Av. Republica 220, Santiago, Chile.
[3] Dipartimento di Fisica, Universita' di Trieste, Via G. B. Tiepolo, 11, 34100 Trieste.
email: jimenez@ieec.uab.es

Abstract. We implement the Single Degenerate (SD) scenario proposed for Type Ia Supernova (SNIa) progenitors in SPH simulations. We analyse the chemical evolution of bulge-type galaxies together with the observed correlations relating SNIa rates with the characteristics of the host galaxy, such as their SFR. The models reproduce the observed signatures shown by [O/Fe] ratios in the Galactic Bulge and the present day SNIa rates. Also, the observed correlation found by Sullivan *et al.* (2006) between SSFR (specific star formation rate) and the SNIa rate per unit of galaxy mass (SSNIaR), naturally arises. This analysis helps to set more stringent constraints to the galaxy formation models and gives some hint on the progenitor problem.

Keywords. galaxies: evolution - galaxies: formation - -galaxies: supernovae -galaxies: numerical simulations

1. Introduction

Galaxy formation models make use of sub-grid treatments to include complex physical processes which cannot be numerically resolved, as for example, the SNIa explosions (Kobayashi 2004; Jiménez *et al.* 2011; Tissera *et al.* 2013). The still uncertain identity of the SNIa progenitors constitutes a major issue for the models since the regulation of both chemical and energetic feedbacks in the gas dynamics and in the overall stellar evolution is affected. Hence, is of utmost importance to confront the model results with observations to learn about galaxy formation and to test the validity of the adopted hypotheses. The information provided by chemical patterns left by the nucleosynthesis products of the stellar populations can provide stringent constraints for galaxy formation models (Tinsley 1979; Matteucci & Greggio 1986; Tissera *et al.* 2014). Additionally, observations relating the SNIa rates of galaxies with the characteristics of the host galaxy as their morphology, colours and SFR can be used as constraints to the progenitors or SNIa (Greggio 2010).

The SD scenario is one of the most popular scenario proposed for explaining the SNIa based on a thermonuclear explosion of a C-O white dwarf exceeding the Chandrasekhar mass through accreting from a non-degenerate companion star, where the mass accretion can assume many configurations. We adopt the SD proposed by Matteucci & Recchi (2001, from now, MR01), where SNIa originate from a binary systems with one C-O WD and a red (super) giant star. The explosion times correspond to the lifetimes of stars in the mass range $0.8 - 8M_\odot$. The first system made of two $8M_\odot$ stars explodes after $\sim 3 - 4 \times 10^7$ yrs from the beginning of star formation.

2. The Simulations

We use the Tree-PM SPH code GADGET-3 (Springel 2005; Scannapieco *et al.* 2005, 2006), including star formation, chemical enrichment by supernova feedback,

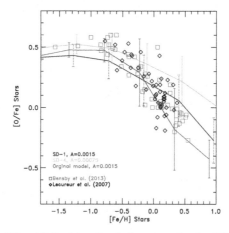

Figure 1. [O/Fe] vs. [Fe/H] exhibited by the bulge stars in the SPH simulated galaxies with the SD scenario by Matteucci & Recchi (2001), with varying A and the Original model. These models are compared to observational [O/Fe] ratios for stars in the Galactic Bulge by Bensby *et al.* (2013) & Lecureur *et al.* (2007).

metal-dependent cooling and a multiphase model for the gas component. At the beginning, the gas component represents $\sim 65\%$ of the total baryonic mass and is distributed in the disc component. The final bulges are formed by a dissipative-dominated component and a central bar concentrated within the inner ~ 3 kpc. The B+D stellar mass are nearly the same for all the simulations $\sim 3.5 \times 10^{10}$ M$_\odot$, typical for a small ellipticals or a bulge of spiral galaxies. Moreover, the analysis of the SFR of the bulges shows that the stars formed during the strong starbursts, where the cold gas is exhausted within ~ 1 Gyr. Hence, the new stars formed for which we follow the chemical abundances are mainly part of the bulge component. In the model of Scannapieco *et al.* (2005, 2006) the lifetimes of the binary system that explode as SNIa are assumed to be randomly distributed within a certain range given by $\tau_{SNIa} = [0.1, 1]$ Gyr. This model is referred as the "Original model" to distinguish it from the implementation of the SD presented here. Thus, considering a larger minimum lifetime for the distribution of the secondary mass, compared to the predictions of the SD scenario.

2.1. Calibrating the SD scenario

The SD scenario involve a free parameter (A), which is associated to the fraction of binary systems of the type necessary to produce SNIa (see MR01 for further details). This value should be adjusted to match the observed SNIa rate at the present time. We use the SNIa rates from spheroidal-dominated galaxies given by Li *et al.* (2011), obtaining good agreement with models: SD-1 (run with A=0.0015, SNIa rates =0.002 SN/yr) and SD-4 (A−0.0075, SNIa rates−0.0016 SN/yr). The Original model (A=0.0015, SNIa rates=0.0027 SN/yr) show rates of similar values. We require the models also to reproduce the knee expected in [O/Fe] vs [Fe/H] distribution. Therefore, we compute the average stellar mass abundance of [Fe/H] and [O/Fe] for the stars in the bulge of each model and compared to Galactic Bulge stars, taken this comparison as indicative. The sample includes dwarfs and subgiant stars from Bensby *et al.* (2013) and Red giants stars from Lecureur *et al.* (2007). Form Fig. 1 it can be seen that the observed data points are best represented by the models SD-4 and SD-1. The former model (SD-4) fits the zero point of the data following the observed trend up to [Fe/H]~ -0.25. Meanwhile, SD-1 matches the slope and passes through the data for larger metallicities ranges. These

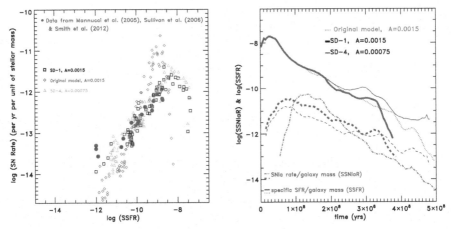

Figure 2. *(Left Panel):* SSFR as a function of the SSNIaR for all the models. The red circles represent combined data from Sullivan *et al.*; Mannucci *et al.*; Smith *et al.*. Notice that these models also fit the observed [O/Fe] ratios coming from the Galactic Bulge and the present-day SNIa rates. *(Right Panel):* The log(SSFR) and log(SSNIaR), evolving with time for all the models.

models predict a long plateau for the [O/Fe] ratio and a knee occurring at high [Fe/H], as observations suggest. However, at high metallicity the slopes of both models do not follow so nicely the observed [O/Fe] ratios. This is perhaps a consequence of yields of Woosley & Weaver (1995) that do not include mass loss from massive stars. The Original model, albeit simple, fits the data in the whole range.

3. SNIa rates and SSFR correlation

By using the SuperNova Legacy Survey (SNLS) galaxy sample Sullivan *et al.* (2006), showed an existing correlation between the specific SFR and the specific SNIa rates (SSFR and SSNIaR, respectively). He also found this correlation for observations of a morphologically classified sample of galaxies in the local universe presented by Mannucci *et al.* (2005). Later, Smith *et al.* (2012) confirmed this correlation, adding the fact that it does not seem to evolve with redshift.

We test if the models can reproduce the observed correlation. For this the SSNIaR and the SSFR are estimated as the galaxy evolves, under the hypothesis that observations might catch galaxies at different stages of evolution (see also Greggio (2010) for a similar approach using analytical models). In the left panel of Fig. 2 we show a combination of the observed data by Smith *et al.* (2012); Sullivan *et al.* (2006); Mannucci *et al.* (2006), and the results for the SD-1, SD-4 and the Original model. Given the reported uncertainties in the zero point of this correlation (see Smith *et al.* 2012), we focus on the study of the slope of this correlation. Consequently, the simulated relations are normalized by an *ad hoc* factor to make the comparison of the slope easier. First, we notice that the observed correlation comes out naturally from the SD models. This is remarkable, since it was not used for their calibration. On the other hand, the Original model, albeit reproducing the SNIa rates and the [O/Fe] ratios, does not follow the trend displayed by the data, showing a different slope.

All the models show that the SSNIaR increases abruptly with high values of SSFR before reaching the expected observed trend. To understand this behaviour we plot in the right panel of Fig. 2 the logarithm of SSFR and the SSNIaR as a function of time. Here it can be seen that the sharp increase of SSNIaR for high SSFR reflects the onset of

SNIa and how quickly it reaches the maximum value. The Original model differs from the SD ones showing a delayed peak with a stronger declination after reaching the maximum value. For all the models, after the maximum is reached, both the SSNIaR and the SSFR decrease establishing the observed correlation. For very low SSFR, there is still a residual SSNIaR, as the generation of these events is delayed in time and the correlation is lost.

Although we are using a simple example of galaxy evolution, it is encouraging that the SD scenario reproduces a clear correlation which agrees remarkably well with observations. On the other hand, in the Original model, the SNIa production is shifted to later times compared to the SD scenario, beginning at $\sim 1 \times 10^8$ yrs. This excludes the so-called prompt type Ia SNe, that are indeed observed, additionally this model fails at reproducing the slope of the observed correlation. In a forthcoming paper we will explore different SNIa progenitor scenarios applying galaxy formation models to understand the nature of the SNIa progenitors.

Acknowledgments

NJ warmly thanks the organizers of the IAU 309 Symposium and the financial aid received. NJ acknowledges support from the European Commission's Framework Programme 7, through the Marie Curie International Research Staff Exchange Scheme LACEGAL (PIRSES-GA-2010-269264). This work has been partially supported by "Proyecto Interno" of UNAB (Chile), PIP Conicet-2009/0305 and PICT Racies-2011/959 (Argentina). Simulations were run in Fenix Cluster (IAFE, CONICET-UBA) and Hal Cluster (UNC).

References

Bensby, T., Yee, J. C., Feltzing, S., *et al.* 2013, *A&A*, 549, A147
Greggio, L. 2010, *MNRAS*, 406, 22
Jiménez, N., Cora, S. A., Bassino, L. P., Tecce, T. E., & Smith Castelli, A. V. 2011, *MNRAS*, 417, 785
Kobayashi, C. 2004, *MNRAS*, 347, 740
Lecureur, A., Hill, V., Zoccali, M., *et al.* 2007, *A&A*, 465, 799
Li, W., Chornock, R., Leaman, J., *et al.* 2011, *MNRAS*, 412, 1473
Mannucci, F., Della Valle, M., & Panagia, N. 2006, *MNRAS*, 370, 773
Mannucci, F., Della Valle, M., Panagia, N., *et al.* 2005, *A&A*, 433, 807
Matteucci, F. & Greggio, L. 1986, *A&A*, 154, 279
Matteucci, F. & Recchi, S. 2001, *ApJ*, 558, 351
Scannapieco, C., Tissera, P. B., White, S. D. M., & Springel, V. 2005, *MNRAS*, 364, 552
Scannapieco, C., Tissera, P. B., White, S. D. M., & Springel, V. 2006, *MNRAS*, 371, 1125
Smith, M., Nichol, R. C., Dilday, B., *et al.* 2012, *ApJ*, 755, 61
Springel, V. 2005, *MNRAS*, 364, 1105
Sullivan, M., Le Borgne, D., Pritchet, C. J., Hodsman, A., & Neill. 2006, *ApJ*, 648, 868
Tinsley, B. M. 1979, *ApJ*, 229, 1046
Tissera, P. B., Beers, T. C., Carollo, D., & Scannapieco, C. 2014, *MNRAS*, 439, 3128
Tissera, P. B., Scannapieco, C., Beers, T. C., & Carollo, D. 2013, *MNRAS*, 432, 3391
Woosley, S. E. & Weaver, T. A. 1995, *ApJS*, 101, 181

Galaxies in 3D across the Universe
Proceedings IAU Symposium No. 309, 2014
B. L. Ziegler, F. Combes, H. Dannerbauer, M. Verdugo, eds.

© International Astronomical Union 2015
doi:10.1017/S1743921314009685

Dissecting galactic winds with the SAMI Galaxy Survey

I-Ting Ho[1] and the SAMI Galaxy Survey Team[2]

[1] Institute for Astronomy, University of Hawaii, 2680 Woodlawn Drive, Honolulu, HI 96822, USA
email: `itho@ifa.hawaii.edu`
[2] Full list of team members is available at `http://sami-survey.org/members`

Abstract. We conduct a case study on a normal star-forming galaxy ($z = 0.05$) observed by the SAMI Galaxy Survey and demonstrate the feasibility and potential of using large integral field spectroscopic surveys to investigate the prevalence of galactic-scale outflows in the local Universe. We perform spectral decomposition to separate the different kinematic components overlapping in the line-of-sight direction that causes the skewed line profiles in the integral field data. The three kinematic components present distinctly different line ratios and kinematic properties. We model the line ratios with the shock/photoionization code MAPPINGS IV and demonstrate that the different emission line properties are caused by major galactic outflows that introduce shock excitation in addition to photoionization. These results set a benchmark of the type of analysis that can be achieved by the SAMI Galaxy Survey on large numbers of galaxies.

Keywords. galaxies: evolution, galaxies: kinematics and dynamics - galaxies: starburst

1. Introduction

Galactic-scale outflows or galactic winds are one of the major "feedback" mechanisms regulating the mass assembly and star-forming activities of galaxies as they evolve over cosmic time (see Veilleux *et al.* 2005 for a review). Despite the import role outflows play in galaxy evolution, the prevalence and magnitudes of outflows are not well constrained. Observations of extra-planar optical emission and blue-shifted absorption lines have shown that starburst-driven winds are common in local galaxies ($z < 0.5$) with extreme star-forming activities with high star formation rate surface densities (e.g., $\Sigma_* \geqslant 0.1$ M$_\odot$ yr^{-1} kpc^{-2}, Heckman 2002; Rupke *et al.* 2005), and ubiquitous at high redshifts where the cosmic star formation rates peak (e.g., $z > 1.4$; Weiner *et al.* 2009; Steidel *et al.* 2010). For the normal star-forming population in the local Universe, detecting outflow signatures in individual galaxies are still difficult with the current spectroscopy surveys, unless with stacking analysis (e.g. Chen *et al.* 2010). Without spatially resolved data on a statistical sample, the prevalence and degree of galactic-scale outflows in normal star-forming galaxies remain largely unclear. The recent integral field spectroscopy survey using the new Sydney-AAO Multi-object Integral field spectrograph (SAMI; Croom *et al.* 2012) on the 3.9-m Anglo-Australian Telescope (i.e. the SAMI Galaxy Survey; Bryant *et al.* 2014; Allen *et al.* 2014) provides a unique opportunity to unravel the wind population in the local Universe ($z \lesssim 0.05$). In Ho *et al.* (2014), we demonstrate that by taking advantage of the high spectral resolution of SAMI (up to $R \approx 4500$) we can unambiguously identify wind galaxies through investigating the emission line ratios and kinematics. In this proceeding, some of the key results are summarized.

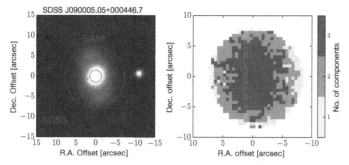

Figure 1. *Left:* SDSS *g, r, i* color composite image of SDSS J090005.05+000446.7 ($z = 0.05386$) over-plotted with footprints of Sloan Digital Sky Survey (SDSS) fiber and SAMI. The SDSS fiber has a diameter of $3''$. The SAMI hexabundle has a circular field of view of $15''$ in diameter. *Right:* Component map describing the number of components required to model the spectral profiles in each SAMI spaxel.

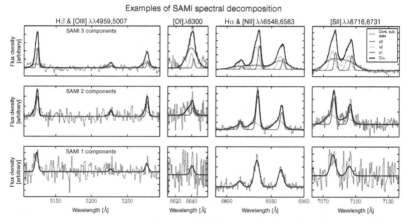

Figure 2. Examples of our spectral decomposition on selected spaxels requiring 3-component (top row), 2-component (middle row) and 1-component (bottom row) fits to describe the spectral profiles. Each panel shows a zoom-in of some key diagnostic emission lines. Continuum subtracted spectra are shown in grey. Best-fitting $c3$, $c2$, and $c1$ are shown in red, orange, and blue, respectively. The best-fitting models, $\sum c_i$, are shown in black. To avoid confusion where the lines are blended, we also show individual line transitions of $c3$ as transparent red lines.

2. Shocks and outflows in SDSS J090005.05+000446.7

In the integral field data of SDSS J090005.05+000446.7 (hereafter, SDSS J0900; left panel of Fig. 1) delivered by the SAMI Galaxy Survey, SDSS J0900 presents skewed line profiles changing with position in the galaxy because of the different kinematic components overlapping in the line-of-sight direction. We fit each spaxel with multiple component Gaussians depending on the complexity of the line profiles. In the most complicated cases, a broad $c3$, an intermediate $c2$ and a narrow $c1$ kinematic components are required to described the spectral profiles. Some examples of our spectral decomposition are shown in Fig. 2, and the right panel of Fig. 1 shows the numbers of components required at different locations in the SAMI field of view.

The different kinematic components present vastly different kinematic properties (see Ho *et al.* 2014) and line ratios. On the classical BPT diagrams (top row of Fig. 3), the emission line ratios of the different components are distinctly different, with the narrow $c1$ component consistent with photoionization, the broad $c3$ component consistent

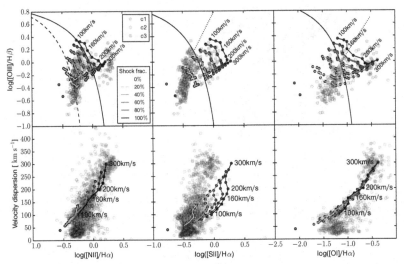

Figure 3. *Top row:* The three classical BPT diagrams (Baldwin *et al.* 1981). *Bottom row:* The three key line ratios versus velocity dispersion. The blue, orange, and red points correspond to the different kinematic components (see Fig. 2). The other color points connected by line segments are shock/photoionization model grids from MAPPINGS (see Ho *et al.* 2014 for details). The line segments connect model grids of certain shock fractions with changing shock velocities. The pure photoionization points (zero shock fraction) are at the bottom left corners (magenta). The different components show distinctly different line ratios and kinematics and the MAPPINGS models well reproduce the observations.

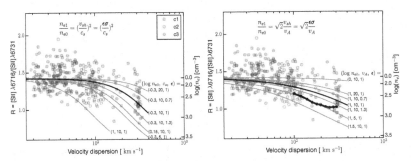

Figure 4. [S II] ratios and electron densities, versus velocity dispersion. The curves are models of isothermal shocks propagating in ISM without magnetic field (left panel) and with magnetic field (right panel). The models providing the best representation of the data are shown as the black curves. The models with different parameters close to those of the black models are shown in grey. The three parameters adopted in each model are labeled next to each curve. The units are cm^{-3} for the pre-shock electron density (n_{e0}), and km s^{-1} for the sound speed (c_s) and Alfvén velocity (v_A). The geometric factor (ϵ) is unit-less. The predictions from the MAPPINGS IV shock models are shown as thick black segments in the right panel.

with low-ionization narrow emission-line regions, and the intermediate *c2* component presenting intermediate line ratios. Clear correlations between the line ratios and velocity dispersion are also obvious in the bottom row of Fig. 3. We model the line ratios using the shock/photoionization code MAPPINGS (Sutherland & Dopita 1993; Dopita & Sutherland 1996; Allen *et al.* 2008; Dopita *et al.* 2013), and we reach remarkable agreement between the data and the models. The elevated line ratios and velocity dispersion of the broad *c3* component are due to the presence of interstellar shocks ($v \approx 200$–300 km s^{-1}) induced by the galactic winds in SDSS J0900. The models demonstrate that the narrow *c1* component is excited by pure photoionization originated from star-forming regions on the disk, and the intermediate *c2* component is excited by a combination of both photoionization and shock excitation.

The presence of interstellar shocks also directly result in the compression of gas, causing the elevation of electron density traceable with the density sensitive line ratio [S II] $\lambda6716$/[S II] $\lambda6731$. The degrees of compression are related to the shock velocities. The elevated electron densities of the broad *c3*, shock-excited component are both observed in our data and consistent with the inferred shock velocities of 200–300 km s^{-1}. Fig. 4 shows two scenarios where the shocks travel in non-magnetized (left panel) and magnetized (right) medium. The high electron densities of *c3* are consistent with the model predictions, demonstrating the importance of shock excitation and the presence of major galactic winds in SDSS J0900.

3. Conclusions

We present the key features to investigate in high-resolution integral field data when major galactic-scale outflows occur. While this case study poses no immediate answer to the prevalence of galactic winds in the local Universe, the large sample size (final sample size of 3,400 galaxies) of the SAMI Galaxy Survey will soon enable an array of studies on galactic-scale outflows and winds in the local Universe (e.g. Fogarty *et al.* 2012).

Acknowledgements

This research makes use of the data products from the SAMI Galaxy Survey†.

References

Allen, J. T., Croom, S. M., Konstantopoulos, I. S., *et al.* 2014, ArXiv e-prints, arXiv:1407.6068
Allen, M. G., Groves, B. A., Dopita, M. A., Sutherland, R. S., & Kewley, L. J. 2008, *ApJS*, 178, 20
Baldwin, J. A., Phillips, M. M., & Terlevich, R. 1981, *PASP*, 93, 5
Bryant, J. J., Owers, M. S., Robotham, A. S. G., *et al.* 2014, ArXiv e-prints, arXiv:1407.7335
Chen, Y.-M., Tremonti, C. A., Heckman, T. M., *et al.* 2010, *AJ*, 140, 445
Croom, S. M., Lawrence, J. S., Bland-Hawthorn, J., *et al.* 2012, *MNRAS*, 421, 872
Dopita, M. A. & Sutherland, R. S. 1996, *ApJS*, 102, 161
Dopita, M. A., Sutherland, R. S., Nicholls, D. C., Kewley, L. J., & Vogt, F. P. A. 2013, *ApJS*, 208, 10
Fogarty, L. M. R., Bland-Hawthorn, J., Croom, S. M., *et al.* 2012, *ApJ*, 761, 169
Heckman, T. M. 2002, in Astronomical Society of the Pacific Conference Series, Vol. 254, Extragalactic Gas at Low Redshift, ed. J. S. Mulchaey & J. T. Stocke, 292
Ho, I.-T., Kewley, L. J., Dopita, M. A., *et al.* 2014, *MNRAS*, in press (arXiv:1407.2411)
Rupke, D. S., Veilleux, S., & Sanders, D. B. 2005, *ApJS*, 160, 115
Steidel, C. C., Erb, D. K., Shapley, A. E., *et al.* 2010, *ApJ*, 717, 289
Sutherland, R. S. & Dopita, M. A. 1993, *ApJS*, 88, 253
Veilleux, S., Cecil, G., & Bland-Hawthorn, J. 2005, *ARA&A*, 43, 769
Weiner, B. J., Coil, A. L., Prochaska, J. X., *et al.* 2009, *ApJ*, 692, 187

† The SAMI Galaxy Survey is based on observation made at the Anglo-Australian Telescope. The Sydney-AAO Multi-object Integral field spectrograph was developed jointly by the University of Sydney and the Australian Astronomical Observatory. The SAMI input catalogue is based on data taken from the Sloan Digital Sky Survey, the GAMA Survey and the VST ATLAS Survey. The SAMI Galaxy Survey is funded by the Australian Research Council Centre of Excellence for All-sky Astrophysics, through project number CE110001020, and other participating institutions. The SAMI Galaxy Survey website is **http://sami-survey.org/**.

Galaxies in 3D across the Universe
Proceedings IAU Symposium No. 309, 2014
B. L. Ziegler, F. Combes, H. Dannerbauer, M. Verdugo, eds.

The Theory of Forming Submillimetre Galaxies

Desika Narayanan

Department of Astronomy and Physics, Haverford College,
370 W Lancaster Ave, Haverford, PA 19041, USA
email: dnarayan@haverford.edu

Abstract. Submillimetre-selected galaxies are the most luminous, heavily star-forming galaxies in the Universe. While this field has exploded observationally over the past decade and a half, theorists have struggled to develop a concordance model for their origin. Here, I review the major theoretical efforts in this field. I then present a newly developed model for the origin of this enigmatic population of galaxies.

Keywords. galaxies: star formation, starburst, evolution, formation

1. Introduction

Since the report of their discovery in back to back *Nature* papers in 1998 (Barger *et al.* 1998; Hughes *et al.* 1998), understanding the origin and cosmological role of 850μm-selected Submillimetre Galaxies (SMGs) has been a source of great interest to both observational and theoretical astrophysicists. The most luminous of these systems are the brighest galaxies in the Universe, and they form as many as a few thousand solar masses per year (compared to the paltry $\sim 1 - 2$ M$_\odot$ yr^{-1} of our own galaxy; see the recent review by Casey, Narayanan & Cooray 2014). These galaxies are heavily dust-enshrouded, and the less luminous subset of these galaxies (that form at just tens to hundreds of solar masses per year) contribute significantly to the cosmic infrared background. SMGs generally reside at redshifts $z \sim 2 - 3$, though have been discovered at redshifts as high as $z \sim 5 - 6$ Riechers *et al.* (2013); Vieira *et al.* (2013). While the exact stellar masses of these systems are argued about in the literature (e.g. Hainline *et al.* 2011; Michałowski *et al.* 2012), it is known that they are generally quite massive, residing in many$\times 10^{12} - 10^{13}$ M$_\odot$ halos at $z \sim 2$,

2. The Theoretical Challenge

Understanding the physical origin of these systems has proven to be a source of great consternation and frustration for galaxy formation theorists. The fundamental difficulty is as follows. The easiest mechanism for forming galaxies with SFRs $\geqslant 1000$ M$_\odot$ yr^{-1} at redshifts $z \sim 2 - 3$ that are as massive as SMGs are via gas-rich galaxy mergers (Narayanan *et al.* 2009, 2010a,b). However, as noted by Davé *et al.* (2010), when examining large cosmological dark-matter only simulations, it becomes rapidly apparent that there are nowhere near enough galaxy mergers at $z = 2 - 3$ to account for the SMG population given the typical duty cycle of SMGs formed in merger simulations (Fakhouri & Ma 2008; Hopkins *et al.* 2010; Hayward *et al.* 2013). At the same time, mergers are an extremely *inefficient* producer of submillimetre-wave photons. For example, when Hayward *et al.* (2011) examined the submllimetre evolution (utilising the dust radiative transfer package SUNRISE; Jonsson, Groves & Cox 2010) of idealised galaxy merger simulations, they found a very weak dependence between the synthetic $850\mu m$ flux density and the star formation rate. This weak powerlaw relation owes to the fact that mergers typically produce rather compact starbursts, and thus heat the dust. As the dust heats, the SED shifts toward the blue, thus giving relatively negligible increase to the submillimetre-band SED. On the other hand, the typical star formation history of galaxies of comparable mass to observed SMGs in cosmological simulations typically peak at redshifts $z = 4 - 6$, and therefore are unable to provide enough ultraviolet photons to be reprocessed into the copious submillimetre luminosity typically seen in SMGs (e.g. Davé, Finlator & Oppenheimer 2011).

This fundamental challenge has resulted in numerous physical models in the literature with a wide range of predicted physical properties of SMGs. While I don't have the luxury of unlimited space to fully delve into the various models, I highly recommend, in my very unbiased opinion, Chapter 10 in the super-awesome review by Casey, Narayanan & Cooray (2014) for more details, and the history of this field as of early 2014. Here, we provide a few anecdotal examples from some of the more well-cited models in the field.

Baugh *et al.* (2005) was one of the first to highlight the central issues surrounding the theoretical challenge in forming SMGs. Utilising a fork of the Durham Semi-Analytic Model (SAM), these authours found that in order to account for SMGs in their cosmological model (while, importantly, simultaneously matching both the high-redshift Lyman Break Galaxy population, as well as the present epoch K-band luminosity function), that SMGs needed to owe their origin to roughly 22% major mergers, and 77% minor (mass ratio $< 1/3$) mergers (González *et al.* 2011). A notable distinction in this model was that the stellar initial mass function was flat for starbursting systems. This model has since been updated, and while a top-heavy stellar IMF is still necessary, the magnitude of the slope is much less extreme (Cowley *et al.* 2014).

Cosmological hydrodynamic simulations have similarly had a difficult time in fully matching the observed properties of SMGs. Owing to a lack of spatial resolution of most hydrodynamic simulations in the early 2000s (typically 3-5 kpc), to date no cosmological hydrodynamic simulation has performed bona fide radiative transfer to identify galaxies as SMGs. The most extensive study utlising this modeling technique was performed by Davé *et al.* (2010). These authours examined a 100 Mpc h^{-1} volume, and took the ansatz that the most heaviliy star forming galaxies in the simulation were the SMGs. In order to then match the observed abundances of SMGs, this required SMGs to have a typical SFR of $\sim 180 - 600$ M$_\odot$ yr^{-1}. This model was able to match the inferred stellar masses, metallicities, gas fractions and clustering measurements of SMGs. The typical star formation rates were a factor 2-3 lower in these models than typically inferred from observations, though these could potentially be reconciled by uncertainties in the assumed stellar IMF of high-z starbursts (Narayanan & Davé 2012a,b). Importantly, in these models, almost none of the galaxies were undergoing a major merger.

Finally, taking a hybrid approach to understand the origin of SMGs, in a series of papers, Hayward *et al.* and Narayanan *et al.* combined a large number of hydrodynamic simulations of idealised galaxy discs and galaxy mergers (with a large range of both galaxy merger mass ratios and galaxy masses) with dust radiative transfer calculations (Hayward *et al.* 2011, 2012, 2013; Narayanan *et al.* 2009, 2010b,a). These authours then convolved the calculated $850\mu m$ duty cycles of these galaxies with observed stellar mass functions and galaxy-galaxy merger rates derived from cosmological dark matter only simulations to infer the theoretical number counts. While these models found success in matching the observed abundances of SMGs, they were unable to capture the realistic environment of SMGs in formation, owing to the idealised nature of the galaxy evolution simulations.

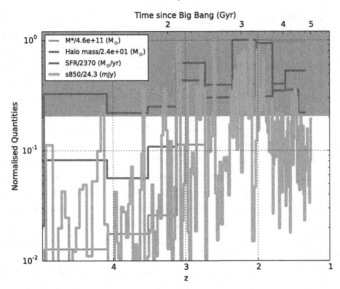

Figure 1. Evolution of physical quantities and our synthetic 850μ m flux density with time. The orange shaded region shows the typical selection criteria for SMGs.

3. A New Picture for SMG Formation

Motivated by these numerical experiments, we have conducted a series of cosmological hydrodynamic zoom simulations of galaxy formation (Hopkins *et al.* 2014). The goal with these simulations is to simulate the cosmic environment surrounding galaxies that could potentially represent SMGs, while still maintaining relatively high spatial resolution. We first conducted a large box/low resolution dark-matter only simulation; we then identified a halo of interest, and re-simulated this at high resolution, and with gas included. Motivated by clustering measurements of SMGs (Hickox *et al.* 2012), we selected a massive halo with a $(z = 0)$ mass of $M_{\rm halo} = 6 \times 10^{13}$ $M_\odot$. The hydrodynamics were simulated utilising the code GIZMO (Hopkins 2014). Important to this study are the inclusion of various forms of stellar feedback (including radiative momentum deposition, energy deposition from supernova and stellar winds, and HII region feedback; see Hopkins *et al.* (2014) for details).

In order to calculate the inferred photometric properties of these hydrodynamic simulations, we have developed a new dust radiative transfer package, POWDERDAY. This code projects the particle data from the hydrodynamic simulation onto an octree adaptive grid, and then calculates the stellar SEDs utilising FSPS (Conroy, Gunn & White 2009). The stellar radiation fields are then propagated through the dusty interstellar medium utilising HYPERION (Robitaille 2011), and subsequently analysed. Threaded throughout the code is YT (Turk *et al.* 2011), facilitating the individual codes playing nice with one another.

Figure 1 shows our main results. The galaxy has a rising star formation history (SFH), which peaks at redshifts $z \approx 2 - 3$. The SFH peaks late compared to most models of galaxies of this mass owing to the stellar feedback in the simulations: feedback from early star formation drives gaseous bubbles which provides a reservoir to be banked for later star formation. By this time, the ISM is sufficiently enriched with metals that the galaxy may be viewed as a submillimetre-luminous system.

The morphology of this system is extremely clumpy. As the central galaxy evolves, numerous subhaloes bombard the centre, providing more fuel for star formation. This is

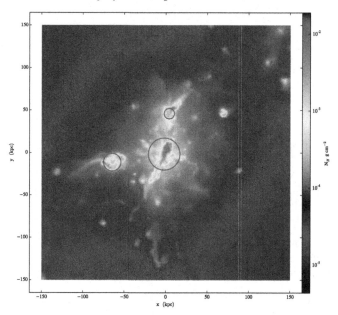

Figure 2. Gas surface density projection plot of an SMG at $z \sim 2$. The red circles correspond to individual dark matter halos with the sizes corresponding to their virial radii. The panel is 300 kpc on a side. SMGs will typically break up into multiple emission peaks in our model.

evident in Figure 2, where we show an example snapshot of one of our model galaxies, with individual subhalos denoted (with size corresponding to their virial radius). Our model therefore predicts that high-resolution observations of SMGs will reveal multiple emission sources, rather than one (or even two merging) galaxies. The galaxy has stellar masses and star formation rates comparable to those observed. This picture is notably different from a classical merger-driven scenario. At the same time, the long duty cycle and general abundance of halos of this mass suggest that this sort of model can account for the observed abundance of high-z SMGs.

Acknowledgements

DN thanks Helmut Dannerbauer for sending him a frantic email on the final day of conference registration reminding him to register to attend this very interesting meeting! DN also thanks the Thyssen-Bornemisza Art Company in Vienna for hosting an epic party on the final night of the conference. Partial support for DN was provided by NSF grants AST-1009452, AST-1442650 and NASA grant HST-AR-13906.001.

References

Barger A. J., Cowie L. L., Sanders D. B., Fulton E., Taniguchi Y., Sato Y., Kawara K., & Okuda H., 1998, *Nature*, 394, 248

Baugh C. M., *et al.*, 2005, *MNRAS*, 356, 1191

Casey C. M., Narayanan D., & Cooray A., 2014, *Physics Reports*, 541, 45

Conroy C., Gunn J. E., & White M., 2009, *ApJ*, 699, 486

Cowley W. I., Lacey C. G., Baugh C. M., & Cole S., 2014, arXiv/1406.0855

Davé R., Finlator K., & Oppenheimer B. D., 2011, arXiv/1108.0426

Davé R., Finlator K., Oppenheimer B. D., Fardal M., Katz N., Kereš D., & Weinberg D. H., 2010, *MNRAS*, 404, 1355

Fakhouri O. & Ma C.-P., 2008, *MNRAS*, 386, 577

González J. E., Lacey C. G., Baugh C. M., & Frenk C. S., 2011, *MNRAS*, 413, 749

Hainline L. J., Blain A. W., Smail I., Alexander D. M., Armus L., Chapman S. C., & Ivison R. J., 2011, *ApJ*, 740, 96

Hayward C. C., Jonsson P., Kereš D., Magnelli B., Hernquist L., & Cox T. J., 2012, *MNRAS*, 424, 951

Hayward C. C., Kereš D., Jonsson P., Narayanan D., Cox T. J., & Hernquist L., 2011, *ApJ*, 743, 159

Hayward C. C., Narayanan D., Kereš D., Jonsson P., Hopkins P. F., Cox T. J., & Hernquist L., 2013, *MNRAS*, 428, 2529

Hickox R. C., *et al.*, 2012, *MNRAS*, 421, 284

Hopkins P. F., 2014, arXiv/1409.7395

Hopkins P. F., Kereš D., Oñorbe J., Faucher-Giguère C.-A., Quataert E., Murray N., & Bullock J. S., 2014, *MNRAS*, 445, 581

Hopkins P. F., Younger J. D., Hayward C. C., Narayanan D., & Hernquist L., 2010, *MNRAS*, 402, 1693

Hughes D. H., *et al.*, 1998, *Nature*, 394, 241

Jonsson P., Groves B. A., & Cox T. J., 2010, *MNRAS*, 186

Michałowski M. J., Dunlop J. S., Cirasuolo M., Hjorth J., Hayward C. C., & Watson D., 2012, *A&A*, 541, A85

Narayanan D., Cox T. J., Hayward C. C., Younger J. D., & Hernquist L., 2009, *MNRAS*, 400, 1919

Narayanan D. & Davé R., 2012a, *MNRAS*, 423, 3601

Narayanan D. & Davé R., 2012b, arXiv/1210.6037

Narayanan D. *et al.*, 2010a, *MNRAS*, 407, 1701

Narayanan D., Hayward C. C., Cox T. J., Hernquist L., Jonsson P., Younger J. D., & Groves B., 2010b, *MNRAS*, 401, 1613

Riechers D. A., *et al.*, 2013, *Nature*, 496, 329

Robitaille T. P., 2011, *A&A*, 536, A79

Turk M. J., Smith B. D., Oishi J. S., Skory S., Skillman S. W., Abel T., & Norman M. L., 2011, *ApJS*, 192, 9

Vieira J. D. *et al.*, 2013, *Nature*, 495, 344

Galaxies in 3D across the Universe
Proceedings IAU Symposium No. 309, 2014
B. L. Ziegler, F. Combes, H. Dannerbauer, M. Verdugo, eds.

© International Astronomical Union 2015
doi:10.1017/S1743921314009703

Kinematics of superdense galaxies in clusters

Alessia Moretti[1], Bianca Poggianti[2], Daniela Bettoni[2], Michele Cappellari[3], Giovanni Fasano[2] and the WINGS team

[1]Department of Physics and Astronomy, University of Padova,
Vicolo dell'osservatorio 3, 35127 Padova, Italy
email: alessia.moretti@unipd.it

[2]INAF - Padova Astronomical Observatory
Vicolo dell'osservatorio 2, 35127 Padova, Italy

[3]Dep. of Astrophysics, University of Oxford, Denys Wilkinson Building, Keble Road, Oxford,
OX1 3RH, UK

Abstract. We present the first results obtained by analyzing the detailed kinematics of a sub-sample of 9 massive and compact galaxies found in the WINGS survey. The observed galaxies are very old (both luminosity and mass- weighted age are on average $\geqslant$ 10 Gyr), while they resemble more typical galaxies in the other characteristics. The total M/L ratio is determined using as free parameters the anisotropy β and the galaxy inclination i.

Keywords. galaxies: elliptical and lenticular, cD - galaxies: evolution - galaxies: formation

1. Sample selection and analysis

Valentinuzzi *et al.*, 2010 found in the WINGS sample of local cluster of galaxies ($0.04 < z < 0.07$) a significant number of massive and compact galaxies ($3 \times 10^{10} M_\odot < M < 4 \times 10^{11} M_\odot, \Sigma_{50} > 3 \times 10^9 M_\odot/kpc^2$), that have not been found in other local surveys (e.g. SDSS). Our results show that selecting already passive galaxies at high z means selecting the most compact ones. Assuming they do not resume the star formation activity at later times, the high-z sizes should be compared with the sizes of the oldest (in terms of their LW-age) low-z galaxies, to avoid a significant progenitor bias. Taking into account this progenitor bias Poggianti *et al.*, 2013b found that for compact galaxies in low density environments a significant size evolution for these compact objects is not needed. In order to validate our stellar masses, derived using fiber spectroscopy and our spectrophotometric tool SINOPSIS (Fritz *et al.*, 2007,2011), we obtained two sets of IFU observations for a subsample of 17 WINGS SDGs. 9 of them have been observed with G-MOS @GEMINI and 8 of them with VIMOS (purple and green dots, respectively, in Fig. 1, top left panel).

We present here WINGS SDGs that have been observed with GEMINI G-MOS with an exposure time of 2700 s, using the B600 grism in the 1-slit mode. The FoV is $3.5'' \times 5''$. The wavelength coverage goes from 4000 to 6900 Å. We derived the stellar kinematics by using the pPXF algorithm by Cappellari & Emsellem (2004), that works in regions where S/N=10 (defined by a Voronoi tessellation). In our case we used as templates a set of MILES SSPs with 16 values of ages and 6 metallicities. Fig. 1, bottom panel shows the velocity map for one of our galaxies. We derived the luminosity-weighted angular momentum λ_R following Emsellem *et al.*, 2007, and found that most SDGs are fast rotators (Fig. 1, top right panel). We then started evaluating the dynamical mass using the Jeans Anysotropic MGE models by Cappellari 2008. This model requires an accurate estimate of the galaxy light decomposition, that is made by superimposing a set of multiple gaussian profiles (MGE expansion). We then used a self-consistent JAM

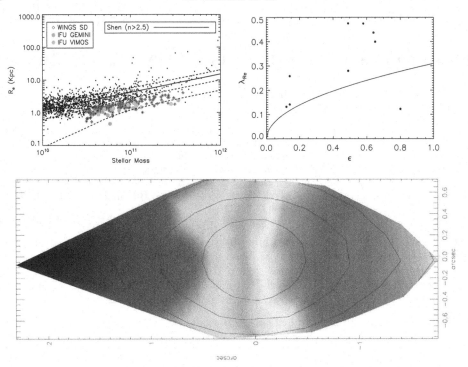

Figure 1. Top left: Mass-size relation for WINGS cluster members. Black lines are the local (SDSS) relations. Red dots are WINGS SDGs. Green and purple symbols are IFU candidates. Top right: Luminosity weighted angular momentum (λ_R) within R_e versus ellipticity. The line divides fast rotators (FR) from slow rotators (SR). Bottom panel: example of the velocity map derived for our SDGs.

model (see Cappellari *et al.*, 2013, Paper XV), where the underlying assumption is that the mass distribution follows the light one derived from the MGE fit, to evaluate the M/L ratio. This model has two non linear parameters (the vertical anisotropy β and the galaxy inclination i). The M/L ratio is then derived imposing the match with the observed V_{rms}. The derived M/L is the total one, and not only the stellar one. We are currently deriving the M/L JAM for our galaxies with the aim of estimating the dark matter fraction inside R_e (by comparing it with the stellar M/L derived from spectrophotometry) and the possible influence of the adopted IMF.

References

Cappellari M., 2008, *MNRAS*, 390, 71
Cappellari M., Emsellem E., 2004, *PASP*, 116, 138
Cappellari M., *et al.*, 2013, *MNRAS*, 432, 1709
Emsellem E., *et al.*, 2007, *MNRAS*, 379, 401
Fritz J., *et al.*, 2007, *A&A*, 470, 137
Fritz J., *et al.*, 2011, *A&A*, 526, A45
Poggianti B. M., *et al.*, 2013, *ApJ*, 777, 125
Valentinuzzi T., *et al.*, 2010, *ApJ*, 712, 226

Galaxies in 3D across the Universe
Proceedings IAU Symposium No. 309, 2014
B. L. Ziegler, F. Combes, H. Dannerbauer, M. Verdugo, eds.

© International Astronomical Union 2015
doi:10.1017/S1743921314009715

A 3D view of the Hydra I galaxy cluster core - I. Kinematic substructures

Michael Hilker[1], Carlos Eduardo Barbosa[1,2], Tom Richtler[3], Lodovico Coccato[1], Magda Arnaboldi[1] and Claudia Mendes de Oliveira[2]

[1]European Southern Observatory, Garching, Germany, email: mhilker@eso.org
[2]Universidade de Sao Paulo, Sao Paulo, Brazil
[3]Universidad de Concepción, Concepción, Chile

Abstract. We used FORS2 in MXU mode to mimic a coarse 'IFU' in order to measure the 3D large-scale kinematics around the central Hydra I cluster galaxy NGC 3311. Our data show that the velocity dispersion field varies as a function of radius and azimuthal angle and violates point symmetry. Also, the velocity field shows similar dependence, hence the stellar halo of NGC 3311 is a dynamically young structure. The kinematic irregularities coincide in position with a displaced diffuse halo North-East of NGC 3311 and with tidal features of a group of disrupting dwarf galaxies. This suggests that the superposition of different velocity components is responsible for the kinematic substructure in the Hydra I cluster core.

Keywords. galaxies: elliptical and lenticular, cD - galaxies: kinematics and dynamics - galaxies: halos - galaxies: individual (NGC 3311)

1. Introduction, target and method

The mass determination of central cluster galaxies based on kinematical data normally uses simplified assumptions of virial equilibrium and spherical symmetry. The round appearance of these galaxies and the smooth distribution of hot X-ray emitting gas might suggest that these assumptions are justified. On the other hand, it has been shown that the halos of massive ellipticals grow by a factor of about 4 in mass since z=2 (van Dokkum *et al.* 2010). The mass growth is dominated by the accretion of low mass systems (minor mergers). Thus, one might expect that accretion events leave kinematical signatures in the phase space of the outer stellar population especially of central cluster galaxies.

Here we present the vivid case of the central giant elliptical of the Hydra I cluster, NGC 3311. This early-type galaxy dominated cluster is regarded as dynamically evolved. Recent photometric and kinematical studies of the diffuse stellar light, planetary nebulae and globular clusters in the core of Hydra I, however, have shown that 1) NGC 3311 exhibits a steeply rising velocity dispersion profile (Ventimiglia *et al.* 2010, Richtler *et al.* 2011), 2) the velocity dispersion profiles differ from each other in different azimuthal directions, as judged from longslit analyses (Ventimiglia *et al.* 2011, Richtler *et al.* 2011), and 3) the diffuse light is not centered around NGC 3311's main spheroid, but it is displaced towards the North-East by about 15 kpc (Arnaboldi *et al.* 2012).

In order to find kinematic signatures of these substructures and to disentangle the past and present active assembly history of the Hydra I cluster core, we used FORS2 in MXU mode (ESO programme 088.B-0448, PI: T. Richtler) to mimic a coarse 'IFU'. Our novel approach is to place short slits in an onion shell-like pattern around NGC 3311 to measure its 3D large scale kinematics out to 3 effective radii. Sky slits were positioned far outside of NGC 3311's main body and bright halo. The borders of the 'IFU' spaxels

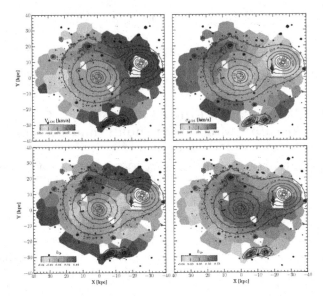

Figure 1. Upper left: Radial velocity map. **Upper right:** Velocity dispersion map. **Lower left:** Skewness (h3) map. **Lower left:** Kurtosis (h4) map. All maps were smoothed with a locally weighted scatterplot smoothing algorithm (Cleveland 1979). The contours are surface brightness levels of NGC 3311 (in the center) and NGC 3309 (on the right) V-band light between 19 and 23.5 mag/arcsec^2 in steps of 0.5 mag.

were defined via Voronoi tesselation to create kinematic maps (see Fig. 1). The S/N of our final spectra ranges from >20 to 2 from the inner to the outer radii.

2. Results and conclusions

The main results of our analysis, shown in Fig. 1, are: 1) there are pronounced azimuthal variations both in radial velocity and velocity dispersion as a function of galactocentric distance, explaining the above mentioned discrepancies from previous longslit results; 2) in the North-East there are significant small scale variations in both radial velocity and velocity dispersion data points; 3) the very large velocity dispersion values in some parts of the cluster core ($\sigma > 500$ km/s) probably point to a superposition of kinematical substructures; and 4) the displaced diffuse stellar halo around NGC 3311 coincides with regions of positive h3 and h4 values, also evidence for more than one velocity component in the stellar halo.

We conclude that the stellar halo around NGC 3311 in the core of Hydra I is still forming and is not in dynamical equilibrium. Probably, a group of infalling dwarf galaxies and their tails are responsible for the kinematical substructures. The general lesson is that one has to take care when inferring the properties of the central dark matter halo around NGC 3311 from kinematical data.

References

Arnaboldi, M., Ventimiglia, G., Iodice, E., Gerhard, O., & Coccato, L., 2012, *A&A*, 545, 37
Cleveland, W. S., 1979, *JASA*, 74, 829
Richtler, T., Salinas, R., Misgeld, I., *et al.* 2011, *A&A*, 531, 119
van Dokkum, P. G., Whitaker, K. E., Brammer, G., *et al.* 2010, *ApJ*, 709, 1018
Ventimiglia, G., Gerhard, O., Arnaboldi, M., & Coccato, L., 2010, *A&A*, 520, 9
Ventimiglia, G., Arnaboldi, M., & Gerhard, O., 2011, *A&A*, 528, 24

Galaxies in 3D across the Universe
Proceedings IAU Symposium No. 309, 2014
B. L. Ziegler, F. Combes, H. Dannerbauer, M. Verdugo, eds.
© International Astronomical Union 2015
doi:10.1017/S1743921314009727

A 3D view of the Hydra I cluster core- II. Stellar populations

Carlos Eduardo Barbosa[1,2]†, Magda Arnaboldi[1], Michael Hilker[1], Lodovico Coccato[1], Tom Richtler[3] and Cláudia Mendes de Oliveira[2]

[1]European Southern Observatory, Garching, Germany
[2]IAG-USP, São Paulo, SP, Brazil
[3]Universidad de Concepción, Concepción, Chile

Abstract. Several observations of the central region of the Hydra I galaxy cluster point to a multi-epoch assembly history. Using our novel FORS2/VLT spectroscopic data set, we were able to map the luminosity-weighted age, [Fe/H] and [α/Fe] distributions for the stellar populations around the cD galaxy NGC 3311. Our results indicate that the stellar populations follow the trends of the photometric substructures, with distinct properties that may aid to constrain the evolutionary scenarios for the formation of the cluster core.

Keywords. galaxies: elliptical and lenticular, cD - galaxies: clusters: individual (Hydra I) - galaxies: stellar content

1. Introduction

Although previously considered a system in equilibrium, observations of the Hydra I cluster core, dominated by NGC 3311, have shown instead a complex, ongoing process of assembly of matter. Recently, Arnaboldi *et al.* (2012) reported the existence of an faint surface brightness offset envelope. In part I of our work (see Hilker at al. in this proceedings), we have shown that the envelope substructure have line-of-sight velocity distributions distinct from the surrounding regions, which is also related to a group of infalling galaxies.

Coccato *et al.* (2011) has shown that the stellar content in the region of the offset envelope is different from the other regions, thus attesting that stellar populations can be used to trace and constrain the processes involved in galaxy disruption and in the build up of the massive cD galaxies. However, a spatially resolved study to observe the shape and extent of the stellar population features was still missing. Therefore, here we aim to further demonstrate that the Hydra I core is indeed actively forming using stellar populations as tracers of the past evolution of the system.

2. Data and Methods

We have used our novel FORS2/VLT spectroscopic data set (programme 88.B-448; PI: Richtler), i.e. short slits placed in an onion shell-like pattern onto NGC 3311 to mimic a coarse 'IFU'. The data set consists of high resolution optical spectra for 130 positions out to ∼30 kpc (∼3 effective radii). We modelled seven absorption line features in the Lick/IDS system with the alpha-enhanced models of Thomas *et al.* (2011) using a Monte Carlo Markov Chains method in order to obtain uncertainties and to access whether we can break the well known metallicity-age degeneracy.

---

† email: carlos.barbosa@usp.br

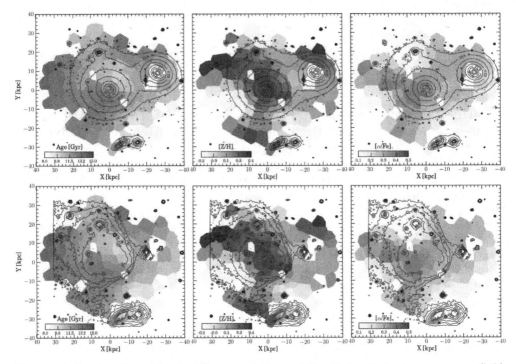

Figure 1. Luminosity-weighted stellar populations of the Hydra I cluster core: ages (left), metallicity (centre) and alpha-elements content (right). Upper and lower panels display contours for the V-band image and the offset envelope respectively, as in Arnaboldi *et al.* (2012).

3. Results and Conclusion

We present results for the luminosity-weighted ages, metallicities, and alpha element abundances in Fig. 1. The offset envelope exhibits populations which are slightly younger, more metal-rich and have lower alpha abundances. These results are consistent with the "two stages" formation scenario of central galaxies: the central parts form first "in situ", possibly in a quasi monolithic collapse with violent starbursts. The outer regions then grow by accreting less massive systems with extended periods of star formation.

Acknowledgements

Based on observations made with ESO Telescopes at the La Silla Paranal Observatory. CEB and CMdO are grateful to the São Paulo Research Foundation (FAPESP) funding (Procs. 2011/21325-0 and 2012/22676-3). TR acknowledges support from FONDECYT project Nr. 1100620, the BASAL Centro de Astrofísica y Tecnologías Afines (CATA) PFB-06/2007, and a visitorship at ESO/Garching.

References

Arnaboldi M., Ventimiglia G., Iodice E., Gerhard O., & Coccato L., 2012, *A&A*, 545, A37
Coccato L., Gerhard O., Arnaboldi M., & Ventimiglia G., 2011, *A&A*, 533, A138
Thomas D., Maraston C., & Johansson J., 2011, *MNRAS*, 412, 2183

Galaxies in 3D across the Universe
Proceedings IAU Symposium No. 309, 2014
B. L. Ziegler, F. Combes, H. Dannerbauer, M. Verdugo, eds.
© International Astronomical Union 2015
doi:10.1017/S1743921314009739

Understanding the transformation of spirals to lenticulars

Evelyn J. Johnston, Alfonso Aragón-Salamanca and Michael R. Merrifield

School of Physics and Astronomy, University of Nottingham, University Park, Nottingham, NG7 2RD, UK
email: ejohnsto@eso.org

Abstract. By studying the individual star-formation histories of the bulges and discs of lenticular (S0) galaxies, it is possible to build up a sequence of events that leads to the cessation of star formation and the consequent transformation from the progenitor spiral. In order to separate the bulge and disc stellar populations, we spectroscopically decomposed long-slit spectra of Virgo Cluster S0s into bulge and disc components. Analysis of the decomposed spectra shows that the most recent star formation activity in these galaxies occurred within the bulge regions, having been fuelled by residual gas from the disc. These results point towards a scenario where the star formation in the discs of spiral galaxies are quenched, followed by a final episode of star formation in the central regions from the gas that has been funnelled inwards through the disc.

1. Introduction

Lenticular (S0) galaxies are often seen as an endpoint in the evolution of spiral galaxies since they display the same discy morphologies as spirals, but contain older stellar populations. In order to transform a spiral into an S0, the star-formation in the disc must be truncated and the bulge luminosity enhanced (Christlein & Zabludoff 2004). Therefore, to understand how these two phenomena occur, we need to study the individual star-formation histories of the bulges and discs.

2. Disentangling the Star Formation Histories of the Bulge and Disc

A sample of 21 S0s from the Virgo Cluster were used in this study, with B-band magnitudes in the range of -17.3 to -22.3 and inclinations above 40 degrees to reduce contamination from misclassified ellipticals. A long-slit spectrum was obtained along the major axis of each galaxy using Gemini-GMOS with typical exposure times of $\sim 2 - 3$ hours. The reduced spectra were decomposed using the spectroscopic bulge–disc decomposition technique of Johnston *et al.* (2012), in which light profile of the galaxy in each wavelength bin was separated into bulge and disc components by fitting a Sérsic bulge plus exponential disc model. Having found the best fit and obtained the decomposition parameters for each wavelength bin, the total light from each component was then calculated by integration, and plotted against wavelength to create two one-dimensional spectra representing purely the bulge and disc light.

The ages, metallicities and Mgb/⟨Fe⟩ ratios of the bulges and discs were measured using the strengths of the absorption features in the decomposed spectra. It can be seen in Fig. 1 that the bulges contain systematically younger and more metal rich stellar populations than their associated discs, suggesting that a final star-formation event occurred in the bulge region during the transformation. Additionally, a comparison of the Mgb/⟨Fe⟩

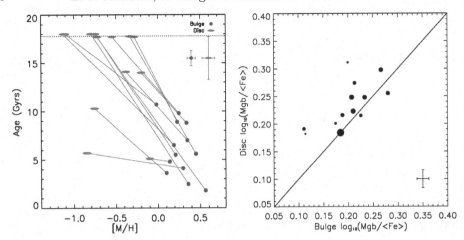

Figure 1. *Left*: The relative, light-weighted ages and metallicities of the bulges and discs of S0s in the Virgo Cluster. The lines link bulges and discs from the same galaxy. *Right*: Comparison of the Mgb/⟨Fe⟩ ratios for the bulges and discs. Errors are shown on the right of each plot.

ratios for the bulges and discs, also shown in Fig. 1, reveals a correlation, indicating that their star-formation histories are linked.

3. Conclusions

We find that during the transformation from a spiral to an S0, gas is stripped from the disc gently enough that no significant amounts of star formation were triggered there. At some point, a final star-formation event occurs in the bulge region, resulting in younger, more metal rich stellar populations there. The most likely origin of the gas that fuelled this star formation is from the associated disc, which appears to be true from the clear correlation in the chemical enrichment of the the bulges and discs. Together, these results present a scenario for the transformation of spirals to S0s in which gas is stripped gently from the disc, while the residual gas is channelled in towards the centre of the galaxy. Eventually, this gas induces a final episode of star formation within the bulge, which uses up all the remaining gas and boosts the luminosity of the bulge relative to the disc. The whole galaxy then fades to an S0.

Clearly, further information on the transformation can be obtained if the decomposition was carried out in two dimensions using integral field unit (IFU) data. The up-coming IFU Mapping of Nearby Galaxies at APO (MaNGA, Bundy *et al.* in preparation) survey in SDSS-IV promises the resolution and wide field of view that will answer the remaining questions about how the bulge and disc star-formation histories are linked.

Acknowledgements

EJ acknowledges support from the STFC, the IAU and the RAS.

References

Christlein D. & Zabludoff A. I., 2004, *ApJ*, 616, 192
Côté P. *et al.*, 2004, *ApJS* , 153, 223
Johnston, E. J., Aragón-Salamanca, A., Merrifield, M. R., & Bedregal, A. G. 2012, *MNRAS*, 422, 2590

Galaxies in 3D across the Universe
Proceedings IAU Symposium No. 309, 2014
B. L. Ziegler, F. Combes, H. Dannerbauer, M. Verdugo, eds.
© International Astronomical Union 2015
doi:10.1017/S1743921314009740

Abundant molecular gas and inefficient SF in intra-cluster regions of a ram pressure stripped tail

P. Jáchym[1], M. Sun[2], F. Combes[3], L. Cortese[4] and J. D. P. Kenney[5]

[1] Astronomical Institute, Academy of Sciences of the Czech Republic, Boční II 1401, 14100, Prague, Czech Republic, email: jachym@ig.cas.cz

[2] Department of Physics, University of Alabama in Huntsville, 301 Sparkman Drive, Huntsville, AL 35899, USA

[3] Observatoire de Paris, LERMA, 61 Av. de l'Observatoire, 75014 Paris, France

[4] Centre for Astrophysics & Supercomputing, Swinburne University of Technology, Mail H30, P. O. Box 218, Hawthorn, VIC 3122, Australia

[5] Department of Astronomy, Yale University, 260 Whitney Ave., New Haven, CT 06511, USA

Abstract. For the first time in any ram pressure stripped galaxy, we detect large amounts of cold molecular gas in the X-ray bright, and star forming tail of ESO 137-001 in the Norma cluster. We find very low star formation efficiency in the stripped gas, similar to values found in the outer spiral disks where however molecular gas is mostly undetected. The results were recently published in Jáchym *et al.* (2014).

1. Introduction

Ram pressure of the intra-cluster medium (ICM) can strip cold, star-forming interstellar matter (ISM) reservoir from late-type galaxies in clusters (Gunn & Gott 1972), and thus quench their star formation (SF). The stripped ISM forms one-sided 'cometary' tails that have been revealed in many galaxies in HI, diffuse Hα, young stars (seen either in Hα or UV), and X-rays (see Jáchym *et al.* 2014, for references). It is now understood that gas-stripped tails observed in different wavelengths are manifestations of the stripped ISM that mixed with the surrounding ICM. Part of the stripped gas can get heated to temperatures between that of HI and the ICM, while other part of the (denser) gas can efficiently cool and form cold, molecular clouds, and thus stars.

2. Results

With the APEX telescope, we searched for cold molecular gas, as traced by CO(2-1) emission, in the iconic ram pressure stripped galaxy ESO 137-001 located in the central part of the Norma cluster (A3627). This galaxy has an 80 kpc long, bright, double structure X-ray tail, and a 40 kpc Hα tail with over 30 HII regions (Sun *et al.* 2006, 2007, 2010). Fig. 1 shows the galaxy, together with the APEX CO(2-1) spectra obtained in four pointings (marked with circles). Clearly, CO emission is detected in all positions, including the most distant one at $\sim$40 kpc downstream in the tail, where also Hα and X-ray emission have a local peak. It is the first time that cold molecular gas is found at large distances downstream in a ram pressure stripped tail. This observation thus brings a new insight into the fate of the stripped ISM and its evolution in the intracluster space.

The corresponding total H_2 mass in the tail is $\gtrsim 10^9$ $M_\odot$, assuming a Galactic value of the CO-to-H_2 conversion factor. This is an amount similar to the total mass of the X-ray emitting gas. Together, they nearly account for the gas mass missing from the galaxy. The

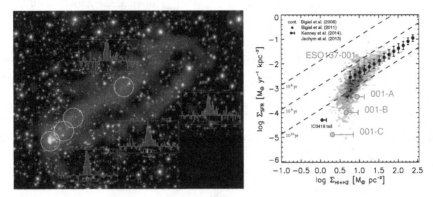

Figure 1. Left: Observed APEX positions on a composite X-ray *Chandra* (blue) / Hα SOAR (red) image of ESO 137-001. The FWHM of the CO(2-1) beam is 25"≈8.4 kpc. Right: SFR surface density as a function of total gas surface density in ESO 137-001 and the three observed tail regions 001-A, 001-B, and 001-C. The filled circles correspond to H_2 column density, while the 'error bars' show ATCA HI upper limits. Figure adapted from Bigiel *et al.* (2008). From Jáchym *et al.* (2014).

Hα-based star formation rates (SFRs) averaged over the APEX beams (following Kennicutt & Evans 2012) are for the observed pointings placed into the Schmidt-Kennicutt-type plot in Fig. 1. While in the main body of the galaxy, the measured molecular gas depletion time is well consistent with average values found in nearby galaxies, clearly, the location of the tail points indicates much lower efficiency of star formation (SFE) in the stripped gas than is typical in star-forming ISM in the inner parts of spiral disks.

3. Conclusions

The measured low SFE ($\tau_{\mathrm{dep,H_2}} > 10^{10}$ yr) in the tail of ESO 137-001 suggests that most of the stripped gas does not form stars but ultimately joins the ICM. Similarly low SFEs are found in the outer disks of spiral galaxies, where however, HI is likely the dominant component of the ISM and CO is mostly undetected (e.g., Bigiel *et al.* 2010). This might be due to low volume density of the stripped gas, and/or ram pressure induced shocks that may increase thermal and turbulent pressure of the stripped ISM. The CO-bright tail of ESO 137-001 thus represents a special environment of star formation.

The detection of abundant molecular gas in the X-ray bright tail of ESO 137-001 suggests that in massive, high-ICM-pressure clusters (such as Norma), the conditions are right for the mixing of the stripped ISM with the surrounding ICM to produce observable levels of individual gas components. This motivates further multi-wavelength studies of ram pressure stripped galaxies in rich clusters that would allow us to reveal more of the stripped ISM and thus to learn about its fate and evolution in the intra-cluster space.

Acknowledgements

We acknowledge support from the project M100031203 of the Academy of Sciences of the Czech Republic.

References

Bigiel, F., Leroy, A., Walter, F., *et al.* 2010, *AJ*, 140, 1194
Bigiel, F., Leroy, A., Walter, F., *et al.* 2008, *AJ*, 136, 2846

Gunn, J. E. & Gott, III, J. R. 1972, *ApJ*, 176, 1
Jáchym, P., Combes, F., Cortese, L., *et al.* 2014, *ApJ*, 792, 11
Kennicutt, R. C. & Evans, N. J. 2012, *ARA&A*, 50, 531
Sun, M., Donahue, M., Roediger, E., *et al.* 2010, *ApJ*, 708, 946
Sun, M., Donahue, M., & Voit, G. M. 2007, *ApJ*, 671, 190
Sun, M., Jones, C., Forman, W., *et al.* 2006, *ApJ*, 637, L81

Galaxies in 3D across the Universe
Proceedings IAU Symposium No. 309, 2014
B. L. Ziegler, F. Combes, H. Dannerbauer, M. Verdugo, eds.

© International Astronomical Union 2015
doi:10.1017/S1743921314009752

The Cosmic Skidmark: witnessing galaxy transformation at z = 0.19

David N. A. Murphy

Instituto de Astrofísica, Facultad de Física,
Pontificia Universidad Católica de Chile
email: dmurphy@astro.puc.cl

Abstract. We present an early-look analysis of the "Cosmic Skidmark". Discovered following visual inspection of the Geach, Murphy & Bower (2011) SDSS Stripe 82 cluster catalogue generated by ORCA (an automated cluster algorithm searching for red-sequences; Murphy, Geach & Bower 2012), this $z = 0.19$ 1.4L* galaxy appears to have been caught in the rare act of transformation while accreting onto an estimated 10^{13}–$10^{14} h^{-1} M_\odot$-mass galaxy group. SDSS spectroscopy reveals clear signatures of star formation whilst deep optical imaging reveals a pronounced 50 kpc cometary tail. Pending completion of our ALMA Cycle 2 and IFU observations, we show here preliminary analysis of this target.

Keywords. galaxies: evolution - galaxies: clusters: general

1. Introduction

Study of galaxy populations reveal clearly bimodal properties such as morphology, the colour-magnitude distribution (CMD) and star formation rates (see e.g. Kauffmann *et al.* 2006). The origin of this dichotomy appears to be driven mainly by galaxy environment; compared to those in the low-density field, spiral galaxies in clusters exhibit suppressed star formation, smaller cold-gas reservoirs and truncation of their gaseous disks (Boselli & Gavazzi 2006). The evolution of the CMD into a progressively pronounced "red sequence" (Bower, Lucey & Ellis 1992) at early times, indications of a transitional CM "green valley" and a lack of high-mass blue spirals point to the transformation of late-type spirals into early-type ellipticals within cluster environments. Examples of such transformations (most recently Jáchym *et al.* 2014) are valuable case-studies that deepen our understanding of galaxy evolution. To this end, we present a preliminary look at the "Cosmic Skidmark" - an example of galaxy transformation in action.

2. Data

The Cosmic Skidmark investigation consists of three datasets. The first is with deep (7hr) *griz*-band SOAR Adaptive Module (SAM; Tokovinin *et al.* 2008) photometry (shown in Figure 1 *left*. We compliment this with 14hrs of VLT/VIMOS IFU data covering a 27"x27" FOV over 3837Å–7430Å rest-frame at a spaxel resolution of 0.67"x0.67". Finally, we await completion of a 3hr ALMA Band 3 scan to measure tracers of neutral gas from reservoirs potentially fuelling the sites of strong star-formation. The latter two datasets are partially complete. The z=0.19 redshifting of spectral features permits IFU measurement of [OII] line emission in addition to ALMA mapping of CO(1-0) and CN(1-0) over one pointing covering the target galaxy, group members and environs to a 3σ CO line sensitivity of 0.34mJy.

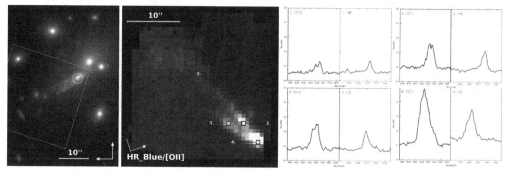

Figure 1. (*left*) SAM *gri*-composite imaging of the Cosmic Skidmark along with the VI-MOS/IFU footprint and SDSS spectrographic fiber position. (*middle*) Blue-grism VLT/VIMOS IFU datacube centred on observed-frame [OII] emission (λ_{obs} 4420Å). Samples of spectra corresponding to the numbered spaxels are shown *right*.

3. Preliminary findings and outlook

Figure 1 (*centre*) shows a datacube for the high-resolution VPH blue VLT/VIMOS grism centred on the wavelength corresponding to observed-frame [OII] line emission. Spectra for the numbered spaxels can also be found in Figure 1 (*right*). With three key spectral indicators ([OII],[OIII],Hβ) we find clear indications of star formation in the galaxy core (spaxel 4), in agreement with SDSS fiber spectroscopy. Star formation, evident throughout the cometary plume, traces out the dual-tail structure (spaxels 2,3) alluded to in optical imaging. We find evidence for line emission out to 50 kpc from the galaxy core (spaxel 1). Inspection of deep *g*-band SAM imaging near this spaxel reveals a number of small galaxies and clumps clustered downstream at the tail-end of the plume. With *g-r* colours consistent with the group red-sequence suggesting a similar redshift to the Cosmic Skidmark, they may possibly have been disturbed by it's passage through the group. Future work will focus on probing this site, in particular two diffuse (*g*~25.2mag/arcsec^2) tidal dwarf-scale sources seen stretching between two brighter sources. VIMOS star-formation maps can be connected to ALMA CO(1-0), CN(1-0) and SiO(3-2) line emission mapping of molecular gas reservoirs and will permit measurement of the star formation efficiency, kinematical differences between the star-forming and cold-gas, gas depletion timescales and outflow signatures. A detailed analysis of this target will appear in Murphy *et al.* (2014, in preparation).

Acknowledgements

The author acknowledges support through FONDECYT grant 3120214.

References

Boselli A. & Gavazzi G., 2006, *PASP*, 118, 517

Bower R. G., Lucey J. R., & Ellis R. S., 1992, *MNRAS*, 254, 601

Geach J. E., Murphy D. N. A., & Bower R. G., 2011, *MNRAS*, 413, 3059

Jáchym P., Combes F., Cortese L., Sun M., & Kenney J. D. P., 2014, *ApJ*, 792, 11

Kauffmann G., Heckman T. M., Brinchmann G., Charlot J., Tremonti S., White C., & Brinkmann S. D. M., 2006, *MNRAS*, 367, 1394

Murphy D. N. A., Geach J. E., & Bower R. G., 2012, *MNRAS*, 420, 1861

Tokovinin A., Tighe R., Schurter P., Cantarutti R., Martinez N., Mondaca M., & Montane E., 2008, *SPIE*, 7015

Galaxies in 3D across the Universe
Proceedings IAU Symposium No. 309, 2014
B. L. Ziegler, F. Combes, H. Dannerbauer, M. Verdugo, eds.
© International Astronomical Union 2015
doi:10.1017/S1743921314009764

Galaxy Growth at Early Times
from 3D Studies

N. M. Förster Schreiber and the SINS/zC-SINF & KMOS³ᴰ teams

Max-Planck-Institut für extraterrestrische Physik,
Giessenbachstrasse 1, D-85748 Garching, Germany
email: forster@mpe.mpg.de

Abstract. Spatially- and spectrally-resolved studies have proven very powerful in exploring the processes driving baryonic mass assembly and star formation at redshifts $z \sim 1-3$, the heyday of massive galaxy formation. This contribution presents selected key results from near-infrared integral field spectroscopic studies of $z \sim 1-3$ galaxies, including from the SINS/zC-SINF and KMOS³ᴰ surveys with SINFONI and KMOS at the ESO Very Large Telescope, highlights synergies with observations at other wavelengths, and discusses some of the implications for early galaxy evolution from mass assembly to feedback processes.

1. The new picture of galaxy evolution and the role of near-IR IFU surveys

In the emerging *"equilibrium growth model,"* early galaxies are fed by smooth accretion from their halos and minor mergers, replenishing their gas reservoirs and fueling their star formation (e.g., Dekel *et al.* 2009; Lilly *et al.* 2013). The balance between accretion and star formation maintains galaxies on a tight "main sequence" (MS) in stellar mass ($M_\star$) vs star formation rate (SFR), with ~ 0.3 dex scatter, whose evolution reflects the decrease in cosmic SFR density from its peak around $z \sim 2$ to the winding down epochs at $z \leqslant 1$ (e.g., Elbaz *et al.* 2007; Rodighiero *et al.* 2011; Whitaker *et al.* 2012). This scenario is further empirically motivated by the evolution of molecular gas mass fractions and the dominance of disk-like systems among MS star-forming galaxies (SFGs) out to $z \sim 2.5$ (e.g., Förster Schreiber *et al.* 2009; Wuyts *et al.* 2011; Tacconi *et al.* 2013; Wisnioski *et al.* 2014). The growth of galaxies thus appears to be tightly regulated, until they reach $M_\star \sim 10^{11} M_\odot$ where their star formation is rapidly quenched (e.g., Peng *et al.* 2010).

Spatially-resolved kinematics, star formation, and ISM conditions of $z \sim 1-3$ galaxies from observations of rest-optical line emission with sensitive near-IR integral field units (IFUs) at $8-10$m-class telescopes have provided some of the key empirical evidence in support of the above picture — highlighting the importance of internal dynamical processes such as violent gravitational instabilities in gas-rich disks, and of stellar and AGN feedback. The richness of information provided by such "3D studies" allows one to map the gas motions, the sites of on-going star formation, and spatial variations in gas-phase metallicity and nebular excitation across galaxies. Detailed line profile analysis can further reveal multiple emission line components and, in particular, the broad emission signature of outflowing gas. With IFUs, not only the kinematics but also the location, size, and geometry of outflows can be determined, all essential parameters to derive mass ejection rates and mass loading that are unconstrained from traditional rest-UV studies.

Obtaining such detailed but crucial information of faint distant galaxies requires a substantial investment of on-source integration per target. Near-IR IFUs at large ground-based telescopes such as SINFONI at the Very Large Telescope (VLT), OSIRIS on Keck II, and NIFS on Gemini-North, are all single-object IFUs. In the decade up to

2013, published studies using these instruments assembled data of ~ 300 galaxies at $z \approx 1-4$ in ~ 300 nights. "SINS/zC-SINF," carried out with SINFONI, provided the largest such survey with 110 SFGs at $z \sim 1.5-3$ observed in natural seeing ($4-5$ kpc resolution) and 35 of them followed up at higher resolution ($1-2$ kpc) with adaptive optics (AO; Förster Schreiber *et al.* 2006, 2009; Genzel *et al.* 2006, 2008, 2011; Mancini *et al.* 2011; Newman *et al.* 2013). SINS/zC-SINF probes the bulk of the $z \sim 2$ SFG population over two orders of magnitude in stellar mass and SFR ($M_\star \sim 3 \times 10^9 - 3 \times 10^{11}$ M$_\odot$, SFR $\sim 10-800$ M$_\odot$yr^{-1}). Other near-IR IFU surveys in the range $1 < z < 3.5$ have targeted galaxies with various selection criteria, probing somewhat different mass/luminosity ranges and/or focussing on different redshift intervals, and include among others "MASSIV" (e.g., Épinat *et al.* 2009, 2012; Contini *et al.* 2012), "AMAZE" and "LSD" (e.g., Gnerucci *et al.* 2011a,b), "WiggleZ" (Wisnioski *et al.* 2011, 2012), "sHiZELs" (Swinbank *et al.* 2012a,b), the "BX/BM" OSIRIS samples (Law *et al.* 2009, 2012; Wright *et al.* 2009). A small but important set of strongly lensed $z > 1$ galaxies was also observed with near-IR IFUs, reaching an effective resolution on sub-kpc scales in the source plane and even as small as $100-200$ pc in a few cases (Stark *et al.* 2008; Jones *et al.* 2010a,b, 2013; Yuan *et al.* 2011, 2012; Wuyts *et al.* 2014a).

With the new near-IR KMOS instrument at the VLT, featuring 24 IFUs deployable over a $7'$-diameter field (Sharples *et al.* 2013), survey efficiency is now considerably boosted. Samples of a few 1000's of distant galaxies may now be collected in the same amount of ~ 300 nights. Moreover, the effective cost of accessing fainter galaxies in so far poorly explored regimes of galaxy parameters is much reduced — useful data of even a handful of such objects with single-IFUs implies prohibitively large time allocations. KMOS thus opens up new opportunities for progress through more complete and homogeneous censuses. Aside from improved statistics, advances in our physical understanding of the processes driving galaxy evolution will continue to require *sensitivity, high resolution,* and *synergies with multi-wavelength observations.* High S/N is needed to measure reliably weaker emission lines and details of line profiles. AO-assisted observations are necessary to resolve objects down to the 1kpc-scales of galaxy components (bulges, star-forming clumps). Information about the distribution of the bulk of stellar populations (tracing past evolution) and of the cold molecular gas (the fuel of on-going/future star formation) is essential for a full and coherent picture, and to constrain timescales and scaling relations.

The next sections summarize selected key results from near-IR IFU studies, with an emphasis on the SINS/zC-SINF survey and highlights from the 1st-year data of the KMOS^{3D} survey. A comprehensive review covering near-IR single-object IFU studies is given by Glazebrook (2013).

2. Kinematics and structure of SFGs, and properties of disks at z $\sim 1-3$

The SINS/zC-SINF and other near-IR IFU surveys of kinematics revealed that $> 50\%$ of $z \sim 1-3$ SFGs are fairly regular rotation-dominated disks, whereas a minority consists of disturbed (major) merging systems or more compact, velocity dispersion-dominated objects (e.g., Shapiro *et al.* 2008; Förster Schreiber *et al.* 2009; Law *et al.* 2009; Jones *et al.* 2010b; Gnerucci *et al.* 2011a; Épinat *et al.* 2012; Newman *et al.* 2013). The rotation velocity $v_{\rm rot}$ scales roughly linearly with galaxy size, consistent with centrifugally-supported baryonic disks of constant angular momentum parameter within virialized dark matter halos. The rest-optical light and stellar mass maps from high-resolution HST near-IR imaging of the SINS/zC-SINF AO sample also show a majority of disk-like profiles and reveal a significant bulge-like component in the most massive galaxies (Förster Schreiber *et al.* 2011a; Tacchella *et al.* 2014). Concurring evidence for the prevalence of high-z disks has come from HST-based studies of rest-UV/optical morphologies and stellar mass distributions of large mass-selected samples out to $z \sim 2.5$, which showed that MS SFGs typically have a disk-like stellar structure with a systematic increase of bulge-to-disk ratio towards high masses (e.g. Wuyts *et al.* 2011; Lang *et al.* 2014). These results from kinematics and stellar structure provided some of the compelling evidence that smoother mass accretion modes onto galaxies and internal dynamical processes within galaxies play an important — if not dominant — role in growing galaxies.

High-z disks were found to be characterized by remarkably high intrinsic gas velocity dispersions of $\sigma_0 \sim 30 - 100 \, \mathrm{km \, s^{-1}}$, large gas-to-baryonic mass fractions of $\sim 30\% - 50\%$, and luminous kpc-sized star-forming clumps (e.g., Genzel *et al.* 2006, 2008; Förster Schreiber *et al.* 2006, 2009, 2011b; Law *et al.* 2009; Cresci *et al.* 2009; Jones *et al.* 2010b; Wisnioski *et al.* 2011; Swinbank *et al.* 2012a; Newman *et al.* 2013). At $z \sim 2$, the velocity dispersions are $5 - 10$ times higher than in present-day disks, implying that high-z disks are dynamically hotter and geometrically thicker, plausibly as a consequence of enhanced accretion, gas fractions, and SFRs. Elevated velocity dispersions were also measured from high-resolution IRAM/Plateau de Bure interferometric observations of CO emission in $z \sim 1 - 2$ star-forming disks, confirming that it is a property of the entire ISM and not just of the ionized gas layer (Tacconi *et al.* 2013, e.g.,). These properties, along with trends of increasing central dynamical mass fraction and bulge-to-disk ratio with galaxy mass and evolutionary stage, and possibly older clump ages at smaller galactocentric distances (e.g., Genzel *et al.* 2008, 2011; Förster Schreiber *et al.* 2011b; Wuyts *et al.* 2012; Tacchella *et al.* 2014), are consistent with theoretical arguments and numerical simulations of turbulent gas-rich disks in which giant star-forming clumps result from violent gravitational instabilities and bulges form via efficient secular processes on timescales < 1 Gyr (e.g., Bournaud *et al.* 2007, 2014; Ceverino *et al.* 2012; Dekel & Burkert 2014). Bulge growth in high-z disks can lead to "morphological quenching" as the central stellar spheroidal component stabilizes the gas-rich disk against fragmentation (e.g., Martig *et al.* 2009). Dynamical analysis of the SINS/zC-SINF AO sample revealed increasingly centrally peaked Toomre Q values well above unity in the inner $2 - 3$ kpc of the most massive galaxies, where star formation is suppressed and a massive stellar bulge is present — tantalizing evidence for inside-out gravitationally-driven quenching acting at $z \sim 2$ (Genzel *et al.* 2014a).

3. Vigorous feedback from star formation and AGN

Major breakthoughs emerged from the high quality, S/N, and resolution SINFONI+AO data from the SINS/zC-SINF survey. Strong, spatially-extended ionized gas outflows were discovered via a broad FWHM $\sim 450 \, \mathrm{km \, s^{-1}}$ Hα+[NII] component underneath the narrower emission tracing star formation, often associated with individual bright clumps in the disks (Genzel *et al.* 2011; Newman *et al.* 2012a,b). This discovery pinned down the roots of the ubiquitous galactic-scale winds at high-z so far only observed on > 10 kpc scales (e.g., Shapley *et al.* 2003; Weiner *et al.* 2009). Shock excitation plausibly due to these *star formation-driven outflows* is detected around bright clumps in the best-case, highest S/N disk (Newman *et al.* 2012a). The winds appear to set at a SFR surface density $\sim 1 \, \mathrm{M_\odot yr^{-1} kpc^{-2}}$, ten times higher than the break-out threshold in nearby starbursts and attributed to enhanced gas pressure in $z \sim 2$ disks. Outflow rates from clumps and from the galaxies reach up to several times their SFRs — the first empirical constraint at $z > 1$ on mass loading, a key parameter in theoretical models of star formation-driven feedback. Such vigorous feedback may disrupt intensely star-forming

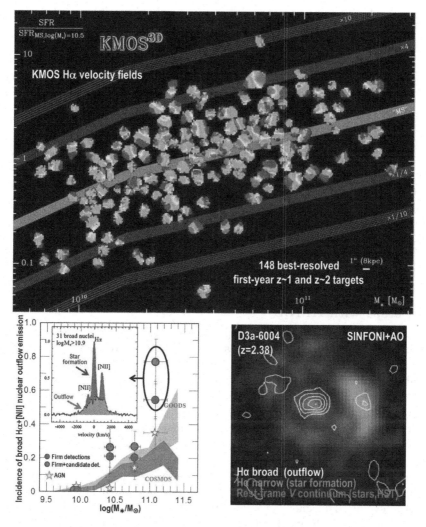

Figure 1. *Top:* Hα velocity fields for the 148 best-resolved $z \sim 1$ and $z \sim 2$ targets among the 246 galaxies observed in the first year of the KMOS^{3D} survey with the multi-IFU VLT/KMOS instrument (adapted from Wisnioski *et al.* 2014). All galaxies are shown on the same angular scale and are plotted within 0.2 dex of their location in the stellar mass versus SFR plane (with SFRs normalized to the SFR of the "main sequence" (MS) at the median $z \approx 0.9$ and ≈ 2.3 of the respective samples). Blue to red colors correspond to blue- to redshifted velocities relative to the systemic velocity of each galaxy. At least 70% of all the galaxies observed exhibit rotation-dominated disk kinematics, robustly confirming previous results from smaller samples observed with single-object IFUs. *Bottom left:* Incidence as a function of galaxy stellar mass of nuclear AGN-driven outflows, as revealed by the presence of a broad FWHM $\sim 1000 - 2000$ kms^{-1}, high-excitation component detected in Hα, [NII], and [SII] emission originating from the central few kpc of $z \sim 1 - 2.5$ of galaxies (adapted from Genzel *et al.* 2014b). There is a sharp onset of this incidence around $\log(M_\star/M_\odot) \sim 10.9$ (large blue/red circles) among the sample of > 100 galaxies analyzed (mostly from the KMOS^{3D} and SINS/zC-SINF surveys), reaching $\sim 2/3$ above this mass and at least as high as the fraction of luminous AGN from independent classical AGN indicators (star symbols and shaded areas). *Bottom right:* Deep SINFONI+AO observations have resolved the broad emission associated with nuclear AGN-driven outflows in the most massive $z \sim 2$ SINS/zC-SINF targets, and imply typical intrinsic sizes of $2 - 3$ kpc. One example is shown here, with the galaxy D3a-6004 (from Förster Schreiber *et al.* 2014). The white contours, green colors, and red colors correspond to the distribution of the broad emission tracing the nuclear outflow, the narrow Hα emission tracing star formation, and the rest-optical light from the bulk of stars. This galaxy has a significant stellar bulge and a prominent star-forming ring, and dynamical stability analysis is suggest it may also be undergoing morphological quenching (Genzel *et al.* 2014a).

gas-rich clumps before they migrate as bound entities to the galaxy centers (e.g., Genel *et al.* 2012; Wuyts *et al.* 2012).

The SINFONI+AO data further revealed evidence of *nuclear AGN-driven outflows* in all of the most massive SINS/zC-SINF galaxies ($\log(M_\star/M_\odot) > 10.9$), all of which are near-MS disks hosting a significant bulge (Förster Schreiber *et al.* 2014). Although classical AGN diagnostics indicate half of these galaxies host an AGN, none is bolometrically important. The broad emission in the inner few kpc of these massive SFGs has a larger FWHM $\sim 1000 - 1500$ kms^{-1} and higher [NII]/H$\alpha > 0.5$; inferred mass outflow rates are comparable or larger than the SFRs in the centers of the galaxies. The detection of this broad emission component in Hα *and* in the forbidden [NII] and [SII] lines, together with the intrinsic extent of $2 - 3$ kpc and the velocity width, are telltale signatures of an outflow driven by a Type 2 AGN. While energetic AGN-driven outflows have been reported at high-z in very luminous but rare QSOs, submillimeter, and radio galaxies (e.g., Nesvadba *et al.* 2011; Harrison *et al.* 2012; Cano-Díaz *et al.* 2012; Brusa *et al.* 2014), the SINS/zC-SINF survey has now detected nuclear AGN-driven winds in *typical* massive SFGs. The frequency of these AGN-driven outflows at the high-mass end suggests they have a large duty cycle and, given the inferred outflow rates, may contribute to shutting down star formation by efficiently expelling gas out of the inner regions of the galaxies.

4. Next steps with KMOS

With KMOS, it is now possible to expand near-IR IFU studies to larger, more homogeneous and complete samples. Our team has undertaken KMOS^{3D}, a comprehensive multi-year survey of Hα+[NII]+[SII] emission of 600+ $z = 0.7 - 2.7$ mass-selected galaxies (see also contribution by E. Wuyts *et al.*, this volume). KMOS^{3D} capitalizes on the 3D-HST and CANDELS HST Treasury Surveys, providing unique multi-wavelength data and large mass-limited samples with accurate redshifts. The survey will enable us to establish firmly the connection between galaxy kinematics and stellar structure, determine the role of star formation and AGN in governing stellar mass growth, test systematically the imprint of feedback on galaxy formation efficiency, and explore the influence of environment. By observing similarly selected samples and targeting the same spectral diagnostics over the full $0.7 < z < 2.7$ range, KMOS^{3D} will allow us to track consistently the evolution in galaxy properties over 5 Gyrs all across the peak epochs of cosmic star formation activity. First-year results from deep observations of nearly 250 targets, split in about equal-sized samples at median $z \approx 0.9$ and $z \approx 2.3$, show that our strategy emphasizing unbiased selection and sensitivity is successful at providing high quality data of individual galaxies and extending into regimes unexplored so far by near-IR IFU surveys. These results: (1) confirm with greater robustness the prevalence of kinematically-identified disks among SFGs at $\log(M_\star/M_\odot) > 9.5$ and both $z \sim 1$ and $z \sim 2$, and show a decrease by a factor of ≈ 2 of the intrinsic dispersion σ_0 consistent with expectations from the evolution of cold gas mass fractions in the framework of marginally-stable gas-rich disks (Wisnioski *et al.* 2014, and E. Wuyts *et al.* this volume); (2) strengthen the ubiquity of nuclear AGN-driven outflows in massive high-z SFGs from a larger sample (with a 6-fold increase at the highest masses) that clearly reveals a fairly sharp onset at $\log(M_\star/M_\odot) \sim 10.9$ of the incidence of this phenomenon, reaching about 2/3 above this mass — a fraction at least as high as that of luminous AGN identified in the X-ray, optical, infrared, and radio (Genzel *et al.* 2014b); (3) set tighter constraints, in combination with data from SINS/zC-SINF and from LUCI multi-object spectroscopy at the Large Binocular Telescope, on the slope and evolution of the mass-metallicity relation in the range $0.7 < z < 2.7$, and the (lack of) dependence on SFR at fixed z (Wuyts *et al.* 2014b, and E. Wuyts *et al.* this volume).

We can look forward to a great many new findings over the next few years from our KMOS[3D] and other surveys undertaken with KMOS (e.g., Sobral *et al.* 2013; Stott *et al.* 2014) in conjunction with high resolution AO-assisted IFU follow-ups and multi-wavelength information — a new era in near-IR IFU studies has begun!

References

Bournaud, F., Elmegreen, B. G., & Elmegreen, D. M. 2007, *ApJ*, 670, 237

Bournaud, F., Perret, V., Renaud, F., *et al.* 2014, *ApJ*, 780, 57

Brusa, M., Bongiorno, A., Cresci, G., *et al.* 2014, *MNRAS*, in press (arXiv:1409.1615)

Cano-Díaz, M., Maiolino, R., Marconi, A., *et al.* 2012, *A&A*, 537, L8

Ceverino, D., Dekel, A., Mandelker, N., *et al.* 2012, *MNRAS*, 420, 3490

Contini, T., Garilli, B., Le Fèvre, O., *et al.* 2012, *A&A*, 539, A91

Cresci, G., Hicks, E. K. S.., Genzel, R., *et al.* 2009, *ApJ*, 697, 115

Dekel, A., Sari, R., & Ceverino, D. 2009, *ApJ*, 703, 785

Dekel, A. & Burkert, A. 2014, *MNRAS*, 438, 1870

Elbaz, D., Daddi, E., Le Borgne, D., *et al.* 2007, *A&A*, 468, 33

Épinat, B., Contini, T., Le Fèvre, O., *et al.* 2009, *A&A*, 504, 789

Épinat, B., Tasca, L., Amram, P., *et al.* 2012, *A&A*, 539, A92

Förster Schreiber, N. M., Genzel, R., Lehnert, M. D., *et al.* 2006, *ApJ*, 645, 1062

Förster Schreiber, N. M., Genzel, R., Bouché, N., *et al.* 2009, *ApJ*, 706, 1364

Förster Schreiber, N. M., Shapley, A. E., Erb, D. K., *et al.* 2011a, *ApJ*, 731, 65

Förster Schreiber, N. M., Shapley, A. E., Genzel, R., *et al.* 2011b, *ApJ*, 739, 45

Förster Schreiber, N. M., Genzel, R., Newman, S. F., *et al.* 2014, *ApJ*, 787, 38

Genel, S., Naab, T., Genzel, R., *et al.* 2012, *ApJ*, 745, 11

Genzel, R., Tacconi, L. J., Eisenhauer, F., *et al.* 2006, *Nature*, 442, 786

Genzel, R., Burkert, A., Bouché, N., *et al.* 2008, *ApJ*, 687, 59

Genzel, R., Newman, S. F., Jones, T., *et al.* 2011, *ApJ*, 733, 101

Genzel, R., Förster Schreiber, N. M., Lang, P., *et al.* 2014a, *ApJ*, 785, 75

Genzel, R., Förster Schreiber, N. M., Rosario, D., *et al.* 2014b, *ApJ*, in press (arXiv:1406.0183)

Glazebrook, K., 2013, Publ. Astron. Soc. Australia, 30, 56

Gnerucci, A., Marconi, A., Cresci, G., *et al.* 2011a, *A&A*, 528, A88

Gnerucci, A., Marconi, A., Cresci, G., *et al.* 2011b, *A&A*, 533, A124

Harrison, C. M., Alexander, D. M., Swinbank, A. M., *et al.* 2012, *MNRAS*, 426, 1073

Jones, T. A., Ellis, R. S., Jullo, E., & Richard, J. 2010a, *ApJ*, 725, L176

Jones, T. A., Swinbank, A. M., Ellis, R. S., Richard, J., & Stark, D. P. 2010b, *MNRAS*, 404, 1247

Jones, T. A., Ellis, R. S., Richard, J., & Jullo, E. 2013, *ApJ*, 765, 48

Lang, P., Wuyts, S., Somerville, R. S., *et al.* 2014, *ApJ*, 788, 11

Law, D. R., Steidel, C. C., Erb, D. K., *et al.* 2009, *ApJ*, 697, 2057

Law, D. R., Shapley, A. E., Steidel, C. C., *et al.* 2012, *Nature*, 487, 338

Lilly, S. J., Carollo, C. M., Pipino, A., Renzini, A., & Peng, Y. 2013, *ApJ*, 772, 119

Mancini, C., Förster Schreiber, N. M., Renzini, A., *et al.* 2011, *ApJ*, 743, 86

Martig, M., Bournaud, F., Teyssier, R., & Dekel, A. 2009, *ApJ*, 707, 250

Nesvadba, N. P. H.., Polletta, M., Lehnert, M. D., *et al.* 2011, *MNRAS*, 415, 2359

Newman, S. F., Shapiro Griffin, K., Genzel, R., *et al.* 2012a, *ApJ*, 752, 111

Newman, S. F., Genzel, R., Förster Schreiber, N. M., *et al.* 2012b, *ApJ*, 761, 43

Newman, S. F., Genzel, R., Förster Schreiber, N. M., *et al.* 2013, *ApJ*, 767, 104

Peng, Y., Lilly, S. J., Kovac, K., *et al.* 2010, *ApJ*, 721, 193

Rodighiero, G., Daddi, E., Baronchelli, I., *et al.* 2011, *ApJ*, 739, L40

Shapiro, K. L., Genzel, R., Förster Schreiber, N. M., *et al.* 2008, *ApJ*, 682, 231

Shapley, A. E., Steidel, C. C., Pettini, M., & Adelberger, K. L. 2003, *ApJ*, 588, 65

Sharples, R., Bender, R., Agudo Berbel, A., *et al.* 2013, The Messenger, 151, 21

Sobral, D., Swinbank, A. M., Stott, J. P., *et al.* 2013, *ApJ*, 779, 139

Stark, D. P., Swinbank, A. M., Ellis, R. S., *et al.* 2008, *Nature*, 455, 775

Stott, J. P., Sobral, D., Swinbank, A. M., *et al.* 2014, *MNRAS*, 443, 2695

Swinbank, A. M., Smail, I., Sobral, D., *et al.* 2012a, *ApJ*, 760, 130

Swinbank, A. M., Sobral, D., Smail, I., *et al.* 2012b, *MNRAS*, 426, 935

Tacchella, S., Lang, P., Carollo, C. M., *et al.* 2014, *ApJ*, submitted

Tacconi, L. J., Neri, R., Genzel, R., *et al.* 2013, *ApJ*, 768, 74

Weiner, B. J., Coil, A. L., Prochaska, J. X., *et al.* 2009, *ApJ*, 692, 187

Whitaker, K. E., van Dokkum, P. G., Brammer, G., & Franx, M. 2012, *ApJ*, 754, L29

Wisnioski, E., Förster Schreiber, N. M., Wuyts, S., *et al.* 2014, *ApJ*, submitted (arXiv:1409.6791)

Wisnioski, E., Glazebrook, K., Blake, C., *et al.* 2011, *MNRAS*, 417, 2601

Wisnioski, E., Glazebrook, K., Blake, C., *et al.* 2012, *MNRAS*, 422, 3339

Wright, S. A., Larkin, J. E., Law, D. R., *et al.* 2009, *ApJ*, 699, 421

Wuyts, S., Förster Schreiber, N. M., van der Wel, A., *et al.* 2011, *ApJ*, 742, 96

Wuyts, S., Förster Schreiber, N. M., Genzel, R., *et al.* 2012, *ApJ*, 753, 114

Wuyts, S., Förster Schreiber, N. M., Nelson, E. J., *et al.* 2013, *ApJ*, 779, 135

Wuyts, E., Rigby, J. R., Gladders, M. D., & Sharon, K. 2014a, *ApJ*, 781, 61

Wuyts, E., Kurk, J., Förster Schreiber, N. M., *et al.* 2014b, *ApJ*, 789, L40

Yuan, T.-T., Kewley, L. J., Swinbank, A. M., Richard, J., & Livermore, R. C. 2011, *ApJ*, 732, L14

Yuan, T.-T., Kewley, L. J., Swinbank, A. M., & Richard, J. 2012, *ApJ*, 759, 66

Galaxies in 3D across the Universe
Proceedings IAU Symposium No. 309, 2014
B. L. Ziegler, F. Combes, H. Dannerbauer, M. Verdugo, eds.
© International Astronomical Union 2015
doi:10.1017/S1743921314009776

Blowin' in the wind: both 'negative' and 'positive' feedback in an outflowing quasar at $z \sim 1.6$

Giovanni Cresci

INAF - Osservatorio Astrofisco di Arcetri, largo E. Fermi 5, 50127, Firenze, Italy
email: gcresci@arcetri.astro.it

Abstract. Quasar feedback in the form of powerful outflows is invoked as a key mechanism to quench star formation, preventing massive galaxies to over-grow and producing the red colors of ellipticals. On the other hand, some models are also requiring 'positive' AGN feedback, inducing star formation in the host galaxy through enhanced gas pressure in the interstellar medium. However, finding observational evidence of the effects of both types of feedback is still one of the main challenges of extragalactic astronomy, as few observations of energetic and extended radiatively-driven winds are available. We present SINFONI near infrared integral field spectroscopy of XID2028, an obscured, radio-quiet $z = 1.59$ QSO, in which we clearly resolve a fast (1500 km/s) and extended (up to 13 kpc from the black hole) outflow in the [OIII] lines emitting gas, whose large velocity and outflow rate are not sustainable by star formation only. The narrow component of Hα emission and the rest frame U band flux show that the outflow position lies in the center of an empty cavity surrounded by star forming regions on its edge. The outflow is therefore removing the gas from the host galaxy ('negative feedback'), but also triggering star formation by outflow induced pressure at the edges ('positive feedback'). XID2028 represents the first example of a host galaxy showing both types of feedback simultaneously at work.

Keywords. Galaxies: active - Galaxies: evolution - ISM: jets and outflows

1. Introduction

We have recently completed two follow-up programs on obscured quasars pre-selected for being in a significant outflowing phase. The first consists in X-Shooter observations of a sample of X-ray selected obscured QSOs at z ~ 1.5, selected from the XMM-COSMOS survey on the basis of their observed red colors (R-K> 4.5) and high MIR and X-ray to optical flux ratio (X/O> 10, see Brusa *et al.* 2010). The presence of outflowing material identified by [OIII]λ4958,5007 kinematics was indeed detected in 6 out of 8 sources, confirming the efficiency of a selection based on X-ray to optical to NIR colors in isolating such objects undergoing a 'blow-out' phase (Brusa *et al.* 2014).

We also recently completed a SINFONI program on a sample of AGN pre-selected to have high accretion rates, where radiatively-driven winds are supposed to be more effective since they likely originate from the acceleration of disk outflows by the AGN radiation field, and a combination of moderate obscuration observed in the X-rays and high Eddington ratio where outflows or transient absorption are expected to happen (Fabian *et al.* 2008). Interestingly, XID2028 at $z = 1.59$ was selected as an obscured, outflowing QSOs using both criteria described above.

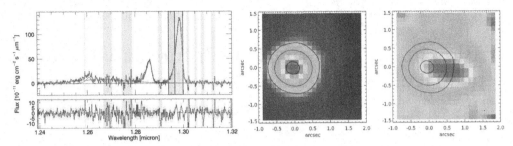

Figure 1. [OIII] in XID2028. *Left panels:* the J band SINFONI spectrum of XID2028, integrated in a region of 8×8 spaxels (1" × 1") around the QSO. The observed spectrum is shown in black, the different broken power-law components in the fit for each line (Hβ, large BLR Hβ, [OIII]λ4959,5007) are shown in blue, while their sum is shown in red. The shaded regions show the location of sky lines, that were excluded from the fit. The red and blue box show the intervals in wavelength in which the maps shown in the right panels are integrated. *Right Panels:* [OIII]λ5007 channel maps obtained integrating the continuum subtracted SINFONI datacube on the line core (*left*, red box) and on the blue wing (*right*, blue box). The contours on the line core, marking the position of the central QSO, are shown in black in both panels. The fully resolved, extended blue wing due to the outflow is extended up to 1.5", i.e. 13 projected kpc from the QSO position. North is up and East is left, 1" corresponds to 8.5 kpc.

2. Results

The integrated J band SINFONI spectrum of XID2028 is shown in Fig. 1: an asymmetric, prominent blue wing is clearly visible for all the emission lines, already suggesting the presence of outflowing material towards the observer. In fact, the forbidden [OIII] emission lines are an ideal tracer of extended outflowing ionized gas, as cannot be produced in the high-density, sub-parsec scales typical of AGN broad-line regions (BLR). To assess the spatial extent of the blueshifted wing, in Fig. 1 we also show the continuum subtracted SINFONI datacube collapsed on the spectral channels corresponding to the line core and to the line wing. The prominence of an extended, fully resolved outflow is evident from the image of the blue wing in the west-east direction. The blue wing is originating on the QSO position, as marked from the line core black contours, and it is extending up to 1.5", i.e. 13 projected kpc, from the center.

We have fitted separately the spectrum of each spatial spaxel in the field of view using a broken power law, in order to measure and map the velocity of the [OIII] line wing. We measure strongly blueshifted velocities in the outflow region, with velocities as high as $v_{10} = -1500\ km/s$ towards our line of sight, where v_{10} is the velocity at the 10th percentile of the overall emission-line profile. The detected velocity are far too high to be due to rotational motions in the host galaxy. Moreover, in case of ordered rotation we expect the line width to be peaked at the center of the galaxy (e.g. Cresci *et al.* 2009), while in this case the line width is larger in the blue wing region.

We used the Hβ emission in the outflow region to derive an estimate of the ionized outflowing mass. Depending on a number of assumptions (see Cresci *et al.* 2014), we estimate a mass outflow rate $\dot{M}_{out} \sim 300\ M_\odot/yr$ for the ionized component only, which may correspond to a total outflow rate of $\dot{M}_{out} \gtrsim 1000\ M_\odot/yr$ when accounting for a neutral and molecular component. This would translate to a mass loading factor $\gtrsim 3$, suggesting that such an energetic outflow is not sustainable by star formation only.

We fitted the integrated H+K SINFONI spectrum in the Hα region with a single broad Gaussian, finding a $FWHM = 3600\ km/s$. Although a single Gaussian represents a good fit on the wings of the emission line, a weaker but significant narrow component is present in the fit residuals. The redshift $z = 1.594$ of this narrow component ($FWHM =$

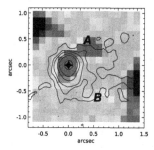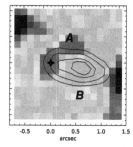

Figure 2. Narrow Hα map obtained integrating the single broad Gaussian Hα fit residuals on the spectral channels $1.7015 < \lambda < 1.7047$ μm. In the *left panel* the HST-ACS rest frame U band contours are superimposed in black. The same pattern is obtained by these two independent tracers of star formation in the host galaxy, with two additional clumps of star formation (marked with A and B) elongated at the west of the QSO (marked with a star). In the *right panel* the blue wing contours from Fig. 1, tracing the outflow position, are plotted for comparison. A clear anti-correlation between the outflow location and the star formation tracers suggests that the outflowing material is sweeping the gas along the outflow core ('negative feedback'), while is compressing the gas at its edges inducing star formation at the locations marked as *A* and *B* on the map ('positive feedback').

320 km/s) is consistent with the systemic redshift of the host galaxy (Brusa *et al.* 2010), and it is interpreted as due to the star formation in the host galaxy.

We map the spatial extent of the Hα narrow component by fitting in each spaxels of the datacube the single Gaussian derived from the integrated spectrum, keeping the width and centroid fixed and leaving the normalization free to vary. We then integrate on the spectral channels corresponding to the narrow component in the integrated spectrum ($1.7015 < \lambda < 1.7047$ μm) the residual map obtained subtracting the Gaussian fit to the datacube. The result is illustrated in Fig. 2, where we show that the star formation in the host is not symmetrical distributed: most of the star formation activity is concentrated in the nucleus and in just two additional clumps, elongated to the west away from the QSO in two almost parallel branches around the location of the outflow (marked as A and B, detected in an integrated spectrum at 9σ and 4σ respectively). We derive a limit of $log([NII]/H\alpha) < -1.1$ over the regions A and B. The narrow components of Hβ and [OIII]λ5007 are undetected across all the field of view, making impossible to place the two regions on classical AGN-star forming galaxies BPT spectral diagnostic. Nonetheless, the limit derived for [NII]/Hα is already supporting a star forming origin of the line emission, as no known AGN shows such low values of this ratio (Kauffmann *et al.* 2003). For comparison, we measure $log([NII]/H\alpha) = -0.4$ for the narrow components in the nuclear region, fully consistent with AGN ionization.

The same peculiar spatial pattern is also found in the available HST-ACS F814W imaging, sampling the rest frame U band at the redshift of the source (Fig. 2, left panel). Although at least part of the U emission may be due to scattered light from the QSO, the U band is also sensitive to the light emitted by young, massive stars. Therefore, the similarity of the shape of the U-band image with that of the narrow Hα emission suggests that the U continuum is also dominated by star formation that covers a horseshoe region around the outflow. Interestingly, the outflow position is coincident with the cavity between these two star forming regions (Fig. 2) both in the narrow Hα residuals maps as well as in the ACS rest frame U image. The star formation activity is therefore heavily suppressed in the core of the outflow, where the fast expanding gas is able to sweep away the gas needed to sustain the assemble of new stars. This represents a clear example of *'negative feedback'* in action, showing the powerful outflow expelling most of the gas and quenching the star formation along its route in the host galaxy. On the other hand, we also find evidence that enhanced, triggered star forming activity is detected in the two off-center, elongated regions surrounding the bulk of the outflow, marked as A and B in Fig. 2. The causal connection between the outflows and the star forming regions is supported by the highly asymmetric shape of the star forming regions, that are extending out of the galaxy exactly along both edges of the outflow, while no other tail or star forming clump is detected in the rest of the host galaxy: the integrated flux of the narrow Hα line emission in the rest of the disk is $\sim 40\%$ of regions A and B alone.

3. Conclusions

These observations suggest that we are also witnessing *'positive feedback'* in the host galaxy with the gas clouds compressed by the outflow-driven shock, which drives turbulent compression especially at the outflow edges. Such 'positive feedback' has been invoked in recent years to explain the correlation between AGN luminosities and nuclear star formation rates as well as between black hole accretion rate and star formation rate in AGN (Ishibashi *et al.* 2012, 2014; Silk 2013, Zinn *et al.* 2013). Despite its possible importance, the few available observational evidences of such feedback were till now ascribed to an handful of extreme, powerful jets in radio-loud galaxies (Feain *et al.* 2007, Crockett *et al.* 2012). Moreover, the outflow-induced star formation has been usually found not in the AGN host galaxy, but in a companion satellite aligned along the radio axis (Croft *et al.* 2006, Elbaz *et al.* 2009, Kramer *et al.* 2004).

Therefore, XID2028 represents the first direct detection of outflow induced star formation in a radio quiet AGN, as well as the first example of both types of feedback simultaneously at work in the same galaxy. As the quasar is observed at the peak epoch of galaxy and black hole assembly, where we expect to have the maximum influence of feedback on galaxy evolution, the data presented demonstrate that both 'positive' and 'negative' AGN feedback are a crucial ingredient to shape the evolution of galaxies, by regulating the star formation in the host and driving the BH-galaxy coevolution.

References

Brusa, M., Civano, F., Comastri, A., *et al.* 2010, *ApJ*, 716, 348
Brusa, M., *et al.* 2014, *MNRAS*, submitted
Cresci, G., Hicks, E. K. S., Genzel, R., *et al.* 2009, *ApJ*, 697, 115
Crockett, R. M., Shabala, S. S., Kaviraj, S., *et al.* 2012, *MNRAS*, 421, 1603
Croft, S., van Breugel, W., de Vries, W., *et al.* 2006, *ApJ*, 647, 1040
Elbaz, D., Jahnke, K., Pantin, E., Le Borgne, D., & Letawe, G. 2009, *A&A*, 507, 1359
Fabian, A. C., Vasudevan, R. V., & Gandhi, P. 2008, *MNRAS*, 385, L43
Feain, I. J., Papadopoulos, P. P., Ekers, R. D., & Middelberg, E. 2007, *ApJ*, 662, 872
Ishibashi, W. & Fabian, A. C. 2012, *MNRAS*, 427, 2998
Ishibashi, W. & Fabian, A. C. 2014, arXiv:1404.0908
Kauffmann, G., Heckman, T. M., Tremonti, C., *et al.* 2003, *MNRAS*, 346, 1055
Klamer, I. J., Ekers, R. D., Sadler, E. M., & Hunstead, R. W. 2004, *ApJ*, 612, L97
Silk, J. 2013, *ApJ*, 772, 112
Zinn, P.-C., Middelberg, E., Norris, R. P., & Dettmar, R.-J. 2013, *ApJ*, 774, 66

Galaxies in 3D across the Universe
Proceedings IAU Symposium No. 309, 2014
B. L. Ziegler, F. Combes, H. Dannerbauer, M. Verdugo, eds.

© International Astronomical Union 2015
doi:10.1017/S1743921314009788

The Evolution of Resolved Kinematics and Metallicity from Redshift 2.7 to 0.7 with LUCI, SINS/zC-SINF and KMOS^{3D}

Eva Wuyts and the SINS/zC-SINF and KMOS^{3D} Teams[‡]

Max-Planck-Institut für extraterrestrische Physik,
Giessenbachstr. 1, D-85741 Garching, Germany
email: `evawuyts@mpe.mpg.de`

Abstract. The KMOS^{3D} survey will provide near-IR IFU observations of a mass-selected sample of $\sim$600 star-forming galaxies at $0.7 < z < 2.7$ with the K-band Multi Object Spectrograph (KMOS) at the VLT. We present kinematic results for a first sample of $\sim$200 galaxies, focusing on the evolution of the gas velocity dispersion with redshift. Combined with existing measurements, we find an approximate $(1+z)$ evolution from $z \sim 4$ to the present day, which can be understood from the co-evolution of the gas fraction and specific star formation rate (sSFR) in the the equilibrium picture of galaxy evolution.

We combine the KMOS^{3D} sample with data from the LUCI and SINFONI spectrographs, as well as multi-wavelength HST imaging from CANDELS, to address the relations between stellar mass, SFR, and the [N II]/Hα flux ratio as an indicator of gas-phase metallicity for a sample of 222 star-forming galaxies. We find a constant slope at the low-mass end of the mass-metallicity relation and can fully describe its redshift evolution through the evolution of the characteristic turnover mass where the relation begins to flatten at the asymptotic metallicity. At a fixed mass and redshift, our data do not show a correlation between the [N II]/Hα ratio and SFR.

Keywords. galaxies: formation - galaxies: evolution - galaxies: high-redshift

1. Introduction

The KMOS^{3D} GTO survey (P. I.: N. M. Förster Schreiber/D. J. Wilman) takes advantage of the multiplex capabilities of the KMOS deployable IFU system at the VLT to observe the Hα and [N II] emission of a mass-selected sample of $\sim$600 star-forming galaxies (SFGs) at $0.7 < z < 2.7$ in the COSMOS, GOODS-South and UDS deep fields. Targets are selected from the rest-frame optical 3D-HST Treasury Survey (Brammer *et al.* 2012), which reduces the historical bias of previous rest-frame UV-selected IFU studies towards blue star-forming galaxies. The crucial synergy of KMOS^{3D} with the CANDELS multi-wavelength HST imaging allows to connect the properties of the ionised gas (kinematics, SF, metallicity and outflows) with the stellar structure and populations, as well as galaxy environment.

[‡] Jaron Kurk[1], Natascha M. Förster Schreiber[1], Reinhard Genzel[1], Emily Wisnioski[1], Kaushala Bandara[1], Stijn Wuyts[1], Alessandra Beifiori[1], Ralf Bender[1], Gabriel B. Brammer, Andreas Burkert, Peter Buschkamp[1], C. Marcella Carollo, Jeffrey Chan[1], Ric Davies[1], Frank Eisenhauer[1], Max Fabricius, Matteo Fossati[1], Sandesh K. Kulkarni[1], Yoshi Fudamoto[1], Philipp Lang[1], Simon J. Lilly, Dieter Lutz[1], Chiara Mancini, Trevor Mendel[1], Ivelina G. Momcheva, Thorsten Naab, Erica J. Nelson, Alvio Renzini, David Rosario[1], Roberto P. Saglia[1], Stella Seitz, Ray M. Sharples, Amiel Sternberg, Sandro Tacchella, Linda J. Tacconi[1], Pieter van Dokkum, David J. Wilman[1]

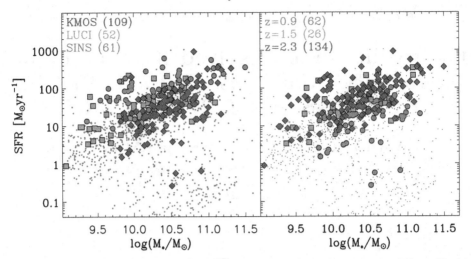

Figure 1. Location of the combined KMOS^{3D}, LUCI and SINS sample in the SFR-stellar mass plane. Targets are color-coded by instrument in the left panel and redshift in the right panel. The numbers in parentheses note the number of sources in each survey or redshift bin. As a reference, the small gray dots represent the total mass-selected galaxy population at $0.7 < z < 2.7$ in the CANDELS/3D-HST fields.

2. The redshift evolution of the velocity dispersion

The first year of KMOS^{3D} observations has targeted 223 SFGs, 106 galaxies in YJ-band at median $z = 0.9$ and 117 galaxies in K-band at median $z = 2.3$ (Figure 1), and has detected Hα emission for 80% of them. For all detected targets, we have carried out the standard kinematic extraction to produce maps of Hα flux, velocity and velocity dispersion as well as 1D velocity and dispersion profiles along the kinematic axis. When these 1D profiles are well-defined, we fit an exponential disk model and measure the dispersion in the outer regions of the galaxy, at or outside the turnover radius, to minimize the impact of beam-smearing in the inner regions. We find an average dispersion of 50 km/s at $z = 2.3$ and 25 km/s at $z = 0.9$ (Wisnioski *et al.* 2014, in prep.). Figure 2 combines the KMOS^{3D} dispersion results with literature data for a range of different surveys and shows a continuous $(1 + z)$ decline in dispersion from $z \sim 4$ towards the present day. This can be physically understood starting from the Toomre stability criterium, where for a critical disk $\sigma(z) \sim v_{\rm rot} f_{\rm gas}(z)$ (Genzel *et al.* 2008). The redshift evolution of the gas fraction $f_{\rm gas}(z)$ depends on the evolution of the depletion time and specific star formation rate. Adopting recent scaling relations from the literature (Whitaker *et al.* 2014; Tacconi *et al.* 2013; Genzel *et al.* 2014a), the shaded gray area in Figure 2 shows the predicted velocity dispersion evolution for a range of rotational velocities $v_{\rm rot} = 100 - 250$ km/s (see Wisnioski *et al.* 2014 (in prep.) for more details). The agreement with the data supports the equilibrium picture of galaxy evolution, where the increased turbulence in high redshift disks is driven by their higher gas fractions.

3. The redshift evolution of the mass-metallicity relation

We measure the average [N II]/Hα line ratio for all KMOS^{3D} targets from an integrated, velocity-shifted 1D spectrum within the maximal elliptical aperture along the kinematic axis. To increase the statistics and redshift coverage of this metallicity study, we include existing LUCI long-slit and SINFONI IFU spectroscopic data (Förster Schreiber *et al.* 2009; Mancini *et al.* 2011), adding 26 SFGs at $z \sim 1.5$ and 87 SFGs at $z \sim 2.3$

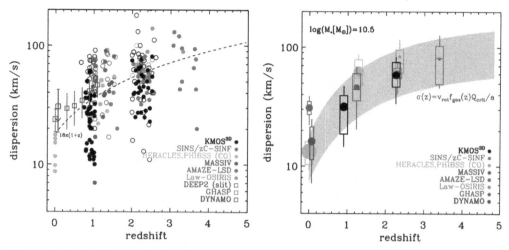

Figure 2. *Left:* Galaxy velocity dispersion measurements from the literature at $z = 0 - 4$ from molecular and ionized gas emission. KMOS^{3D} results at $z = 0.9$ and $z = 2.3$ are shown by black circles. The dashed line shows a simple $(1 + z)$ evolution scaled to overlap with the data. *Right:* The gray band shows the expected dispersion based on the Toomre criterium and the redshift evolution of the gas fraction (see text). It agrees well with the mean (filled circles) and 50% and 90% distributions (boxes and vertical lines, respectively) of the data points shown on the left.

(Figure 1). Our analysis is carried out as much as possible in terms of the directly observable [N II]/Hα line ratio to avoid systematics associated with the choice of strong-line metallicity calibration (Kewley & Ellison 2008); when necessary, we employ the linear calibration by Pettini & Pagel (2004).

Figure 3 shows the [N II]/Hα ratios as a function of stellar mass, 2σ upper limits are included for the 17% of targets for which [N II] was not detected. We identify 18 AGNs in our sample from X-ray and radio data, infrared colors, and rest-UV spectroscopy (Genzel *et al.* 2014a). Additionally, Förster Schreiber *et al.* (2014) and Genzel *et al.* (2014a) report evidence for AGN-driven outflows in the nuclear regions of 20 targets included in our combined sample from broad (FWHM$\gtrsim$1000 km/s) emission components with enhanced [N II]/Hα line ratios. We stack our galaxies at $z = 0.9$ and $z = 2.3$, both with and without the AGN-contaminated sources. The right panel of Figure 3 shows that the 17% AGN contamination of our sample does not significantly affect the derived mass-metallicity relation (MZR). We find good agreement with literature studies at the high mass end (Erb *et al.* 2006; Zahid *et al.* 2014b; Steidel *et al.* 2014). The different slopes at the low-mass end are likely due to differences in sample selection, especially the inclusion of red, dusty objects. This needs further investigation.

We provide fits to the MZR based on the parameterization introduced by Zahid *et al.* (2014a),

$$12 + \log(O/H) = Z_0 + \log\left[1 - \exp\left(-\left[\frac{M_*}{M_0}\right]^\gamma\right)\right] \quad (3.1)$$

where Z_0 corresponds to the asymptotic metallicity, M_0 is the characteristic turnover mass where the relation begins to flatten and γ is the power-law slope at stellar masses $\ll M_0$. Within the uncertainties, Z_0 and γ are constant, such that redshift evolution of the MZR depends solely on the evolution of the characteristic turnover mass (Wuyts *et al.* 2014). For a fixed $Z_0 = 8.69$ and $\gamma = 0.4$, we can describe the redshift evolution of the characteristic turnover mass as $\log(M_0/M_\odot) = (8.86\pm0.05)+(2.92\pm0.16)\log(1+z)$. The

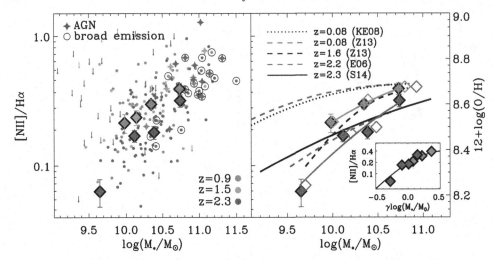

Figure 3. [N II]/Hα detections and 2σ upper limits vs. stellar mass, color-coded by redshift. AGNs identified from classic X-ray, etc., indicators and broad nuclear AGN-driven outflows are indicated with four-pointed stars and black circles, respectively. Large colored diamonds indicate the stacked [N II]/Hα ratios in three, one, and four bins of stellar mass for the $z = 0.9$, $z = 1.5$, and $z = 2.3$ redshift slices, excluding the AGN-contaminated sources. In the right panel, the open diamonds show the stacked results when all galaxies are included. The blue and red solid lines correspond to our best-fit MZRs. The inset shows the stacked [N II]/Hα ratios as a function of $\gamma \log(M_*/M_0)$, which follow a universal relation.

relation between metallicity and stellar mass scaled by M_0 thus becomes independent of redshift, as can be seen in the right panel of Figure 3.

Finally, we investigate the importance of SFR in the mass-metallicity relation by stacking our sample at $z = 0.9$ and $z = 2.3$ in two bins of SFR for each mass bin. We do not find a correlation between the [N II]/Hα ratio and SFR at fixed mass and redshift (Wuyts *et al.* 2014). Both metallicity and SFR evolve with redshift, but the lack of correlation between these two parameters at fixed mass and redshift, suggests that both processes are not necessarily causally related.

References

Brammer, G. B., *et al.* 2012, *ApJS*, 200, 13
Erb, D. K., *et al.* 2006, *ApJ*, 644, 813
Förster Schreiber, N. M., *et al.* 2009, *ApJ*, 706, 1364
Förster Schreiber, N. M., *et al.* 2014, *ApJ*, 787, 38
Genzel, R., *et al.* 2008, *ApJ*, 687, 59
Genzel, R., *et al.* 2014a, arXiv:1406.0183
Genzel, R., *et al.* 2014b, arXiv:1409.1171
Kewley, L. J. & Ellison, S. L. 2008, *ApJ*, 681, 1183
Mancini, C., *et al.* 2011, *ApJ*, 743, 86
Pettini, M. & Pagel, B. E. J. 2004, *MNRAS*, 348, L59
Steidel, C. C., *et al.* 2014, arXiv:1405.5473
Tacconi, L. J., *et al.* 2013, *ApJ*, 768, 74
Whitaker, K. E., *et al.* 2014, arXiv:1407.1843
Wuyts, E., *et al.* 2014, *ApJ*, 789, L40
Zahid, H. J., *et al.* 2014a, *ApJ*, 791, 130
Zahid, H. J., *et al.* 2014b, *ApJ*, 792, 75

Galaxies in 3D across the Universe
Proceedings IAU Symposium No. 309, 2014
B. L. Ziegler, F. Combes, H. Dannerbauer, M. Verdugo, eds.

© International Astronomical Union 2015
doi:10.1017/S174392131400979X

Resolved Spectroscopy of Gravitationally Lensed Galaxies at $z \simeq 2$

Tucker Jones

Department of Physics, University of California, Santa Barbara
Santa Barbara, CA 93106, USA
email: tajones@physics.ucsb.edu

Abstract. Spatially resolved spectroscopy is even more powerful when combined with magnification by gravitational lensing. I discuss observations of lensed galaxies at $z \simeq 2$ with spatial resolution reaching 100 parsecs. Near-IR integral field spectroscopy reveals the kinematics, distribution and physical properties of star forming regions, and gas-phase metallicity gradients. Roughly two thirds of observed galaxies are isolated systems with coherent velocity fields, large velocity dispersion, multiple giant star-forming regions, and negative gas-phase metallicity gradients, suggestive of inside-out growth in gravitationally unstable disks. The remainder are undergoing mergers and have shallower metallicity gradients, indicating mixing of the interstellar gas via gravitational interaction. The metallicity gradients in isolated galaxies are consistent with simulations using standard feedback prescriptions, whereas simulations with enhanced feedback predict shallower gradients. These measurements therefore constrain the growth of galaxies from mergers and star formation as well as the regulatory feedback.

Keywords. galaxies: evolution - galaxies: high-redshift - galaxies: kinematics and dynamics - ISM: abundances - gravitational lensing: strong

1. Introduction

The advent of integral field spectroscopy (IFS) with 8–10 meter telescopes has led to tremendous understanding of the internal structure of high redshift galaxies, particularly at $z \simeq 2-3$ where bright nebular emission lines are redshifted to favorable near-infrared wavelengths. Two-dimensional spectral mapping of emission lines reveals the kinematics and distribution of star formation while multiple line ratios can be used to infer the gas-phase metallicity (e.g., Jones *et al.* 2013). However, typical galaxies at $z \gtrsim 2$ are poorly sampled by ground-based instruments, even when using adaptive optics (AO). A promising solution is to observe gravitationally lensed galaxies with high areal magnification factors. Early results showed that lensing can be utilized to achieve physical resolution as fine as 100 pc (Nesvadba *et al.* 2006; Stark *et al.* 2008), and a spectacular example from Jones *et al.* (2013) is shown in Figure 1.

2. Kinematics and Morphology

Samples of lensed galaxies have revealed that a majority of typical galaxies at $z \simeq 2-3$ exhibit coherent rotation, while approximately 1/3 are undergoing major mergers (Jones *et al.* 2010a, 2013). These results are consistent with surveys of the most massive galaxies which are sufficiently sampled even without the benefit of lensing (e.g., Förster Schreiber *et al.* 2009). But unlike local disk galaxies, the velocity dispersions are very high in *all* star forming galaxies observed with IFS at $z \gtrsim 2$, typically $\sigma \simeq 50$–100 km s^{-1} and often exceeding the circular rotation velocity. A further discovery was that nearly all star forming galaxies at $z \gtrsim 2$ are resolved into multiple giant "clumps" of star

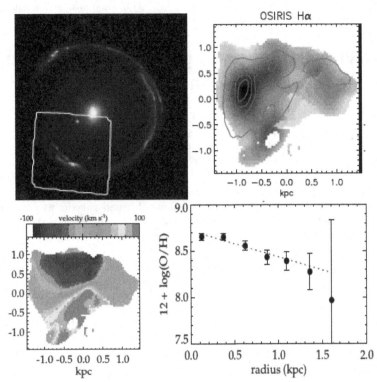

Figure 1. This figure shows an example of the data quality obtained with AO-fed IFS observations of lensed galaxies. *Top left:* HST image of the gravitational lens system SDSS J1148+1930. The blue arc is a star forming galaxy at $z = 2.38$, lensed into a nearly complete Einstein ring. The white box shows the region observed with Keck/OSIRIS and AO. *Top right:* morphology of SDSS J1148+1930, reconstructed in the source plane using the gravitational lens model. Grayscale shows the Hα emission and contours show rest-frame UV continuum from HST imaging. Both quantities trace recent unobscured star formation, revealing giant star forming regions with sizes of several hundred parsec. These are also evident as discrete blue clumps of emission in the upper left image, which not be resolved without the combination of gravitational lensing and AO. *Bottom left:* two-dimensional velocity field derived from Hα emission centroids in each spatial pixel, reconstructed in the source plane. The galaxy exhibits a regular rotating velocity field with peak-to-peak amplitude $\sim$150 km s^{-1}. *Bottom right:* radial gas-phase metallicity gradient derived from the ratios of strong emission lines (Hα, [N II], Hβ, [O III]). The metallicity is approximately solar in the center, and decreases to $\sim 1/5$ solar in the outer regions.

formation when observed with sufficient spatial resolution (Jones *et al.* 2010a; Livermore *et al.* 2012). Lensing has enabled measurements of their sizes which are found to be comparable to giant H II regions in local starburst galaxies (typically several hundred pc in diameter; Swinbank *et al.* 2009; Jones *et al.* 2010a). Simple calculations show that the kinematic properties and giant clumps are both naturally explained by a scenario in which star formation occurs in gravitationally unstable thick disks: kinematics indicate marginal gravitational instability (with Toomre parameters $Q_T \simeq 1$) which should lead to collapse on scales consistent with observed clump sizes (Genzel *et al.* 2008; Jones *et al.* 2010a). Livermore *et al.* (2012) further demonstrated that the properties of clumps in high redshift galaxies are consistent with the same scaling relations as local star forming disks, given the increased gas fractions (e.g., Tacconi *et al.* 2013). To summarize, the data suggest that a majority of galaxies at $z \simeq 2 - 3$ are evolving in situ with star formation occurring predominantly in gas-rich, gravitationally unstable thick disks.

3. Metallicity Gradients

More recent work has focused on measuring the time evolution of metallicity gradients to constrain models of galaxy evolution. Disk galaxies in the local universe have negative radial gradients in their gas-phase metallicity, with higher metallicities in the central regions (e.g., Vila-Costas & Edmunds 1992). The presence of gradients is explained by inside-out growth of galactic disks, i.e., the disk scale length increases at later times. Evolution of the gradient slope is a sensitive probe of recent dynamical history (e.g., Rupke *et al.* 2010) and of stellar feedback (e.g., Gibson *et al.* 2013; Anglés-Alcázar *et al.* 2014): gravitational interactions and galactic outflows can rapidly redistribute the metal-enriched interstellar medium over large physical scales, flattening any radial gradients.

The first measurement of a metallicity gradient at high redshift found a steep negative slope which would not have been identified without lensing magnification (Jones *et al.* 2010b). Subsequent observations of non-lensed galaxies at $z = 1$–4 have found mostly flat gradients with a significant fraction of positive slopes, but these data may suffer from a number of potential systematic effects related to the metallicity indicators and coarse ~ 5 kpc spatial resolution (Yuan *et al.* 2013). It is clear that $\lesssim 1$ kpc resolution is essential to recover accurate gradients even for large galaxies at $z \simeq 2$, and the best means of achieving this is with lensed galaxies (Yuan *et al.* 2013). Such samples are currently small but are growing rapidly thanks to dedicated efforts with Keck/OSIRIS+AO (Figure 1; Jones *et al.* 2013) and with HST grism spectroscopy via the GLASS survey (Schmidt *et al.* 2014; Jones *et al.* 2014 *in prep*). Figure 2 summarizes the high-resolution measurements presently available. Isolated (non-interacting) lensed galaxies have steeper gradients on average compared to $z \simeq 0$ descendants (Jones *et al.* 2013), in reasonable agreement with cosmological simulations using standard feedback models.

4. Discussion

High resolution measurements of kinematics and star formation are now available for large samples of galaxies at $z \simeq 2$, and their properties are reasonably well explained. However metallicity gradient measurements remain sparse and different samples do not agree. As an example, non-lensed galaxies in Figure 2 have systematically flatter gradients than the lensed sample. This may be due to insufficient spatial resolution, or it could be a real trend induced by selection effects: the non-lensed galaxies are systematically much larger in size which is known to correlate with flatter physical gradient slopes (e.g., Sánchez *et al.* 2014). Larger, reliable samples spanning a range of galaxy properties are clearly needed to understand the evolution of metallicity gradients and the implications for dynamical evolution and feedback. Such samples are within reach of ongoing surveys of lensed galaxies with Keck and HST, and this promises to be an interesting avenue of research in the near future.

Acknowledgements

It is a pleasure to thank the symposium organizers for their effort. This proceeding represents a summary of various work pertinent to the symposium; it would not have been possible without the efforts of numerous collaborators. Readers are referred to the relevant publications for more details: Jones *et al.* (2010a,b, 2013); Schmidt *et al.* (2014); Jones *et al.* (2014 *in prep*), and the references cited here.

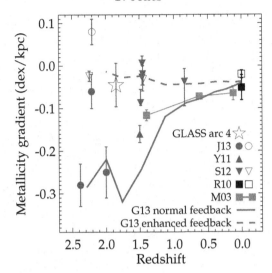

Figure 2. This figure shows the evolution of metallicity gradients with redshift. Measurements of gravitationally lensed galaxies with $\lesssim$ 300 pc resolution are shown in red (GLASS arc 4: Jones *et al.in prep*; J13: Jones *et al.* 2013; Y11: Yuan *et al.* 2011), non-lensed galaxies with ~1 kpc resolution are shown in blue (S12: Swinbank *et al.* 2012), an average of local gradients are shown in black (R10: Rupke *et al.* 2010), and the Milky Way metallicity gradient evolution measured from planetary nebulae is shown in green (M03: Maciel *et al.* 2003). Solid and hollow symbols denote isolated disks and interacting (or merging) galaxies, respectively, to illustrate that interacting galaxies have flatter gradients on average in the lensed and $z = 0$ samples. We additionally show results of two different feedback schemes in otherwise identical simulations of a galaxy similar to those shown in this figure (G13: Gibson *et al.* 2013). Isolated lensed galaxies at high redshift are more consistent with the "normal" feedback model.

References

Anglés-Alcázar, D., Davé, R., Özel, F., & Oppenheimer, B. D. 2014, *ApJ*, 782, 84

Förster Schreiber, N. M., Genzel, R., Bouché, N., *et al.* 2009, *ApJ*, 706, 1364

Genzel, R., Burkert, A., Bouché, N., *et al.* 2008, *ApJ*, 687, 59

Gibson, B. K., Pilkington, K., Brook, C. B., Stinson, G. S., & Bailin, J. 2013, *A&A*, 554, A47

Jones, T. A., Swinbank, A. M., Ellis, R. S., Richard, J., & Stark, D. P. 2010, *MNRAS*, 404, 1247

Jones, T., Ellis, R., Jullo, E., & Richard, J. 2010, *ApJ*, 725, L176

Jones, T., Ellis, R. S., Richard, J., & Jullo, E. 2013, *ApJ*, 765, 48

Livermore, R. C., Jones, T., Richard, J., *et al.* 2012, *MNRAS*, 427, 688

Maciel, W. J., Costa, R. D. D., & Uchida, M. M. M. 2003, *A&A*, 397, 667

Nesvadba, N. P. H., Lehnert, M. D., Eisenhauer, F., *et al.* 2006, *ApJ*, 650, 661

Rupke, D. S. N., Kewley, L. J., & Chien, L.-H. 2010, *ApJ*, 723, 1255

Sánchez, S. F., *et al.* 2014, *A&A*, 563, A49

Schmidt, K. B., *et al.* 2014, *ApJ*, 782, L36

Stark, D. P., Swinbank, A. M., Ellis, R. S., *et al.* 2008, *Nature*, 455, 775

Swinbank, A. M., Webb, T. M., Richard, J., *et al.* 2009, *MNRAS*, 400, 1121

Swinbank, A. M., Sobral, D., Smail, I., Geach, J. E., Best, P. N., McCarthy, I. G., Crain, R. A., & Theuns, T. 2012, *MNRAS*, 426, 935

Tacconi, L. J., Neri, R., Genzel, R., *et al.* 2013, *ApJ*, 768, 74

Vila-Costas, M. B. & Edmunds, M. G. 1992, *MNRAS*, 259, 121

Yuan, T.-T., Kewley, L. J., Swinbank, A. M., Richard, J., & Livermore, R. C. 2011, *ApJ*, 732, L14

Yuan, T.-T., Kewley, L. J., & Rich, J. 2013, *ApJ*, 767, 106

Galaxies in 3D across the Universe
Proceedings IAU Symposium No. 309, 2014   © International Astronomical Union 2015
B. L. Ziegler, F. Combes, H. Dannerbauer, M. Verdugo, eds.   doi:10.1017/S1743921314009806

Probing Individual Star Forming Regions Within Strongly Lensed Galaxies at z > 1

Matthew B. Bayliss[1], Jane R. Rigby[2], Keren Sharon[3], Michael D. Gladders[4] and Eva Wuyts[5]

[1]Dept. of Physics, Harvard University, 17 Oxford St., Cambridge, MA, 02138
email: `mbayliss@cfa.harvard.edu`
[2]Obs. Cosmology Lab, NASA Goddard Space Flight Center, Greenbelt, MD, 20771
[3]Dept. of Astronomy, the Univ. of Michigan, 1085 S. University Ave., Ann Arbor, MI, 48109
[4]Dept. of Astronomy & Astrophysics, Univ. of Chicago, 5640 S. Ellis Ave., Chicago, IL, 60637
[5]MPE, Postfach 1312, Giessenbachstr., D-85741 Garching, Germany

Abstract. Star formation occurs on physical scales corresponding to individual star forming regions, typically of order ~100 parsecs in size, but current observational facilities cannot resolve these scales within field galaxies beyond the local universe. However, the magnification from strong gravitational lensing allows us to measure the properties of these discrete star forming regions within galaxies in the distant universe. New results from multi-wavelength spectroscopic studies of a sample of extremely bright, highly magnified lensed galaxies are revealing the complexity of star formation on sub-galaxy scales during the era of peak star formation in the universe. We find a wide range of properties in the rest-frame UV spectra of individual galaxies, as well as in spectra that originate from different star forming regions within the same galaxy. Large variations in the strengths and velocity structure of Lyman-alpha and strong P Cygni lines such as C IV, and MgII provide new insights into the astrophysical relationships between extremely massive stars, the elemental abundances and physical properties of the nebular gas those stars ionize, and the galactic-scale outflows they power.

Keywords. gravitational lensing: strong — high-redshift galaxies — galaxies: star formation

1. Introduction

The rest-frame ultraviolet spectra of galaxies contain a wealth of information about the population of massive stars, the properties of the nebular gas those stars ionize, and the galactic-scale outflows they power. We can therefore use rest-UV spectra to constrain the famous "galactic feedback" that drives metals into the intergalactic medium and to better understand the role of this feedback in shutting down future star formation. A key science goal for 20–30 m telescopes is to understand this feedback process, but until the next generation of telescopes are built, there are only two ways to obtain rest-UV diagnostics for typical star-forming galaxies at redshifts above z ~1.5. The first is to stack low-quality spectra of many galaxies (e.g., Shapley *et al.* 2003). This results in a good sampling of the average properties of star forming galaxies, but removes the possibility of understanding the variations in the observable properties and how changes in one or more observables affects some or all of the others. The second way is to target galaxies that have been highly amplified by gravitational lensing. In previously published good signal-to-noise (S/N) rest-frame UV spectra for four such bright gravitationally lensed galaxies there are numerous strong features that trace the properties of massive stars and outflowing gas (Pettini *et al.* 2002, Finkelstein *et al.* 2009, Quider *et al.* 2009, 2010). All four of these galaxies have remarkably similar P Cygni profiles, yet each has very different outflow properties – we do not yet know what is a "normal" outflow at z > 1.

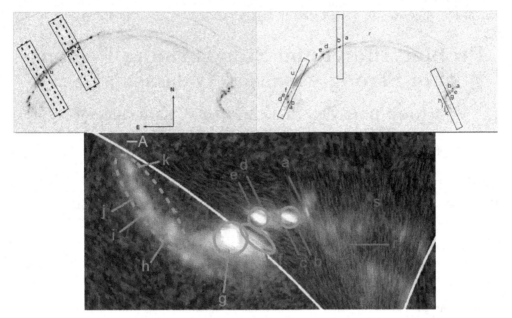

Figure 1. Slit positions for MagE observations of RCS2 J0327-1326. The top grayscale images show individual emission regions/knots in the arc labeled with lowercase letters. Solid and dashed lines indicate the positions of the MagE slit along the arc. Slits indicated by solid and dashed lines on the top left depict the slit positions for observations taken at two different slit widths, where the wider-slit was used for observations during times of worse seeing. The bottom panel shows a reconstruction of the giant arc in the source plane from Sharon *et al.* (2012), with the same emission knots – each with a physical size of $\lesssim$100 pc – are labeled with lowercase letters. The solid cyan ellipsoids indicate the four knots for which we have good S/N spectra.

At the heart of the problem is the fact that the fundamental physical scale of star formation and stellar outflows is not the scale of a single galaxy, but rather the scale of individual star forming regions within galaxies. However, strong gravitational lensing provides uniquely powerful views of distant galaxies, and in the most extreme high-magnification cases we are able to spatially resolute structures on $\lesssim$100 pc scales within galaxies at z > 1. In this proceeding we present a few highlights of the first results from an ongoing effort to obtain good S/N rest-frame UV spectra of individual star forming regions within distant galaxies. Here we will focus on two source in particular – RCS2 J0327-1326 and SGAS J1050+0017.

2. RCS2 J0327-1326

RCS2 J0327-1326 is a spectacular strongly lensed galaxy at $z = 1.704$ that forms a giant arc extending $\sim$38" along the sky (Wuyts *et al.* 2010, Sharon *et al.* 2012). This galaxy has been well-studied at NIR wavelengths, which samples the rest-frame optical (Rigby *et al.* 2011, Whitaker *et al.* 2014, Wuyts *et al.* 2014). We have also been conducting a follow-up campaign to acquire good S/N moderate resolution optical spectra of this arc with the MagE spectrograph on the Magellan-II (Clay) telescope. Due to the high magnification it is possible to place standard spectroscopic slits at different positions that map to distinct star forming regions with sizes $\lesssim$100 pc in the source galaxy (Figure 1).

Spectra of each of four star forming knots within RCS2 J0327-1326 exhibit strong differences in the strength and structure of many prominent rest-UV features. In particular, the C IV and Mg II P Cygni lines, which result from outflowing gas, show large variations

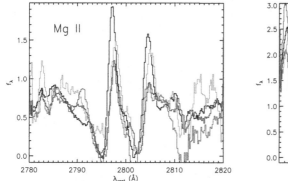 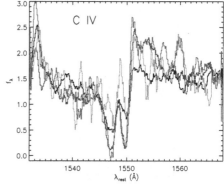

Figure 2. The Mg II 2796,2803 (left) and C IV 1548,1551 (right) P Cygni lines as observed in four different $\lesssim 100$ pc scale star forming regions within RCS2 J0327-1326. Both panels are consistent in their use of four different colors to indicate the four distinct star forming regions.

between the different knots (Figure 2). It is also notable that there is no positive correlation between the strength of the C IV and Mg II lines, indicating that these two lines are generated by different mechanisms and/or in different physical locations. These data also informed the observed lack of correlation between Ly-α and Mg II P Cygni emission in a sample of strongly lensed galaxies (Rigby *et al.* 2014), which further supports a physical picture in which Mg II emission traces stellar wind driven outflows, possibly providing a diagnostic measure of the radiative transfer mechanisms in those outflows.

3. SGAS J1050+0017

This bright lensed galaxy at $z = 3.625$ was published by Bayliss *et al.* (2014), in which the authors analyzed optical through NIR (rest-frame UV through optical) spectra of the source, as well as imaging data spanning 0.4 through 4.5 microns. This multi-wavelength analysis had difficultly explaining the properties of the integrated spectra of the giant arc, including very strong and narrow P Cygni features, as well as differences between the ionization parameter measurements. Here we show a new analysis of the Gemini/GMOS-North spectra of this arc in which we isolate and individually extract the emission from two distinct $\sim$100-200 pc star forming knots along the arc (see Figures 1 & 2 in Bayliss *et al.* 2014). The spectra clear exhibit differences between the two distinct knots, with the second knot, plotted in red, having a much weaker C IV feature, as well as weaker He II emission (Figure 3). These spectra provide direct evidence that the conditions of the ionizing radiation field and the population of massive stars vary significantly across different regions at $z = 3.625$.

4. Conclusions

Looking at individual star forming regions within the most highly magnified giant arcs we see significant variations along different lines of sight, including variable strength and profile shape of structure of P Cygni wind lines, with no evidence for correlation between C IV and Mg II P Cygni features, and indications of significant differences in the ionizing radiation field. These results reinforce the argument that the astrophysics of the interstellar medium (ISM) in distant star forming galaxies is a complex business.

Bright lensed galaxies extending out to z $\sim$ 5 have recently become common place (e.g., Bayliss *et al.* 2010, Koester *et al.* 2010, Bayliss *et al.* 2011b, Gladders *et al.* in prep), and

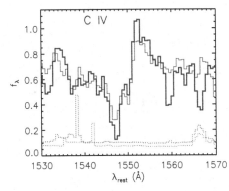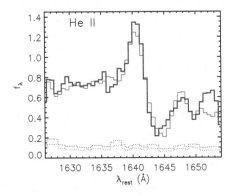

Figure 3. The C IV 1548,1551 (left) P Cygni lines and He II 1640 (right) emission line as seen in distinct star forming regions (indicated by different colors) within SGAS J1050+0017.

these lensed galaxies typically reside at z ~ 2, and therefore sample the epoch of peak star formation in the universe (Bayliss *et al.* 2011a, Bayliss 2012). The highest magnification systems drawn from these large samples provide a new opportunity to leave single-object analysis behind and study the UV spectra of individual star forming regions within many z > 1 galaxies. Such observations would allow us to characterize the relationship between local signatures of the winds of massive stars and their interaction with their surrounding ISM. These data will, in turn, reveal important new information about the astrophysics of the radiative transfer in the ISM in regions of prolific star formation.

Acknowledgements

This work is based on observations from the Magellan-II (Clay) and the Gemini-North Telescopes. Supported was provided by the National Science Foundation through Grant AST-1009012, by NASA through grant HST-GO-13003.01.

References

Bayliss, M. B., Wuyts, E., Sharon, K., *et al.* 2010, *ApJ*, 720, 1559
Bayliss, M. B., Gladders, M. D., Oguri, M., *et al.* 2011, *ApJ*, 727, L26
Bayliss, M. B., Hennawi, J. F., Gladders, M. D., *et al.* 2011, *ApJS*, 193, 8
Bayliss, M. B. 2012, *ApJ*, 744, 156
Bayliss, M. B., Rigby, J. R., Sharon, K., *et al.* 2014, *ApJ*, 790, 144
Finkelstein, S. L., Papovich, C., Rudnick, G., *et al.* 2009, *ApJ*, 700, 376
Koester, B. P., Gladders, M. D., Hennawi, J. F., *et al.* 2010, *ApJ*, 723, 73
Pettini, M., Rix, S. A., Steidel, C. C., *et al.* 2002, *ApJ*, 569, 742
Quider, A., Shapley, A., Pettini, M., Steidel, C. C., & Stark, D. 2010, *MNRAS*, 402, 1467
Quider, A., Pettini, M., Shapley, A., & Steidel, C. C. 2009, *MNRAS*, 398, 1263
Rigby, J. R., Wuyts, E., Gladders, M. D., Sharon, K., & Becker, G. D. 2011, *ApJ*, 732, 59
Rigby, J. R., Bayliss, M. B., Gladders, M. D., *et al.* 2014, *ApJ*, 790, 44
Shapley, A. E., Steidel, C. C., Pettini, M., & Adelberger, K. L. 2003, *ApJ*, 588, 65
Sharon, K., Gladders, M. D., Rigby, J. R., *et al.* 2012, *ApJ*, 746, 161
Whitaker, K. E., Rigby, J. R., Brammer, G. B., *et al.* 2014, *ApJ*, 790, 143
Wuyts, E., Barrientos, L. F., Gladders, M. D., *et al.* 2010, *ApJ*, 724, 1182
Wuyts, E., Rigby, J. R., Gladders, M. D., & Sharon, K. 2014, *ApJ*, 781, 61

Galaxies in 3D across the Universe
Proceedings IAU Symposium No. 309, 2014
B. L. Ziegler, F. Combes, H. Dannerbauer, M. Verdugo, eds.

© International Astronomical Union 2015
doi:10.1017/S1743921314009818

Mapping and resolving galaxy formation at its peak epoch with Mahalo-Subaru and Gracias-ALMA

**Tadayuki Kodama[1,2,3], Masao Hayashi[1], Yusei Koyama[4],
Ken-ichi Tadaki[1], Ichi Tanaka[2], Rhythm Shimakawa[2,3],
Tomoko Suzuki[1,3] and Moegi Yamamoto[1,3]**

[1]Optical and Infrared Astronomy Division, National Astronomical Observatory of Japan,
Mitaka, Tokyo 181-8588, Japan
email: `t.kodama@nao.ac.jp`

[2]Subaru Telescope, National Astronomical Observatory of Japan, 650 North A'ohoku Place,
Hilo, HI 96720, USA

[3]Department of Astronomical Science, Graduate University for Advanced Studies, Mitaka,
Tokyo 181-8588, Japan

[4]Institute of Space Astronomical Science, Japan Aerospace Exploration Agency Sagamihara,
Kanagawa, 252-5210, Japan

Abstract. MAHALO-Subaru (MApping HAlpha and Lines of Oxygen with Subaru) project aims to investigate how the star forming activities in galaxies are propagated as a function of time, mass, and environment. It employs a unique set of narrow-band filters on MOIRCS/Subaru to search for Ha emitters associated to the proto-clusters or in narrow redshift slices in the general field. We have shown not only filamentary/clumpy structures of all the proto-clusters but also very high star formation activities therein especially at $z > 2$. HST images from the CANDELS survey have revealed that nearly half of the Hα emitters in the field at $z \sim 2$ have clumpy structures. Among them, "red dusty clumps" are preferentially found at or near the mass center of galaxies. Therefore, they are probably linked to the formation of bulge component. To explore physical states and the mode of star formation of those forming galaxies, we have started Gracias-ALMA project in full coordination with the Mahalo-Subaru project. We will resolve molecular gas contents and dusty star formation within these galaxies, and tell whether clumps are formed by gravitational instability of gas rich disks, and whether bulges are formed by clump migration or through galaxy-galaxy mergers.

Keywords. galaxies: clusters, galaxies: formation, galaxies: evolution

1. Mahalo-Subaru: Mapping star formation at its peak epoch

It is widely recognised that the cosmic star formation rate (SFR) density and the AGN number density both show peaks at $z \sim 2$ (e.g. Hopkins & Beacom 2006). Also galaxy formation should be dependent on environment. Our previous Subaru observations have revealed that the star formation activity in the cluster core at $z \sim 1.5$ is as high as that in the field, and that the peak of star forming activity is shifted outwards of clusters as time progresses (e.g. Hayashi *et al.* 2010; Koyama *et al.* 2011). We want to understand how the star forming activities in high density regions at high redshifts are intrinsically biased. how they are externally affected by surrounding environments, how the peak activity is shifted outwards to lower density environments with time, and what physical processes are involved and responsible for these phenomena.

To address those critical issues, we have been conducting the "Mahalo-Subaru" project (MApping HAlpha and Lines of Oxygen with Subaru; Kodama *et al.* 2013), a large

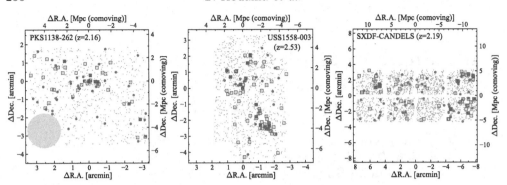

Figure 1. 2-D maps of two proto-clusters, USS1558–003 at $z = 2.53$ (Hayashi *et al.* 2012), and PKS1138–262 at $z = 2.16$ (Koyama *et al.* 2013), and a redshift slice ($z = 2.09$) of a general field, SXDF-UDS-CANDELS (Tadaki *et al.* 2013). The red filled circles and blue open squares indicate HAE candidates associated to the proto-cluster separated at $J - K_s) = 2.3$.

part of which was conducted as a Subaru open-use intensive program. Our targets are clusters/proto-clusters at $0.4 < z < 3.13$, and an un-biased general field SXDF-UDS-CANDELS (4 redshift slices at $2.19 < z < 3.63$) (Table 1). We employ unique sets of narrow-band (NB) filters on the two wide-field cameras, Suprime-Cam (optical; $34' \times 27'$) and MOIRCS (NIR; $7' \times 4'$). Most of the MOIRCS NB filters are designed and manufactured specifically to our targets, and their wavelengths perfectly match to redshifted Hα or [OIII] lines from our $z > 1.5$ targets. The Suprime-Cam NB filters in the z'-band are also used to search for [OII] emitters at $z = 1.46$–1.62. The line-of-sight velocity range that fall within the filter FWHM with respect to the cluster center is optimal (± 1000-3000km/s).

Using these NB filters, we have successfully identified lots of star-forming galaxies very efficiently in narrow redshift slices associated to the clusters or in the general field, which show nebular emission lines from ionized star-forming regions. We are sure about their membership to the clusters by the presence of emission lines and their SEDs based on colour-colour diagrams or photometric redshifts. We have completed most of the imaging observations under excellent conditions. Exposure times were typically 1–1.5hrs per broad-band, and 3–4hrs per narrow-band for $z \gtrsim 1.5$ targets. Figure 1 shows three best examples; two proto-clusters associated to the radio galaxies and the general field. We have discovered complex substructures traced by Hα emitters (HAEs), in the vicinity of, and physically associated the radio galaxies. It clearly indicates that these systems are still in a vigorous assembly phase. Star formation activity is also very intense even in the core regions of the proto-clusters. This is in contrast to lower redshift clusters where star formation activities are very low and the peak area of star formation activity is shifted outwards. It shows that cluster formation takes place in an inside-out fashion (Fig. 2).

It is notable that some of the HAEs show very red colours in $J - K_s$ (>2.3 in Vega-mag), indicating that they are dusty star forming galaxies (SFGs). Interestingly, those red HAEs tend to favor high density regions such as clumps and filaments along the structures. Therefore we expect that these are the key populations under the influence of environmental effects (Koyama *et al.* 2013).

These proto-clusters are very likely the sites where early-type galaxies, which will eventually dominate a rich cluster by the present-day, were just in their formation phase. Therefore, they serve us unique, excellent targets to investigate in detail the physical mechanisms of galaxy formation and its early evolution in dense environment.

Table 1. A target list of the Mahalo-Subaru survey. S-Cam means Suprime-Cam.

environ-ment	target	z	line	λ (μm)	camera	NB-filter	status
Low-z cluster	CL0024+1652	0.395	Hα	0.916	S-Cam	NB912	Kodama+'04
	CL0939+4713	0.407	Hα	0.923	S-Cam	NB921	Koyama+'11
	CL0016+1609	0.541	Hα	1.011	S-Cam	NB1006	not yet
	RXJ1716.4+6708	0.813	Hα	1.190	MOIRCS	NB1190	Koyama+'10
			[OII]	0.676	S-Cam	NA671	observed
High-z cluster	XCSJ2215−1738	1.457	[OII]	0.916	S-Cam	NB912/921	Hayashi+'10
	4C65.22	1.516	Hα	1.651	MOIRCS	NB1657	Koyama+'14
	CL0332−2742	1.61	[OII]	0.973	S-Cam	NB973	observed
	ClGJ0218.3−0510	1.62	[OII]	0.977	S-Cam	NB973	Tadaki+'12
Proto-cluster	PKS1138−262	2.156	Hα	2.071	MOIRCS	NB2071	Koyama+'12
	4C23.56	2.483	Hα	2.286	MOIRCS	CO	Tanaka+'11
	USS1558−003	2.527	Hα	2.315	MOIRCS	NB2315	Hayashi+'12
	MRC0316−257	3.130	[OII]	2.539	MOIRCS	NB1550	not yet
			[OIII]	2.068	MOIRCS	NB2071	not yet
General field	GOODS-N (70 arcmin^2)	2.19	Hα	2.094	MOIRCS	NB2095	Tadaki+'11
			[OII]	1.189	MOIRCS	NB1190	observed
	SXDF-CANDELS (92 arcmin^2)	2.19	Hα	2.094	MOIRCS	NB2095	Tadaki+'13
		2.53	Hα	2.315	MOIRCS	NB2315	Tadaki+'13
		3.17	[OIII]	2.093	MOIRCS	NB2095	Suzuki+'14
		3.63	[OIII]	2.317	MOIRCS	NB2315	Suzuki+'14

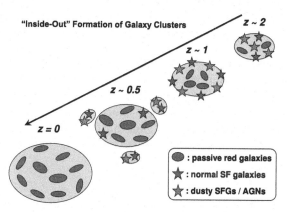

Figure 2. Schematic diagram of the inside-out evolution of clusters over cosmic times.

Deep optical-NIR HST images with high spatial resolution (0.1–0.16″) are available in the SXDF-UDS-CANDELS field (Grogin *et al.* 2011). From those images we find that 40% of our HAEs at $z \sim 2$ show clumpy structures. Interestingly, massive clumpy galaxies tend to have a "red dusty clump" at or near the mass center of galaxies. Those clumpy galaxies may originate either from subgalaxy mergers or fragmentation of gas rich turbulent disks fed by cold streams of gas along the surrounding filamentary structures (Dekel *et al.* 2009). The red clumps may correspond to the merger of those clumps at galactic centers after they migrate to the centers due to dynamical friction (Bournaud *et al.* 2013). We may be witnessing the site of bulge formation in action. To test this hypothesis, we need to spatially resolve these clumpy galaxies and map out star formation activities within the galaxies. Such central dusty starburst, if present, must be identified.

We have also obtained deep spectra of HAE candidates in the two proto-clusters (PKS1138 and USS1558) with MOIRCS on Subaru to study physical properties of the SFGs in high density regions at high redshifts. The data have not only spectroscopically confirmed the membership of 63 galaxies (70% of the observed targets) (Shimakawa *et al.* 2014a), but also provided us with their gaseous metallicities (Shimakawa *et al.* 2014b). We find that protocluster galaxies are more chemically enriched than those of field galaxies at a given stellar mass in the range of $M_\star \lesssim 10^{11}$ $M_\odot$. There are two possible scenarios to acount for this interesting result. One is that due to higher gas pressure of intra-cluster medium, the gas once ejected from galaxies by outflows may fall back again and be recycled for further star formation, leading to more chemically enriched systems. The other is that metal poor gas that are loosely bounded to outer regions of galaxies may be selectively more stripped off due to tidal interaction or ram-pressure stripping, leading to higher averaged metallicity (Shimakawa *et al.* 2014b). Since the gas fraction for a given metallicity is expected to be different between the two scenarios, accurate measurement of molecular gas content by radio observations such as with ALMA will be the key to idetify the physical mechanism behind the environmental dependence of chemical evolution.

2. Gracias-ALMA: Resolving gas and dust contents in SFGs

Although we have identified many vigorously SFGs in our targets, we do not know the physical processes that are actually acting on the galaxies with the optical-NIR data alone. We therefore conduct the Gracias-ALMA project that is fully coordinated with the Mahalo-Subaru project. Gracias-ALMA stands for "GRAphing CO Intensity And Submm with ALMA". ALMA will provide us with (1) molecular gas content and (2) amount of hidden dusty star formation, and characterize the properties of SFGs. Furthermore, with ALMA, we can easily resolve individual galaxies on an angular scale of $0.1''$ which corresponds to 0.8 kpc at $z = 2.5$. Therefore we will be able to know where in galaxies dusty star formation and the dense molecular gas are distributed. For example, if the molecular gas content of the central red clump in the above clumpy galaxies is observed with ALMA, we can investigate the mode of star formation or star formation efficiency there which is expected to be different during a burst phase, providing a crucial test for the hypothesis of clumpy galaxy – bulge formation connection.

References

Bournaud, F., Perret, V., Renaud, F., *et al.* 2013, *ApJ*, 780, 57
Dekel, A., Birnboim, Y., Engel, G., *et al.* 2009, *Nature*, 457, 451
Grogin, N. A., Kocevski, D. D., Faber, S. M., *et al.* 2011, *ApJS*, 197, 35
Hayashi, M., Kodama, T., Koyama, Y., *et al.* 2010, *MNRAS*, 402, 1980
Hayashi, M., Kodama, T., Tadaki, K., *et al.* 2012, *ApJ*, 757, 15
Hopkins, A. M. & Beacom, J. F. 2006, *ApJ*, 651, 142
Kodama, T., Balogh, M. L., Smail, I., *et al.* 2004, *MNRAS*, 354, 1103
Kodama, T., Tanaka, I., Hayashi, M., *et al.* 2013, *ASPC*, 476, 37
Koyama, Y., Kodama, T., Shimasaku, K., *et al.* 2010, *MNRAS*, 403, 1611
Koyama, Y., Kodama, T., Nakata, F., *et al.* 2011, *ApJ*, 734, 66
Koyama, Y., Kodama, T., Tadaki, K., *et al.* 2013, *MNRAS*, 428, 1551
Koyama, Y., Kodama, T., Tadaki, K., *et al.* 2014, *ApJ*, 789, 18
Shimakawa, R., Kodama, T., Tadaki, K., *et al.* 2014a, *MNRAS*, 441, L1
Shimakawa, R., Kodama, T., Tadaki, K., *et al.* 2014b, arXiv1406.5219
Suzuki, T., *et al.* 2014, in preparation
Tadaki, K., Kodama, T., Koyama, Y., *et al.* 2011, *PASJ*, 63, 437
Tadaki, K., Kodama, T., Ota, K., *et al.* 2012, *MNRAS*, 423, 2617
Tadaki, K., Kodama, T., Tanaka, I., *et al.* 2013, *ApJ*, 778, 114
Tanaka, I., De Breuck, C., Kurk, J., *et al.* 2011, *PASJ*, 63, 415

Galaxies in 3D across the Universe
Proceedings IAU Symposium No. 309, 2014
B. L. Ziegler, F. Combes, H. Dannerbauer, M. Verdugo, eds.
© International Astronomical Union 2015
doi:10.1017/S174392131400982X

An Excess of Dusty Starbursts at $z = 2.2$

Helmut Dannerbauer

Department for Astrophysics, University of Vienna,
Türkenschanzstr. 17, 1180 Vienna, Austria

Abstract. Searching for massive, dusty starbursts offers the great opportunity to trace galaxy overdensities and thus the cosmic web in the distant universe. I will present our APEX-LABOCA 870 μm imaging of the protocluster field of the radio galaxy MRC1138−262 — the so-called Spiderweb Galaxy — at z = 2.16, uncovering a large number of so-called submm galaxies (SMGs). The number counts already indicate an excess of SMGs compared to blank fields. Based on an exquisite multi-wavelength dataset, I will show that a large fraction of these massive, dusty starbursts are physically associated with the protocluster at z = 2.16. Finally, I will discuss both the properties of this starburst overdensity and their individual members.

Keywords. galaxies: individual: MRC1138−262 — galaxies: clusters: individual: MRC1138−262 — galaxies: high redshift — cosmology: observations — Infrared: galaxies — submillimeter: galaxies

1. Motivation

Understanding when and how did present-day galaxy cluster form at high redshifts have been the science driver for the extensive search, especially at optical and near-infrared wavelengths, for protoclusters of galaxies in the distant universe in the past decade. Powerful high-redshift radio galaxies (HzRG; see for more details the review by Miley & De Breuck 2008) are considered to be the most promising signposts of massive clusters in formation. The VLT survey of Lyα emitters (LAEs), Hα emitters (HAEs), Lyman Break Galaxies and Extremely Red Objects in seven fields containing radio galaxies at redshifts up to 5.2 provided in almost all cases evidence for galaxy overdensities associated with the central galaxy (e.g., Kurk *et al.* 2000, 2004ab; Pentericci *et al.* 2000; Venemans *et al.* 2007; Overzier *et al.* 2006; Miley *et al.* 2004).

2. Dusty Star Forming Galaxies

However, we note that these optical/NIR techniques mainly trace (rather low-mass) galaxies with unobscured star formation, making up only 50% of the cosmic star formation activity (Dole *et al.* 2006). In the last decade (sub)millimeter surveys have revolutionized our understanding of the formation and evolution of galaxies, by revealing an unexpected population of high-redshift, dust-obscured galaxies which are forming stars at a tremendous rate. Submm galaxies (SMGs; see the review by Blain *et al.* 2002), first discovered by Smail *et al.*, (1997), have intense star formation, with rates of a few hundred to several thousands solar masses per year. These dusty starbursts are massive (e.g., Genzel *et al.* 2003; Greve *et al.* 2005), most probably the precursors of present-day ellipticals (e.g., Ivison *et al.* 2013) and excellent tracers of mass density peaks and thus of protoclusters. Studying SMGs offer us an unique opportunity to explore episodes of bursting star formation in a critical epoch of galaxy formation.

3. Galaxy Clusters

From the collapse of the first highest density peaks of matter density fluctuations (after inflation) arose galaxy clusters. These are the nodes of the cosmic web, formed by the inhomogeneously assembled mass along filaments. As the first structures to be collapsed, galaxy clusters have to be seen as the earliest starting fingerprint of galaxy formation and evolution. These structures grew hierarchically through the merging and accretion of smaller units of galaxy halos. These high density regions provide a remarkable environment for investigating the physical processes responsible for the triggering and suppression of star formation and AGN activity. Enhanced star formation activity is observed in large scale structures beyond $z = 1.5$ (Hilton *et al.* 2010; Tran *et al.* 2010). We now know that the wide spread star formation activity (both in the cluster center and the outer parts) is the by-product of the build up of distant clusters whereas local clusters devoid star formation in their centers (e.g., Koyama *et al.* 2010). To summarize, there is strong evidence for star formation being triggered in galaxies that fall into the cluster potential and are affected by the cluster environment, although the details of the physical process are not yet clear. But did this also happen to the most massive galaxies in todays clusters which formed rapidly in the early universe?

A prominent feature of colour-magnitude diagrams of (local) clusters is the red-sequence. Red-sequence galaxies are dominating the core of galaxy clusters and are massive, passively evolving, early-type galaxies. The red-sequence is found in clusters up to $z = 1.5$ (e.g., Rosati *et al.* 2009; Santos *et al.* 2011), and its tightness suggests that they are formed in a short time scale beyond $z = 2$. When and how the progenitors of the red-sequence are formed in clusters is one of the most hotly debated topic. Scenarios for their ancestors include in situ formation or accretion/falling into the cluster potential. SMGs with short time scales of building up the bulk of their stellar mass are promising candidates for being the ancestors of red sequence galaxies.

The limited mapping speed in the (sub)mm wavelength regime (between 870 m and 1.2 mm) does not allow to scan extremely large blank fields in order to find systematically overdensities of SMGs. Only a handful of serendipitous discoveries of single SMGs physically associated to galaxy overdensities traced at other wavelengths (*et al.* Capak *et al.* 2011; Daddi *et al.* 2009; Riechers *et al.* 2010; Walter *et al.* 2012) are known. At larger mm-wavelengths large/all sky surveys such as with the South Pole Telescope or the Planck satellite discover clumps which could be high-z clusters but due to the coarse spatial resolution of these telescopes individual cluster members cannot be detected (see e.g. Clements *et al.* 2014). Targeting a few protocluster fields, and finding overdensities of SMGs in comparison to random fields (e.g., Stevens *et al.* 2003, 2010; De Breuck *et al.* 2004; Greve *et al.* 2007) beyond $z = 1.5$ has already demonstrated the huge potential of submm observations to trace large scale structures. However, in none of these cases the obligatory and time consuming identification work was properly done for the individual sources and presumably cluster members.

These first results imply that a complete census of the cluster members is only possible if source populations selected both from the optical/near-infrared and far-infrared/submm are considered. E.g. the so-called Lyα emitters and Hα emitters are expected to be relatively low, medium mass objects with SFRs between few to hundred solar masses per year whereas dusty starbursts are massive and intense star forming galaxies. Follow-up with Herschel of so-called Planck-clumps (high-z cluster candidates) finds star formation rate densities based on the FIR orders of magnitude higher than in the field (Clements *et al.* 2014) and thus demonstrated the importance of searching for dusty starbursts in known cluster fields selected from optical/near-infrared wavelengths.

4. MRC1138-262

In order to verify the hypothesis that SMGs enable us to reveal overdensities of massive galaxies with intense on-going star formation, we observed the field of a protocluster at $z = 2.16$ with APEX-LABOCA at 870 μm (Dannerbauer *et al.* 2014). This protocluster associated with the high-redshift radio galaxy MRC1138-262 at $z = 2.16$, the so-called Spiderweb Galaxy, is one of the best studied fields so far. Lyα and Hα imaging/spectroscopy of this field reveal an excess of LAEs and HAEs compared to blank fields (Kurk *et al.*, 2000; Pentericci *et al.*, 2000; Kurk *et al.*, 2004a,b; Hatch *et al.*, 2011). The spatial distribution of LAEs and HAEs provides strong evidence that the central galaxy lies in a forming cluster (Pentericci *et al.* 2000; Kurk *et al.* 2000, 2004ab). Two attempts to search for submillimeter overdensities in this field are known. Using SCUBA, Stevens *et al.* (2003) report the (tentative) excess of SMGs, and spatial extension of the submillimeter emission of the HzRG MRC1138-262. However, I note that the field of view of SCUBA has only a diameter of 2 arcmin ($\sim$1 Mpc at $z = 2.16$), and thus the reported SMG excess is based on very small numbers. Rigby *et al.* (2014) present Herschel SPIRE observations of a large field ($\sim$400 arcmin^2), centered on the HzRG. They report an excess of SPIRE 500 μm sources but found no filamentary structure in the far-infrared as seen in the rest-frame optical (Kurk *et al.* 2004a; Koyama *et al.* 2013). However, in both cases no counterpart identification was attempted for the individual sources. In addition, Valtchanov *et al.* (2013) report the serendipitous discovery of an overdensity of SPIRE sources 7 armin south of the protocluster. Based on the modified blackbody derived redshift distribution, incorporating the color information as well as the SED shape, they conclude that the majority of the 250 μm sources in the overdensity are likely to be at a similar redshift. With the available scarce multi-wavelength data they cannot exclude the attractive possibility that the overdensity is within the same structure as the Spiderweb at $z = 2.2$.

With our 40 hours APEX LABOCA observations, we extended drastically previous SCUBA observations by Stevens *et al.* (2003) focusing on the HzRG and the surrounding region of 2 arcmin diameter. We detected a large number (16) of SMGs down to a 3σ peak flux of 3mJy/beam (Fig, 1), roughly up to a factor four more than expected from the blank field surveys at 870 μmm as e.g. LESS, the LABOCA survey of the Extended Chandra Deep Field South (ECDFS) at these wavelengths (e.g., Weiss *et al.* 2009). This excess is consistent with an excess of Herschel SPIRE 500 μm sources in the same field of the radio galaxy reported by Rigby *et al.* (2014). Based on an exquisite multi-wavelength database, including VLA 1.4 GHz radio and infrared observations, we investigate whether these sources are members of the protocluster structure at $z \approx 2.2$.

Using Herschel PACS+SPIRE and Spitzer MIPS photometry, we derive reliable far-infrared photometric redshifts for all sources. Our VLT ISAAC and SINFONI near-infrared spectra confirm that four of these SMGs have redshifts of $z \approx 2.2$ (Kurk *et al.* in prep.) with consistent FIR-photo-z. We also present evidence that another SMG in this field, earlier detected at 850 μm, has a counterpart exhibiting Hα (HAE229) and CO(1-0) emission at $z = 2.15$ (Emonts *et al.* 2013; Dannerbauer *et al.* in prep.). Including the radio galaxy and two SMGs with FIR-photometric redshifts at z = 2.2, we conclude that at least eight submm sources are part of the protocluster at $z = 2.16$ associated with the radio galaxy MRC1138-262 (see Fig. 1 and 2). Strikingly, these eight sources are concentrated within a region of 2 Mpc (the typical size of clusters in the local universe) and are distributed within the filaments traced by the Hα emitters at $z \approx 2.2$. Thus, as predicted by theories (e.g., De Lucia *et al.* 2004), massive dusty starbursts are indeed tracing the cosmic web. Their location within the filamentary structure would support

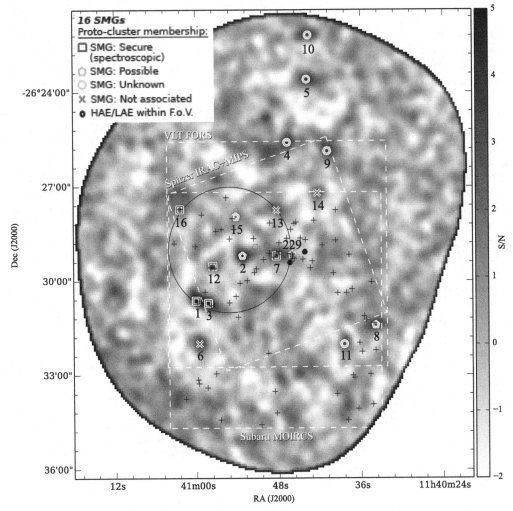

Figure 1. We show the location of 16 SMGs extracted from our LABOCA map of the field of MRC1138 on top of the LABOCA signal-to-noise map (courtesy from Dannerbauer *et al.* 2014). The large circle (solid line) has a diameter of ∼240″ (corresponding to a physical size of 2 Mpc) and shows the region where all seven SMGs with $z = 2.2$ are located. The SMG overdensity is a factor four higher than compared to blank fields (Weiss *et al.* 2009).

the picture that these dusty starbursts are falling into the cluster (center). We even detected SMGs separated between 5 to 10 Mpc from the HzRG, the cluster center, which could trace a further filament (see Fig. 1) but with the current data their membership cannot be verified.

For the 2 Mpc central region, the excess of SMGs is a factor of four higher than the well-known structure of six SMGs at z = 1.99 in GOODS-N distributed over 7x7 Mpc^2 (Blain *et al.* 2004; Chapman *et al.* 2009). Blain *et al.* (2004) report an association of five sources in the HDF-North. All five SMGs have spectroscopically measured redshifts of z = 1.99 (see also Chapman *et al.* 2009). This is the largest blank field association known so far. Chapman *et al.* (2009) report an apparently less significant overdensity of UV-selected galaxies at the same redshift and region of the sky. The protocluster MRC1138−262 at z = 2.2 is securely traced by galaxy populations probing different mass

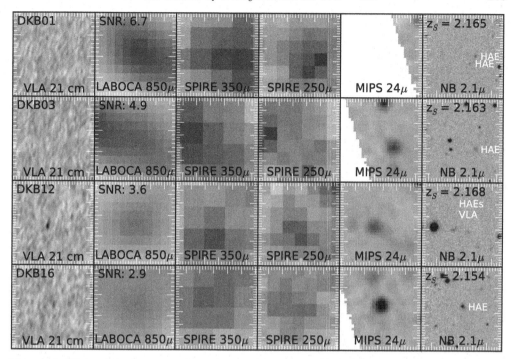

Figure 2. VLA 1.4 GHz image, LABOCA 870 μm, SPIRE 350 μm, SPIRE 250 μm, MIPS 24 μm and MOIRCS Hα images of LABOCA sources. The large white circles represent the size of the LABOCA beam ($\sim 11''$ diameter). Small white circles are VLA and/or HAE sources.

ranges, star formation and degree of obscuration including LAEs, HAEs and SMGs. The data excludes the cluster membership for three SMGs. For the remaining six SMGs we do not have enough data to make a robust judgement on their membership. We measure a star formation rate density SFRD ~ 1500 M$_\odot$ yr^{-1} Mpc^{-3}, four magnitudes higher compared to the global SFRD at this redshift in the field, similar to results obtained by Clements *et al.* (2014) and thus demonstrating the importance of searching for dusty starbursts in known cluster fields selected from optical/near-infrared wavelengths.

5. Conclusions

Our results demonstrate that submillimeter observations enable us to reveal overdensities of massive, dusty starbursts and is key to complete our census of protocluster member populations. We conclude that systematic and detailed investigations of (proto)clusters in the early universe at submillimeter wavelengths are feasible. Finally, future sensitive subarcsecond resolution observations with mm-interferometers such as ALMA will allow us to complete the characterization of the 16 SMGs discovered by LABOCA.

Acknowledgements

I would like to thank all the scientists coming to our symposium, thus making through their participation this event happening.

References

Blain, A. W., Smail, I., Ivison, R. J., Kneib, J.-P., & Frayer, D. T. 2002, *Phys. Rep.*, 369, 111
Blain, A. W., Chapman, S. C., Smail, I., & Ivison, R. 2004, *ApJ*, 611, 725
Capak, P. L., Riechers, D., Scoville, N. Z., *et al.* 2011, *Nature*, 470, 233
Chapman, S. C., Blain, A., Ibata, R., *et al.* 2009, *ApJ*, 691, 560
Clements, D., *et al.*, 2014, *MNRAS*, 439, 1139
Daddi, E., Dannerbauer, H., Stern, D., *et al.* 2009a, *ApJ*, 694, 1517
Emonts, B. H. C., Feain, I., Röttgering, H. J. A., *et al.* 2013, *MNRAS*, 430, 3465
Dannerbauer, H., Kurk, J. D., De Breuck, C., *et al.* 2014, *A&A*, in press
De Breuck, C., Bertoldi, F., Carilli, C., *et al.* 2004, *A&A*, 424, 1
De Lucia, G., Poggianti, B. M., Aragón-Salamanca, A., *et al.* 2004, *ApJ*, 610, L77
Dole, H., Lagache, G., Puget, J.-L., *et al.* 2006, *A&A*, 451, 417
Greve, T. R., Bertoldi, F., Smail, I., *et al.* 2005, *MNRAS*, 359, 1165
Greve, T. R., Stern, D., Ivison, R. J., *et al.* 2007, *MNRAS*, 382, 48
Hatch, N. A., Kurk, J. D., Pentericci, L., *et al.* 2011b, *MNRAS*, 415, 2993
Hilton, M., Lloyd-Davies, E., Stanford, S. A., *et al.* 2010, *ApJ*, 718, 133
Ivison, R. J., Swinbank, A. M., Smail, I., *et al.* 2013, *ApJ*, 772, 137
Koyama, Y., Kodama, T., Shimasaku, K., *et al.* 2010, *MNRAS*, 403, 1611
Kurk, J. D., Röttgering, H. J. A., Pentericci, L., *et al.* 2000, *A&A*, 358, L1
Kurk, J. D., Pentericci, L., Röttgering, H. J. A., & Miley, G. K. 2004a, *A&A*, 428, 793
Kurk, J. D., Pentericci, L., Overzier, R. A., Röttgering, H. J. A., & Miley, G. K. 2004b, *A&A*, 428, 817
Miley, G. K., Overzier, R. A., Zirm, A. W., *et al.* 2006, *ApJ*, 650, L29
Miley, G. & De Breuck, C. 2008, *A&A Rev.*, 15, 67
Overzier, R. A., Miley, G. K., Bouwens, R. J., *et al.* 2006, *ApJ*, 637, 58
Pentericci, L., Kurk, J. D., Röttgering, H. J. A., *et al.* 2000, *A&A*, 361, L25
Rigby, E., *et al.*, 2014, *MNRAS*, 437, 1882
Riechers, D. A., Capak, P. L., Carilli, C. L., *et al.* 2010, *ApJ*, 720, L131
Rosati, P., Tozzi, P., Gobat, R., *et al.* 2009, *A&A*, 508, 583
Santos, J. S., Fassbender, R., Nastasi, A., *et al.* 2011, *A&A*, 531, L15
Smail, I., Ivison, R. J., & Blain, A. W. 1997, *ApJ*, 490, L5
Stevens, J. A., Ivison, R. J., Dunlop, J. S., *et al.* 2003, *Nature*, 425, 264
Stevens, J. A., Jarvis, M. J., Coppin, K. E. K., *et al.* 2010, *MNRAS*, 405, 2623
Tran, K.-V. H., Papovich, C., Saintonge, A., *et al.* 2010, *ApJ*, 719, L126
Valtchanov, I., *et al.* 2013, *MNRAS*, 436, 2505
Venemans, B. P., Röttgering, H. J. A., Miley, G. K., *et al.* 2007, *A&A*, 461, 823
Walter, F., Decarli, R., Carilli, C., *et al.* 2012, *Nature*, 486, 233
Weiß, A., Kovács, A., Coppin, K., *et al.* 2009, *ApJ*, 707, 1201

Galaxies in 3D across the Universe
Proceedings IAU Symposium No. 309, 2014
B. L. Ziegler, F. Combes, H. Dannerbauer, M. Verdugo, eds.

© International Astronomical Union 2015
doi:10.1017/S1743921314009831

A molecular scan in the Hubble Deep Field North

Roberto Decarli[1], Fabian Walter[1], Chris Carilli[2] and Dominik Riechers[3]

[1]Max Planck Institut für Astronomie, Königstuhl 17, D-69117, Heidelberg, Germany
email: `decarli@mpia.de`
[2]NRAO, Pete V. Domenici Array Science Center, P. O. Box O, Socorro, NM, 87801, USA
[3]Cornell University, 220 Space Sciences Building, Ithaca, NY 14853, USA

Abstract. Our understanding of galaxy evolution has traditionally been driven by pre-selection of galaxies based on their broad-band continuum emission. This approach is potentially biased, in particular against gas-rich systems at high-redshift which may be dust-obscured. To overcome this limitation, we have recently concluded a blind CO survey at 3mm in a region of the Hubble Deep Field North using the IRAM Plateau de Bure Interferometer. Our study resulted in 1) the discovery of the redshift of the bright SMG HDF850.1 ($z = 5.183$); 2) the discovery of a bright line identified as CO(2-1) arising from a BzK galaxy at $z = 1.785$, and of other 6 CO lines associated with various galaxies in the field; 3) the detection of a few lines (presumably CO(3-2) at $z \sim 2$) with no optical/NIR/MIR counterparts. These observational results allowed us to expand the parameter space of galaxy properties probed so far in high-z molecular gas studies. Most importantly, we could set first direct constraints on the cosmic evolution of the molecular gas content of the universe. The present study represents a first, fundamental step towards an unbiased census of molecular gas in 'normal' galaxies at high-z, a crucial goal of extragalactic astronomy in the ALMA era.

Keywords. galaxies: evolution - galaxies: ISM - ISM: evolution

1. Introduction

Most of our current understanding of the formation of galaxies is based on blank optical and near-IR deep observations of their stellar continua, and to less extent of their ionized gas and the dust continuum emission: The luminosities and colors of stellar continua are used to identify high-z galaxies and to gauge their stellar contents and star formation rates (e.g., Steidel *et al.* 2004; Le Fèvre *et al.* 2005; Lilly *et al.* 2007). Narrow-band observations unveiled the presence of a population of star-forming galaxies with bright line emission arising from the ionized gas, but faint stellar continua (e.g., Ouchi *et al.* 2009, 2010). Wide field observations of the dust continuum gauge the obscured component of the star formation process (see review by Casey *et al.* 2014). The observational milestone resulting from all these investigations is the 'Lilly–Madau' plot, which traces the evolution of star formation rate (SFR) density in the universe as a function of cosmic time. These searches targeted star formation processes in act (through photoionized gas tracers or dust emission) or in the past (through stellar masses and ages). On the other hand, observations of the molecular gas, which is the fuel for star formation, have been limited to follow–up studies of galaxies that have been pre–selected from optical/NIR deep surveys, or to rare, extreme sources like bright sub-mm galaxies. Color–selection techniques (e.g., BzK, BMBX) have revealed significant samples of gas–rich, star forming galaxies at $z = 1.5$ to 2.5 ($M_{H_2} = 10^{11}$ M$_\odot$, $M_* = 10^{10}$ M$_\odot$, SFR = 100 M$_\odot$ yr^{-1}, Daddi *et al.* 2008, 2010a,b; Greve *et al.* 2010; Tacconi *et al.* 2010, 2013; Genzel *et al.* 2010, 2011, 2014;

Decarli *et al.* 2014a). However, optical/NIR selections are likely missing gas-dominated and/or obscured galaxies, especially at intermediate to high redshift.

2. A molecular scan in the HDF-N

In order to overcome this limitation, we performed a volume–limited census of the molecular gas content in primeval galaxies, through a blind CO search over a large part of the 3mm transparent window of the atmosphere (79–115 GHz). The pointing center (RA=12:36:50.300, Dec=+62:12:25.00, J2000.0) was chosen to include the bright sub-mm galaxy HDF850.1 (Hughes *et al.* 1998). The primary beam was $41''$–$60''$ in diameter. The typical beam size in our molecular line scan is $3'' \times 2.7''$ ($\sim$25 kpc at $z = 2$, and roughly constant at any $z \gtrsim 1$). The final cubes have a typical rms of 0.3 mJy beam^{-1} in 90 km s^{-1} channels, or $L'_{\mathrm{lim}} = (4-8) \times 10^9$ K km s^{-1} pc^2, assuming a typical line width of 300 km s^{-1}, and by requiring a 3.5-σ line detection (see Decarli *et al.* 2014b).

Our collapsed cube resulted in the deepest 3mm continuum map available to date, with a 1-σ rms of 8.5 μJy beam^{-1} (Decarli *et al.* 2014b). More importantly, we discovered various CO lines:

i- We observed 2 lines arising from the bright sub-mm galaxy HDF850.1. We identified them as CO(5-4) and CO(6-5), which placed the source at z=5.183, as confirmed by follow-up observations of the CO(2-1) and [CII] lines (Walter *et al.* 2012; Neri *et al.* 2014). This was the first spectroscopic redshift measurement for this iconic source (14 years after its discovery by Hughes *et al.* 1998).

ii- We discovered 7 CO lines spatially associated with optical/NIR counterparts. One of these galaxies is a massive ($M_* = 2.5 \times 10^{11}$ M$_\odot$) star-forming (38 M$_\odot$ yr^{-1}) BzK galaxy at z=1.784. A similarly bright CO line, on the other hand, is associated with a galaxy with $\sim$1 order of magnitude smaller stellar mass at z=2.044 (Decarli *et al.* 2014b). This example shows that large molecular gas reservoirs can be found in galaxies with very different properties at $z \sim 2$ (Decarli *et al.* 2014b; Walter *et al.* 2014).

iii- Other line candidates were found in a spatial position which shows absolutely no optical/NIR/MIR counterparts, despite the exquisite depth of the available ancillary data. The most remarkable example, dubbed ID.18, has a flux of 0.58 ± 0.11 Jy km s^{-1} and is detected in independent data blocks. It peaks at 112.593 GHz. We argue that the line is CO(3-2) at $z = 2.071$. A second transition, CO(4-3), is tentatively detected (at $\lesssim$4-σ level) at the same position in our 2mm follow-up observations, thus confirming the line identification.

3. CO luminosity functions and molecular gas content

We compare our results with the empirical predictions by Sargent *et al.* (2014) and with semi-analytical cosmological models by Lagos *et al.* (2011) and Obreschkow *et al.* (2009a,b), as well as with the observed CO luminosity function at $z = 0$ as measured by Keres *et al.* (2003). We place our constraints in the following way: We bin the CO blind detections in terms of CO luminosity, and we normalize to the volume of the universe in each redshift interval. For each luminosity bin, we place a lower limit corresponding to our secure detections (i.e., the lines that we managed to confirm via independent follow-up observations) and upper limits that represent the case where all line candidates are in fact real. Our blind detections probe a CO luminosity range close to the 'knee' of the predicted CO luminosity functions (Walter *et al.* 2014).

We convert the CO luminosity functions into a cosmic molecular content, $\rho_{\mathrm{H}2}(z)$. This is the amount of molecular gas in galaxies per unit comoving volume. The luminosity

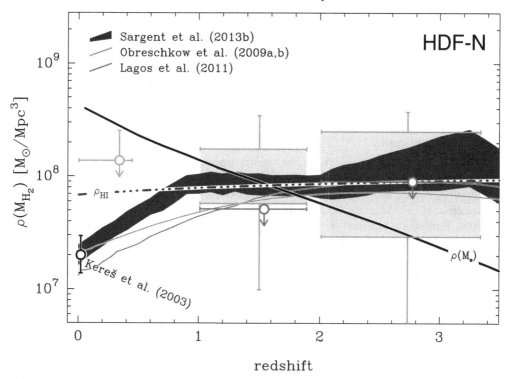

Figure 1. The evolution of the cosmic H_2 mass density, $\rho_{H2}(z)$ based on predictions from semi–analytical comological models (Obreschkow *et al.* 2009a, 2009b, Lagos *et al.* 2011) as well as the empirical predictions by Sargent *et al.* (2014). The blue–shaded area shows only the contribution of our blind detections to $\rho(M_{H2})$, *not* corrected/extrapolated for a population of undetected sources at lower or higher L'_{CO}, while the limits from the CO stacks over a H-band flux-limited sample of galaxies are shown as red points and arrows. For a comparison, the evolution of the cosmic neutral gas mass density (ρ_{HI}) and of the stellar mass density (ρ_*) are also plotted. Figure taken from Walter *et al.* (2014).

functions from empirical predictions or semi-analytical models are converted into H_2 mass via standard Galactic correction factors, and then integrated over the whole L_{CO} range. On the other side, we do not attempt to correct for sources not detected in our scan at both lower and higher L'_{CO} luminosities, given the unknown shape of the luminosity function. A comparison to empirical predictions of $\rho_{H2}(z)$ shows that the securely detected sources in our molecular line scan already provide significant contributions to the predicted $\rho_{H2}(z)$ in the redshift bins $\langle z \rangle \approx 1.5$ and $\langle z \rangle \approx 2.7$. Accounting for galaxies with CO luminosities that are not probed by our observations results in cosmic molecular gas densities $\rho_{H2}(z)$ that are higher than current predictions (see Fig. 1 and Walter *et al.* 2014). We note however that the current uncertainties (in particular the luminosity limits, number of detections, as well as cosmic volume probed) are significant.

4. Conclusions

Our molecular line scan of the HDF-N allows us to place first direct limits on CO luminosity function and on the molecular gas density in 'normal' galaxies at high redshift without any pre-selection based on optical/NIR/MIR wavelength observations. Our data shows that galaxies close to the peak of star formation activity ($z = 1 - 3$) have much higher molecular gas content than galaxies at $z = 0$. This evolution matches or even exceeds the one predicted by our current understanding of galaxy properties. We note however that the current uncertainties in our precursor study are significant, and that current models can thus not be ruled out given the available data. The emerging capabilities of the Jansky Very Large Array (JVLA) and of the Atacama Large (Sub–)Millimeter Array (ALMA) will enable similar molecular deep field studies to much deeper levels and larger areas (e.g., da Cunha *et al.* 2013).

Acknowledgements

This IAU proceeding is based on the results presented in Decarli *et al.* (2014b) and Walter *et al.* (2014). We thank our collaborators M. Aravena, E. Bell, F. Bertoldi, D. Colombo, P. Cox, E. da Cunha, E. Daddi, D. Downes, M. Dickinson, R. Ellis, L. Lentati, R. Maiolino, K. M. Menten, R. Neri, H.-W. Rix, M. Sargent, D. Stark, B. Weiner, and A. Weiß. This study is based on observations with the IRAM Plateau de Bure Interferometer (PdBI). IRAM is supported by INSU/CNRS (France), MPG (Germany) and IGN (Spain). Support for RD was provided by the DFG priority program 1573 'The physics of the interstellar medium'.

References

Carilli, C. & Walter, F. 2013, *ARA&A*, 51, 105
Casey, C., *et al.* 2014, *PhR*, 541, 45
Daddi, E., *et al.* 2008, *ApJ*, 673, L21
Daddi, E., *et al.* 2010a, *ApJ*, 713, 686
Daddi, E., *et al.* 2010b, *ApJ*, 714, L118
da Cunha, E., *et al.* 2013, *ApJ*, 765, 9
Decarli, R., *et al.* 2014a, *ApJ*, 780, 115
Decarli, R., *et al.* 2014b, *ApJ*, 782, 79
Genzel, R., *et al.* 2010, *MNRAS*, 407, 2091
Genzel, R., *et al.* 2011, *ApJ*, 733, 101
Genzel, R., *et al.* 2014, arXiv:1409.1171
Greve, T., *et al.* 2010, *ApJ*, 719, 483
Hughes, D. H., *et al.* 1998, *Nature*, 394, 241
Keres, D., Yun, M. S., & Young, J. S. 2003, *ApJ*, 582, 659
Lagos, C. D. P., *et al.* 2011, *MNRAS*, 418, 1649
Le Feèvre, O., *et al.* 2005, *Nature*, 437, 519
Lilly, S. J., *et al.* 2007, *ApJ Suppl.*, 172, 70
Neri, R., *et al.* 2014, *A&A*, 562, 35
Obreschkow, D., *et al.* 2009a, *ApJ*, 702, 1321
Obreschkow, D., *et al.* 2009b, *ApJ*, 698, 1467
Ouchi, M., *et al.* 2009, *ApJ*, 706, 1136
Ouchi, M., *et al.* 2010, *ApJ*, 723, 869
Sargent, M. T., *et al.* 2014, *ApJ*, 793, 19
Steidel, C. C., *et al.* 2004, *ApJ*, 604, 534
Tacconi, L., *et al.* 2010, *Nature*, 463, 781
Tacconi, L., *et al.* 2013, *ApJ*, 768, 74
Walter, F., *et al.* 2012, *Nature*, 486, 233
Walter, F., *et al.* 2014, *ApJ*, 782, 79

Galaxies in 3D across the Universe
Proceedings IAU Symposium No. 309, 2014   © International Astronomical Union 2015
B. L. Ziegler, F. Combes, H. Dannerbauer, M. Verdugo, eds.   doi:10.1017/S1743921314009843

Cold streams: detectability, relation to structure and characteristics

Tobias Goerdt

Department for Astrophysics, University of Vienna,
Türkenschanzstr. 17, 1180 Vienna, Austria
email: tobias.goerdt@univie.ac.at

Abstract. Cold gas streaming along the dark-matter filaments of the cosmic web is predicted to be the major provider of resources for disc buildup and star formation in massive galaxies in the early universe. We use hydrodynamical simulations to study to what extent these cold streams are traceable in the extended circum-galactic environment of galaxies via Ly alpha emission, Ly alpha absorption and selected low ionisation metal absorption lines. We predict the strength of the absorption signal produced by the streams and find that it is consistent with observations in high redshift galaxies. The characteristics of the Ly alpha emission of our simulated galaxies are similar in luminosity, morphology and extent to the observed Ly alpha blobs, with distinct kinematic features. We analyse the characteristics of the cold streams in simulations and present scaling relations for the amount of infall, its velocity, distribution and its clumpiness and compare our findings with observations.

Keywords. cosmology: theory – galaxies: evolution – galaxies: formation – galaxies: high redshift – intergalactic medium — galaxies: ISM

1. Introduction

Our understanding of how galaxies form has changed substantially in recent years. A decade ago it was thought that galaxies collect their baryons through diffuse gas, spherically symmetrically falling into dark matter haloes and being shock-heated as it hits the gas residing in the halos, the so-called hot mode accretion. Whether the gas eventually settles into the equatorial plane, forming a galactic disk was depending on the mass of the dark halo. Below a critical mass, the gas could cool efficiently, forming a disk galaxy, while for larger masses the cooling time would be longer than the Hubble time, leading to structures that resemble galaxy clusters with a large baryon fraction in the hot, diffuse intergalactic gas component. Recent theoretical work and simulations (Dekel *et al.* 2009) however have shown that at high redshift ($z \gtrsim 2$), galaxies acquire their baryons primarily via cold streams of relatively dense and pristine gas with temperatures around 10^4 K that penetrate through the diffuse shock-heated medium, the so-called cold mode accretion. These streams peak in activity around redshift 3.

N-body simulations suggest that about half the mass in dark-matter haloes is built-up smoothly, suggesting that the baryons are also accreted semi-continuously as the galaxies grow. Hydrodynamical cosmological simulations also show rather smooth gas accretion, including mini-minor mergers with mass ratios smaller than 1:10, that brings in about two thirds of the mass (Dekel *et al.* 2009). The massive, clumpy and star-forming discs observed at $z \sim 2$ may have been formed primarily via the smooth and steady accretion provided by the cold streams, with a smaller contribution by major merger events.

2. Absorption line profiles

In this section† we consider the Lyα and metal line absorption that occurs as UV light emitted by the central galaxy is absorbed by gas in the circum-galactic environment. This is defined to be situated in the spherical zone from just outside the galactic disk to around the virial radius R_v. Observations of absorption against the central galaxy itself have the advantage of being able to discriminate between inflows and outflows because the absorption may be assumed to occur in foreground material only. Radiation emitted or scattered from behind the galaxy is blocked by the galaxy itself. However, such absorption do not provide spatial information about the distance from the galaxy centre as they are all by definition at an impact parameter $b = 0$ from the galaxy centre.

Along a given sight-line the gas has a varying density, temperature and radial velocity as function of radial position r from the central galaxy. The convolutions of the different densities and radial velocities in the gas along the line of sight are the main ingredients to compute an absorption line profile. For a fair comparison to the observations, we mimic a Gaussian point spread function. It has a beam-size (= FWHM η of the Gaussian) of 4 kpc.

The likelihood of a detection of a cold stream while looking at a single galaxy from a single direction without averaging is for Mg II: an inflow >150 km s^{-1} with an EW > 0.2 Å in 1.3 % of all observations. These values should be achievable by future observations. We produce stacked spectra using our simulations by summing up the line profiles of several thousand different directions for each of the three galaxies and stack them together. We determine the absorption line profile for a spherical shell between an outer radius and an inner radius. The predicted metal line absorption profiles appear tiny. Probably the most suitable lines for the purpose of detecting cold streams in absorption are C II and Mg II since they have the strongest signal. Out of those Mg II is closer to being observed with the needed sensitivity and resolution.

3. Cold streams as Lyα blobs

Hydrodynamical cosmological simulations robustly demonstrate that massive galaxies at high redshifts were fed by cold gas streams, inflowing into dark-matter haloes at high rates along the cosmic web. In this section‡ we show that these streams should be observable as luminous Lα sources, with elongated irregular structures stretching for distances of over 100 kpc. The release of gravitational potential energy by the instreaming gas as it falls into the halo potential well is the origin of the Lα luminosity. The predicted Lα emission morphologies and luminosities make such streams likely candidates for the sources of observed high redshift Lα blobs. Most of the gas in the cold streams is at temperatures in the range $(1 - 5) \times 10^4$K, for which the Lα emissivity is maximised.

Using state-of-the-art cosmological AMR simulations with 70 pc resolution and cooling to below 10^4K, and applying a straightforward analysis of Lα emissivity due to electron impact excitation, we produced maps of Lα emission from simulated massive galaxies at $z \sim 3$. We computed the average Lα luminosity per given halo mass and the predicted luminosity function of these extended Lα sources, with an uncertainty of ± 0.4 dex. The properties of the individual images resemble those of the observed LABs in terms of morphology and kinematics. The predicted luminosity function is close to the observed LAB luminosity function. The simulated LABs qualitatively reproduce the

† published as Goerdt *et al.* (2012)
‡ published as Goerdt *et al.* (2010)

observed correlations between the global LAB properties of Lα luminosity, isophotal area and linewidth.

4. Relation to structure

We know that gas and, for our purposes more interestingly, dark matter haloes hosting satellite galaxies enter a host halo via smooth accretion streams (Dekel *et al.* 2009). Suppose now that galaxies and their satellite systems form in the focal point of a number m of randomly oriented cold streams. What is then the probability that a certain fraction of satellites lies within a thin plane around the central galaxy of the host halo? 15 out of Andromeda galaxy's 27 satellites are within a plane with 200 kpc radius and 12.6 kpc vertical rms scatter. In order to work out the likelihood to have of order 56% of satellites in a plane as thin as 3.8 degree we have performed Monte Carlo experiments, drawing randomly orientated streams and calculating the probabilities of having a given ratio of subhaloes within a plane of certain thickness†.

Since we assumed that all the streams pass trough the galaxy centre, two streams will always lie on a perfect plane. The '2 / m' values therefore are always at probability $P = 1.0$. For a three stream scenario we would expect at least 67 % of the satellites to lie within one plane. Since two, three or four stream configurations are most commonly seen in cosmological simulations, we can already say that there is nothing surprising about the configuration seen around the Andromeda galaxy. On the contrary, the fact that so little satellites lie on a thin plane rules out a two stream or a three stream scenario for the Andromeda galaxy, unless secular evolution or other processes move some of the satellites out of that plane. Turning the argument around, the fact that we observe a large fraction of satellites in a thin plane can be seen as indirect observational evidence for the cold mode accretion stream scenario proposed by Dekel *et al.* (2009) and that scattering events are not efficient, at least for the Local Group.

5. Streaming rates into high-z galaxies: smooth flows and mergers

In this section we look at the inflow rates of cold accretion streams, at their average inflow velocities, at the distribution of inflow rates as well as the role of mergers versus smooth flows. For this purpose we analyse the clumpiness of the gas streams and use the sharp peaks of inflow in the $\dot{M}(r)$ profiles as well as a new and unique clump finder. We evaluate each clump mass and estimate a mass ratio for the expected merger, ignoring further mass loss in the clump on its way in. We find: (1) The velocity profiles for the cold streams are by and large constant and show no sign of free-fall at all. (2) The constant infall velocity follows power law relations with respect to the halo mass and to the redshift. (3) The inflow distribution can be described by a double-Gaussian distribution (i.e. the sum of two Gaussians). (4) Most of the amount of inflow (>75%) is entering the halo at low inflow rates (i.e. smooth accretion) and only a small portion (<25%) of the inflow is coming in at high inflow rates (massive merger events). (5) Merger rates follow for a given mass and redshift bin of galaxies a robust double-logarithmic behaviour (i.e. the convolution of two logarithmic functions).

Interestingly, observer also found a double-Gaussian distribution when looking at the specific star formation rate distribution at fixed stellar mass in observations. They interpreted this as contributions from main-sequence and starburst activity. The analogy between their and our findings indicates that connections might exist between smooth

† published as Goerdt & Burkert (2013)

accretion and main-sequence star formation on the one hand as well as accretion through merger events and starburst activity star formation on the other side.

6. Conclusions

We conclude that the signatures of cold inflows are subtle, and when stacked are overwhelmed by the outflow signatures. Our predicted Lyα line absorption profiles agree with the observations, while the stacked metal line absorption from the inflows is much weaker than observed in the outflows. The single-galaxy line profiles predicted here will serve to compare to single-galaxy observations.

Our results support the idea that the observed LABs are direct detections of the cold steams that drive the evolution of massive galaxies at high redshifts. Even though the observed LABs are sometimes associated with central sources that are energetic enough to power the observed Lα emission, such as starbursts and AGNs, these central sources are very different from each other in the different galaxies, and in many LABs they are absent altogether. The gravitational heating associated with the inflowing cold streams is a natural mechanism for driving the extended Lα cooling radiation observed as LABs, and this extended Lα emission is inevitable in most high-redshift galaxies.

We conclude that the special spatial alignment of Andromeda's satellite galaxies can naturally be explained by cold stream accretion and simple geometry. Hydrodynamical simulations naturally produce coplanar structures of satellites via cold mode accretion streams. Satellite alignments around galaxies can be seen as indirect observational evidence for the cold stream paradigm and provides important information about the accretion history of galaxies.

We conclude that gas is flowing into a galaxy's halo mainly as smooth accretion flows with only a minority ($\sim 25\%$) coming in via merger events such as clumps or small satellite galaxies. The velocity profile of those streams is, contrary to what might be expected, constant and not free-falling. The potential energy of the gas which is lost on its way in must be dissipated by other mechanisms, such as Lyα radiation (Goerdt et al. 2010). The analogousness of the equations used to describe the distribution of the gas inflow in simulations or to describe the specific star formation rate distribution in observations (the double-Gaussian decomposition) indicates a connection might exist between smooth accretion and main-sequence star formation on the one hand side as well as a connection between accretion through merger events and starburst activity star formation on the other hand.

Acknowledgements

Tobias Goerdt is a Lise Meitner fellow. This work was supported by FWF project number M 1590-N27, the Spanish *Ministerio de Economía y Competitividad (MINECO)* via the projects AYA2009-13875-C03-02 and AYA2012-31101 and via the JdC subprogramme JCI-2011-10289.

References

Dekel, A. *et al.* 2009, *Nature*, 457, 451

Goerdt, T., Dekel, A., Sternberg, A., Ceverino, D., Teyssier, R., & Primack, J. R., 2010, *MNRAS*, 407, 613

Goerdt, T., Dekel, A., Sternberg, A., Gnat, O., & Ceverino, D., 2012, *MNRAS*, 424, 2292

Goerdt, T. & Burkert, A., 2013, arXiv:1307.2102

Galaxies in 3D across the Universe
Proceedings IAU Symposium No. 309, 2014
B. L. Ziegler, F. Combes, H. Dannerbauer, M. Verdugo, eds.

© International Astronomical Union 2015
doi:10.1017/S1743921314009855

Galaxy Mass Assembly with VLT & HST and lessons for E-ELT/MOSAIC

François Hammer, Hector Flores and Mathieu Puech

GEPI, Observatoire de Paris
5 Place Jules Janssen F-92 195 Meudon, France
email: `francois.hammer@obspm.fr`

Abstract. The fraction of distant disks and mergers is still debated, while 3D-spectroscopy is revolutionizing the field. However its limited spatial resolution imposes a complimentary HST imagery and a robust analysis procedure. When applied to observations of IMAGES galaxies at $z = 0.4$-0.8, it reveals that half of the spiral progenitors were in a merger phase, 6 billion year ago. The excellent correspondence between methodologically-based classifications of morphologies and kinematics definitively probes a violent origin of disk galaxies as proposed by Hammer *et al.* (2005). Examination of nearby galaxy outskirts reveals fossil imprints of such ancient merger events, under the form of well organized stellar streams. Perhaps our neighbor, M31, is the best illustration of an ancient merger, which modeling in 2010 leads to predict the gigantic plane of satellites discovered by Ibata *et al.* (2013). There are still a lot of discoveries to be done until the ELT era, which will open an avenue for detailed and accurate 3D-spectroscopy of galaxies from the earliest epochs to the present.

Keywords. galaxies: spiral - galaxies: evolution - galaxies: formation -galaxies: Local Group

1. Methodology to probe distant galaxy kinematics

Galaxies assemble their mass through mergers and gas accretion but the balance between these mechanisms is still in debate. Spatially resolved motions are currently observed (Flores *et al.* 2006; Forster-Schreiber *et al.* 2006; Contini *et al.* 2012) at 3 to 7 kpc scales, for galaxies up to $z = 2.5$. Imagery with the HST-ACS reveals sub-kpc details, and galaxy spectral energy distribution is currently retrieved from UV to near-IR (Spitzer) and sub-mm (ALMA) wavelengths. However observations lead to different, sometimes contradictory results (Glazebrook 2013), which could require to investigate further, e.g., the various methodologies for distinguishing a disk from a merger.

The first difficulty in 3D spectroscopy, especially at $z > 1$† is to gather complete, mass-selected samples of galaxies. Secondly, distant galaxies have small sizes, with stellar half-light diameters decreasing from 2 to 0.6 *arcsecond* from $z = 0.55$ to $z = 2$ (for stellar masses larger than $10^{10} M\odot$, see Dahlen *et al.* 2007). Fortunately ionized gas is currently more extended than the continuum (see Figure 4 of Puech *et al.* 2010) and the [OII] emission diameters at $z - 0.65$ average to 2.8 ± 0.8 arc second that is covered by ~ 3 spatial resolution elements for a 0.8-1 arc sec FWHM PSF. At $z = 2$, $H\alpha$ emissions are on average, resolved by the same number of elements: the smaller galaxy sizes are compensated by the better seeing (0.4-0.5 arc second FWHM) in K band‡, and the expected larger ionized gas to continuum size ratio. The determination of rotation curves

---

† Selecting $z > 1$ galaxies using visible filters strongly biases the selection to actively forming, young and low-mass galaxies. Moreover, linking $z > 1$ galaxies to their descendants is problematic given the possibility of merger events during the elapsed time, i.e., 8 billion years.

‡ Observations using adaptive optics are useful to zoom in bright regions in the galaxy cores (e.g., to probe AGN or central outflows, see Genzel *et al.* 2014; Forster-Schreiber *et al.* 2014), but

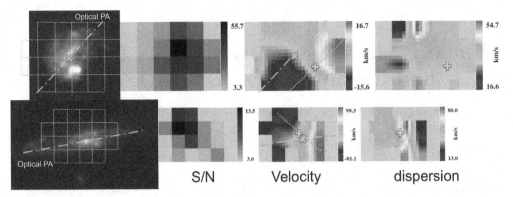

S/N   Velocity   dispersion

Figure 1. IMAGES galaxies J033239.72-275154.7 at z = 0.4151 (top) and J033213.06-274204.8 at z = 0.4215 (bottom), with the scale provided by the size of one IFU micro lens, 0.5 arc sec. From left to right: rbg colors using v, i and z filter from HST/ACS, S/N, velocity field and velocity dispersion maps. PA_{opt} and PA_{VF} are indicated by dot-dash and dot lines, respectively. Cross and plus signs mark the optical center (c_{opt}) and the velocity field center (c_{VF}), allowing to apply the morpho-kinematic classification procedure.

requires more than 7 resolution elements per optical radius (and better 10, e.g., Bosma 1978), which can be reached by only few massive and extended distant galaxies, or later by the future ELT 3D spectrographs such as MOSAIC.

With such conditions, interpreting kinematics may appear to be complex though it remains the unique technique for differentiating, e.g., rotating disks from mergers, while smaller perturbations (lopsidedness, warps and bars) are hardly or not detected (Shapiro et al. 2008). One needs to elaborate a comprehensive, efficient methodology. Rotating disks possess remarkable geometrical properties, which render them very easy to recognize. Mergers can lead to strong misalignments between (ionized) gas and old star distributions, rotating disks do not. Analysis of kinematics requires a priori a good knowledge of the mass center, the PA and the inclination of a supposed disk (Krajnovic et al. 2006). Complimentary deep space-based imagery at red rest-frame wavelengths is thus mandatory to recover the above properties and then to interpret distant galaxy kinematics. The classification procedure for a rotating disk requires to verify, within the observational uncertainties, whether or not:

(a) the mass center coincides to the velocity field center;

(b) the difference in PA between red imagery and velocity field is lower than 20 degrees (see Epinat et al. 2012 and references therein and Garcia-Lorenzo et al. 2014);

(c) the 'kinemetry' (see Shapiro et al. 2008) or the 'σ centering' (see Flores et al. 2006) are consistent with a dominant rotation motion.

The three steps are mandatory to be verified, and they require mass center, PA and inclination, respectively. The 'kinemetry' method uses an harmonic decomposition of the kinematical maps based on asymmetries. The 'σ centering' method uses the beam smearing effect: the large and sharp gradient between extremal velocities ($\sim 2 \times V_{max}$) is almost unresolved, and when convolved with random motions, the σ map unavoidably shows a prominent peak at the center of a rotating disk. We notice that 'kinemetry' can be applied on only a tiny fraction of z ∼ 2 galaxies (15 among 62 galaxies, Forster-Schreiber et al. 2009) and cannot distinguish turbulent disks (as those simulated by Bournaud et al. 2010) from mergers (see Perret 2014).

the outskirts observed with very small spaxels are affected by very low S/N and surface-brightness dimming.

Figure 1 illustrates the three above criteria with two galaxies from IMAGES (see e.g., Yang *et al.* 2008). In the top panels, stellar mass and velocity field centers are off by 0.8 arc second, which is much larger than the 0.35 arc second uncertainty, obtained from the quadratic convolution of the astrometry (0.25 arc second) and the line peak location (0.25 arc second) precisions. Interestingly PA_{VF} is almost aligned with PA_{opt}, and without imagery, the system could have been confusedly classified as a rotating disk. However this system would have not passed the "σ centering" test, as shown in the top-left panel of Figure 1 (see the σ peak far offset from the velocity field center). In fact this system has been fully analyzed and modeled as a 3:1, near completion merger by Peirani *et al.* (2009) explaining the giant star-bursting bar, the offset center and the small velocity gradient for a 35 degree inclined galaxy. Figure 1 (bottom) shows a galaxy for which morphology resembles well that of an almost edge-on, well defined spiral galaxy. Kinematics is essential here: after examining the S/N map, the most illuminated spaxels are well outside, at the top-left of the edge-on disk. Moreover PA_{VF} is at ~ 50 degrees from the PA_{opt}, which definitively discards a rotating disk classification. Both examples illustrate how kinematics and imagery are insufficient when used separately, while combined together, they provide a very powerful technique, the morpho-kinematical classification. If not applied (e.g., without high resolution imagery), one can derive only an upper limit on the rotating disk fraction (see Contini *et al.*, same volume).

2. The IMAGES survey

The explicit goal of IMAGES is to gather enough constraints on $z = 0.4$-0.8 galaxies for linking them directly to their local descendants. Its selection is limited by an absolute J-band magnitude ($M_J(AB) < -20.3$), a quantity relatively well linked to the stellar mass (Yang *et al.* 2008), leading to a complete sample of 63 emission line galaxies with $W_0([OII]) > 15$ A and $M_{stellar} > 1.5\ 10^{10}\ M_\odot$. The set of measurements includes HST imagery (one to three orbits in b, v, i and z), spatially-resolved kinematics from VLT/GIRAFFE (from 8 to 24hrs integration times), deep VLT/FORS2 observations (two × three hours with two grisms at $R = 1500$) and deep Spitzer observations to sample mid and near-IR. IMAGES galaxies have been observed in four different fields of view to avoid cosmological variance effects. It is hardly affected by cosmological dimming, because the imaging depth ensures the detection of the optical disk of a MW-like galaxy after being redshifted to $z \sim 0.5$.

According to the Cosmological Principle, progenitors of present-day giant spirals have properties similar to those of galaxies having emitted their light ~ 6 Gyr ago. Figure 2 presents the results of a morphological analysis of 143 distant galaxies for which depth, rest-frame filters, spatial resolution and selection are strictly equivalent to what has been done for the Hubble Sequence derived from SDSS galaxies (Delgado-Serrano *et al.* 2010). The past Hubble Sequence shown in Figure 2(left) includes the IMAGES galaxies as well as a complimentary sample of distant galaxies with $W_0([OII]) \leqslant 15$ A (Delgado-Serrano *et al.* 2010). Statistical distributions (stellar mass, star formation, etc.) of both low and high redshift galaxies have been verified to be consistent with the results of large surveys. The methodology for classifying the morphologies follows a semi-automatic decision tree, which uses as templates the well known morphologies of local galaxies that populate the Hubble sequence, including the color of their sub-components. For example a blue nucleated galaxy is classified peculiar. Such a conservative method is the only way for a robust morphological classification, and interestingly, it confirms the van den Bergh (2002)'s results.

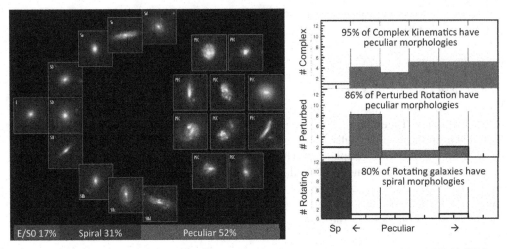

Figure 2. *Left*: The past Hubble Sequence, six billion years ago, based on 143 IMAGES
galaxies, and for which each stamp represents 5% of the galaxies.

However star formation may affect the morphological appearance, and the morphological classification has to be compared to the spatially-resolved kinematics. The classification of velocity fields is ranging from rotating disk, perturbed rotation (i.e., σ peak offset) or complex kinematics (Flores *et al.* 2006; Yang *et al.* 2008). Figure 2 (see also Neichel *et al.* 2008) evidences that peculiar morphologies correlate very well with anomalous kinematics and vice versa.

Is that correlation preserved when using automatic classification methods based on e.g., concentration and asymmetry of the galaxy luminosity profile? The answer is negative (Neichel *et al.* 2008), and these methods overestimate the number of spirals by a factor of two, a problem already identified by Conselice *et al.* (2005). Automatic classification methods are interesting and appealing because they can be applied to very large number of galaxies. However they provide too simplistic results to interpret complex objects like distant galaxies. Their limitations in distinguishing peculiar from spiral morphologies lead to a far larger systematic than the Poisson statistical uncertainty for a sample of one hundred of objects (Delgado-Serrano *et al.* 2010), making very large samples treated this way not very useful.

3. The observed frequency of mergers in the past histories of spirals

The remarkable agreement between morphological and kinematical classifications implies that dynamical perturbations of the gaseous component at large scales are linked to the peculiar morphological distribution of the stars. This suggests a common process that profoundly modified the overall gas and star properties. Minor mergers (mass ratio well above five) can affect locally the dispersion map (Puech *et al.* 2007) while they cannot affect the large scale rotational field over several tens of kpc†. Outflows provoked by stellar feedback are not observed in the IMAGES galaxies that are generally forming stars at moderate rate (Rodrigues *et al.* 2012). Internal fragmentation is limited because less than

---

† This is entirely true for very minor mergers (mass ratio larger than 10) while intermediate mergers may affect significantly the galaxy properties during short phases. Due to their lower impact and their longer duration (Jiang *et al.* 2008), they are considerably less efficient than major mergers in distorting morphologies and kinematics or they do it in a somewhat sporadic way (Hopkins *et al.* 2008).

20% of the IMAGES sample show clumpy morphologies (Puech 2010) while associated cold gas accretion tends to vanish in massive halos at z<1, with rates <1.5 $M_\odot$/yr at z ∼ 0.7 (Keres *et al.* 2009). In fact in many simulations of cold flows, the clumps are merging galaxies containing gas, stars and dark matter (see e.g. Danovich *et al.* 2011), and for which mergers are easily identified by large jumps in angular momentum acquisition (see e.g., Kimm *et al.* 2011). Major mergers appear to be the mechanism explaining peculiar morphologies as well as kinematically unrelaxed galaxies. They are indeed unique in bringing enough angular momentum to explain the large scatter of the mass-velocity (Tully-Fisher) relation (Flores *et al.* 2006; Puech *et al.* 2008; Covington *et al.* 2010). This explanation indeed solves the spin catastrophe (see also Maller, Dekel, & Somerville 2002), i.e. the longstanding problem that disks formed in isolation are too small and with too small angular momentum (Navarro & Steinmetz 2000).

This has led Hammer *et al.* (2009) to test and then successfully model all the unrelaxed IMAGES galaxies as consequences of major mergers using hydrodynamical simulations (GADGET2 and ZENO). Using a comparison of their morpho-kinematics properties to those from a grid of simple major merger models (Barnes 2002), they provided convincing matches in about two-thirds of the cases. It results that a third of z = 0.4-0.8 galaxies are in one or another merger phase. Letting aside the bulge-dominated galaxies (E/S0) which fraction does not evolve (see Figure 2), it means that six billion years ago half of present-day spirals were involved in a major merger phase (Hammer *et al.* 2009).

Why IMAGES is finding so many spiral progenitors experiencing a major merger phase? In fact the morpho-kinematic technique used in IMAGES is found to be sensitive to all merger phases, from pairs to the post-merger relaxation phase. The merger rate associated with different phases has been calculated (Puech *et al.* 2012), and found to match perfectly predictions by state-of-the-art ΛCDM semi-empirical models (Hopkins *et al.* 2010) with no particular fine-tuning. Thus, both theory and observations predict an important impact of major mergers for progenitors of present-day spiral galaxies and the whole Hubble sequence could be just a vestige of merger events (Hammer *et al.* 2009). Nowadays all recent cosmological simulations report the formation of late-type disks after major mergers (Font *et al.* 2011; Brook *et al.* 2011; Keres *et al.* 2012; Guedes *et al.* 2011; Aumer *et al.* 2013).

4. Relics of ancient mergers in the nearby Universe

If many spiral galaxies have experienced a major merger, it should have left fossil imprints into their halo. Mergers are often producing gigantic tidal tails from which many stars are captured by the halo remnant gravitational potentials (Font *et al.* 2006; Wang *et al.* 2012; Hammer *et al.* 2010, 2013). Because individual stars are not affected by dynamical friction neither by the halo hot gas, they form stellar streams as illustrated in Figure 3, which can be persistent for a Hubble time.

The outskirts of nearby, isolated spiral galaxies often show low-surface brightness stellar streams (Martinez-Delgado *et al.* 2010), which are likely relics of mergers. An explanation from recent minor mergers seems unlikely because of the stream red colors and of the absence of any residual core. The whole NGC5907 galaxy (disk, bulge and thick disk) and associated loops have been successfully modeled by assuming a 3:1 gas-rich major merger (see Figure 3 and Wang *et al.* 2012). Our nearest neighbor, M31 shows a classical bulge and a high halo metallicity suggesting a major merger origin (van den Bergh 2005; Kormendy 2013). The later provides a robust explanation of the stellar Giant Stream, which could be made of tidal tail stars captured by the galaxy gravitational potential after the fusion time (Hammer *et al.* 2010). Stars of the Giant Stream have ages older

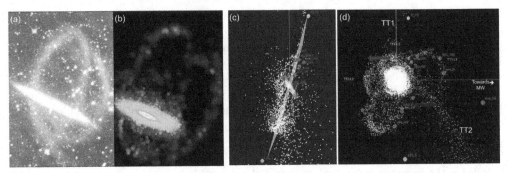

Figure 3. *(a)*: Deep observations of NGC 5907 in the visible light showing the thin disk with the gigantic loops (from Martinez-Delgado *et al.* 2008). *(b)*: GADGET2 simulation (2.2M particles) of the stellar loops, 5.3 Gyrs after a 3:1 merger (first passage, 6.8 Gyrs ago, $f_{gas}^{initial} = 60\%$, $f_{baryons}^{initial} = 17\%$, see Wang *et al.* 2012). *(c)*: Stellar particles associated to an ancient 3:1 merger at M31 location (GADGET2, 8M particles see Hammer *et al.* 2010, 2013; Fouquet *et al.* 2012), and forming a plane seen almost edge-on, coinciding with the observed satellite plane (almost seen edge-on, in blue color) found by Ibata *et al.* (2013). *(d)*: Same as (c) but rotated by 90 degrees to reveal the loops system in the loop plane. Most M31's satellite plane (it also includes the recently found CasIII, see Martin *et al.* 2013) are found in the loop system and their observed motions follow the predictions from the loop model (see red and blue arrows).

than 5.5 Gyr (Brown *et al.* 2007), and a 3:1 gas-rich merger may reproduce it as well as the M31 disk, bulge, 10 kpc ring & thick disk, assuming the interaction and fusion have occurred 8.75±0.35 and 5.5±0.5 Gyr ago, respectively.

In fact, the merger model of M31 of Hammer *et al.* (2010) predicted the disk of satellites discovered by Ibata *et al.* (2013). Spatial locations of these M31 satellites are drawing two loop systems (see Figure 3d and Hammer *et al.* 2013) similar to those found around NGC5907, i.e., typical signatures of ancient major mergers. Besides providing a prediction, the M31 merger model is the only one able to account for the satellite plane thinness, its apparent rotation (see the arrows in Figure 3d) and the excess of dwarves towards the MW (see Hammer *et al.* 2013 for details).

Both MW and M31 possess fossil satellite systems, which might be linked together because the Ibata *et al.* (2013) disk of satellites is pointing to the MW. Not surprisingly it is also a prediction of the modeling of M31 as an ancient merger, and the main associated tidal tail may have reached the MW (Fouquet *et al.* 2012; Hammer *et al.* 2013). This scenario could be at the origin of the vast plane of satellites observed around the MW (Pawlowski *et al.* 2011), and in such a case many Local Group dSphs would be relics of tidal dwarves, and then without dark matter as verified by Yang *et al.* (2014).

5. Discussion & Conclusions

The "merger hypothesis" (Toomre & Toomre 1972) for elliptical galaxies interprets them as the product of gas-poor or successive major mergers. It appears more and more plausible that spiral galaxies could also result from mergers of gas-rich galaxies (Hammer *et al.* 2005, 2009). Re-formation of disks after major mergers has been yet described in 2002 (Barnes 2002), assuming a small gas fraction (12%) in the progenitors†. With much larger gas fraction a rebuilt disk could be more prominent and after subsequent star formation, it could dominate the remnant galaxy. Gas fractions in galaxies at earliest

---

† the rebuilding disk scenario (Hammer *et al.* 2005, 2009) is also supported by the very recent discovery of rotating molecular disks by ALMA, in nearby merger remnants (Ueda *et al.* 2014).

epochs indicate values that exceed 50% at z~ 1.5-2 (Rodrigues *et al.* 2012), allowing even the formation of late type spirals with small B/T values (Hopkins *et al.* 2010).

Perhaps the numerous bulges with low Sersic indices (often called pseudo-bulges, see Kormendy 2013 and references therein) found in local galaxies are at odd with a major merger history. Forming pseudo bulge galaxies is however expected through ancient gas-rich mergers (Keselman & Nusser 2012; Hammer *et al.* 2012), though further investigations are necessary. The scatter of the $M_{stellar} - SFR$ relation for emission line galaxies could be too narrow (0.3 dex, Noeske *et al.* 2007) for accommodating major mergers, which are expected to show strong and short living bursts during the fusion phase. However the fusion phase is unlikely an instantaneous episode (see e.g., Figure 8 of Hopkins *et al.* 2012 for mergers of MW-mass galaxies) and they are indeed consistent with duration of 0.4 Gyr for SFR variations by a factor 4 (or 2 σ). Indeed the shape of the SFR peak depends on the treatment of the overall feedback (see Hopkins, same volume and Hopkins *et al.* 2012). Interestingly, observations (from IMAGES, see Figure 4 of Puech *et al.* 2012), which average orbits, gas fractions and mass ratios are also fully consistent with the $M_{stellar} - SFR$ scatter (Puech *et al.* 2014).

Because 3D spectroscopy is photon starving, its future is linked to the ELTs. The combination of the *JWST* and of the multi-IFUs MOSAIC on the E-ELT will be uniquely powerful for investigations into the physics underway as the first galaxies were forming (near or at the time when the Universe was only partially reionized). The potential performances are impressive, i.e. detection of Lyman-α emission line at $z \sim 9$ (for $J_{AB} = 29$) and the study of rest-frame UV absorption lines in a brighter galaxy at $z \sim 7$ ($J_{AB} = 26$). In addition to the larger collecting area of the E-ELT, MOSAIC will provide further gains compared to the first-light instruments envisaged for the GMT and TMT. Specifically, the use of MOAO-corrected IFUs will provide excellent image quality for spectroscopy over wide fields, enabling the study of spatially-resolved physical properties for large samples of high-redshift galaxies, together with optimum subtraction of the sky background.

Acknowledgements

We warmly thank both the IMAGES and MOSAIC teams for their support and crucial achievements. We are grateful to the organizers for their invitation at such an excellent Conference and their very nice welcome in the beautiful city of Vienna.

References

Abazajian, K. N., *et al.* 2009, *ApJS*, 182, 543
Aumer, M., White, S. D. M., Naab, T., Scannapieco, C., 2013 *MNRAS*, 434, 3142
Barnes, J. E., 2002 *MNRAS*, 333, 481
Bosma, A., 1978, PhD Thesis
Bournaud, F., Elmegreen, B. G., Teyssier, R., *et al.* 2010 *MNRAS*, 409, 1088
Brook, C. B., Stinson, G., Gibson, B. K., *et al.*, 2011 *MNRAS*, 419, 771
Brown, T. M., *et al.* 2007, *ApJ*, 658, L95
Conselice C. J., Bundy K., Ellis R. S., *et al.* 2005, *ApJ*, 628, 160
Contini, T., Garilli, B., Le Fvre, O., *et al.* 2012, *A&A*, 539, 91
Covington, M. D., Kassin, S. A., Dutton, A. A., *et al.* 2010, *ApJ*, 710, 279
Dahlen, T., Mobasher, B., Dickinson, M., *et al.* 2007, *ApJ*, 654, 172
Danovich, M., Dekel, A., & Hahn, O. and Teyssier, R., 2012, *MNRAS*, 422, 1732
Delgado-Serrano, R., Hammer, F., Yang, Y. B., *et al.*, 2010 *A&A*, 509, 78
Epinat, B., Tasca, L., Amram, P., *et al.* 2012 *A&A*, 539, 92
Flores, H., Hammer, F., Puech, M., *et al.* 2006 *A&A*, 455, 107
Font, A. S., Johnston, K. V., & Bullock, J. S. Robertson, B. E., 2006, *ApJ*, 646, 886

Font, A. S., McCarthy, I. G., Crain, R. A., *et al.* 2011, *MNRAS*, 416, 2802

Forster-Schreiber, N. M., Genzel, R., Lehnert. M., *et al.* 2006, *ApJ*, 645, 1062

Forster-Schreiber, N. M., Genzel, R., Bouche, N., *et al.* 2009, *ApJ*, 706, 1364

Forster Schreiber, N. M., Genzel, R., Newman, S. F., *et al.* 2014, *ApJ*, 787 38

Fouquet, S., Hammer, F., Yang, Y., Puech, M., & Flores, H., 2012, *MNRAS*, 427, 1769

Garcia-Lorenzo, B., *et al.* 2014, *A&A*, in press (arXiv:1408.5765)

Genzel, R., Forster Schreiber, N. M., Rosario, D., *et al.* 2014, *ApJ*, in press (arXiv:1406.0183)

Glazebrook, K., 2013, PASA, 30, 56

Guedes, J., Callegari, S., Madau, P., *et al.* 2011 *ApJ*, 742, 76

Hammer, F., Flores, H., Elbaz, D., *et al.* 2005, *A&A*, 430, 115

Hammer, F., Puech, M., Chemin, L., *et al.*, 2007 *ApJ*, 662, 322

Hammer, F., Flores, H., Puech, M., Yang, Y. B., Athanassoula, E., *et al.* 2009, *A&A*, 507, 1313

Hammer, F., Yang, Y. B., Wang, J. L., *et al.* 2010, *ApJ*, 725, 542

Hammer, F., Yang, Y. B., Fouquet, S., *et al.* 2013, *MNRAS*, 431, 2543

Hammer, F., *et al.*, 2012, Modern Physics Letters A, 33, 1230034 (arXiv:1110.1376)

Ibata, R., Lewis, G., Conn, A. R., *et al.* 2013, Nature, 493, 62

Hopkins, P. F., Hernquist, L., Cox, T. J., *et al.*, 2008, *ApJ*, 688, 757

Hopkins, P. F., Bundy, K., Croton, D., *et al.*, 2010, *ApJ*, 715, 202

Hopkins, P. F., Cox, T. J., Hernquist, L., *et al.*, 2012, *MNRAS*, 430, 1901

Jiang, C. Y., Jing, Y. P., Faltenbacher, A., *et al.*, 2008 *ApJ*, 675, 1095

Keres, D., Katz, N., Fardal, M., *et al.*, 2009 *MNRAS*, 395, 160

Keres, D., Vogelsberger, M., Sijacki, D., *et al.* 2012 *MNRAS*, 425, 2027

Keselman, J. A., Nusser, A., 2012 *MNRAS*, 424, 1232

Kimm, T., Devriendt, J., & Slyz, A, 2011, *MNRAS* submitted (arXiv:1106.0538)

Kormendy, J. 2013, In Secular Evolution of Galaxies, XXIII Canary Islands Winter School of
    Astrophysics, ed. J. Falcón-Barroso & J. H. Knapen (Cambridge: Cambridge Univ. Press)

Krajnovic, D., Cappellari, M., de Zeeuw, T. , Copin, Y., 2006 *MNRAS*, 366, 787

Maller A. H., Dekel A., Somerville R., 2002, *MNRAS*, 329, 423

Martin, N *et al.*, 2013, *ApJ*, 772, 15

Martinez-Delgado, D., *et al.* 2008, *ApJ*, 689, 184

Martínez-Delgado, D., *et al.*, 2010, *AJ*, 140, 962

Navarro, J. & Steinmetz, M., 2000, *ApJ* 528, 607

Neichel, B., *et al.* 2008, *A&A*, 484, 159

Noeske, K. G., *et al.* 2007, *ApJ*, 660, L43

Pawlowski, M. S., Kroupa, P., & de Boer, K. S. 2011, *A&A*, 532, A118

Peirani S., Hammer F., Flores H., Yang Y., Athanassoula E., 2009, *A&A*, 496, 51

Perret, V., 2014, PhD Thesis

Puech, M., Hammer, F., Flores, H., Neichel, B., Yang, Y., & Rodrigues, M. 2007, *A&A*, 476,
    21

Puech, M., *et al.* 2008, *A&A*, 484, 173

Puech M., 2010 *MNRAS*, 406, 535

Puech, M., Hammer, F., Flores, H., *et al.*, 2010, *A&A* 510, 68

Puech, M., Hammer, F., Hopkins, P. F., *et al.* 2012, *ApJ*, 753, 128

Puech, M., Hammer, F., Rodrigues, M., *et al.* 2014, *MNRAS*, 443, 49

Rodrigues, M., Puech, M., Hammer, F., *et al.*, 2012 *MNRAS*, 421, 2888

Shapiro, K. L., Genzel, R., Forster-Schreiber, N. M., *et al.*, 2008 *ApJ*, 682, 231

Toomre, A. & Toomre, J., 1972, *ApJ*, 178, 623

Ueda, J., Iono, D., Yun, M. S., *et al.*, 2014, *ApJS*, accepted (arXiv:1407.6873)

van den Bergh, S., 2002, *Pub. Astron. Soc. Pacific*, 114, 797

van den Bergh, S., 2005, in *The Local Group as an Astrophysical Laboratory*, ed. M. Livio &
    T. M. Brown (Cambridge: Cambridge Univ. Press), P.1-15

Wang, J., Hammer, F., Athanassoula, E., *et al.* 2012, *A&A*, 538, A121

Yang, Y., Flores, H., Hammer, F., *et al.* 2008, *A&A*, 477, 789

Yang, Y., Hammer, F., Fouquet, S., *et al.* 2014, *MNRAS*, 442, 2419

Galaxies in 3D across the Universe
Proceedings IAU Symposium No. 309, 2014
B. L. Ziegler, F. Combes, H. Dannerbauer, M. Verdugo, eds.

© International Astronomical Union 2015
doi:10.1017/S1743921314009867

Oxygen abundances of zCOSMOS galaxies at $z \sim 1.4$ based on five lines and implications for the fundamental metallicity relation

Christian Maier[1], Simon J. Lilly[2], Bodo L. Ziegler[1] and zCOSMOS team

[1]Department for Astrophysics, University of Vienna,
Türkenschanzstr. 17, 1180 Vienna, Austria
email: `christian.maier@univie.ac.at`
[2]Institute of Astronomy, ETH Zurich, 8093 Zurich, Switzerland

Abstract. A relation between the stellar mass M and the gas-phase metallicity Z of galaxies, the MZR, is observed up to higher redshifts. It is a matter of debate, however, if the SFR is a second parameter in the MZR. To explore this issue at $z > 1$, we used VLT-SINFONI near-infrared (NIR) spectroscopy of eight zCOSMOS galaxies at $1.3 < z < 1.4$ to measure the strengths of four emission lines: Hβ, [OIII]λ5007, Hα, and [NII]λ6584, additional to [OII]λ3727measured from VIMOS. We derive reliable O/H metallicities based on five lines, and also SFRs from extinction corrected Hαmeasurements. We find that the MZR of these star-forming galaxies at $z \approx 1.4$ is lower than the local SDSS MZR by a factor of three to five, a larger change than reported in the literature using [NII]/Hα-based metallicities from individual and stacked spectra. Correcting N2-based O/Hs using recent results by Newman *et al.* (2014), also the larger FMOS sample at $z \sim 1.4$ of Zahid *et al.* (2014) shows a similar evolution of the MZR like the zCOSMOS objects. These observations seem also in agreement with a non-evolving FMR using the physically motivated formulation of the FMR from Lilly *et al.* (2013).

1. Introduction

In the local universe, the metallicity at a given mass also depends on the SFR of the galaxy (e.g., Mannucci *et al.* 2010), i.e., the SFR appears to be a "second-parameter" in the mass-metallicity relation (MZR). Mannucci *et al.* (2010) claimed that this local $Z(M, SFR)$ is also applicable to higher redshift galaxies, and coined the phrase fundamental metallicity relation (FMR) to denote this *epoch-invariant* $Z(M, SFR)$ relation. Lilly *et al.* (2013) showed that a $Z(M, SFR)$ relation is a natural outcome of a simple model of galaxies in which the SFR is regulated by the mass of gas present in a galaxy.

Most studies of the Z(M,SFR) at $z \sim 1.4$ were based on small samples or samples with limited spectroscopic information, producing contradictory results. Except for the $z \sim 1.4$ study of Maier *et al.* (2006), the derived O/H metallicities in the literature at $z \sim 1.4$ have been based just on [NII]/Hα, via the N2-method (Pettini & Pagel 2004). However, Newman *et al.* (2014) found that the MZR at high z determined using the N2-method might be up to a factor of three times too high in terms of metallicity, when the effects of both photoionization and shocks are not taken into consideration.

2. Results

We have carried out NIR follow-up spectroscopy with VLT-SINFONI of eight massive star-forming galaxies at $1.3 < z < 1.4$, selected from the zCOSMOS-bright sample (Lilly *et al.* 2009) based on their [OII]λ3727emission line observed with VIMOS. Observations

282   C. Maier *et al.*

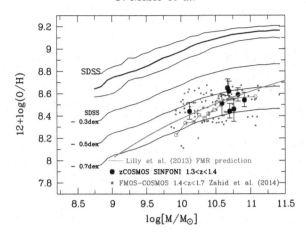

Figure 1. The median local MZR and 1σ values of local SDSS galaxies from Tremonti *et al.* (2004) are shown by the three upper black lines, while the three thinner lower black lines show the median SDSS MZR shifted downward by 0.3, 0.5 and 0.7 dex, respectively. The Zahid *et al.* (2014) N2-based O/Hs are corrected for the N2-calibration issues discussed in Newman *et al.* (2014), and shown as red dots for the individual measurements, and as open red circles for the O/Hs derived from the stacked spectra. The observed MZR at $z \approx 1.4$ is lower than the local SDSS MZR by a factor of three to five, and is also in agreement with the FMR prediction of Lilly *et al.* (2013).

with SINFONI in two bands (J to observe the H$\beta$and [OIII]λ5007, and H to observe H$\alpha$and [NII]λ6584) were performed between November 2013 and January 2014. SFRs, masses and O/H metallicities were measured as described in Maier *et al.* (2014).

In Fig. 1 we compare the MZR of galaxies at $1.3 < z < 1.7$ with the relation of SDSS. Stellar masses are converted to or computed with a Salpeter (1955) IMF, O/Hs are computed or converted to the Kewley & Dopita (2002) calibration. The Zahid *et al.* (2014) N2-based O/Hs were converted to the Kewley & Dopita (2002) calibration and then systematically corrected to 0.4 dex lower metallicities, in agreement with recent results of Newman *et al.* (2014) who found that the MZR at $z > 1$ determined using the N2-method might be $2 - 3$ times too high in terms of metallicity. We also overplot (as a magenta line) the Lilly *et al.* (2013) prediction of a non-evolving Z(M,SFR), the case represented by the dashed lines in Fig. 7 of Lilly *et al.* (2013). The $z \sim 1.4$ observational data shown in Fig. 1 are in agreement with this FMR prediction.

References

Kewley, L. J. & Dopita, M. A. 2002, *ApJS*, 142, 35
Lilly, S. J., Le Brun, V., Maier, C., *et al.* 2009, *ApJS*, 184, 218
Lilly, S. J., Carollo, C. M., Pipino, A., *et al.* 2013, *ApJ*, 772, 119
Maier, C., Lilly, S. J., Carollo, C. M., *et al.* 2006, *ApJ*, 639, 858
Maier, C., Lilly, S. J, Ziegler, B. L., *et al.* 2014, *ApJ*, 792, 3
Mannucci, F., Cresci, G., Maiolino, R., *et al.* 2010, *MNRAS*, 408, 2115
Newman, S. F., Buschkamp, P., Genzel, R., *et al.* 2014, *ApJ*, 781, 21
Pettini, P. & Pagel, B. E. J.. 2004, *MNRAS*, 348, 59
Salpeter, E. E. 1955, *ApJ*, 121, 161
Tremonti, C. A., Heckman, T. M., Kauffmann, G., *et al.* 2004, *ApJ*, 613, 898
Zahid, H. J., Kashino, D., Silverman, J. D., *et al.* 2014, *ApJ*, 792, 75

Galaxies in 3D across the Universe
Proceedings IAU Symposium No. 309, 2014
B. L. Ziegler, F. Combes, H. Dannerbauer, M. Verdugo, eds.

© International Astronomical Union 2015
doi:10.1017/S1743921314009879

Investigating the evolution of merger remnants from the formation of gas disks

Junko Ueda[1,2,3] Daisuke Iono[1,4], Min S. Yun[5], Alison F. Crocker[6],
Desika Narayanan[7], Shinya Komugi[1], Daniel Espada[1,8], Bunyo
Hatsukade[1], Hiroyuki Kaneko[9], Yuichi Matsuda[1], Yoichi Tamura[10],
David J. Wilner[3], Ryohei Kawabe[1,2,4] and Hsi-An Pan[4,11,12]

[1] National Astronomical Observatory of Japan, 2-21-1 Osawa, Mitaka,Tokyo, 181-8588, Japan
email: junko.ueda@nao.ac.jp
[2] Department of Astronomy, School of Science, The University of Tokyo,
7-3-1 Hongo, Bunkyo-ku, Tokyo 133-0033, Japan
[3] Harvard-Smithsonian Center for Astrophysics, 60 Garden Street, Cambridge, MA 02138, USA
[4] Department of Astronomical Science, The Graduate University for Advanced Studie
(SOKENDAI), 2-21-1 Osawa, Mitaka, Tokyo 181-8588, Japan
[5] Department of Astronomy, University of Massachusetts, Amherst, MA 01003, USA
[6] Ritter Astrophysical Research Center, University of Toledo, Toledo, OH 43606, USA
[7] Department of Astronomy, Haverford College, 370 Lancaster Ave, Haverford, PA 19041, USA
[8] Joint ALMA Observatory, Alonso de Córdova 3107, Vitacura, Santiago 763-0355, Chile
[9] Graduate School of Pure and Applied Sciences, University of Tsukuba,
1-1-1 Tennodai, Tsukuba, Ibaraki 305-8577, Japan
[10] IoA, The University of Tokyo, 2-21-1 Osawa, Mitaka,Tokyo, 181-0015, Japan
[11] Nobeyama Radio Observatory, 462-2 Minamimaki, Minamisaku, Nagano 384-1305, Japan
[12] Dept. of Physics, Faculty of Science, Hokkaido Univ., Kita-ku, Sapporo 060-0810, Japan

Abstract. Our new compilation of interferometric CO data suggests that nuclear and extended molecular gas disks are common in the final stages of mergers. Comparing the sizes of the molecular gas disk and gas mass fractions to early-type and late-type galaxies, about half of the sample show similar properties to early-type galaxies, which have compact gas disks and low gas mass fractions. We also find that sources with extended gas disks and large gas mass fractions may become disk-dominated galaxies.

Keywords. galaxies: evolution - galaxies: formation - galaxies: interactions - galaxies: ISM - galaxies: kinematics and dynamics - radio lines: galaxies

---

Major mergers of two disk galaxies are widely believed to provide a way to form a spheroid-dominated early-type galaxy (ETG; e.g., Barnes & Hernquist 1992). Contrary to the classical scenario, recent simulations with more realistic gas physics have shown that some mergers will reform extended gas disks and evolve into disk-dominated late-type galaxies (LTGs; e.g., Springel & Hernquist 2005). In order to check this scenario, we investigate interferometric CO maps of 37 optically-selected merger remnants, which are taken from the merger remnant sample (Rothberg & Joseph 2004). New maps were obtained toward 27/37 sources with ALMA, CARMA, and SMA, and the CO detection rate is 74% (20/27). We find that 80% (24/30) of the sources with robust CO detections show kinematical signatures of the molecular gas disk in their velocity fields (Figure 1).

We compare the properties of molecular gas and stellar components in the merger remnants with ETGs and LTGs to investigate the evolution of merger remnants. The majority of the merger remnants shows a compact molecular gas disk relative to the stellar component. Unless the disks grow significantly, for example from the return of ejected molecular gas or tidal HI gas, the sources will likely evolve into ETGs. We tentatively

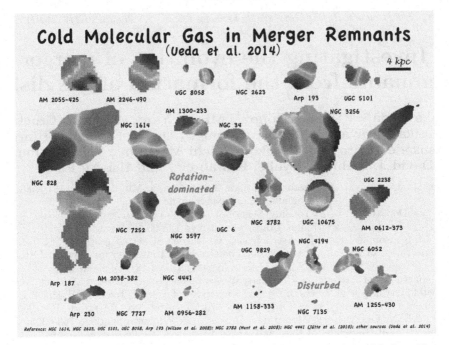

Figure 1. The CO velocity fields of 30 merger remnants (References of the CO data: Hunt *et al.* 2008; Wilson *et al.* 2008; Jütte *et al.* 2010; Ueda *et al.* 2014). The CO data products are publicly released on the website (http://alma-intweb.mtk.nao.ac.jp/~jueda/data/mr/data.html).

suggest that sources with extended gas disks and large gas mass fractions may become disk-dominated LTGs, if there are no further mechanisms to transport the molecular gas toward the central region thereby decreasing the disk size.

Acknowledgements

We thank L. K. Hunt, S. García-Burillo, E. Juette, B. Rothberg, and C. D. Wilson for kindly providing their published data. This paper has made use of the following ALMA data: ADS/JAO.ALMA#2011.0.00099.S and ALMA Science Verification data: ADS/JAO.ALMA#2011.0.00002.SV. ALMA is a partnership of ESO (representing its member states), NSF (USA) and NINS (Japan), together with NRC (Canada) and NSC and ASIAA (Taiwan), in cooperation with the Republic of Chile. The Joint ALMA Observatory is operated by ESO, AUI/NRAO and NAOJ.

This research was supported, in part, by a grant from the Hayakawa Satio Fund awarded by the Astronomical Society of Japan. J. U. is financially supported by a Research Fellowship from the Japan Society for the Promotion of Science for Young Scientists.

References

Barnes, J. E. & Hernquist, L. 1992, *ARA&A*, 30, 705
Hunt, L. K., Combes, F., García-Burillo, S., *et al.* 2008, *A&A*, 482, 133
Jütte, E., Aalto, S., & Hüttemeister, S. 2010, *A&A*, 509, A19
Rothberg, B. & Joseph, R. D. 2004, *AJ*, 128, 2098
Springel, V. & Hernquist, L. 2005, *ApJ*, 622, L9
Ueda, J., Iono, D., Yun, M. S., *et al.* 2014, *ApJS*, 214, 1
Wilson, C. D., Petitpas, G. R., Iono, D., *et al.* 2008, *ApJS*, 178, 189

Galaxies in 3D across the Universe
Proceedings IAU Symposium No. 309, 2014
B. L. Ziegler, F. Combes, H. Dannerbauer, M. Verdugo, eds.

© International Astronomical Union 2015
doi:10.1017/S1743921314009880

Molecular gas content in typical L^* galaxies at $z \sim 1.5 - 3$

Miroslava Dessauges-Zavadsky[1], Michel Zamojski[1], Daniel Schaerer[1],
Françoise Combes[2], Eiichi Egami[3], A. Mark Swinbank[4],
Johan Richard[5], Panos Sklias[1], Tim D. Rawle[6], Jean-Paul Kneib[7],
Frédéric Boone[8] and Andrew Blain[9]

[1] Observatoire de Genève, Université de Genève,
51 Ch. des Maillettes, 1290 Versoix, Switzerland
email: miroslava.dessauges@unige.ch
[2] Observatoire de Paris, LERMA, 61 Avenue de l'Observatoire, 75014 Paris, France
[3] Steward Observatory, University of Arizona,
933 North Cherry Avenue, Tucson, AZ 85721, USA
[4] Institute for Computational Cosmology, Durham University,
South Road, Durham DH1 3LE, UK
[5] CRAL, Observatoire de Lyon, Université Lyon 1,
9 Avenue Ch. André, 69561 Saint Genis Laval Cedex, France
[6] ESAC, ESA, PO Box 78, Villanueva de la Canada, 28691 Madrid, Spain
[7] Laboratoire d'Astrophysique, Ecole Polytechnique Fédérale de Lausanne (EPFL),
Observatoire de Sauverny, 1290 Versoix, Switzerland
[8] CNRS, IRAP, 14 Avenue E. Belin, 31400 Toulouse, France
[9] Department of Physics & Astronomy, University of Leicester,
University Road, Leicester LE1 7RH, UK

Abstract. To extend the molecular gas measurements to typical L^* star-forming galaxies (SFGs) at $z \sim 1.5 - 3$, we have observed CO emission for five strongly-lensed galaxies selected from the *Herschel* Lensing Survey. The combined sample of our L^* SFGs with CO-detected SFGs at $z > 1$ from the literature shows a large spread in star formation efficiency (SFE). We find that this spread in SFE is due to variations of several physical parameters, primarily the specific star formation rate, but also stellar mass and redshift. An increase of the molecular gas fraction ($f_{\rm gas}$) is observed from $z \sim 0.2$ to $z \sim 1.2$, followed by a quasi non-evolution toward higher redshifts, as found in earlier studies. We provide the first measure of $f_{\rm gas}$ of $z > 1$ SFGs at the low-stellar mass end between $10^{9.4} < M_*/{\rm M}_\odot < 10^{9.9}$, which shows a clear $f_{\rm gas}$ upturn.

Keywords. gravitational lensing: strong - galaxies: high-redshift - ISM: molecules

1. Introduction

Improvements in the sensitivity of the IRAM Plateau de Bure Interferometer (PdBI) have made it possible to start getting a census of the molecular gas content in star-forming galaxies near the peak of the cosmic star formation activity. However, the sample of CO-detected objects at $z = 1 - 3$ is still small and mostly confined to the high star formation rate and high stellar mass end of main-sequence SFGs (Daddi *et al.* 2010; Tacconi *et al.* 2010, 2013; Saintonge *et al.* 2013). We could extend the dynamical range of star formation rates and stellar masses of SFGs with observationally constrained molecular gas contents below SFR < 40 M$_\odot$ yr^{-1} and $M_* < 2.5 \times 10^{10}$ M$_\odot$ by observing five strongly-lensed SFGs with the PdBI and 30 m telescope, selected from the *Herschel* Lensing Survey (Egami *et al.* 2010). We hence build up a sample of L^* galaxies at $z \sim 1.5 - 3$ with CO measurements (Dessauges-Zavadsky *et al.* 2014) to which we add the well-known strongly-lensed MS 1512-cB58 and Cosmic Eye (Baker *et al.* 2004; Coppin *et al.* 2007).

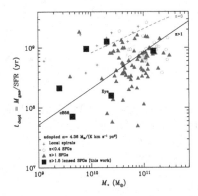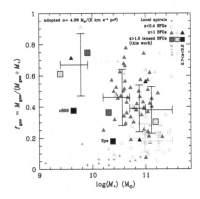

Figure 1. *Left.* Trend for a molecular gas depletion timescale increase with stellar mass. *Right.* Stellar mass dependence of the molecular gas fraction, an upturn is observed at the low-M_* end.

2. L^* galaxies in the context of galaxies with CO measurements

The combined sample of L^* SFGs with CO-detected SFGs at $z > 1$ from the literature shows a large spread in SFE with a dispersion of 0.33 dex, such that SFE now extend well beyond the low values of local spirals and overlap the distribution of $z > 1$ sub-mm galaxies. What drives this large spread in SFE or equivalently in molecular gas depletion timescale ? We find that it is due to variations of primarily the specific star formation rate, but also stellar mass and redshift. Correlations of the SFE with the offset from the main-sequence and the compactness of the starburst are less clear. The increase of the molecular gas depletion timescale with M_* now revealed by the low stellar mass SFGs at $z > 1$ (Fig. 1), and also observed at $z = 0$, is opposed to the constant molecular gas depletion timescale generally admitted and refutes the linear Kennicutt-Schmidt relation.

3. Molecular gas fraction of high-redshift galaxies

Various physical processes at play in the evolution of galaxies (e.g., accretion, star formation, and feedback) have direct impact on the molecular gas fraction. Thus, a solid way to test galaxy evolution models is to confront their predictions with the f_{gas} behaviour. We observe an increase of f_{gas} from $z \sim 0.2$ to $z \sim 1.2$, followed by a quasi non-evolution toward higher redshifts, in contrast with the expected steady redshift increase of f_{gas}. At each redshift the gas fraction shows a large dispersion, due to the dependence of f_{gas} on M_*, producing a gradient of increasing f_{gas} with decreasing M_*. We provide the first measure of $\langle f_{gas} \rangle = 0.67 \pm 0.20$ of $z > 1$ SFGs at the low-M_* end $10^{9.4} < M_*/M_\odot < 10^{9.9}$, which shows a clear f_{gas} upturn (Fig. 1). This upturn, predicted by models, has a strength varying with the outflow/feedback/wind which are taking place in the galaxy.

References

Baker A. J., *et al.* 2004, *ApJ*, 604, 125
Coppin K. E. K., *et al.* 2007, *ApJ*, 665, 936
Daddi E., *et al.* 2010, *ApJ*, 713, 686
Dessauges-Zavadsky M., *et al.* 2014, *A&A*, submitted [arXiv:1408.0816]
Egami E., *et al.* 2010, *A&A*, 518, L12
Saintonge A., *et al.* 2013, *ApJ*, 778, 2
Tacconi L. J., *et al.* 2010, *Nature*, 463, 781
Tacconi L. J., *et al.* 2013, *ApJ*, 768, 74

Galaxies in 3D across the Universe
Proceedings IAU Symposium No. 309, 2014
B. L. Ziegler, F. Combes, H. Dannerbauer, M. Verdugo, eds.

© International Astronomical Union 2015
doi:10.1017/S1743921314009892

Jansky VLA S-band view of Hα emitters (HAEs) associated with a protocluster 4C23.56 at z = 2.5

Minju Lee[1,2], Kenta Suzuki[3], Kotaro Kohno[3], Yoichi Tamura[3],
Daisuke Iono[2], Bunyo Hatsukade[2], Kouichiro Nakanishi[2],
Ichi Tanaka[2], Tadayuki Kodama[2], Kenichi Tadaki[2], Soh Ikarashi[2,4],
Junko Ueda[2], Hideki Umehata[3], Toshiki Saito[1,2] and Ryohei Kawabe[2]

[1] Department of Astronomy, The University of Tokyo,
7-3-1 Hongo, Bunkyo-ku, Tokyo 133-0033, Japan
[2] National Observatory of Japan, 2-21-1 Osawa,
Mitaka, Tokyo 181-0015, Japan
email: minju.lee@nao.ac.jp
[3] Institute of Astronomy, The University of Tokyo,
2-21-1 Osawa, Mitaka, Tokyo, 181-0015, Japan
[4] European Southern Observatory, Karl-Schwarzschild-Str. 2,
D-85748 Garching, Germany

Abstract. We present recent results on Karl Jansky Very Large Array (JVLA) deep S-band (2-4 GHz) observation towards a protocluster 4C23.56 at redshift z ∼ 2.5. The protocluster 4C23.56 is known to have a significant over density (∼ 5 times) of star-burst galaxies selected to be Hα line-bright by a Subaru narrow band imaging. Now we have found 25 HAEs associated with the protocluster. These starburst HAEs are likely to become massive ellipticals at z = 0 in a cluster. Various other galaxy populations also reside in this field and the fact makes the field very unique as a tool to understand galaxy formation in a over dense region. Subsequent deep 1100-μm continuum surveys by the ASTE 10-m dish have discovered that several submillimeter bright galaxies (SMGs) coincide with HAEs, suggesting HAEs undergoing dusty starbursts. As star formation rates (SFRs) of HAEs might have been underestimated, we use radio being resistant to dust extinction. We investigate the correlation between $SFR_{1.4GHz}$ and $SFR_{H\alpha}$ for radio index $\alpha = 0.8$ to see if the correlation holds for the sources and to check the number of dusty star forming galaxies. Our final results will allow us to evaluate quantitatively how the galaxy formation channel may be different under the condition of over-densities.

Keywords. galaxies: active – galaxies: clusters: general –surveys – galaxies: evolution – galaxies: formation – galaxies: high-redshift – radio continuum: galaxies – galaxies: starburst

1. Introduction

Clusters of galaxies, being the most massive structures bounded with gravity in the Universe, provide us an exclusive window to investigate the galaxy formation and evolution in the over-dense region. The progenitors of clusters, protoclusters, are unsettled systems having over-densities than the fields, which is becoming present-day clusters, perhaps by merging and intense star formation, although only a handful to them are identified up to now. A protocluster 4C23.56 is at redshift z = 2.48, where a group of HAEs are found from the Subaru narrow band (NB) survey (MAHALO; MApping HAlpha and Lines of Oxygen with Subaru). The total number of the detected HAEs is 25 up to now (Tanaka *et al.*, in prep; Tanaka *et al.* (2011)). The protocluster is very unique in that it has various galaxy populations in addition to HAEs, e.g. distant red galaxies (DRGs; Kajisawa *et al.* (2006)), extremely red objects (EROs; Knopp & Chambers (1997)), mid-infrared sources (Galametz *et al.* (2012), Mayo *et al.* (2012)) and that there are multi-wavelength ancillary data sets available. Sub-mm single dish telescope ASTE/AzTEC revealed that four of SMGs were overlapped with some HAEs (Suzuki *et al.* in prep; Zeballos *et al.* in prep). However, we could barely pin down the counterparts of SMGs at other wavelengths due to the coarse resolution with ASTE compared to other optical/NIR surveys as well as their redshifts. The overlap of SMGs suggests that dusty star bursts may be onset in HAEs making even the Hα emissions be diminished, leading the underestimation of SFRs. So, we targeted a protocluster associated to the radio galaxy 4C23.56 with Karl Jansky Very Large Array (JVLA) at S-band (2-4 GHz).

2. Results and conclusions

We have reached the r.m.s. level to detect HAEs having SFR > 100 - 400 $M_\odot yr^{-1}$ (3σ) and, as a result, the detection number of JVLA counterparts was seven out of 25. We have also applied stacking analysis to reveal the averaged properties of undetected sources but no the features detected at 3σ. The figure 1 shows the correlation between 1.4 GHz radio continuum and Hα emission which was derived from JVLA 3 GHz (this work) and Subaru NB observations where we assumed radio spectral index $\alpha = 0.8$ to convert 3 GHz information to 1.4 GHz luminosity. We speculate five HAEs among the rest are undergoing dusty starbursts, of which SFRs have been underestimated. Otherwise, these would represent AGN. Adding the Spitzer/MIPS 24μm data and PdBI CO(5-4) observation, one HAE is likely to be undergoing a merger with a heavily obscured starburst. Differentiating the (dusty) starburst and the AGN for the

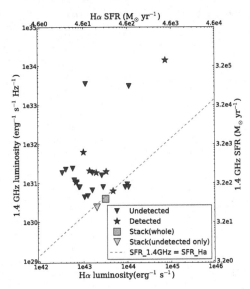

Figure 1. Radio and Hα emission corrleation of a protocluster associated with 4C23.56. The detection rate of HAE counterparts in radio is ~30%. The radio galaxy 4C23.56 itself is on top right of this figure and well-above the correlation representing a radio excess by AGN. All JVLA detections of HAE counterparts, except for one is above the correlation which suggests dusty starbursts of HAEs and/or AGN.

rest will be remained as future works by doing SED fittings and comparison with X-ray observation (Lee *et al.* in prep). Further follow-up observations with ALMA/JVLA will provide more information on the physical properties on each galaxy in this protocluster and illuminate the processes how massive galaxy formation and evolution is affected by its surrounding environment.

References

Galametz, A., Stern, D., De Breuck, C., *et al.* 2012, *ApJ*, 749, 169
Knopp, G. P. & Chambers, K. C. 1997, *ApJS*, 109, 367
Kajisawa, M., Kodama, T., Tanaka, I., Yamada, T., & Bower, R. 2006, *MNRAS*, 371, 577
Mayo, J. H., Vernet, J., De Breuck, C., *et al.* 2012, *A&A*, 539, A33
Tanaka, I., Breuck, C. D., Kurk, J. D., *et al.* 2011, *PASJ*, 63, 415

Galaxies in 3D across the Universe
Proceedings IAU Symposium No. 309, 2014
B. L. Ziegler, F. Combes, H. Dannerbauer, M. Verdugo, eds.

© International Astronomical Union 2015
doi:10.1017/S1743921314009909

Is there a dependence in metallicity evolution on galaxy structures?

U. Kuchner[1], C. Maier[1], B. Ziegler[1], M. Verdugo[1], O. Czoske[1], P. Rosati[2], I. Balestra[3], A. Mercurio[4], M. Nonino[3] and CLASH-, CLASH-VLT-team

[1] University of Vienna, Department of Astrophysics, Türkenschanzstr. 17, 1180 Vienna, Austria
email: ulrike.kuchner@univie.ac.at
[2] University of Ferrara, Department of Physics and Earth Science, [3] INAF, Osservatorio Astronomico di Trieste, [4] INAF, Osservatorio Astronomico di Napoli

Abstract. We investigate the environmental dependence of the mass-metallicty (MZ) relation and its connection to galaxy stellar structures and morphologies. In our studies, we analyze galaxies in massive clusters at z ∼ 0.4 from the CLASH (HST) and CLASH-VLT surveys and measure their gas metallicities, star-formation rates, stellar structures and morphologies. We establish the MZ relation for 90 cluster and 40 field galaxies finding a shift of ∼−0.3 dex in comparison to the local trends seen in SDSS for the majority of galaxies with $logM < 10.5$. We do not find significant differences of the distribution of 4 distinct morphological types that we introduce by our classification scheme (smooth, disc-like, peculiar, compact). Some variations between cluster and field galaxies in the MZ relation are visible at the high mass end. However, obvious trends for cluster specific interactions (enhancements or quenching of SFRs) are missing. In particular, galaxies with peculiar stellar structures that hold signs for galaxy interactions, are distributed in a similar way as disc-like galaxies - in SFRs, masses and O/H abundances. We further show that our sample falls around an extrapolation of the star-forming main sequence (the SFR-$M_\star$ relation) at this redshift, indicating that emission-line selected samples do not have preferentially high star-formation rates (SFRs). However, we find that half of the high mass cluster members ($M_\star > 10^{10} M_\odot$) lie below the main sequence which corresponds to the higher mass objects that reach solar abundances in the MZ diagram.

1. Determination of morphologies and gas abundances

Various studies of the relation between stellar masses and metallicities of star-forming galaxies have outlined their importance for the understanding of cosmic evolution of star-formation rates and galaxy properties. In our investigation we tie in further information of galaxy structures and morphologies across different environments at z ∼ 0.4. Utilizing HST data from the CLASH program (Postman *et al.* 2012), we analyze galaxies in the field of 3 very massive clusters MACS1206, MACS0329 and MACS0416, complemented by Subaru BVRIz imaging which maps the large scale environment of the clusters. We use CLASH-VLT VIMOS spectra over the whole area of 30×30 sq.arcmin to measure accurate star-formation rates and oxygen abundances (gas metallicities) with 5 strong emission lines [OII], Hbeta, [OIII], Halpha and [NII] (see Maier *et al.* 2014). To quantify the stellar structure, we fit the 2D light distribution of galaxies in Subaru images applying the newly developed Galapagos2 algorithms (Häußler *et al.*, 2013) of the MegaMorph project we calculate robust structural parameters over all available Subaru bands simultaneously down to masses of $logM$ ∼8.5 of 90 cluster and 40 field galaxies. To analyze any morphological dependences, we classify the galaxies according to four distinct types. For this, we expand a decision tree (e.g. Nantais *et al.* 2013) to our needs and include the new capabilities of MegaMorph. In short, (B/T)-measurements from 1- and 2-component

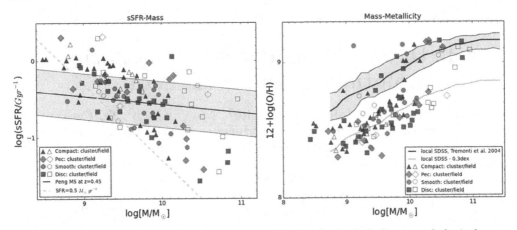

Figure 1. Left: sSFR-mass relation of 130 sample galaxies including morphological information. Right: MZ relation of the same compared to local SDSS measurements.

light profile fittings with varying sets of bulge and disc profiles, as well as color information and visual classifications lead to 4 distinct classes: compact, peculiar, disc-like and objects with smooth, regular structures.

2. Environmental and morphological dependance of the MZ relation

Gas metallicities of galaxies depend on their stellar mass: studies of the mass-metallicity (MZ) relation across a range of redshifts show that galaxies with higher masses form a plateau of higher (solar-like) abundances, whereas towards lower masses a negative slope of the MZ relation is seen. Both the zero point and slope evolve with redshift (Mannucci *et al.* 2009, Maier *et al.* 2014). However, our project is the first to investigate environmental dependencies in very massive clusters ($> 10^{15} M_\odot$). For z~0.4 galaxies, we measure an average evolution of the MZ relation of a factor of 2 but see a shift of the mass threshold below which one finds low metallicities to lower masses for cluster environments. We also see that at z~0.4, galaxies of different types are similarly distributed in the MZ plot, however subtle differences are visible for higher stellar masses and bigger sizes. This region is predominantly occupied by the disc-like star-forming population of our sample, both in the field and cluster. Differences due to environmental effects are more obvious: On average, while the most massive field galaxies of our sample follow the local relation reported by Tremonti *et al.* 2004, a few cluster members exceed solar abundances. This is independent of their morphological type. These are the same galaxies we see below the Main Sequence in the sSFR-Mass diagram. Since peculiar galaxies are found at both high and low metallicities, we can speculate that the time scales for any environmental effects on star formation activity must be short.

References

Maier, C., *et al.* 2014, *ApJ*, 792, 3
Häußler, B., *et al.* 2013, *MNRAS*, 430, 330-369
Mannucci, F., *et al.* 2009, *MNRAS*, 398, 1915
Nantais, J., *et al.* 2013, *A&A*, 555, A5
Postman, M., *et al.* 2012, *ApJS*, 199, 25P
Simard, L., *et al.* 2011, *ApJS*, 196, 11S
Tremonti, C., *et al.* 2004, *ApJ*, 613, 898–913

Galaxies in 3D across the Universe
Proceedings IAU Symposium No. 309, 2014
B. L. Ziegler, F. Combes, H. Dannerbauer, M. Verdugo, eds.

© International Astronomical Union 2015
doi:10.1017/S1743921314009910

Early-type galaxy formation: understanding the role of the environment

Ricardo Demarco[1], Alessandro Rettura[2], Chris Lidman[3], Julie Nantais[1], Yara Jaffe[1] and Piero Rosati[4]

[1] Department of Astronomy, Universidad de Concepción, Casilla 160-C, Concepción, Chile
email: rdemarco@astro-udec.cl
[2] Astronomy Department, California Institute of Technology, 1200 East California Blvd, Pasadena CA 91125, U.S.A.
[3] Australian Astronomical Observatory, 105 Delhi Rd, North Ryde NSW 2113, Australia
[4] Department of Physics and Earth Science, University of Ferrara, Via G.Saragat, 21 - 44122 Ferrara, Italy

Abstract. One of the most characteristic features of galaxy clusters is the so-called "red sequence" (RS) of early-type galaxies. Since these galaxies are, in general, devoid of gas and dust, their red colors are mainly a consequence of their passive nature. However, the physical mechanisms responsible for quenching their star formation, thus originating the RS, are poorly understood. Environmental effects should play a significant role in the formation of the RS by transforming the observed galaxy properties from late to early-type ones. In this respect, we have initiated a KMOS program aimed at studying the kinematical structure of cluster galaxies at $0.8 < z < 1.7$ in an effort to disentangle the physical mechanisms responsible for cluster galaxy evolution and the formation of the RS.

Keywords. galaxies: early-type - galaxies: formation - galaxies: clusters of galaxies

1. Introduction

Observations indicate that the RS begins to form at $z \sim 2.5$ and that by $z \sim 1.4$ is already well established, becoming more apparent in clusters than in the field (Tanaka *et al.* 2005; Kodama *et al.* 2007; Zirm *et al.* 2008; Hilton *et al.* 2009). However, how this actually happens is still unclear. Are early-type galaxies formed directly as such in proto-cluster or field environments through self-quenching mechanisms regulated by their halo mass (nature)? Or do early-type galaxies get into the RS when located in cluster outskirts or group environments in in-falling matter filaments (nurture)? When and where do the effects of ram-pressure stripping and galaxy-galaxy interactions (e.g. harassment, mergers, low-speed encounters; Treu *et al.* 2003) become the dominant way in which galaxies become passive and enter the RS in color-magnitude space? All these processes, whatever their relative importance, should affect the kinematics, spatial distribution and content of the (molecular and ionized) gas component of galaxies. The available observational evidence shows that it is possible, when comparing with numerical models, to establish the most likely cause of any observed kinematical and structural anomaly (Puech *et al.* 2009; Hammer *et al.* 2009). However, these analyses have been done mostly for field galaxies, with systematic and comprehensive studies of the kinematics and detailed structure of baryons in cluster galaxies being still scarce.

Cluster Name	Redshift	Selection	Photometry	Spectroscopy
RXJ0152-13	0.84	X-Ray	ACS/WFC-3, Hawk-I/ISAAC, Spitzer Chandra, VLA, Herschel	yes (FORS2)
XMMJ1229+01	0.98	X-Ray	ACS/WFC-3, Hawk-I/ISAAC	yes
RDCSJ1252-29	1.24	X-Ray	ACS/WFC-3, Hawk-I/ISAAC Spitzer, Chandra	yes (FORS2)
XMMJ2235-25	1.39	X-Ray	ACS/WFC-3, Hawk-I/ISAAC	yes (FORS2)
XMMJ2215-17	1.45	X-Ray	ACS/WFC-3, Hawk-I/ISAAC	yes
ClGJ0218-05	1.62	IR	ACS/WFC-3, Hawk-I/ISAAC	yes

Table 1. Sample of galaxy clusters in our KMOS program. A total of 44 hours of KMOS time has been assigned to this program. Observations are in progress.

2. Datasets and the KMOS observations

Here we aim at exploiting the capabilities of KMOS on the ESO VLT to map out the kinematical structure, spatial distribution and overall properties of the ionized gas components of cluster galaxies at $0.8 < z < 1.7$ (see Table 1). We will search for the presence (or lack) of signatures of the different environmental (galaxy-ICM and galaxy-galaxy) interactions to better understand the physical mechanisms responsible for stellar quenching and the formation of the RS in clusters of galaxies.

The galaxy clusters selected for this program already have extensive spectroscopic and multi-wavelength photometric datasets available (see Table 1). Observational evidence indicates that the most likely place where galaxy transformation and star formation quenching in clusters can occur is their intermediate density regions (e.g. Nantais *et al.* 2013). Star forming galaxies in these regions may show the effects of environmental interactions on their gas content and, therefore, provide key piece of information to understand how galaxies can speed up gas consumption and/or induce gas loss. We focus on the rest-frame Hα emission as a more robust star formation rate (SFR) indicator. We hence select galaxies with [OII] emission to be observed in the YJ- and H-band gratings of KMOS depending on the redshifts. Observations, still in progress, are designed to detect Hα down to a SFR limit of ~ 5 $M_\odot$ yr^{-1} with a signal-to-noise ratio of 10 Å^{-1}, which translates into observing times between 5 and 10 hours (including overheads) depending on the target. The information obtained on the kinematical structure of the hot (ionized) gas component of galaxies will be integrated with that from the cold (molecular) gas structure provided by ALMA (future proposals). All this information will be used to achieve a comprehensive view of cluster galaxy quenching and the formation of the RS.

Acknowledgements
This work is supported by FONDECYT No.1130528.

References
Hammer, *et al.* 2009, *A&A*, 496, 381
Hilton, *et al.* 2009, *ApJ*, 697, 436
Kodama, *et al.* 2007, *MNRAS*, 377, 1717
Nantais, *et al.* 2013, *A&A*, 556, 112
Puech, *et al.* 2009, *A&A*, 493, 899
Tanaka, *et al.* 2005, *MNRAS*, 362, 268
Treu, *et al.* 2003, *ApJ*, 591, 53
Zirm, *et al.* 2008, *ApJ*, 680, 224

Galaxies in 3D across the Universe
Proceedings IAU Symposium No. 309, 2014
B. L. Ziegler, F. Combes, H. Dannerbauer, M. Verdugo, eds.
© International Astronomical Union 2015
doi:10.1017/S1743921314009922

KMOS Clusters and VIRIAL GTO Surveys

D. Wilman[1,2,*], R. Bender[1,2], R. L. Davies[3], J. T. Mendel[2], J. Chan[2],
A. Beifiori[1,2], R. Houghton[3], R. Saglia[2], N. Förster Schreiber[2],
S. Wuyts[2], P. van Dokkum[4], M. Cappellari[3], J. Stott[5], R. Smith[6],
M. Fossati[1,2], S. Kulkarni[2], S. Seitz[1], M. Fabricius[2], R. Sharples[5],
G. Brammer[7], E. Nelson[4], I. Momcheva[4], M. Wegner[1] and I. Lewis[3]

[1] Universitäts-Sternwarte München, Scheinerstr. 1, D-81679 Munich, Germany
[2] Max-Planck-Institut für extraterrestrische Physik, Giessenbachstr. 1, D-85741 Garching,
Germany
[3] Physics Department, University of Oxford, Denys Wilkinson Building, Keble Road, Oxford
OX1 3RH, UK
[4] Department of Astronomy, Yale University, New Haven, CT 06511, USA
[5] Institute for Computational Cosmology, Durham University, South Road, Durham DH1 3LE,
UK
[6] Department of Physics, University of Durham, South Road, Durham DH1 3LE, UK
[7] Space Telescope Science Institute, 3700 San Martin Drive, Baltimore, MD 21218, USA
*email: dwilman@mpe.mpg.de

Abstract. We present the KMOS (**K**-band **M**ulti-**O**bject **S**pectrograph) Cluster and VIRIAL
(**VLT IR IFU A**bsorption **L**ine) Guaranteed Time Observation (GTO) programs. KMOS provides 24 arms each feeding an integral field unit (14×14 spaxels of 0.2″ pixels) for IZ, YJ, H and
K band near infrared (NIR) medium resolution spectroscopy (R ∼ 3500). Targets are selected
from a 7.2′ diameter patrol field. Ultra-deep spectroscopy of ∼ 80 early-type cluster galaxies
(∼ 20hr on source) and ∼ 200 (∼ 10hr on source) early-type field galaxies at $1 < z < 2$ will dramatically improve the situation at $z > 1$ for which measurements of stellar velocity dispersions
and absorption indices are limited to a few, often relatively young passively evolving galaxies
(e.g. Bezanson 2013). In ESO Periods P92 and P93, 15 nights worth of data has been collected
for KMOS-Clusters and 6 nights for VIRIAL: this will be supplemented with more data in
upcoming semesters. All galaxies have multiband HST imaging including existing or upcoming
WFC3 IR imaging, providing stellar mass maps and sizes. Combined with our dispersion measurements, this will allow us to examine the fundamental plane and the dynamical mass of a
large sample of $z > 1$ galaxies for the first time, for both cluster and field galaxies.

Acknowledgements

D. Wilman and M. Fossati acknowledge support from the Deutsche Forschungsgemein-
schaft (DFG) via Project ID 387/1-1. We thank the conference organizers for the opportunity to present this work.

References

Jee, M. J., *et al.*, 2011, *ApJ*, 737, 59
Mullis, C. R., *et al.*, 2005, *ApJ*, 623, L85
Kurk, J., *et al.*, 2009, *A&A*, 504, 331
Bezanson, R., *et al.*, 2013, *ApJ*, 779, L21

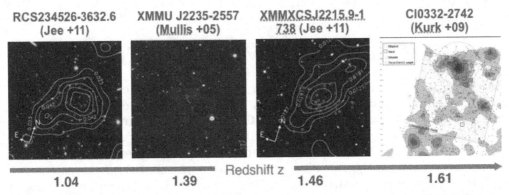

Figure 1. High redshift galaxy clusters observed with KMOS as part of the KMOS Clusters Guaranteed Time program. Well studied clusters at $z > 1$ with known galaxy redshifts, deep photometry and HST imaging are selected for KMOS observations. The panels show existing cluster HST imaging with Xray emission contours (first and third panels) and galaxy density contours (right panel) and are taken from the papers labelled above. Very deep ($\gtrsim 20$ hour) KMOS YJ and IZ band observations of $\gtrsim 20$ galaxies per cluster provides a large sample of $\gtrsim 80$ high redshift cluster galaxies. Most KMOS arms are allocated to brighter red sequence galaxies, with remaining arms allocated to fainter red galaxies or bright blue galaxies with spectroscopic redshifts consistent with the cluster redshift. Clusters which lack high resolution HST imaging in the NIR are soon to be observed with WFC3. All clusters will then have multiband high resolution imaging to complement the KMOS spectroscopy. We have already collected 15 nights worth of data in P92-93 and an optimal data reduction is being developed which maximises S/N and allows us to measure the stellar velocity dispersion for many objects. Fainter objects and indices can be measured for stacked data where the S/N of individual objects is not sufficient.

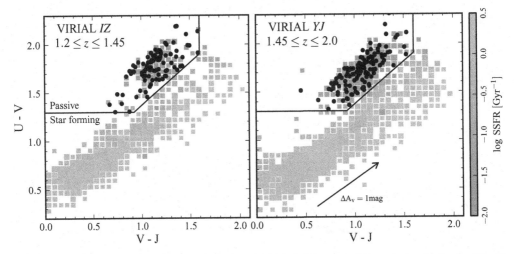

Figure 2. UVJ selection of galaxies for the VIRIAL field survey of passive galaxies in $U - V$ versus $V - J$ space. We select galaxies from the upper left of the diagram (black points correspond to galaxies selected for KMOS observations, bounded by the solid lines). This method separates passively evolving stellar populations from star forming populations including those with high dust extinction, as the dust vector is parallel to the selection threshold (see arrow in right panel). We select galaxies in 2 different redshift ranges, corresponding to the KMOS IZ ($1.2 \leqslant z \leqslant 1.45$, left panel) and YJ ($1.45 \leqslant z \leqslant 2.0$, right panel) grisms. Targets are selected from deep fields with deep CANDELS multiband HST photometry and 3dHST grism spectroscopic redshifts. Observations of $\gtrsim 10$hours for $\gtrsim 200$ galaxies complement those of the KMOS clusters program in terms of environment, depth, selection, redshift range and spectral coverage.

Galaxies in 3D across the Universe
Proceedings IAU Symposium No. 309, 2014
B. L. Ziegler, F. Combes, H. Dannerbauer, M. Verdugo, eds.
© International Astronomical Union 2015
doi:10.1017/S1743921314009934

Spatially resolved Lyman-alpha emission from a virtual dwarf galaxy

Anne Verhamme

Department of Astronomy, University of Geneva,
Chemin des Maillettes 51, CH-1290 Versoix, Switzerland
email: anne.verhamme@unige.ch

Abstract. In the context of the first light of MUSE, Integral Field Unit (IFU) spectrograph of second generation installed recently at VLT, we compute mock IFU Lyman-alpha (lyα) observations of a virtual dwarf galaxy, to help understanding and interpreting forthcoming observations. This study is an extension of the work carried out in Verhamme *et al.* (2012), where we studied the spatially integrated lyα properties of a dwarf galaxy. With the same data, we now investigate the spatial variations of lyα spectra.

Keywords. Radiative transfer – Galaxies: ISM – ISM: structure, kinematics and dynamics

1. Context

The study of spatially resolved lyα properties of galaxies is a *recent* domain: at low redshift, the spatial extend of lyα emission around starforming galaxies is known for less than 10 years (Hayes *et al.* 2005; Östlin *et al.* 2009; Hayes *et al.* 2013). The existence of similar lyα halos around high-redshift galaxies has been reported recently (Steidel *et al.* 2011; Matsuda *et al.* 2012; Momose *et al.* 2014). Newly available IFU facilities will allow to study not only their spatial extent and surface brightness, but also their gas content and kinematics, in lyα. Because of strong radiation transfer effects, models are needed to interpret the forthcoming data.

2. Results

The spatially resolved lyα profiles from our simulated dwarf galaxy vary from pixel to pixel. When the galaxy is seen face-on, the lyα spectrum emerging from the central pixel, directly from the star-forming region, has a usual asymetric, redshifted shape, whereas the surrounding pixels show spectra emerging in the diffuse Lyα halo with non-zero lyα flux on the blue side of the line. When the galaxy is seen edge-on, the 3 central pixels, aligned with the galactic disk, show an unusual double-peaked lyα profile asymetric towards the blue, tracing gas infall; whereas the other pixels show more common double-peaked spectra but with a prominent red peak, tracing gas outflow. However, the number of photons emerging per pixel on the edge-on view is small, the histograms are then noisy and less reliable than face-on.

The lyα spectra emerging from different pixels do not trace the velocity of the gas at the location of escape but they rather seem to trace the bulk velocity field along the path from the lyα sources to the last scattering. Further analyse has to be done to confirm this trend. A strong limitation of this study is the idealised halo in which the galaxy forms, which may be far from producing a realistic cosmological environment.

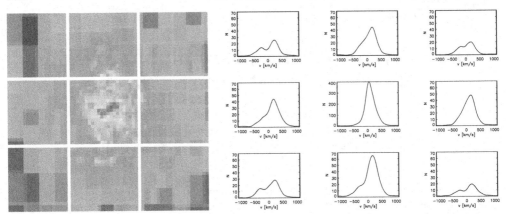

Figure 1. Lyα spectra from 9 pixels of the central 3 × 3 kpc of a dwarf galaxy seen face-on, smoothed at the spectral resolution of MUSE (R=4000). The image on the right shows the correponding ly*alpha* image.

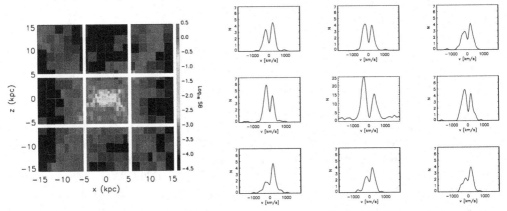

Figure 2. Lyα spectra from a dwarf galaxy seen edge-on, smoothed at MUSE spectral resolution. The image on the right shows the correponding ly*alpha* image divided in 9 pixels from which we extracted the spectra.

References

Verhamme, A., Dubois, Y., Blaizot, J., Garel, T., Bacon, R., Devriendt, J., Guiderdoni, B., & Slyz, A., 2012, *A&A*, 546, 111

Hayes, M., Östlin, G., Mas-Hesse, J. M., Kunth, D., Leitherer, C., & Petrosian, A., 2005, *A&A*, 438, 71

Östlin, G., Hayes, M., Kunth, D., Mas-Hesse, J. M., Leitherer, C., Petrosian, A., & Atek H., 2009, *AJ*, 138, 923

Hayes, M., Östlin, G., Schaerer D., Verhamme, A., Mas-Hesse, J. M., Adamo, A., Atek, H., Cannon, J., Duval, F., Guaita, L., Herenz, C., Kunth, D., Laursen, P., Melinder, J., Orlitova, I., Oti-Floranes, H., & Sandberg, A., 2013, *ApJ*, 765, 27

Steidel, C., Bogoslavljevic, M., Shapley, A., Kollmeier, J., Reddy, N., Erb, D., & Pettini, M., 2011, *ApJ*, 736, 160

Matsuda, Y., Yamada, T., Hayashino, T., Yamauchi, R., Nakamura, Y., Morimoto, N., Ouchi, M., Umemura, M., & Mori, M., 2012, *MNRAS*, 425, 878

Momose, R., Ouchi, M., Nakajima, K., Ono, Y., Shibuya, T., Shimasaku, K., Yuma, S., Mori, M., & Umemura, M., 2014, *MNRAS*, 442, 110

Galaxies in 3D across the Universe
Proceedings IAU Symposium No. 309, 2014
B. L. Ziegler, F. Combes, H. Dannerbauer, M. Verdugo, eds.
© International Astronomical Union 2015
doi:10.1017/S1743921314009946

The relation between the gas, dust and total mass in edge-on spiral galaxies

Flor Allaert

Sterrenkundig Observatorium, Universiteit Gent
Krijgslaan 281, B-9000 Gent, Belgium
email: flor.allaert@ugent.be

Abstract. Each component of a galaxy plays its own unique role in regulating the galaxy's evolution. In order to understand how galaxies form and evolve, it is therefore crucial to study the distribution and properties of each of the various components, and the links between them, both radially and vertically. The latter is only possible in edge-on systems. We present the HEROES project, which aims to investigate the 3D structure of the interstellar gas, dust, stars and dark matter in a sample of 7 massive early-type spiral galaxies based on a multi-wavelength data set including optical, NIR, FIR and radio data.

Keywords. galaxies: spiral - galaxies: evolution - galaxies: formation

1. The HEROES project

The HEROES (HERschel Observations of Edge-on Spirals, Verstappen *et al.* 2013) project was set up as part of three projects to study the distribution of the various components in a total of 32 edge-on galaxies of different Hubble types. The edge-on geometry is crucial for two reasons. Firstly, the dust in edge-on galaxies is visible both in emission (in the FIR/submm) and in absorption (in the optical/NIR). Fitting detailed radiative transfer models to our data using the SKIRT (Baes *et al.* 2003, 2011) and FitSKIRT (De Geyter *et al.* 2013) codes, we determine the dust and stellar distribution in unprecedented detail. Secondly, edge-on galaxies are the only systems that allow to study both the radial and the vertical structure. We will combine our radiative transfer models with kinematical models of the gas content (Allaert *et al.*, in prep.) and dynamical rotation curve decompositions to obtain a full 3D picture of our sample galaxies.

 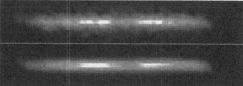

Figure 1. Left: V-band image (top) and FitSKIRT model (bottom) of the HEROES galaxy IC 2531. Right: Observed (top) and model (bottom) total intensity map of the atomic gas in the same galaxy. Note that the physical scales in the left and right images are different.

References

Baes, *et al.* 2003, *MNRAS*, 343, 1081
Baes, *et al.* 2011, *ApJS*, 196, 22
De Geyter, *et al.* 2013, *A&A*, 550, A74
Verstappen, *et al.* 2013, AA&A, 556, 54

Galaxies in 3D across the Universe
Proceedings IAU Symposium No. 309, 2014   © International Astronomical Union 2015
B. L. Ziegler, F. Combes, H. Dannerbauer, M. Verdugo, eds.   doi:10.1017/S1743921314009958

Mirage simulations of the massiv sample

P. Amram[1], V. Perret[1,2], B. Epinat[1], F. Bournaud[3], T. Contini[4],
C. Divoy[4], B. Garilli[5], M. Kissler-Patig[6], O. Le Fevre[1],
C. Lopez-Sanjuan[7], J. Moultaka[4], L. Pairo[5], E. Perez-Montero[8],
J. Queyrel[4], L. Tasca[1], L. Tresse[1] and D. Vergani[9]

[1]LAM, Marseille (F), [2]U. Zurich, (Ch), [3]CEA, SAp, AIM, Saclay(F), [4]IRAP, Toulouse (F),
[5]INAF-IASFBO, Bologna (I), [6]GEMINI, Hilo (USA), [7]FEFCA, Teruel(E), [8]IAC, Granada(E),
[9]INAF-IASF, Milano (I). Email: `philippe.amram@lam.fr`

The MIRAGE sample (Merging & isolated high-redshift AMR galaxies; Perret 2014, PhD dissertation; Perret *et al.* 2014, AA 562, 1) has been built in order to understand the contribution of the merger processes to the mass assembly in the MASSIV (Mass Assembly Survey with SINFONI in VVDS, Contini *et al.* 2012, AA 539, 91) sample. It consists of a sample of idealized simulations based on the RAMSES code; the initial conditions were designed to reproduce the physical properties of the most gas-rich young galaxies. The sample is composed of 20 simulations of mergers exploring the initial parameters of mass and orientation of the disks with a spatial resolution reaching 7 parsecs.

The MIRAGE simulations show (*i*) an absence of star formation bursts in mergers of fragmented and turbulent disks, suggesting a saturation mechanism; (*ii*) that the gas rich clump merging mechanism is able to control the bulge mass growth, to erode the central profile of the dark matter halo and to drive massive gas outflows into the disk plane; (*iii*) whatever the orbital configuration (prograde or retrograde) and whatever the mass ratio between the disks ($1/3/6 - logM = 10.6/10.2/9.8$), a new disk of gas is reconstructed quickly after the merger. This is due to the high gas fraction at the merging event (60%).

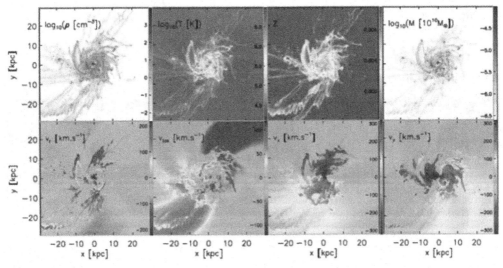

Figure 1. We carry out a comparative study of the MASSIV kinematical data to a set of more than 4000 pseudo-observations at $z = 1.7$ built from simulations of the MIRAGE sample to determine the ability to detect galaxy merger signatures under the observational conditions of the SINFONI instrument. Upper panels from left to right: density, temperature, metallicity and mass maps. Bottom panels from left to right: radial cylindric velocity in the plane of the disk, line-of-sight velocity and the two last ones are projections on the two planes perpendicular to the sky plane (Perret 2014, PhD dissertation).

Galaxies in 3D across the Universe
Proceedings IAU Symposium No. 309, 2014
B. L. Ziegler, F. Combes, H. Dannerbauer, M. Verdugo, eds.

© International Astronomical Union 2015
doi:10.1017/S174392131400996X

The Structure and Kinematics of the ISM in Simulated Star-forming Galaxies

Junichi Baba[1], Kana Morokuma-Matsui[2] and Takayuki R. Saitoh[1]

[1] Earth-Life Science Institute (ELSI), Tokyo Institute of Technology, 2-12-1 Ookayama,
Meguro, Tokyo 152–8551 email: babajn@elsi.jp; babajn2000@gmail.com
[2] Nobeyama Radio Observatory, National Astronomical Observatory of Japan, Minamimaki,
Minamisaku, Nagano, 384-1305, Japan

Abstract. We performed 3D N-body/hydrodynamic simulations of local and high-z ($z \sim 2$) star-forming galaxies (SFGs) to investigate the structure and kinematic properties of the different ISM phases (i.e., ionized, atomic, and molecular gases). We took into account, for the first time, a fully non-equilibrium radiative cooling by solving the non-equilibrium chemistries of atoms (H, He, C and O) and molecules (H_2 and CO), as well as radiative heating, star formation, heating by H_{II} region and supernova explosions.

1. Background and Results

The thermal and kinematic properties of the ISM are crucially important to understand the evolution of galaxies. Although there are some theoretical studies on the ISM properties in local and high-z SFGs, there is no theoretical study on the kinematics of "the different ISM phases". In order to investigate the detailed properties of the ISM in local/high-z SFGs, it is required that the non-equilibrium chemical network and associated cooling/heating processes should be taken into account in galactic-scale hydrodynamic simulations.

We used the N-body/SPH code ASURA-2 (Saitoh & Makino 2010) and solved the hydrodynamics and stellar dynamics of both MW-like ($f_{gas} = 10\%$) and High-z ($f_{gas} = 50\%$ suggested by e.g., Tacconi et al. 2010) SFG models by taken into account "non-equilibrium chemistry" and heating/cooling processes simultaneously. The High-z SFG model succeeded to reproduce the clumpy stellar distribution, as well as giant H_{II} regions, observed in high-z SFGs (e.g., Forster Schreiber et al. 2009). This model predicts that molecular gas associates with massive star clumps and distributes in a thin layer ($\sim$5 pc) as same as MW galaxy. The velocity dispersion of the ionized gas correlates with the star formation rates, but it is much smaller than observed values in high-z SFGs (e.g., Forster Schreiber et al. 2009). This suggests that other energy sources are required to drive the large velocity dispersion of the ionized gas in high-z SFGs.

References

Förster Schreiber, N. M., et al., 2009, *ApJ*, 706, 1364
Saitoh, T. R. & Makino, J., 2010, *PASJ*, 62, 301
Tacconi, L. J., et al., 2010, *Nature*, 463, 781-784

Galaxies in 3D across the Universe
Proceedings IAU Symposium No. 309, 2014
B. L. Ziegler, F. Combes, H. Dannerbauer, M. Verdugo, eds.

© International Astronomical Union 2015
doi:10.1017/S1743921314009971

How to Simulate Galactic Outflows?

Paramita Barai

INAF - Osservatorio Astronomico di Trieste, Via G. B. Tiepolo 11, I-34143 Trieste, Italy
email: pbarai@oats.inaf.it

Abstract. A challenge in cosmological simulations is to formulate a physical model of star-formation (SF) and supernovae (SN) feedback which produces galactic outflows like that widely observed. In several models an outflow velocity (v_{out}) and mass loading factor (η) are input to the sub-resolution recipe. We present results from our MUPPI model, which *uses local properties* of gas, and is able to develop galactic outflows whose properties *correlate with global galaxy properties*, consistent with observations; demonstrating a significant improvement in such work.

Keywords. Cosmology: theory – Methods: Numerical – Galaxies: formation

1. Introduction

We explore our novel *sub-resolution* model for SF and SN feedback, **MUPPI** [**MU**lti-**P**hase **P**article **I**ntegrator] (Murante *et al.* 2010, 2014), embedded in the GADGET-3 code, which uses only local properties of the gas. Unlike popular adaptation in the literature, our model has no input expression of v_{out} and η for SN feedback.

2. Conclusions

We measure properties of SN-driven galactic outflows over redshifts $z = 1 - 5$ in cosmological hydrodynamical simulations (Barai *et al.* 2013, 2014). The MUPPI model generates v_{out} and mass outflow rate ($\dot{M}_{out}$) exhibiting positive correlations with galaxy mass and with the star formation rate (SFR) (Fig. 1). However, most of the relations present a large scatter. The mass-loading is between $0.2 - 10$, with an average $\eta \sim 1$.

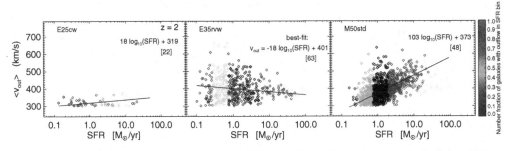

Figure 1. Outflow velocity as a function of SFR of galaxies in three cosmological simulation runs at $z = 2$. The black line is the best-fit relation between v_{out} and $\log_{10}$ (SFR).

Acknowledgements

I acknowledge support from the ERC Starting Grant "cosmoIGM".

References

Barai P. *et al.*, 2013, *MNRAS*, 430, 3213

Barai P., Monaco P., Murante G., Ragagnin A., & Viel M., 2014, submitted to *MNRAS*

Murante G., Monaco P., Giovalli M., Borgani S., & Diaferio A., 2010, *MNRAS*, 405, 1491

Murante G., Monaco P., Borgani S., Tornatore L., Dolag K., & Goz D., 2014, submitted to *MNRAS*

Galaxies in 3D across the Universe
Proceedings IAU Symposium No. 309, 2014
B. L. Ziegler, F. Combes, H. Dannerbauer, M. Verdugo, eds.

© International Astronomical Union 2015
doi:10.1017/S1743921314009983

A low-luminosity type-1 QSO sample: Insight from integral-field spectroscopy

Gerold Busch[1], Semir Smajić[1,2], Lydia Moser[1], Mónica Valencia-S.[1], Jens Zuther[1], Julia Scharwächter[3] and Andreas Eckart[1,2]

[1] I. Physikalisches Institut, Universität zu Köln, Cologne, Germany
email: busch@ph1.uni-koeln.de
[2] Max Planck Institut für Radioastronomie, Bonn, Germany
[3] LERMA, Observatoire de Paris, Paris, France

Abstract. The properties of the host galaxies of quasi-stellar objects (QSOs) are essential for the understanding of the suspected coevolution of central supermassive black holes (BHs) and their host galaxies. In Busch *et al.* (2014), we present a study of 20 low-luminosity type-1 QSOs (LLQSO; Busch *et al.* 2012) that have been selected from the Hamburg/ESO survey for bright UV-excess QSOs ($z \leqslant 0.06$). Performing careful decomposition of deep near-infrared J,H,K images, we found that the observed sources do not follow published $M_{BH} - L_{bulge}$ relations for inactive galaxies, supporting similar results found for type-1 AGN in the optical. This can be explained by overluminous bulges with very young stellar populations or undermassive black holes that are observed in a phase of growth. We use 3d-spectroscopy in the optical and near-infrared as a powerful tool to analyze gas and stellar kinematics, determine gas masses, star formation rate, trace underlying stellar continuum, etc. (e.g., Smajić *et al.* 2014) and thereby constrain possible evolution scenarios. The results will be interpreted in the context of galaxy evolution and particularly the still unknown role of the AGN in this process. Here, we show first results for HE 1029-1831.

Keywords. galaxies: active - galaxies: Seyfert - galaxies: evolution

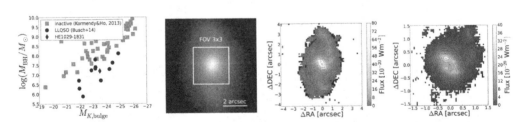

Figure 1. *From left to right:* (1) In the $M_{BH} - L_{bulge}$ relation, the observed LLQSOs clearly lie offset compared to inactive galaxies collected by Kormendy & Ho (2013) (see Busch *et al.* 2014). (2) K-band image of the LLQSO HE 1029-1831 from Fischer *et al.* (2006). (3) Paα map of HE 1029-1831 with a FOV of 8×8 arcsec^2. (4) the same with FOV of 3×3 arcsec^2. The atomic gas is distributed in spiral arms and a clumpy circum-nuclear ring, indicating star formation. This motivates further investigation of the star formation properties in the centers of LLQSOs.

References

Busch G., Zuther J., Valencia-S. M., *et al.*, 2014, *A&A*, 561, A140
Busch G., Zuther J., Valencia-S. M., *et al.*, 2012, PoS(Seyfert2012)060, arXiv:1307.1049
Fischer S., Iserlohe C., Zuther J., *et al.*, 2006, *A&A*, 452, 827
Kormendy J. & Ho L., 2013, *ARA&A*, 51, 511
Smajić S., Moser L., Eckart A., *et al.*, 2014, *A&A*, 567, A119

Galaxies in 3D across the Universe
Proceedings IAU Symposium No. 309, 2014
B. L. Ziegler, F. Combes, H. Dannerbauer, M. Verdugo, eds.

© International Astronomical Union 2015
doi:10.1017/S1743921314009995

Fabry-Perot spectroscopy: a powerful method for detecting superbubbles in galaxy discs

A. Camps-Fariña[1,2], J. Beckman[1,2,3], J. Zaragoza-Cardiel[1,2], J. Font[1,2] and K. Fathi[4]

[1] Instituto de Astrofísica de Canarias, La Laguna, Spain
[2] Departamento de Astrofísica, Universidad de la Laguna, Spain
[3] Consejo Superior de Investigaciones Científicas, Spain
[4] Department of Astronomy, Stockholm University, Stockholm, Sweden

Abstract. We present a new method for the detection and characterization of large scale expansion in galaxy discs based on $H\alpha$ Fabry-Perot spectroscopy, taking advantage of the high spatial and velocity resolution of our instrument (GHαFaS). The method analyses multi-peaked emission line profiles to find expansion along the line of sight on a pixel-by-pixel basis. At this stage we have centred our attention on the large scale structures of expansive gas which show a coherent gradient of velocities from their centres as a result of both bubble shape and projection effect. The results show a wide range of expansion velocities in these superbubbles, ranging from 30-150 km/s, with the expected trend of finding the higher velocities in the more violent areas of the galaxies. We have applied the technique to the Antennae and M83, obtaining spectacular results, and used these to investigate to what extent kinematically derived ages can be found and used to characterize the ages of their massive star clusters.

Keywords. galaxies: superbubbles, galaxies: kinematics, galaxies: evolution

---

GHαFaS is a Fabry-Perot integral field spectrometer, producing a 3.4'x3.4' data cube with a seeing-limited spatial resolution, and velocity resolution of $\sim$7-8 km/s. This allows us to separate different kinematic components of the line profile (normally $H\alpha$) with the aim of detecting expansion.

In order to make this a systematic, galaxy-wide analysis, we need to automate the process of finding and fitting multiple components to the $H\alpha$ profile. To do this we take advantage of the properties of the second derivative of a Gaussian, which tell us that for a combination of Gaussians each contribution lies between two zeros of the second derivative (Goshtasby & O'Neill 1994). The position of the zeros is also related to the offset and width of each Gaussian, so we can estimate the parameters before fitting, to ensure an efficient solution.

We perform this on the entire cube, first detecting the number of components that can be separated in each profile and then fitting them to produce their final parameters. We then search for the expansion signature: two smaller peaks situated symmetrically at each side of the main, brightest peak. We produce new maps by assigning to each pixel the mean expansion velocity (distance from the main peak) and the mean intensity of the secondary peaks. In the poster **Kinematic properties of superbubbles in the Antennae, M83 and Arp 270** we present results obtained aplying this method.

References

Goshtasby, A. & O'Neill W., 1994, *Graph. Models Image Process.*, 56, 281

Galaxies in 3D across the Universe
Proceedings IAU Symposium No. 309, 2014
B. L. Ziegler, F. Combes, H. Dannerbauer, M. Verdugo, eds.

© International Astronomical Union 2015
doi:10.1017/S174392131401000X

Kinematic properties of superbubbles in the Antennae, M83 and Arp 270

A. Camps-Fariña[1,2], J. Beckman[1,2,3], J. Zaragoza-Cardiel[1,2], J. Font[1,2], K. Fathi[4], P. Fabian Velázquez[5] and A. Rodríguez-González[5]

[1] Instituto de Astrofísica de Canarias, La Laguna, Spain
[2] Departamento de Astrofísica, Universidad de la Laguna, Spain
[3] Consejo Superior de Investigaciones Científicas, Spain
[4] Department of Astronomy, Stockholm University, Stockholm, Sweden
[5] Instituto de Ciencias Nucleares, Universidad Nacional de México, D.F, Mexico

Abstract. Superbubbles and large scale expansion in galaxies are important indicators of activity in galaxies: they are formed in starbursts and around active nuclei. Superbubbles can be used to give information about the star-forming region which produced them. We present in-depth results of our study of kinematically detected superbubbles using a method based on Fabry-Perot spectroscopy, which allows us to map regions of expansion across the entire disk of a galaxy. Three objects have been selected for this poster based on the interest of the results they show: two interacting galaxies, the Antennae and Arp270, at different stages of galaxy interaction, and the more isolated galaxy M83. We present the kinematic expansion maps, as well as a census of detected superbubbles and a dynamical study of their properties.

Keywords. galaxies: superbubbles, galaxies: kinematics, galaxies: evolution

We use the method described in the poster **Fabry-Perot spectroscopy: a powerful method for detecting superbubbles in galaxy discs** to derive expansion maps of three galaxies and detect on these the presence of superbubbles. We find 10 superbubbles in the Antennae galaxies (Arp 244), 8 in M83 and 3 in the central zones of the interacting pair Arp 270. We measure their properties: radius, expansion velocity, Hα luminosity, density, mass, kinetic energy, and an approximate age. To assess the validity of our results we use hydrodynamic simulations to reproduce a well-characterised superbubble from its measured properties. The simulation successfully reproduces the bubble, in Fig. 1 we present the expansion map around the bubble and the equivalent measurement from the simulated bubble.

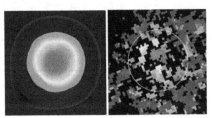

Figure 1. Comparison between a detected superbubble in the Antennae (right) and the equivalent projection of a hydrodynamical simulation done using its measured parameters. Colors represent observed line of sight expansion velocity from maximum in red through lower green-blue.

References

Raga, A. C., Navarro-González, R., & Villagrán-Muniz, M., 2000, *RMxAA*, 36, 67

Galaxies in 3D across the Universe
Proceedings IAU Symposium No. 309, 2014
B. L. Ziegler, F. Combes, H. Dannerbauer, M. Verdugo, eds.

© International Astronomical Union 2015
doi:10.1017/S1743921314010011

Star Formation in the Local Universe from the CALIFA sample: calibration and contribution of disks to the SFR density

Cristina Catalán-Torrecilla[1], Armando Gil de Paz[1], África Castillo-Morales[1], Jorge Iglesias-Páramo[2], Sebastián F. Sánchez[2,3] and CALIFA Collaboration

[1]Departamento de Astrofísica y CC. de la Atmósfera, Universidad Complutense de Madrid, E-28040, Madrid, Spain, email: ccatalan@ucm.es
[2]Instituto de Astrofísica de Andalucía-CSIC, Glorieta de la Astronomía, 18008 Granada, Spain
[3]Instituto de Astronomía,Universidad Nacional Autonóma de México, 04510, México

Abstract. The study of the star formation rate (SFR) is crucial for understanding the birth and evolution of the galaxies (Kennicutt 1998), with this aim in mind, we make use of a well-characterized sample of 380 nearby galaxies from the CALIFA survey that fill the entire color-magnitude diagram in the Local Universe. The availability of wide-field CALIFA IFS ensures a proper determination of the underlying stellar continuum and, consequently, of the extiction-corrected Hα luminosity. We compare our integrated Hα-based SFRs with single and hybrids tracers at other wavelengths found in the literature (Calzetti 2013). Then, we provide a new set of single-band and hybrid calibrators anchored to the extinction-corrected Hα luminosities. In the case of the hybrid calibrators we determine the best fitting a_{IR} coefficients for different combinations of observed (UV or Hα) and dust-reprocessed (22μm or TIR) SFR contributions (where SFR $\propto$ L$_{obs}$ + $a_{IR} \times$ L[IR]). This analysis allow us to provide, for the first time, a set of hybrid calibrations for different morphological types and masses. These are particularly useful in case that the sample to be analyzed shows a different bias in terms of morphology or, more commonly, luminosity or stellar mass. We also study the dependence of this coefficient with color and ionized-gas attenuation. The distributions of a_{IR} values are quite wide in all cases. We found that not single physical property can by itself explain the variation found in a_{IR}.

Finally, we explore the spatial distribution of the SFR by measuring the contribution of disks to the total SFR in the Local Universe. Our preliminary spatially-resolved analysis shows that the disk to total (disk + spheroidal component) SFR ratio is on average $\sim$ 88%. The use of the 2D spectroscopic data is critical to properly determine the Hα luminosity function and SFR density in the Local Universe per galaxy components, the ultimate goal of this project.

Keywords. galaxies: spiral - galaxies: evolution - galaxies: star formation

Acknowledgements

This study makes uses of the data provided by the CALIFA survey. CALIFA would like to thank the IAA-CSIC and MPIA-MPG. C. C.-T. thanks the support of the Spanish *Ministerio de Educación, Cultura y Deporte* by means of the FPU fellowship program.

References

Kennicutt, R. C. 1998, *ARA&A*, 36, 189
Calzetti, D. 2013, *Star Formation Rate Indicators*, ed. J. Falcón-Barroso & J. H. Knapen, 419

Galaxies in 3D across the Universe
Proceedings IAU Symposium No. 309, 2014
B. L. Ziegler, F. Combes, H. Dannerbauer, M. Verdugo, eds.

© International Astronomical Union 2015
doi:10.1017/S1743921314010023

3D-structure of galactic disks

Ekaterina Chudakova

Sternberg Astronomical Institute, Lomonosov Moscow State University,
email: `artenik@gmail.com`

Abstract. We suggest a new photometric method to estimate thicknesses of galactic disks which is working for the galaxies seen under arbitrary inclination. We have applied our method to 45 galaxies from the open SDSS archive and to 57 galaxies from the Spitzer Survey S4G).

Keywords. galaxies: elliptical and lenticular, cD - galaxies: spiral - galaxies: structure

---

We propose a new photometric method to estimate thickness of a galactic disk seen under arbitrary inclination. We stress two profits of our method: it provides an individual estimate for every galaxy and for the galaxies with orientation different from edge-on it allows to confront its vertical structure to the radial scale. We apply our method to two samples of disk galaxies which radial brightness profiles are already known: to the r-band SDSS images of the galaxies which structures are described by Erwin *et al.* (2008) and Gutiérrez *et al.* (2011) and to the 3.6– and 4.5–μm images collected by the Spitzer telescope during the survey S4G. The radial structures of the S4G sample are described by Muñoz-Mateos *et al.* (2013). In the first sample we have obtained very different thickness distributions for the purely exponential-profile galactic disks (Type I) and for the inner disks of antitruncated stellar disks (Type III) (Chudakova & Sil'chenko 2014).

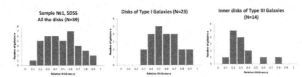

Figure 1. Sample no.1: Distributions of the relative thicknesses q.

In the second sample for most galaxies we have two images: in the 3.6– and in the 4.5–μm bands. Good agreement between relative thicknesses found in the 3.6– and 4.5–μm bands shows us the stability of our method.

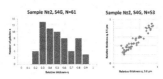

Figure 2. The q distribution and the correlation between the values found in two bands.

References

Erwin, P., Pohlen, M., & Beckman, J. E. 2008, *AJ*, 135, 20
Gutiérrez, L., Erwin, P., Aladro, R., & Beckman, J. E. 2011, *AJ*, 142, 145
Muñoz-Mateos, J. C., Sheth, K., Gil de Paz, A., *et al.* 2013, *ApJ*, 771, aid 59
Chudakova, E. M. & Sil'chenko, O. K., 2014, *Astron. Rep.*, 58, 281

Galaxies in 3D across the Universe
Proceedings IAU Symposium No. 309, 2014
B. L. Ziegler, F. Combes, H. Dannerbauer, M. Verdugo, eds.

© International Astronomical Union 2015
doi:10.1017/S1743921314010035

Metallicity gradients in the Milky Way thick disk as relic of a primordial distribution

Anna Curir[1], Ana Laura Serra[1,2], Mario G. Lattanzi[1], Alessandro Spagna[1], Paola Re Fiorentin[1] and Antonaldo Diaferio[2]

[1]INAF-Osservatorio Astrofisico di Torino
10025 Pino Torinese (Torino), Italy email: curir@oato.inaf.it
[2]Dipartimento di Fisica e Istituto Nazionale di Fisica Nucleare, 10125 Torino, Italy

Abstract. We examine the evolution induced by secular processes of an initial, cosmologically motivated, radial chemical distribution, in a barred and an unbarred disk.

Keywords. galaxies: evolution - galaxies: formation

The simulations presented in Curir *et al.* (2012) describe two Milky Way (MW)-like exponential disks embedded in a suitable dark halo. The radial chemical function injected in the initial configuration of the N-body disk is justified with an inside-out disk formation model (Spitoni & Matteucci 2011).

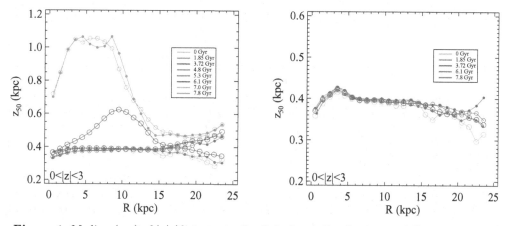

Figure 1. Median (z_{50}) of $|z|$ (distance to the disk plane) distribution at different radii and times for barred (left) and unbarred (right) disks.

We find that this chemical profile does not undergo major transformations after 6 Gyr of dynamical evolution in both disks examined. The final radial $[Fe/H]$ gradients in the solar neighborhood are positive and of the same order as those recently observed in the MW thick disk, in both the barred (Curir *et al.* 2014) and unbarred disks. Therefore the influence of the bar on the evolution of chemical features is negligible. On the other hand, the bar appears to have an important role in provoking a disk thickening and in eliciting a flare in the outer disk (Fig. 1).

References

Curir, A., Lattanzi, M. G., Spagna, A., *et al.* 2012, *A&A*, 545, A133
Curir, A., Serra, A. L., Spagna, A., *et al.* 2014, *ApJ*, 784, L24
Spitoni, E. & Matteucci, F. 2011, *A&A*, 531, A72

Galaxies in 3D across the Universe
Proceedings IAU Symposium No. 309, 2014
B. L. Ziegler, F. Combes, H. Dannerbauer, M. Verdugo, eds.
© International Astronomical Union 2015
doi:10.1017/S1743921314010047

The connection between shape and stellar population in early-type galaxies

Mauro D'Onofrio and the WINGS team

Department of Physics and Astronomy, University of Padova,
Vicolo Osservatorio 3, 35122 Padova, Italy
email: mauro.donofrio@unipd.it

Abstract. We comment on the observed connection between the shape of the photometric profiles and the stellar population content in early-type galaxies.

Keywords. galaxies: elliptical and lenticular, cD - galaxies: evolution - galaxies: formation

1. Link with the Fundamental Plane

Early-type galaxies (ETGs) are virialized stellar systems, so that a relation between their mass and structural properties (radius and velocity dispersion) is expected (the so called Virial Plane, VP). Observations show instead that ETGs share the Fundamental Plane (FP) which appears tilted with respect to the VP. By combining the VP and FP eqs. one can get (see, D'onofrio *et al.* (2013) for details):

$$\log(M/L) - \log(K_V) = (2 - a)\log(\sigma) + b(\langle\mu\rangle_e - M_\odot^\lambda - 21.572) - \log(2\pi G) - c \quad (1.1)$$

where $K_V = 1/k_v k_r$ parametrizes our ignorance of the galaxy structure and dynamics ($\langle V^2\rangle = k_v\sigma^2$ and $\langle R\rangle = k_r R_e$). Eq.1.1 already tell us that a connection between structural parameters and stellar population content must be in place.

The classical interpretation of the FP tilt is based on the hypothesis that $K_V = const$ and $(M/L) \propto M^\alpha$ (see e.g., Cappellari *et al.* 2013). However, D'onofrio *et al.* (2013) have demonstrated that $K_V \propto n$, where n is the Sersic parameter determining the shape of the light profiles of ETGs.

The proof that behind the FP tilt there is the connection between the photometric shape of the light profiles and the stellar population comes from eq.1.1 that can be transformed into a relation between $\log(M/L) - \log(n) - \log(M)$ which has been observed to exist (see, D'onofrio *et al.* 2011).

In conclusion, we claim that the shape of the light profiles and the stellar population in ETGs are connected and this connection is at the origin of the FP tilt. This link between structure and stellar population in ETGS can be explained by the multiple dry merging capture of small galaxies. Evidence is growing in fact that dry merging events produce a change in the Sersic index of galaxies while little effects on total mass and stellar population (M/L) (see e.g., Hiltz *et al.* 2013).

References

Cappellari, M., *et al.* 2013, *MNRAS*, 432, 1709
D'Onofrio, M., *et al.* 2011, *ApJ*, 727, L6
D'Onofrio, M., *et al.* 2013, *MNRAS*, 435, 45
Hiltz, M., *et al.* 2013, *MNRAS*, 429, 2924

Galaxies in 3D across the Universe
Proceedings IAU Symposium No. 309, 2014
B. L. Ziegler, F. Combes, H. Dannerbauer, M. Verdugo, eds.

© International Astronomical Union 2015
doi:10.1017/S1743921314010059

A dust radiative transfer study of the edge-on spiral galaxy NGC 5908

G. De Geyter, M. Baes, P. Camps, J. Fritz and S. Viaene

Sterrenkundig Observatorium, Universiteit Gent, Belgium

Abstract. We present a dust radiative transfer analysis of the edge-on spiral galaxy NGC 5908. In our previous analysis, it was found that the standard assumption of a double-exponential dust distribution resulted in a poor fit. We investigate the possibility of the dust being distributed in one or more rings. The parameters are constrained using FitSKIRT, a code used to automatically determine the best fitting radiative transfer model given a set of observations. We discuss the possible implications of this dust distribution on the predicted spectral energy distribution.

NGC 5908 is an Sb galaxy at a distance 53.4 Mpc and was one of the targets studied in detail in De Geyter *et al.* (2014). The SDSS *g*, *r*, *i* and *z*-band images were used as input for a FitSKIRT (De Geyter *et al.* 2013) radiative transfer model in which the dust was distributed in a standard double-exponential disk. Large scale structures showed up in the residual maps, indicating that this model is a poor representation for the actual distribution of dust in NGC 5908. These features had clear ring-like structures, a feature which is not uncommon in galaxies, e.g. the Sombrero galaxy. We find that a model using two rings, one ring operating as an extended disk with a central cavity and another, narrower, ring describing the overdensity, significantly increases the quality of the fits (see the left-hand panel of Fig. 1).

These models are used as an input for a panchromatic radiative transfer simulation, in which we take both evolved and young stellar populations into account. As can be seen from the right-hand panel of Fig. 1, the model with a double-exponential disk and the two-rings model result in very similar SEDs and both fail to reproduce the observed far-infrared fluxes. Therefore, misclassification of dust rings as dust disk seems to have no significant effect on the predicted far-infrared emission and can not even partially explain the far-infrared deficit found in many dust energy balance studies.

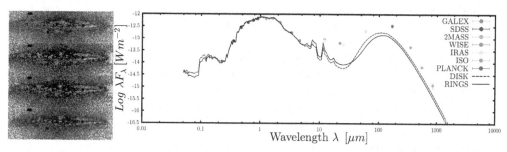

Figure 1. *Left-hand panel:* The residuals between observations and the model using two rings to describe the dust geometry in NGC 5908. *Right-hand panel:* The observed fluxes and SED obtained for the models using a disk and two rings with and without a younger stellar component.

References

De Geyter, G., Baes, M., Fritz, J., & Camps, P. 2013, *A&A*, 550, A74
De Geyter, G., Baes, M., Camps, P., *et al.* 2014, *MNRAS*, 441, 869

Galaxies in 3D across the Universe
Proceedings IAU Symposium No. 309, 2014
B. L. Ziegler, F. Combes, H. Dannerbauer, M. Verdugo, eds.

© International Astronomical Union 2015
doi:10.1017/S1743921314010060

High-resolution, 3D radiative transfer modeling of M51

Ilse De Looze, Jacopo Fritz, Maarten Baes and Sag-2 consortium

Sterrenkundig Observatorium, Universiteit Gent, Krijgslaan 281 S9, B-9000 Gent, Belgium

Abstract. We present a new technique developed to model the radiative transfer (RT) effects in nearby face-on galaxies. The face-on perspective provides insight into the star-forming regions and clumpy structure, imposing the need for high-resolution 3D models to recover the asymmetric stellar and dust geometries observed in galaxies. RT modeling of the continuum emission of stars and its interaction with the embedding dust in a galaxy's interstellar medium enables a self-consistent study of the main dust heating mechanisms in galaxies. The main advantage of RT calculations is the non-local character of dust heating that can be addressed by tracing the propagation of stellar radiation through the dusty galaxy medium.

Keywords. radiative transfer – dust, extinction – galaxies: individual: M 51 galaxies: ISM – infrared: galaxies

1. Model construction

We construct a high-resolution 3D radiative transfer model with the Monte-Carlo code SKIRT (Baes *et al.* 2011) accounting for the absorption, scattering and non-local thermal equilibrium emission of dust in M 51. The 3D distribution of stars is derived from the 2D morphology observed in IRAC 3.6 μm, *GALEX* far-ultraviolet (FUV), Hα and MIPS 24 μm maps, assuming an exponential vertical distribution with a scale height, h_z (based on the observed vertical extent in edge-on galaxies). The dust geometry is constrained through the FUV attenuation. The stellar luminosity, star formation rate and dust mass have been scaled to reproduce the observed stellar spectral energy distribution (SED), FUV attenuation and infrared SED.

2. Conclusions

We demonstrate the capabilities of a new RT modeling technique in the prototype analysis of the grand-design spiral galaxy, M51. The high-resolution 3D model of M51 enables us to study the heating by young (< 100 Myr) and old stars at every infrared (IR) wavelength. Although young stars dominate in mid-IR wavebands, the contribution from old stars becomes non-negligible ($\sim 40\%$) for $\lambda \geqslant 70$ μm, which questions the applicability of IR tracers of the star formation activity in star-forming galaxies such as M51.

Acknowledgements

IDL is a postdoctoral researcher of the FWO-Vlaanderen (Belgium). MB and JF acknowledge the financial support of BELSPO through the PRODEX project "Herschel-PACS Guaranteed Time and Open Time Programs: Science Exploitation" (C90370).

References

Baes, M., Verstappen, J., De Looze, I., *et al.* 2011, *ApJS*, 196, 22

Galaxies in 3D across the Universe
Proceedings IAU Symposium No. 309, 2014
B. L. Ziegler, F. Combes, H. Dannerbauer, M. Verdugo, eds.

© International Astronomical Union 2015
doi:10.1017/S1743921314010072

Decoding 3D Disk Structure and Dynamics Using Doppler Tomography

Arthur D. Eigenbrot and Matthew A. Bershady

Department of Astronomy, University of Wisconsin - Madison,
475 N. Charter St. Madison, WI, 53703, USA
email: `eigenbrot@astro.wisc.edu`

Abstract. We demonstrate a method to measure both rotation curves and 3D ISM structure in edge-on galaxies. Two-dimensional spectral coverage of edge-on galaxies reveals substantial deviations in emission line shapes from a purely gaussian profile that vary with radius and height. Non-gaussianity is quantified using statistical moments to third order. We infer the 3D density distribution by comparing the measured line profiles to synthetic line-of-sight velocity distributions from a suite of three-dimensional galaxy models with different 3D distributions of dust and gas and different rotation curve shapes and amplitudes. We apply this method using multi-position longslit data of nearby edge-on galaxy ESO 435-G25 and find our derived rotation curve matches measured HI rotation from envelope fitting but requires a flared dust disk to accurately describe the radial and vertical trends in the measured statistical moments. Our results are consistent with dynamical expectations for constant pressure support in a disk with exponentially declining surface-density.

Conclusions

Using optical longslit data from SALT we derive a rotation curve consistent HI envelope fitting (Kregel & van der Kruit 2004) despite the effects of extinction and line-of-sight projection. We parametrize rotation by $V(r) = V_{rot} \tanh(r/h_{rot})$ and find derived values of h_{rot} and V_{rot} that are consistent with the results of Andersen & Bershady (2013), implying constant pressure support in an exponentially declining disk surface-density. We acknowledge support from NSF/AST-1009471

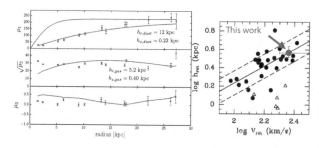

Figure 1. (*left*) The first three statistical moments as a function of radius for data from SALT (points) and the best fitting galaxy model (lines). The model rotation curve is shown as the red curve in the top panel. (*right*) Results from Andersen & Bershady (2013) showing dynamical parameters h_{rot} and V_{rot} for a sample of disk galaxies. ESO 435-G25 is shown as the red point.

References

Andersen, D. R. & Bershady, M. A. 2013, *ApJ*, 768
Andersen, D. R., *et al.* 2006, *ApJS*, 166
Kobulnicky, H. A., *et al.* 2002, *Proceedings of SPIE*, 4841, 1634
Kregel, M. & van der Kruit, P. C. 2004 *MNRAS*, 352, 787-803
Xilouris, E. M., Byun, Y. I., Kylafis, N. D., Paleologou, E. V., & Papamastorakis, J. 1999, *A&A*, 34

Galaxies in 3D across the Universe
Proceedings IAU Symposium No. 309, 2014
B. L. Ziegler, F. Combes, H. Dannerbauer, M. Verdugo, eds.

© International Astronomical Union 2015
doi:10.1017/S1743921314010084

A 3D view of galactic winds in luminous infrared galaxies

P. Martín-Fernández[1], J. Jiménez-Vicente[1,2], A. Zurita[1,2], E. Mediavilla[3,4] and A. Castillo-Morales[5]

[1]Dpto. Física Teórica y del Cosmos, Universidad de Granada, 18071–Granada, Spain
[2]Instituto Carlos I de Física Teórica y Computacional, Granada, Spain
[3]Instituto de Astrofísica de Canarias, 38200–La Laguna, Sta. Cruz de Tenerife, Spain
[4]Dpto. de Astrofísica, Universidad de La Laguna, 38200-La Laguna, Tenerife, Spain
[5]Dpto. de Astrofísica y CC. de la Atmósfera, UCM, 28040–Madrid, Spain
email: pablomartin@ugr.es

Abstract. Galactic winds and outflows are an ubiquitous phenomenon in galaxies with active star formation and/or active nuclei. They constitute the main mechanism for redistributing dust and metals on large scales and are therefore a key ingredient to understand the life cycle of galaxies. Among galaxies, ULIRGs are of particular interest in this context, as they host intense starbursts and are likely to be the dominant star formers at $z > 1$. These objects have been shown to host important winds, but it is not yet known what is the frequency of galactic winds and their properties in galaxies with lower star formation rates (SFR). We are studying galactic winds in a sample of 21 galaxies with different SFRs (including ULIRGs) from observations with the INTEGRAL fiber spectrograph on the 4.2m WHT. In order to be able to address the complex multi–phase nature of the wind phenomenon, we have used the Na I D doublet absorption lines to trace cold gas, and a few emission lines (Hα, [N II] and [S II]) to trace the warmer ionized gas of the wind. The distribution and kinematics of both components in these objects is then analysed. Preliminary results show strong spatial correlation between regions with high non–circular velocities, areas with high star formation activity and regions with two different components in the emission lines. This set of data will help us to characterise the distribution and kinematics of the winds and their relation with the host galaxy type.

Keywords. galaxies: spirals - galaxies: evolution - ISM: jets and outflows

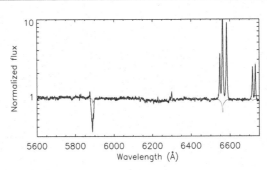

Figure 1. Integrated spectrum of NGC 5394 (solid black) with overlaid fit to the stellar population spectrum (dotted red). In this case, most of the Na I D absorption is produced in the interstellar medium. The subtraction of the synthetic stellar spectra from the total observed spectra yields a spectrum of the nebular component alone for subsequent kinematic analysis.

We acknowledge support from the Spanish "Ministerio de Economía y Competitividad" and "Junta de Andalucía" via grants AYA2011-24728 and FQM-108.

Galaxies in 3D across the Universe
Proceedings IAU Symposium No. 309, 2014
B. L. Ziegler, F. Combes, H. Dannerbauer, M. Verdugo, eds.

© International Astronomical Union 2015
doi:10.1017/S1743921314010096

The formation and build-up of the red-sequence over the past 9 Gyr in VIPERS

Alexander Fritz[1] and the VIPERS Team[2]

[1]INAF - IASF Milano, Via E. Bassini 15, 20133 Milano, Italy
email: `afritz@iasf-milano.inaf.it`
[2]The VIPERS Team: U. Abbas, C. Adami, S. Arnouts, J. Bel, M. Bolzonella, D. Bottini, E. Branchini, A. Burden, A. Cappi, J. Coupon, O. Cucciati, I. Davidzon, G. De Lucia, S. de la Torre, C. Di Porto, P. Franzetti, M. Fumana, B. Garilli, B. R. Granett, L. Guzzo, O. Ilbert, A. Iovino, J. Krywult, V. Le Brun, O. Le Fèvre, D. Maccagni, K. Małek, A. Marchetti, C. Marinoni, F. Marulli, H. J. McCracken, Y. Mellier, L. Moscardini, R. C. Nichol, L. Paioro, J. A. Peacock, W. J. Percival, M. Polletta, A. Pollo, M. Scodeggio, L. A. M. Tasca, R. Tojeiro, D. Vergani, G. Zamorani and A. Zanichelli; Web address: http://vipers.inaf.it

Abstract. We present the Luminosity Function (LF) and Colour-Magnitude Relation (CMR) using ~45000 galaxies drawn from the VIMOS Public Extragalactic Redshift Survey (VIPERS). Using different selection criteria, we define several samples of early-type galaxies and explore their impact on the evolution of the red-sequence (RS) and the effects of dust. Our results suggest a rapid build-up of the RS within a short time scale. We find a rise in the number density of early-type galaxies and a strong evolution in LF and CMR. Massive galaxies exist already 9 Gyr ago and experience an efficient quenching of their star formation at $z = 1$, followed by a passive evolution with only limited merging activity. In contrast, low-mass galaxies indicate a different mass assembly history and cause a slow build-up of the CMR over cosmic time.

Keywords. Galaxy Surveys - Galaxies: evolution - Galaxies: formation - Galaxies: statistics

1. The Evolution of Early-Type Galaxies Since $z = 1.3$

The VIMOS Public Extragalactic Redshift Survey (VIPERS) is mapping out the Universe at $0.5<z<1.2$ using spectroscopic redshifts of 100k galaxies with $I_{AB} < 22.5$ in two sky regions over 24 deg^2. Based on complementary precise UV to IR photometry, this unique dataset provides unprecedented detail on the statistical properties of galaxies and evolution of structure at intermediate redshifts (Guzzo *et al.* 2014; Garilli *et al.* 2014).

We examine the evolution of the Colour-Magnitude Relation (CMR) and Luminosity Function (LF) in VIPERS using ~45000 galaxies between $0.4 \leqslant z \leqslant 1.3$ over 10.32 deg^2 (Fritz *et al.* 2014). Independent from the selection criteria to define early-type galaxies, we find a rise in the number density and a strong evolution in the LF and CMR since $z = 1.3$. Dusty red galaxies show a mild evolution up to $z = 1$ and the effects of dust obscuration become more prominent at $z > 1.3$. Stellar population properties and their modelling favour a rapid build-up of the red-sequence within ~1.5 Gyr. Massive galaxies are already assembled at $z = 1$ and experience a rapid star formation (SF) quenching without significant subsequent dry mergers. Low-massive galaxies indicate a more complex mass assembly with a combination of SF quenching and minor mergers that slowly build-up of the faint- and intermediate-luminosity end of the CMR.

References

Fritz, A., Scodeggio, M., Ilbert, O., *et al.* 2014, *A&A*, 563, A92
Garilli, B., Guzzo, L., Scodeggio, M., *et al.* 2014, *A&A*, 562, A23
Guzzo, L., Scodeggio, M., Garilli, B., *et al.* 2014, *A&A*, 566, A108

Galaxies in 3D across the Universe
Proceedings IAU Symposium No. 309, 2014
B. L. Ziegler, F. Combes, H. Dannerbauer, M. Verdugo, eds.

© International Astronomical Union 2015
doi:10.1017/S1743921314010102

Interacting galaxies in 3D:
Three case studies.

Isaura Fuentes-Carrera[1], Nelli Cárdenas-Martínez[1],
Mónica Sánchez-Cruces[1], Margarita Rosado[2], Philippe Amram[3],
Lorenzo Olguín[4], Jura Borissova[5], Simon Verley[6], Héctor Flores[7],
Lourdes Verdes-Montenegro[8] and Denise Gonçalves[9]

[1]Escuela Superior de Física y Matemáticas, Instituto Politécnico Nacional (ESFM-IPN)
Edificio 9, U. P. Adolfo López Mateos s/n, Col. Zacatenco, 07730 Mexico City, Mexico
email: `isaura@esfm.ipn.mx`
[2]IA-UNAM, Mexico   [3]LAM, France   [4]UNISON, Mexico   [5]Universidad de Valparaíso, Chile
[6]Universidad de Granada, Spain   [7]Observatoire de Paris-Meudon, France   [8]IAA, Spain
[9]Observatorio do Valongo, Rio de Janeiro, Brazil

Abstract. We present 3D observations of three interacting galaxies in order derive their extended kinematics and to trace events such as violent star-formation (SF), mass-transfer, structure perturbation and the presence of energetic sources induced by the interacting processes.

Keywords. galaxies: interacting - galaxies: spiral - galaxies: evolution

1. Introduction

External perturbations and induced star formation. We present scanning Fabry-Perot interferometric observations of a pair of starburst galaxies. We derived the velocity field of each galaxy and compared them with morphologicial features related to SF phenomena.

Kinematics of an "isolated" galaxy. We also study the kinematics of an apparently isolated LIRG galaxy. We find double Hα profiles that could be related to an infalling satellite triggering an intense SF event.

Energetic phenomena in interacting galaxies. Finally we present an IFU study of the surroundings of an elongated ultra-luminous X-ray source in an interacting galaxy pair (Rosado, Ghosh & Fuentes-Carrera 2008) in order to trace particular motions of the ionized gas around this source.

2. Conclusions

3D observations of galaxies enable us to match the motions of stars and gas with different morphological structures. In the case of interacting galaxies these studies allow us to trace the effects of the interacting process in the evolution of these systems.

Acknowledgements

I.F-C. acknowledges financial support from CONACYT project 133520 and SIP project 20144432.

References

Rosado, M., Ghosh, K., & Fuentes-Carrera, I. 2008, *AJ*, 136, 212

Galaxies in 3D across the Universe
Proceedings IAU Symposium No. 309, 2014
B. L. Ziegler, F. Combes, H. Dannerbauer, M. Verdugo, eds.

© International Astronomical Union 2015
doi:10.1017/S1743921314010114

The new method of investigating the orientation of galaxies and their clusters.

Włodzimierz Godłowski[1]*, Paulina Pajowska[1], Piotr Flin[2] and Elena Panko[3]

[1]Institute of Physics, Opole University, Opole, Poland
[2]Institute of Physics, Jan Kochanowski University, Kielce, Poland
[3]Kalinenkov Observatory, Nikolaev State University, Nikolaev, Ukraine

Abstract. We present a new method of analyzing galaxy alignments in clusters originaly proposed by Godowski 2012 and now improved.

Keywords. galaxies:

1. Introduction

Problem of large structure formation is one of the most important problems in modern cosmology and extragalactic astronomy. Classical theories (Peebles 1969; Zeldovich 1970, Efstathiou & Silk 1983) were developed and modified by various researchers (e.g. Lee & Penn 2002; Navarro, Abadi & Steinmetz 2004, Codis *et al.* 2012, Varela *et al.* 2012). Various scenarios of large scale structure formation make different predictions concerning orientation of galaxies in structures, distribution of spins of galaxies and alignment between the brightest galaxy and the major axis of the structure. The ultimate test for a given scenerio would be to test it against observations. A new method of analyzing galaxy alignments in clusters proposed by Godłowski (2012) has now been improved. The distribution of position angles for galaxy major axes has been analyzed, as well as the distribution of the two angles describing the spatial orientation of the galaxy plane, both of which provide information about galaxy angular momenta.

2. Aim

The aim of the project is to investigate the correlation function of orientation of galaxies in clusters and the orientation of clusters themselves.

References

Codis, S., Pichon, C., Devriendt, J., Slyz, A., *et al.* 2012 *Mon. Not. R. Astr. Soc.*, 427, 3320
Efstathiou, G. A. & Silk, J., 1983, *The Formation of Galaxies, Fundamentals of Cosm. Phys.* 9, 1
Godłowski, W. 2012, *ApJ* 747, 7
Lee, J. & Pen, U. 2002, *ApJ*, 567, L111
Navarro, J. F., Abadi, M. G., Steinmetz, M., 2004 *Astrophys. J.* 613, L41
Peebles, P. J. E., 1969 *Astrophys. J.* 155, 393
Varela, J., Betancort-Rijo, J., Trujillo, I., & Ricciardelli, E. 2012, *Astrophys. J.* 744, 82
Zeldovich, B. Ya. 1970, *A&A*, 5, 84

Galaxies in 3D across the Universe
Proceedings IAU Symposium No. 309, 2014
B. L. Ziegler, F. Combes, H. Dannerbauer, M. Verdugo, eds.

© International Astronomical Union 2015
doi:10.1017/S1743921314010126

Construction of luminosity function for galaxy clusters

Włodzimierz Godłowski[1]*, Joanna Popiela[1], Katarzyna Bajan[2], Monika Biernacka[3], Piotr Flin[3] and Elena Panko[4]

[1]Institute of Physics, Opole University, Opole, Poland
[2]Institute of Physics, Pedagogical University, Cracow, Poland
[3]Institute of Physics, Jan Kochanowski University, Kielce, Poland
[4]Kalinenkov Observatory, Nikolaev State University, Nikolaev, Ukraine

Abstract. The luminosity function is an important quantity for analysis of large scale structure statistics, interpretation of galaxy counts (Lin & Kirshner 1996). We investigate the luminosity function of galaxy clusters. This is performed by counting the brightness of galaxies belonging to clusters in PF Catalogue. The obtained luminosity function is significantly different than that obtained both for optical and radiogalaxies (Machalski & Godowski 2000). The implications of this result for theories of galaxy formation are discussed as well.

Keywords. galaxies:

1. Introduction

We construct and study the luminosity function (Lin & Kirshner 1996) of 6168 galaxy clusters. This was performed by counting the brightness of galaxies belonging to clusters in the magnitude range $m3 + 3$ in PF Catalogue (Panko & Flin 2006) taking data for galaxies from MRSS (Ungruhe, Seitter & Durbeck 2003). The obtained luminosity function is significantly different than that obtained both for optical and radiogalaxies (Machalski & Godłowski 2000). Our preliminary result obtained with Condon (1989) method shows that shape of Luminosity Function for galaxy clusters looks like Gauss function rather than a Schechter function.

2. Conclusions

Probable explanation of the diferences in the shape of the luminosity function of galaxy clusters and luminosity function of galaxy and radiogalaxy come from differences in scenarios of formation of galaxies and clusters.

References

Condon, J. J. 1989, *ApJ* 338, 13
Lin, H. & Kirshner, R. P. 1996, *ApJE* 464, 60
Machalski, J. & Godlowski, W. 2000, *A&A* 360, 463
Panko, E. & Flin, P. 2006 *Journal Astron. Data*, 12, 1
Ungruhe, R., Seitter, W. C., & Durbeck, H. W. 2003, *Journal Astronomical Data* 9, 1

Galaxies in 3D across the Universe
Proceedings IAU Symposium No. 309, 2014
B. L. Ziegler, F. Combes, H. Dannerbauer, M. Verdugo, eds.

© International Astronomical Union 2015
doi:10.1017/S1743921314010138

Radio-optical properties of extragalactic populations in the VIPERS Survey

A. Zanichelli[1], L. Gregorini[1,2], M. Bondi[1] and the VIPERS Team*

[1] INAF Istituto di Radioastronomia, via Gobetti 101, I-40129 Bologna, Italy,
[2]Dipartimento di Fisica e Astronomia, Università di Bologna,
via Ranzani 1, I-40126 Bologna, Italy

Abstract. The radio and optical characteristics of faint radio sources are investigated thanks to the VIPERS spectroscopic and photometric data, and radio data from the VLA FIRST survey at 1.4 GHz.

The VIMOS Public Extragalactic Redshift Survey (VIPERS, Guzzo *et al.*, 2014) is an ESO Large Program to measure the redshifts of 100.000 galaxies over $\sim$ 24 deg^2 of sky in the CFHTLS regions. Based on color pre-selection to remove most of the stars and of galaxies at z<0.5, VIPERS is spectroscopically limited to magnitude I$_{AB}$ <22.5 and covers the redshift range 0.5< z <1.2. A large dataset of ancillary data in various photometric bands is available. The first public data release of $\sim$ 57.000 VIPERS redshifts has been done in October 2013 (see http://vipers.inaf.it).

Thanks to VIPERS spectral type classification, the average properties of galaxy types at different cosmic epochs and radio fluxes can be investigated through optical identifications with radio sources in the VLA FIRST catalog (White *et al.*, 1997). For non radio-detected VIPERS galaxies, the stacking of FIRST radio maps allows to push the analysis at very faint radio flux levels.

From the FIRST catalog we classified 1880 radio sources with S$_{1.4GHz}$ $\geqslant$ 0.77 mJy over the VIPERS area. Among these, 1117 have an optical counterpart in the VIPERS photo-metric dataset down to I$_{AB}$ = 24.0, $\sim$80% having either a spectroscopic or a good-quality photometric redshift. The investigation of their radio–optical properties is ongoing.

In parallel, we are conducting the radio stacking analysis for VIPERS galaxies that show no radio emission in the FIRST survey . A preliminary dataset of $\sim$46000 VIPERS galaxies with good-quality spectroscopic redshift and type classification has been used. First tests show that late spirals/irregulars are detected at increasing redshift, as ex-pected if radio emission in these galaxies is dominated by star formation, while red ellipticals/early spirals are detected at any z, suggesting a probable different origin for radio emission.

The final sample will include the full VIPERS spectroscopic set, allowing us to investi-gate the different properties of the optical classes at very faint flux limits compared with that of radio–emitting galaxies, and to follow their evolution with redshift.

References

Guzzo, L., *et al.* 2014, A&A 566, 108
White, R.,L., *et al.* 1997, ApJ 475, 479

* **The VIPERS Team**: U. Abbas, C. Adami, S. Arnouts, J. Bel, J. Blaizot, M. Bolzonella, D. Bottini, E. Branchini, A. Burden, A. Cappi, J. Coupon, O. Cucciati, I. Davidzon, G. De Lucia, S. de la Torre, C. Di Porto, L. Eardley, M. Elmo, P. Franzetti, A. Fritz, M. Fumana, B. Garilli, B. Granett, L. Guennou, L. Guzzo (PI), O. Ilbert, A. Iovino, E. Jullo, J. Krywult, V. Le Brun, O. Le Fevre, D. Maccagni, A. Marchetti, F. Marulli, K. Malek, C. Marinoni, H. J. McCracken, Y. Mellier, L. Moscardini, T. Moutard, R. Nichol, J. Peacock, W. Percival, M. Polletta, A. Pollo, C. Schimd, M. Scodeggio, V. Scottez, O. Solarz, A. Szalay, L. Tasca, R. Tojeiro, D. Vergani, M. Wilson, A. Zanichelli, G. Zamorani.

Galaxies in 3D across the Universe
Proceedings IAU Symposium No. 309, 2014
B. L. Ziegler, F. Combes, H. Dannerbauer, M. Verdugo, eds.

© International Astronomical Union 2015
doi:10.1017/S174392131401014X

Dust as a tracer of gas in galaxies

Brent Groves, Eva Schinnerer and KINGFISH team

Max Planck Institute for Astronomy, Königstuhl 17, 69117 Heidelberg, Germany

Abstract. We use a sample of 36 galaxies to study empirical relations between Herschel infrared (IR) luminosities and the total mass of the interstellar gas (H_2 + HI). Such a comparison provides a simple empirical relationship without introducing the uncertainty of dust model fitting. We find tight correlations, and provide fits to these relations, between Herschel luminosities and the total gas mass integrated over entire galaxies, with the tightest, almost linear, correlation found for the longest wavelength data (SPIRE500). However, we find that accounting for the gas-phase metallicity (affecting the dust-to-gas ratio) is crucial when applying these relations to low-mass, and presumably high-redshift, galaxies. When examining these relations as a function of galactocentric radius, we find the same correlations, albeit with a larger scatter, up to radius of $r \sim 0.7 r_{25}$ (containing most of a galaxy's baryonic mass). The tight relations found for the bulk of the galaxy's baryonic content suggest that total gas masses of disk-like (non-merging/ULIRG) galaxies can be inferred from far-infrared continuum measurements in situations where only the latter are available. This work is to appear in Groves *et al.* (2014).

Keywords. galaxies:ISM–infrared:galaxies

1. Summary

The evolution of the gas mass density over cosmic time plays a direct role in the evolution of galaxies through its link to the star formation rate. As dust and gas are intimately associated, the dust infrared continuum may provide a more feasible way to determine the total gas mass of galaxies at high redshift. To demonstrate the use of the IR continuum as a gas tracer we have compared the total gas mass and IR luminosity of a well-studied sample of 36 nearby galaxies (average distance 10 Mpc) observed with the KINGFISH (IR), THINGS (HI), and HERACLES ($CO_{(2-1)}$) surveys. These galaxies sample the peak of the mass function, with stellar masses from $\sim 10^{6.5}$ to $\sim 10^{10.5} M_\odot$, and, unlike the local merger-induced ULIRGs, these galaxies have disk-like morphologies that are likely more representative of the main-sequence galaxies at high redshift observed in current and future deep sub-mm surveys.

We find a strong correlation between the total gas mass and IR luminosity in all Herschel bands, with the strongest and tightest correlation found for the longest wavelength (SPIRE 500μm). The gas to luminosity ratio, M_{gas}/L_{500}, is found to increase with decreasing sub-mm luminosity, which can be ascribed to the declining metallicity associated with the lower galaxy masses of the low luminosity galaxies. This declining metallicity will lead to a lower dust to gas ratio, and therefore higher gas to dust luminosity ratio. . To minimize the effects of metallicity, we also fit only galaxies with stellar masses greater than $10^9 M_\odot$ in our sample, which also results in a significantly reduced dispersion. For the SPIRE 500 band we find that a linear relation is a reasonable approximation, and find that $\log(M_{gas}/L_{500}) = 28.5 [M_\odot/L_\odot]$, with a dispersion of only 0.118 dex.

Reference

Groves, B., Schinnerer, E., Leroy, A., *et al.* 2014, *ApJ*, accepted

Galaxies in 3D across the Universe
Proceedings IAU Symposium No. 309, 2014
B. L. Ziegler, F. Combes, H. Dannerbauer, M. Verdugo, eds.

© International Astronomical Union 2015
doi:10.1017/S1743921314010151

Gas properties in the disc of NGC 891 from *Herschel* far-infrared spectroscopy

Thomas M. Hughes

Astronomical Observatory, Ghent University, Belgium
email: thomas.hughes@ugent.be

Abstract. We investigate the physical properties of the interstellar gas in the nearby edge-on spiral galaxy NGC 891, using *Herschel* PACS/SPIRE observations of the most important far-infrared (FIR) cooling lines – [CII] 158 μm, [NII] 122, 205 μm, [OI] 63, 145 μm, and [OIII] 88 μm – obtained as part of the Very Nearby Galaxy Survey (P. I.: C. D. Wilson). We compare our observations to the predictions of a photo dissociation region (PDR) model to determine the gas density, n, and the strength of the incident FUV radiation field, G_0, on a pixel-by-pixel basis. The majority of PDRs in NGC 891's disc exhibit properties similar to the physical conditions found in the spiral arm and inter-arm regions of the face-on M51 galaxy. We estimate a stronger FUV field in the far north-eastern side than compared to the rest of the disc.

Keywords. galaxies: individual: NGC 891 - galaxies: spiral - ISM: lines and bands

1. Importance of optical depth effects

We determine the gas properties from the FIR cooling line and TIR emission, estimated from 24, 100 and 160 μm maps (see Hughes *et al.* 2014), using the Kaufman *et al.* (2006) PDR model. Radial trends in n and G_0 found in face-on galaxies, e.g. M51 (Parkin *et al.* 2013), are only reproduced by correcting the [OI] 63 μm line flux to account for increasing optical thickness towards the center (Fig. 1), implying such effects must be considered when interpreting observations of highly inclined systems (Hughes *et al.* submitted).

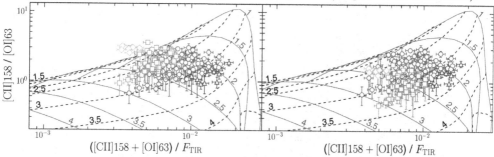

Figure 1. The [CII]/[OI]63 versus ([CII]+[OI]63)/$F_{\rm TIR}$ ratio diagnostic diagram, in which we superimpose our adjusted observations onto grid-lines of constant log n (dashed lines) and log G_0 (solid lines) determined from the PDR model. The points correspond to pixels covering the nucleus (diamonds), mid-plane (squares) and at increasing radial distances (circles/triangles). We contrast a constant (*left*) and varying correction (*right*) to the [OI] 63 μm emission.

References

Hughes, T. M., Baes, M., Fritz, J., *et al.* 2014, *A&A*, 565, 4
Hughes, T. M., Foyle, K., Schirm, M. R. P., *et al.* 2014, *A&A*, submitted
Kaufman, M. J., Wolfire, M. G., & Hollenbach, D. J. 2006, *ApJ*, 644, 283
Parkin, T. J., Wilson, C. D., Schirm, M. R. P., *et al.* 2013, *A&A*, 550, 114

Galaxies in 3D across the Universe
Proceedings IAU Symposium No. 309, 2014
B. L. Ziegler, F. Combes, H. Dannerbauer, M. Verdugo, eds.

© International Astronomical Union 2015
doi:10.1017/S1743921314010163

Metallicity gradients of galaxies in the *Herschel* Reference Survey

Thomas M. Hughes

Astronomical Observatory, Ghent University, Belgium
email: thomas.hughes@ugent.be

Abstract. We introduce a pilot project to measure metallicitiy gradients for a sample of twenty nearby galaxies drawn from the *Herschel* Reference Survey (HRS), representative of normal, star-forming spiral galaxies. We have obtained optical spectroscopic observations using the Very Large Telescope with the FORS2 instrument in multi-object mode, targeting individual HII and star-forming regions across the galaxy discs (P. I.: L. Cortese). From the ratios of the strong emission lines, we estimate the local gas-phase oxygen abundance and construct metallicity gradients. Combining these new data with *Herschel* PACS/SPIRE far-infrared photometric observations and HI 21 cm line maps, to trace the cold dust and gas respectively, will allow the study of the relationships between stars, gas, dust and metals on sub-kiloparsec scales.

Keywords. cosmology: observations galaxies: spiral galaxies: evolution

1. Preliminary reduction of VLT/FORS2 observations

In Fig. 1, we show two example spectra from slits covering HII regions in NGC 5669 (aka HRS 319; Boselli *et al.* 2010), extracted with the ESO Reflex pipeline. Using emission lines observed in each spectra at each position in our twenty galaxies, we apply the method of Hughes *et al.* (2013) and calibrations of Kewley & Ellison (2008) to estimate gas-phase oxygen abundance gradients. We will then investigate, amongst other things, how strong metallicity gradients may bias global metallicity estimates derived from the integrated spectra of gas-stripped galaxies (see Hughes *et al.* 2013; Boselli *et al.* 2013).

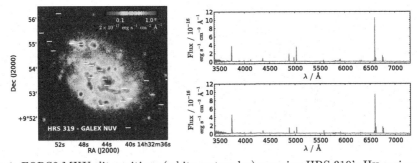

Figure 1. FORS2 MXU slit positions (white rectangles) covering HRS 319's HII regions, traced here via *NUV* emission (*left*), are presented alongside two example optical spectra extracted from the observations using the ESO Reflex standard data reduction pipeline (*right*).

References

Boselli, A., Eales, S., Cortese, L., Bendo, G., & Chanial, P. *et al.*. 2010, *PASP*, 122, 261
Boselli, A., Hughes, T. M., Cortese, L., Gavazzi, G., & Buat, V., 2013, *A&A*, 550, 114
Hughes, T. M., Cortese, L., Boselli, A., Gavazzi, G., & Davies, J. I., 2013, *A&A*, 550, 115
Kewley, L. & Ellison, S., 2008, *ApJ*, 681, 1183

Galaxies in 3D across the Universe
Proceedings IAU Symposium No. 309, 2014
B. L. Ziegler, F. Combes, H. Dannerbauer, M. Verdugo, eds.

© International Astronomical Union 2015
doi:10.1017/S1743921314010175

Survey of lines in M 31: [CII] as SFR tracer at ~ 50 pc scales

M. J. Kapala[1], K. Sandstrom[1,2] and B. Groves[1]

[1] Max Planck Institut für Astronomie, Königstuhl 17, D-69117 Heidelberg, Germany
email: kapala@mpia.de, [2] Steward Observatory, Tucson, USA

The [CII] 158 μm line is typically the brightest far-IR emission line from star-forming galaxies. As such, this line is a possible tracer of star-formation, but to do so we need the relative contributions of different ISM phases. Using high physical resolution observations of the [CII] 158 μm line from *Herschel* PACS in five 3'×3' field in M 31 and optical IFU spectra from PPaK and ancillary IR data, we are able to spatially separate out the ISM phases (Kapala *et al.* subm.). We find that a large fraction of [CII] emission in M 31 arises from diffuse gas (~20–90%), with a sub-linear relation of [CII]–SFR at ~50 pc scales. However, on ~kpc scales, the observed empirical [CII]–SFR relation is in agreement with other extragalactic studies. The observed flattening of the fitted slope to the [CII]–SFR at ~50 pc scales might be explained by possible contributions to ISM gas heating by older stellar populations (ie. [CII] tracing longer timescales) and/or leaked photons from HII regions. The issue of leaked photons from HII regions should go away when averaged over larger scales (>500 pc).

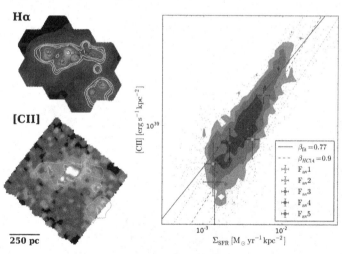

Figure 1: **Left:** Hα and [CII] emission maps (Hα contours overlaid on both maps), in one of the Fields. **Right:** [CII]-SFR relation on small scales – grey points and number density plot (5.2" pixels ~20 pc; 3σ cut). Solid blue lines are fits to the individual pixels. Red dashed line - the relation found by Herrera-Camus et al. (subm.) for the KING-FISH sample at ~kpc res. (dashed-dotted lines - 1σ scatter around their fit)

Acknowledgements

We acknowledge support from the DLR Grant 50OR1115.

Herschel is an ESA space observatory with science instruments provided by European-led Principal Investigator consortia and with important participation from NASA.

Reference

Kapala, M. J., *et al.*, subm., *ApJ*

Galaxies in 3D across the Universe
Proceedings IAU Symposium No. 309, 2014
B. L. Ziegler, F. Combes, H. Dannerbauer, M. Verdugo, eds.

© International Astronomical Union 2015
doi:10.1017/S1743921314010187

A far-IR and optical 3D view of the starburst driven superwind in NGC 2146

Kathryn Kreckel[1], Lee Armus[2], Brent Groves[1], Mariya Lyubenova[3], Tanio Diaz-Santos[2] and Eva Schinnerer[1]

[1]MPIA [2]Caltech [3]Kapteyn Institute

Galaxy outflows are a vital mechanism in the regulation of galaxy evolution through feedback and enrichment. NGC 2146, a nearby infrared luminous galaxy (LIRG), presents evidence for outflows along the disk minor axis in all gas phases (ionized, neutral atomic and molecular). We present new far-IR Herschel imaging and spectroscopy of this galaxy from the Key Insights on Nearby Galaxies: a Far-Infrared Survey with Herschel (KING-FISH) project, as well as new optical integral field unit spectroscopy, to map the kinematics and gas excitation in the central 5 kpc and trace the dust distribution (Kreckel et al. 2014). We observe an increased velocity dispersion in the [OI] 62 um, [OIII] 88 um, [NII] 122 um and [CII] 158 um fine-structure lines that is spatially coincident with shocked gas above and below the disk. Unhampered by extinction, the far-IR lines trace the outflow to the base of the superwind at the disk center, and we discuss the potential for using [CII] as a tracer of outflows such as this in high redshift systems with ALMA. The stellar kinematics are decoupled from the disk rotation seen in all gas phases, which we attribute to a merger that has not produced a fully elliptical morphology.

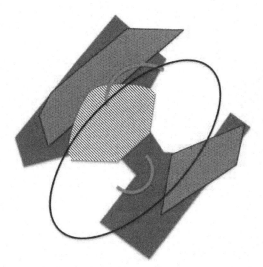

Figure 1. Diagram of outflow indicators in NGC 2146, relative to the stellar disk (black oval). Previous identification of the superwind in soft X-ray emission from hot gas (white hatched) and CO molecular outflows (purple arcs) are consistent with the far-IR [CII] emission from atomic gas in the cone walls (solid blue) and optically identified shock excitation (red dotted).

Reference

Kreckel, K., Armus, L., Groves, B., *et al.* 2014, *ApJ*, 790, 26

Galaxies in 3D across the Universe
Proceedings IAU Symposium No. 309, 2014
B. L. Ziegler, F. Combes, H. Dannerbauer, M. Verdugo, eds.

© International Astronomical Union 2015
doi:10.1017/S1743921314010199

A Radio-Optical Study of Resolved Star Formation in SAMI Galaxies

Sarah Leslie[1,3], Lisa Kewley[1], Elaine Sadler[2,3] and Julia Bryant[2,3,4]

[1]Research School of Astronomy & Astrophysics, Australian National University, Cotter Road, Weston, ACT 2611, Australia
email: u5022102@anu.edu.au

[2]Sydney Institute for Astronomy (SIfA), School of Physics, The University of Sydney, NSW 2006, Australia

[3]ARC Centre of Excellence for All-sky Astrophysics (CAASTRO)

[4]The Australian Astronomical Observatory, PO Box 915, North Ryde, NSW, 1670, Australia

Abstract. With integral field spectroscopic data from the the Sydney-AAO Multi-object Integral-field spectrograph (SAMI) survey and the VLA, we will study the relationship between star formation (as traced by Hα emission) and the radio continuum emission within galaxies with the aim of better understanding the intricacies of local scaling relations.

Keywords. galaxies: star formation, radio continuum: galaxies, galaxies: magnetic fields

1. Project Summary

Despite many decades of study, the relation between radio continuum emission and the star-formation rate (SFR) in galaxies is still not well understood. Previous studies have indicated that the tight correlation between radio and far infrared (FIR) flux used to derive local scaling relations between the radio continuum power and SFR breaks down for faint, low metallicity, low mass galaxies. To better understand these scaling relations across a range of stellar masses, we will combine resolved radio continuum information with the spatially resolved information on gas and stellar processes derived from optical integral field spectroscopy (IFS). We will draw our IFS data for galaxies spanning a wide range of stellar masses from the first massively multiplexed IFS survey, the Sydney-AAO Multi-object Integral-field spectrograph (SAMI) galaxy survey. Our radio data are from the VLA Faint Images of the Radio Sky at Twenty-centimeters (FIRST) survey, supplemented by our own deeper VLA observations. In this project we aim to:

• Compare the spatial distribution of star-forming regions in galaxies with the brightness profile of the radio emission to better understand the radio emission as a SFR tracer within galaxies. In particular we will also

    ○ study low mass star-forming galaxies which fall off the radio-SFR relations for larger galaxies and

    ○ investigate whether the mix of thermal and non-thermal radio emission in star-forming galaxies changes with galaxy stellar mass and metallicity.

• Understand the role played by the ordered magnetic field in galaxies, and test whether the dominant mechanism for radio emission is different in dwarf galaxies which often lack this ordered magnetic field.

Acknowledgements

This research is supported by the Australian Research Council Centre of Excellence for All-sky Astrophysics (CAASTRO), through project number CE110001020.

Galaxies in 3D across the Universe
Proceedings IAU Symposium No. 309, 2014
B. L. Ziegler, F. Combes, H. Dannerbauer, M. Verdugo, eds.

© International Astronomical Union 2015
doi:10.1017/S1743921314010205

Dust in FIR-bright ADF-S galaxies

K. Małek[1,2], A. Pollo[2,3], T. T. Takeuchi[1], V. Buat[4], D. Burgarella[4] and M. Malkan[5]

[1]Department of Particle and Astrophysical Science, Nagoya University, Furo-cho, Chikusa-ku, 464-8602 Nagoya, Japan email: malek@cft.edu.pl
[2]National Centre for Nuclear Research, ul. Hoża 69, 00-681 Warszawa, Poland
[3]The Astronomical Observatory, Jagiellonian University, ul. Orla 171, 30-244 Kraków, Poland
[4]Laboratoire d'Astrophysique de Marseille, OAMP, Universite Aix-Marseille, CNRS, 38 rue Frerdeic Joliot-Curie, 13388 Marseille, cedex 13, France
[5]Department of Physics and Astronomy, University of California, Los Angeles, CA 90024, USA

Abstract. Multiwavelength Spectral Energy Distributions (SEDs) of far-infrared (FIR) galaxies detected in the AKARI South Ecliptic Poles Survey (ADF-S) allow to trace differences between [Ultra]-Luminous Infrared Galaxies ([U]LIRGS) and other types of star-forming galaxies (SF).

Keywords. galaxies: infrared - galaxies: evolution - galaxies: spectral energy distribution

1. Results

The ADF-S provides the highest quality FIR image of the extragalactic Universe. With its four photometric bands (65, 90, 140, and 160 μm) it mapps a wide area of 12 sq^2. We cross-corelated ADF-S catalog with public databases (Małek *et al.* 2013) and used the CIGALE SED fitting code (Noll *et al.* 2009) to measure the physical parameters of the ADF-S sources.

We created average SEDs (Fig. 1), normalized all SEDs at rest frame 90 μm, and divided them into:

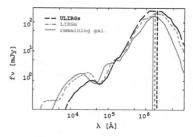

Figure 1. The average SEDs normalized at 90μm.

ULIRGs (17 galaxies), LIRGs (31 galaxies), and the remaining galaxies (82 objects). We notice a significant shift in the peak λ of the dust emission in the FIR and a different ratio between luminosities in the optical and IR parts of these three types of galaxies (Małek *et al.* 2014). It means that [U]LIRGs contain cooler dust that SF galaxies, and that the ratio between luminosities in the optical and IR parts of the spectra increases with the dust luminosity.

Acknowledgements

AP and KM have been supported by the National Science Centre (UMO-2012/07/B/ST9/04425 and UMO-2013/09/D/ST9/04030). KM was supported by the Strategic Young Researcher Overseas Visits Program for Accelerating Brain Circulation No. R2405.

References

Małek, K., Pollo, A., Takeuchi, T. T., *et al.* 2013, *EPS, Special Issue Cosmic Dust V*, 65, 1101
Małek, K., Pollo, A., Takeuchi, T. T., *et al.* 2014, *A&A*, 562, id.A15
Noll, S., Burgarella, D., Giovannoli, E., *et al.* 2009, *A&A*, 507, 1793

Galaxies in 3D across the Universe
Proceedings IAU Symposium No. 309, 2014
B. L. Ziegler, F. Combes, H. Dannerbauer, M. Verdugo, eds.

© International Astronomical Union 2015
doi:10.1017/S1743921314010217

The impact of the orientation on the lobe asymmetry in 3C328 radio galaxy

Andrzej Marecki and Mateusz Ogrodnik

Toruń Centre for Astronomy, Faculty of Physics, Astronomy, and Informatics,
Nicolaus Copernicus University, Toruń, Poland
email: `amr@astro.uni.torun.pl`

Abstract. In the double-lobed Fanaroff-Riley type-II radio galaxy 3C328, one lobe is devoid of a hotspot. We carried out the VLBA observations of the central component of 3C328 and we found there a jet pointing towards the lobe without a hotspot. We suggest that this is because the nucleus of 3C328 remained in off-state for some time in the recent past. No jets were produced during that period so both hotspots faded out. However, under the assumption that one lobe is much farther from the observer than the other, the hotspot of the far-side lobe can be perceived as not decayed yet due to the light-travel lag. Since activity of the nucleus has been restored, we obtain a proof that it is recurrent. It follows that the orientation of the jet is not a paradox.

Normally, both lobes of Fanaroff-Riley (FR) type-II radio galaxies are terminated with hotspots. In FR II radio galaxy 3C328, however, only one lobe has a hotspot whereas the other is clearly devoid of a hotspot (Fig.1). We carried out the Very Long Baseline Array (VLBA) observations of the central component of 3C328 and we found there a jet (Fig.2) oriented towards the southern lobe, i.e. the one with no hotspot.

At first sight, it is puzzling why, despite the presence of the jet seemingly fueling the southern lobe, it appears as a relic. We propose the following interpretation of this apparent paradox. Assume that the nucleus of 3C328 remained quiescent for some time in the recent past. Since no jets were produced in that period, the hotspots were not energized. As a result, they *both* faded out. To explain why they are asymmetric anyway, we assume is that the northern lobe is much farther from us than the southern one hence the hotspot of the former is perceived as not dispersed yet due to the light-travel lag. The jet was relaunched after a period of quiescence, it has therefore no causal connection with the relic it points at. Given that one-sided jets are always beamed towards the observer, the southern lobe must be on the near side in agreement with our second assumption.

This reasoning is similar to that used when explaining the Laing-Garrington effect (Laing 1988; Garrington *et al.* 1988), i.e. the correlation between the jet orientation and the degree of polarization of radio lobes. What we present here is an analogue of Laing-Garrington effect but, instead of an asymmetry of polarization (that perhaps could also be present in 3C328), we observe an asymmetry of its lobes – the near-side lobe appears "older" than the far-side lobe. Such kind of asymmetry may emerge only because of the episodic nature of galactic activity. The images of 3C328 we perceive are thus a record of two subsequent transitions in its nucleus: from radio-loud to radio-quiet and back to radio-loud state. This way we obtain a proof that the activity of 3C328 must be recurrent.

References

Garrington, S. T., Leahy, J. P., Conway, R. G., & Laing, R. A. 1988, *Nature*, 331, 147
Laing, R. A. 1988, *Nature*, 331, 149
Machalski, J., Condon, J. J. 1983, *AJ*, 88, 143

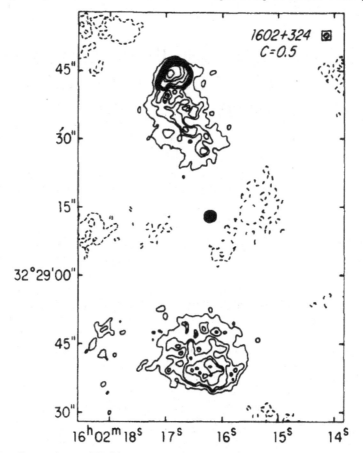

Figure 1. Very Large Array (VLA) image of 3C328 at 1465 MHz (Machalski & Condon 1983)

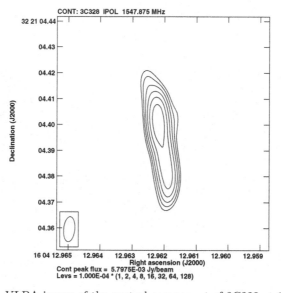

Figure 2. VLBA image of the central component of 3C328 at 1548 MHz

Galaxies in 3D across the Universe
Proceedings IAU Symposium No. 309, 2014
B. L. Ziegler, F. Combes, H. Dannerbauer, M. Verdugo, eds.
© International Astronomical Union 2015
doi:10.1017/S1743921314010229

New empirical metallicity calibrations: Joint analysis of CALIFA data and literature T_e-based measurements

R. A. Marino[1], F. F. Rosales-Ortega[2], S. F. Sánchez[3], A. Gil de Paz[1] and the CALIFA team

[1]CEI Campus Moncloa, UCM-UPM, Departamento de Astrofísica y CC. de la Atmósfera,
Facultad de CC. Físicas, Universidad Complutense de Madrid, Avda. Complutense s/n, 28040
Madrid, Spain. email: ramarino@ucm.es
[2]Instituto Nacional de Astrofísica, Óptica y Electrónica, Puebla, México.
[3]Instituto de Astronomía, Universidad Nacional Autonóma de México, D. F.

Abstract. In Marino *et al.* (2013) we provide revisited empirical calibrations for the oxygen abundances in HII regions based on the O3N2 and N2 indicators. This work is based on the most comprehensive compilation of both T_e-based and multiple strong-line (ONS-based) ionized-gas abundance measurements in external galaxies to date in terms of all statistical significance, quality, and coverage of the parameters space. Our dataset compiles the T_e-based abundances of 603 HII regions extracted from the literature but also includes new measurements from the CALIFA survey. We also present a comparison between our revisited calibrations with a total of 3423 additional CALIFA HII complexes with abundances derived using the ONS calibration. The O3N2 and N2 indicators can be empirically applied to derive oxygen abundances calibrations from either direct-abundance determinations with random errors of 0.18 and 0.16, respectively, and they show shallower abundance dependencies and statistically significant offsets compared to the classical calibrations (as the one of Pettini & Pagel (2004)).

Keywords. galaxies: abundances - galaxies: evolution - ISM: abundances - (ISM): HII regions
- techniques: spectroscopic

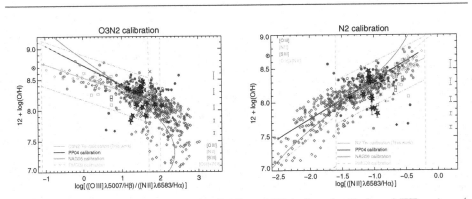

Figure 1. Oxygen abundance versus the O3N2 and N2 indices for T_e-based HII regions (see more details in Marino *et al.* (2013)).

References

Marino, R. A., Rosales-Ortega, F. F., Sánchez, S. F., *et al.* 2013, *A&A*, 559, A114
Pettini, M. & Pagel, B. E. J. 2004, *MNRAS*, 348, L59

Galaxies in 3D across the Universe
Proceedings IAU Symposium No. 309, 2014
B. L. Ziegler, F. Combes, H. Dannerbauer, M. Verdugo, eds.

© International Astronomical Union 2015
doi:10.1017/S1743921314010230

The interaction of the outflow with the molecular disk in the Active Galactic Nucleus of NGC 6951

D. May, J. E. Steiner, T. V. Ricci, R. B. Menezes and I. S. Andrade

Instituto de Astronomia Geofísica e Ciências Atmosféricas, Universidade de São Paulo
Rua do Matão 1226, Cidade Universitária, São Paulo, SP CEP 05508-090, Brazil
email: dmay@usp.br

Abstract. Context: we present a study of the central 200 pc of NGC 6951, in the optical and NIR, taken with the Gemini North Telescope integral field spectrographs, with resolution of $\sim 0''.1$ Methods: we used a set of image processing techniques, as the filtering of high spatial and spectral frequencies, Richardson-Lucy deconvolution and PCA Tomography (Steiner *et al.* 2009) to map the distribution and kinematics of the emission lines. Results: we found a thick molecular disk, with the ionization cone highly misaligned.

Keywords. galaxies: individual (NGC 6951)

1. Introduction

NGC 6951 hosts a Seyfert 2 nucleus, and it is at a distance of 24.1 Mpc ($1'' = 117$ pc), with and inclination of $i = 46°$. It has a radio compact nuclear component, with a position angle of 156°. The HST/ACS image of the ionized gas shows a central elongated structure seen in Hα+[N II], with a similar PA. The orientation of the jets in AGNs and the galaxy disk/torus are uncorrelated, and its non-detection suggests that it is confined in a small region because of its large misalignment and interaction with the ISM.

2. Results and conclusions

If we overlap the images for the ionized gas from the HST and the average image of the H$_2$ molecular lines, we see that the outflow is misaligned with respect the molecular gas, suggesting some kind of interaction. The new detected molecular structure is an edge-on disk of H$_2$, with radius of $\sim$47 pc and $\sim$10 pc of thickness and PA $= 124°$. The radial velocity range is -40 to +40 km s^{-1}, with velocity dispersion of 40 ± 4 km s^{-1}.

The H$_2$ has a larger velocity dispersion in the direction of the ionization cone, likely associated with the turbulence induced by the radio jet. Based on the H$_2$ line ratios, we conclude the excitation mechanism is mainly due to shocks. This is explained as a "digging process" that the jet inflicts on the disk, ejecting some of the molecular gas.

Acknowledgements

The author acknowledges to FAPESP, under grants 2011/19824-8 (DMN).

Reference

Steiner, J. E., Menezes, R. B., Ricci, T. V., & Oliveira, A. S. 2009, *MNRAS*, 395, 64

Galaxies in 3D across the Universe
Proceedings IAU Symposium No. 309, 2014
B. L. Ziegler, F. Combes, H. Dannerbauer, M. Verdugo, eds.

© International Astronomical Union 2015
doi:10.1017/S1743921314010242

Linking star formation and galaxy kinematics in the massive cluster Abell 2163

Veronica Menacho and Miguel Verdugo

Institut f̈r Astrophysik, University of Vienna, Türkenschanzstr. 17, 1180 Vienna, Austria
email: veronica.menacho@univie.ac.at & miguel.verdugo@univie.ac.at

Abstract. The origin of the morphology-density relation is still an open question in galaxy evolution. It is most likely driven by the combination of the efficient star formation in the highest peaks of the mass distribution at high-z and the transformation by environmental processes at later times as galaxies fall into more massive halos. To gain additional insights about these processes we study the kinematics, star formation and structural properties of galaxies in Abell 2163 a very massive ($\sim$4×10^{15} M$_\odot$, Holz & Perlmutter 2012) merging cluster at z=0.2.

We use high resolution spectroscopy with VLT/VIMOS to derive rotation curves and dynamical masses for galaxies that show regular kinematics. Galaxies that show irregular rotation are also analysed to study the origin of their distortion. This information is combined with stellar masses and structural parameters obtained from high quality CFHT imaging. From narrow band photometry (2.2m/WFI), centered on the redshifted Hα line, we obtain star formation rates.

Although our sample is still small, field and cluster galaxies lie in a similar Tully-Fisher relation as local galaxies. Controlling by additional parameters like SFRs or bulge-to-disk ratio do not affect this result. We find however that $\sim$50% of the cluster galaxies display irregular kinematics in contrast to what is found in the field at similar redshifts ($\sim$30%, Böhm *et al.* 2004) and in agreement with other studies in clusters (e.g. Bösch *et al.* 2013, Kutdemir *et al.* 2010) which points out to additional processes operating in clusters that distort the galaxy kinematics.

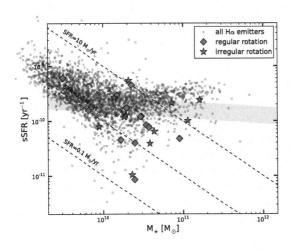

Figure 1. Star forming galaxies in A2163 selected by their Hα emission (small dots). We show in this diagram the cluster galaxies for which we studied their kinematics. Diamonds mark galaxies with regular rotation wheras stars those with irregular kinematics. Dashed lines show constant SFRs whereas the gray area mark the expected sSFR-M$_\star$ relation for this redshift. The most massive galaxy in our sample is a large edge-on spiral clearly expericing ram-pressure stripping as evidenced by its kinematics and trailing star forming regions.

References

Böhm, A., Ziegler, B. L., Saglia, R. P., *et al.* 2004, *A&A*, 420, 97
Bösch, B., Böhm, A., Wolf, C., *et al.* 2013, *A&A*, 554, A97
Holz, D. E. & Perlmutter, S. 2012, *ApJ*, 755, L36
Kutdemir, E., Ziegler, B. L., Peletier, R. F., *et al.* 2010, *A&A*, 520, A109

Galaxies in 3D across the Universe
Proceedings IAU Symposium No. 309, 2014
B. L. Ziegler, F. Combes, H. Dannerbauer, M. Verdugo, eds.

© International Astronomical Union 2015
doi:10.1017/S1743921314010254

Disentangling the stellar populations of the counter-rotating stellar disc in NGC 5719

L. Morelli[1,2], L. Coccato[3], E. M. Corsini[1,2], E. Dalla Bontà[1,2], F. Bertola[1,2], D. Vergani[4] and L. Buson[2]

[1] Dipartimento di Fisica e Astronomia G. Galilei, Università di Padova, vicolo dell'Osservatorio 3, I-35122 Padova, Italy
email: lorenzo.morelli@unipd.it
[2] INAF-Osservatorio Astronomico di Padova, vicolo dell'Osservatorio 2, I-35122 Padova, Italy
[3] European Southern Observatory, Karl-Schwarzschild-Straße 2, D-85748 Garching bei München, Germany
[4] INAF, Osservatorio Astronomico di Bologna, via Ranzani 1, I-40127 Bologna, Italy

Abstract. We present the stellar populations properties of the interacting spiral NGC 5719, which is known to host two cospatial counter-rotating stellar discs .

1. Introduction, analisys and results

The presence of stars counter-rotating with respect to other stars and/or gas has been detected in several disc galaxies and is commonly interpreted as the end result of a retrograde acquisition of external gas and subsequent star formation (Corsini 2014). This picture can be directly tested in the NGC 5719/13 galaxy pair. NGC 5719 is an almost edge-on Sab galaxy with a prominent skewed dust lane at a distance of 23.2 Mpc.

The integral-field spectroscopic observations were carried out in service mode with the Very Large Telescope at the European Southern Observatory in Paranal. The HR blue grism covering the spectral range 4150-6200 Å and the 0.67 arcsec per fibre resolution were used. The instrumental spectral resolution was equivalent to 115 km s^{-1} (FWHM).

At each position in the field of view, the observed galaxy spectrum was decomposed into the contributions of the spectra of two stellar and one ionized-gas components. Therefore we measured separately the kinematics and line strengths of the Lick indices of the two stellar counter-rotating components. We also derived the kinematic of the gaseous component. Finally, we modeled the data of each stellar component with single stellar population models that account for the [α/Fe] overabundance.

We find that the stellar counter-rotating component is younger, less rich in metals, more α-enhanced, and less luminous with respect the main stellar body of the galaxy and it is kinematically associated with the ionized-gas disc (Coccato *et al.* 2011). These findings prove the scenario where gas was accreted first by NGC 5719 on to a retrograde orbit from the large reservoir available in its neighbourhoods as a result of the interaction with its companion NGC 5713, and subsequently fuelled the in situ formation of the counter-rotating stellar disc.

References

Coccato, L., Morelli, L., Corsini, E. M., *et al.* 2011, *MNRAS*, 412, L113
Corsini, E. M. 2014, *Astronomical Society of the Pacific Conference Series*, 486, 51

Galaxies in 3D across the Universe
Proceedings IAU Symposium No. 309, 2014
B. L. Ziegler, F. Combes, H. Dannerbauer, M. Verdugo, eds.

© International Astronomical Union 2015
doi:10.1017/S1743921314010266

Gas accretion history of galaxies at $z \sim 0 - 2$: Comparison of the observational data of molecular gas with the mass evolution model of galaxies

Kana Morokuma[1], Junichi Baba[2], Kazuo Sorai[3] and Nario Kuno[1,4]

[1]Nobeyama Radio Observatory (NRO), NAOJ, email: kana.matsui@nao.ac.jp [2]Tokyo Institute of Technology, [3]Hokkaido University, [4]University of Tsukuba

Abstract. We found stellar mass-dependent evolution of galactic molecular gas fractions ($f_{\mathrm{mol}} = \frac{M_{\mathrm{mol}}}{M_\star + M_{\mathrm{mol}}}$, M_{mol}: molecular gas mass, $M_\star$: stellar mass) where less massive galaxies have decreased f_{mol} from $z = 1$ whereas massive galaxies have already had low f_{mol} until $z = 1$. Comparison of the observed quantities (f_{mol}, optical and near infra-red [NIR] colors, specific star formation rate [sSFR = SFR/$M_\star$]) with mass evolution models suggests that less massive galaxies had high f_{mol} at $z = 1$ thanks to recent gas accretion.

Keywords. galaxies: evolution - galaxies: ISM - ISM: molecules

Measuring f_{mol} of galaxies at $0 < z < 1$ is essential for understanding the evolution of galactic disks. It is because f_{mol} is closely related to mass assembly history of galaxies (gas accretion and star formation). Additionally, optical and NIR surveys revealed that local disk galaxies have acquired their $M_\star$ and developed their disks at $z < 1$ (van Dokkum *et al.* 2013). We observed 12 disk galaxies at $z \sim 0.1\text{-}0.2$ with the 45-m telescope at NRO and detected the CO emissions from 8 galaxies. Combined with literature CO data of different redshifts, we find that the CO detected galaxies follow $M_\star - f_{\mathrm{mol}}$ evolution suggested from the statistical study based on the indirect method (Popping *et al.* 2012). Comparison with theoretical predictions revealed a discrepancy from the observations, suggesting that the baryon processes (star formation, H_2 prescription, feedbacks) adopted in those studies bear improvements. Comparing the observed optical and NIR colors, sSFR and f_{mol} with mass evolution model, we find stellar mass dependence of gas accretion, mass loading factor and depletion time of gas. These suggest that less massive galaxies, possible progenitors of disk galaxies at $z = 0$, had high f_{mol} at $z = 1$ thanks to the gas accretion and have consumed gas to form galactic disks from $z = 1$ to $z = 0$.

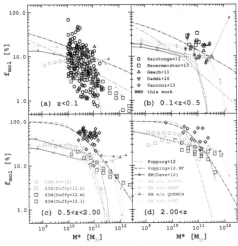

Figure 1. Comparison between observed and theoretically predicted $M_\star - f_{\mathrm{mol}}$.

References

van Dokkum, P. G., *et al.* 2013, *ApJ*, 771, 35
Popping, G., *et al.* 2012, *MNRAS*, 425, 2386
Fu, J., *et al.* 2012, *MNRAS*, 424, 2701
Duffy, A. R., *et al.* 2012, *MNRAS*, 420, 2799
Dave, R., *et al.* 2012, *MNRAS*, 421, 98
Tacconi, L. J., *et al.* 2013, *ApJ*, 768, 74

Galaxies in 3D across the Universe
Proceedings IAU Symposium No. 309, 2014
B. L. Ziegler, F. Combes, H. Dannerbauer, M. Verdugo, eds.

© International Astronomical Union 2015
doi:10.1017/S1743921314010278

High-resolution 3D dust radiative transfer in galaxies with DART-Ray

Giovanni Natale[1,2], Cristina C. Popescu[1,2], Richard. J. Tuffs[2], Victor P. Debattista[1] and Meiert W. Grootes[2]

[1] Jeremiah Horrocks Institute, University of Central Lancashire, Preston, PR1 2HE, UK
[2] Max Planck Institute für Kernphysik, Saupfercheckweg 1, D-69117 Heidelberg, Germany

Abstract. DART-Ray is a 3D ray-tracing dust radiative transfer (RT) code that can be used to derive stellar and dust emission maps of galaxy models and simulations with arbitrary geometries. In addition to the previously published RT algorithm, we have now included in DART-Ray the possibility of calculating the stocastically heated dust emission from each volume element within a galaxy. To show the capabilities of the code, we performed a high-resolution (26 pc) RT calculation for a galaxy N-body+SPH simulation. The simulated galaxy we considered is characterized by a nuclear disc and a flocculent spiral structure. We analysed the derived galaxy maps for the global and local effects of dust on the galaxy attenuation as well as the contribution of scattered radiation to the predicted observed emission. In addition, by performing an additional RT calculation including only the stellar volume emissivity due to young stellar populations (SPs), we derived the contribution to the total dust emission powered by young and old SPs. Full details of this work will be presented in a forthcoming publication.

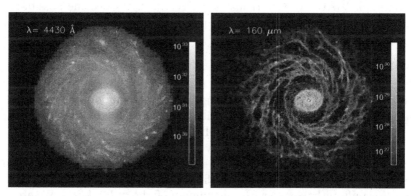

Figure 1. Predicted face-on maps for a galaxy Nbody+SPH simulation. The left panel shows the stellar emission map at 4430 Å(within the B-band) and the right panel the dust emission map at 160 μm. The units of the values aside of the colour bar are erg/(s Å sr pc^2).

References

Cole, D. *et al.* 2014, submitted
Natale, G., Popescu, C. C., Tuffs, R. J., & Semionov, D. 2014, *MNRAS*, 438, 3137
Natale, G. *et al.* 2014, in prep.
Popescu, C. C., Tuffs, R. J., Dopita, M. A., Fischera, J., *et al.* 2011, *A&A*, 527, 109

Galaxies in 3D across the Universe
Proceedings IAU Symposium No. 309, 2014
B. L. Ziegler, F. Combes, H. Dannerbauer, M. Verdugo, eds.

© International Astronomical Union 2015
doi:10.1017/S174392131401028X

Detailed stellar and gaseous kinematics of M31

Michael Opitsch[1,2,3], Maximilian Fabricius [1,2], Roberto Saglia[1,2], Ralf Bender[1,2] and Michael Williams[1,4]

[1] Max Planck Institute for Extraterrestrial Physics, Gießenbachstr., 80748 Garching, Germany
email: `mopitsch@mpe.mpg.de`
[2] University Observatory Munich, Scheinerstr. 1, 81679 Munich, Germany
[3] Excellence Cluster Universe, Boltzmannstr. 2, 85748 Garching, Germany
[4] Department of Astronomy, Columbia University, 550 West 120th Street, NY10027 New York, USA

Abstract. We have collected optical integral field spectroscopic data for M31 with the spectrograph VIRUS-W that result in kinematic maps of unprecedented detail. These reveal the presence of two kinematically distinct gas components.

1. Introduction

Due to its proximity M31 is an ideal target to investigate the kinematics and dynamics of a spiral galaxy in high detail. However, its large angular extent complicates the collection of spectroscopic data for the whole galaxy. With the arrival of the integral field spectrograph VIRUS-W (Fabricius *et al.*, 2008), it has become possible to obtain high-quality two-dimensionally distributed spectra over a large field of view.

2. Conclusions

Our data cover the bulge completely and sample the disk along six different position angles, reaching approximately one scalelength along the major axis (Courteau *et al.*, 2011). We fit the line-of-sight velocity distribution of the stars with pPXF (Cappellari, Emsellem, 2004) and the one of the Hβ, [OIII] and [NI] emission lines with GANDALF (Sarzi *et al.*, 2006). While the stellar velocity field is fairly regular, the gas emission lines in a large fraction of our covered region show double peaks, pointing at two kinematically distinct gas components. The velocity of the first component reaches 300 km/s, the second component is significantly slower with a maximum of about 140 km/s. We are currently testing whether the kinematics of the two components can be be explained by either a warp or the presence of a secondary disk at higher inclination.

Acknowledgements

This research was supported by the DFG cluster of excellence Origin and Structure of the Universe.

References

Cappellari, M. & Emsellem, E. 2004, *PASP*, 116, 138
Courteau, S., *et al.* 2011, *ApJ*, 739, 20
Fabricius, M. H. *et al.* 2008, *SPIE*, 7014, 234
Sarzi, M. *et al.*, 2006 *MNRAS*, 366, 1151

Galaxies in 3D across the Universe
Proceedings IAU Symposium No. 309, 2014
B. L. Ziegler, F. Combes, H. Dannerbauer, M. Verdugo, eds.

© International Astronomical Union 2015
doi:10.1017/S1743921314010291

Evidence for Shocks and Increased SFE in the Lyman Alpha Reference Sample

Johannes Puschnig, Matthew Hayes and Göran Östlin

Department of Astronomy, Oskar Klein Centre, Stockholm University
email: johannes.puschnig@astro.su.se

We report on first molecular gas observations in the Lyman Alpha Reference Sample (LARS; Hayes *et al.* 2013, 2014, Östlin *et al.* 2014), which were performed using the 45m telescope at the Nobeyama Radio Observatory (NRO). The beamsize at the observed ^{12}CO (1–0) emission is $\approx 15''$ corresponding to $\approx 12kpc$ at the given redshift of $z \approx 0.04$. We detected strong ^{12}CO emission in LARS 3 (Arp 238), marginally detected LARS 8 (SDSS 1250+0734) and derived an upper limit for LARS 9 (IRAS 08208+2816).

Our data analysis reveals that LARS 8 and 9 are critically offset from the Kennicutt-Schmidt law leading to an increase in star formation efficiency (SFE). We note that the metallicity in the presented galaxies is sub-solar with a value of [12+log(O/H)]$\approx$ 8.4. Thus, the applicability of CO as a molecular gas tracer is limited and using a constant Milky Way conversion factor $\alpha_{CO} = 4.3$ is most likely not adequate (see green markers in fig.). Therefore, we also make use of a

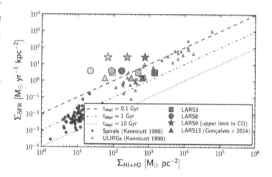

metallicty-dependent α_{CO} which was calculated based on the findings of Magdis *et al.* (2011) and relies on the quite tight correlation between the gas-to-dust ratio and the metallicity. However, even after applying this Z-dependent term, the SFE remains offset by a factor of $\approx$5 (see red markers in fig.) compared to star forming disks.

We interpret the enhanced SFE as a result of shock heating in the ISM introduced by tidal forces. The external ISM pressure exceeding the dynamical pressure of molecular clouds, triggers their collapse and subsequent fragmentation. As shown by Zubovas *et al.* (2014), in such environments of high pressure, new stars preferentially form in massive and compact clusters located in the centre of the initial cloud rather than along cloud filaments. This is supported by our HST images which show tidal features as a result of merging processes and the existence of dense and supermassive stellar associations.

In the case of LARS 09, this scenario is supported by our Herschel/PACS spectroscopy, because the observed [CII] 158 μm line is too strong to be explained by cooling in photo-dissociation regions (PDRs) from star formation only. Most likely, the galaxy contains large amounts of shock-heated gas undergoing turbulent energy dissipation.

In summary, there is evidence for an increased SFE due to shock heating in some LARS galaxies, but uncertainties due to the conversion factor can still affect the result. In a next step, we will resolve the issue by probing higher CO transitions and other molecular lines, which then enable us to compar e the observations to PDR models in order to constrain the underlying physical conditions.

References

Hayes, M., Östlin, Duval, F., *et al.* 2014, *ApJ*, 782, 6

Magdis, G. E., Daddi, E., Elbaz, D., *et al.* 2011, *ApJ*, 740, 15

Zubovas, K., Sabulis, K., Naujalis, R. 2014 *MNRAS*, 442, 2837

Hayes, M., Östlin, G., Schaerer, D., *et al.* 2013, *ApJL*, 765, L27

Östlin, G., Hayes, M., Duval, F., *et al.* 2014, *ApJ*, in prep.

Galaxies in 3D across the Universe
Proceedings IAU Symposium No. 309, 2014
B. L. Ziegler, F. Combes, H. Dannerbauer, M. Verdugo, eds.

© International Astronomical Union 2015
doi:10.1017/S1743921314010308

Stellar Mass Maps for S⁴G

Miguel Querejeta, Sharon E. Meidt, Eva Schinnerer and S⁴G

Max-Planck-Institut für Astronomie, Königstuhl 17, D-69117 Heidelberg, Germany
email: querejeta@mpia.de

Abstract. We present stellar mass maps for the S⁴G sample based on imaging at 3.6 μm that we correct for the presence of non-stellar emission using an ICA technique. Our dust-free images can be readily converted into stellar mass maps, and this important legacy dataset will be made public through IRSA.

Keywords. galaxies: photometry — galaxies: structure — galaxies: evolution

1. Context and method

The Spitzer Survey of Stellar Structure in Galaxies (S⁴G) consists of imaging for 2352 nearby galaxies (D<40Mpc) in the 3.6 and 4.5 μm IRAC bands, an optimal window to trace stellar mass. However, contamination from non-stellar sources (PAH, hot dust) can be locally significant here, severely biasing the derived mass distributions (Meidt *et al.* 2012). We have developed an automatic strategy to identify the old stellar light, based on Independent Component Analysis (ICA), which retains full 2D structural information (Querejeta *et al.* 2014; Fig. 1). As shown by Meidt *et al.* (2014), this dust-corrected 3.6 μm flux can be accurately transformed into stellar mass with even a constant M/L, given the properties of stellar populations at this wavelength.

2. Results

Our final ICA-corrected sample excludes galaxies with low signal-to-noise ($S/N < 10$) and those with colors $[3.6]-[4.5] < 0$ (already consistent with old stellar populations). We find that the contamination from dust (which results in globally red colors $[3.6]-[4.5] > 0$) varies with Hubble type, reaching as high as 40%, and typically in the range 20-30% for spiral galaxies. This likely stems from the dependence of the dust emission on SFR and stellar mass: galaxies with high specific star formation rates (SSFR=SFR/Mass) are clearly associated with the highest dust contributions. With our final mass maps, we calibrate a first-order relationship between [3.6]-[4.5] and dust contamination fraction that can be used to determine stellar masses at 3.6 μm in the absence of more information.

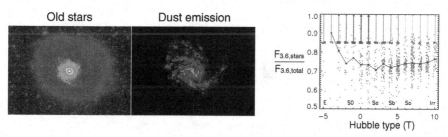

Figure 1. *Left:* ICA map of old stellar flux for NGC4254, next to the identified dust emission. *Right:* fraction of total flux contributed by old stars as a function of Hubble type.

Galaxies in 3D across the Universe
Proceedings IAU Symposium No. 309, 2014
B. L. Ziegler, F. Combes, H. Dannerbauer, M. Verdugo, eds.
© International Astronomical Union 2015
doi:10.1017/S174392131401031X

Gas inflow and AGN-driven outflow in M51

M. Querejeta[1], S. E. Meidt[1], E. Schinnerer[1], S. García-Burillo[2], J. Pety[3], A. Hughes[1], D. Meier[4], F. Bigiel[5], K. Kreckel[1] and G. Blanc[6]

[1]Max-Planck-Institut für Astronomie, Königstuhl 17, D-69117 Heidelberg, Germany
[2]OAN, Madrid, Spain; [3]IRAM, Grenoble, France; [4]New Mexico Tech, USA; [5]University of Heidelberg, Germany; [6]Carnegie Institution, USA. Email: querejeta@mpia.de

Abstract. We study the feeding and feedback of the nucleus of M51 by considering gravitational torques, responsible for gas inflow, in relation to the local distribution of dense gas.

Keywords. galaxies: AGN — galaxies: structure — galaxies: evolution

1. Context

Feedback from AGN is often invoked as a mechanism that regulates star formation. In spite of its cosmological relevance, the way it operates in practice is poorly understood, apart from the basic idea that it must somehow remove the fuel of star formation (cold dense gas) from the disc. The low inclination ($\sim 22°$) and proximity of M51 (7.6 Mpc) make it an ideal target to directly link the feeding of the AGN with its feedback.

2. Torques driving gas inwards

Gravitational torques add or remove angular momentum from the gas ($\tau = dL/dt$), thus effectively driving it outwards or inwards. We use our ICA-corrected $3.6\mu m$ mass map (Querejeta *et al.* 2014) and CO(1-0) imaging from PAWS (Pety *et al.* 2013) to measure the gravitational torques in M51. These imply clear gas inflow to fuel nuclear activity over the entire central region, at an integrated rate of $\sim 20 M_\odot/yr$ inside 20".

3. Molecular outflow

A redshifted and possible blueshifted peak in the CO PAWS velocity field are likely signatures of a molecular outflow. We have obtained new Plateau de Bure (PdBI) observations of the nucleus in three dense gas tracers: HCN, HCO$^+$ and HNC. An off-centered, lopsided dense gas feature along the radio jet is clearly visible in the morphology and kinematics. Interestingly, HCN shows more extreme velocities than CO in this region, suggesting a high degree of jet-ISM interaction.

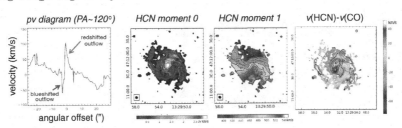

Figure 1. Position-velocity diagram in the center of the PAWS CO map (PA$\sim$ 120°), pointing to the outflow. HCN maps exhibit a significant velocity difference with respect to CO.

Galaxies in 3D across the Universe
Proceedings IAU Symposium No. 309, 2014
B. L. Ziegler, F. Combes, H. Dannerbauer, M. Verdugo, eds.

© International Astronomical Union 2015
doi:10.1017/S1743921314010321

Near-IR Integral Field Spectroscopy of the central region of NGC 5929

Rogemar A. Riffel[1], Thaisa Storchi-Bergmann[2] and Rogério Riffel[2]

[1] Universidade Federal de Santa Maria, Departamento de Física/CCNE, 97105-900, Santa Maria, RS, Brazil.
email: `rogemar@ufsm.br`

[2] Universidade Federal do Rio Grande do Sul, Instituto de Física, CP 15051, Porto Alegre 91501-970, RS, Brazil.
email: `thaisa@ufrgs.br; riffel@ufrgs.br`

Abstract. We present two-dimensional (2D) near-infrared spectra of the inner 300×300 pc^2 of the Seyfert 2 galaxy NGC 5929 at a spatial resolution of $\sim$20 pc obtained with the Gemini Near infrared Integral Field Spectrograph (NIFS). We present 2D maps for the emission line flux distributions and kinematics and report the discovery of a linear structure $\sim$300 pc in extent and of $\sim$50 pc in width oriented perpendicular to the radio jet, showing broadened emission-line profiles.

While over most of the field the emission-line profiles have full-widths-at-half-maximum (FWHM) of $\sim$210 km/s, at a linear structure perpendicular do the radio jet the emission-line FWHMs are twice this value, and are due to two velocity components, one blueshifted and the other redshifted relative to the systemic velocity. We attribute these velocities to an outflow from the nucleus which is launched perpendicular to the radio jet. We reported the detection of this peculiar outflow in Riffel, Storchi-Bergmann & Riffel (2014a), where more details of the analysis can be found. Since, NGC 5929 has a Type 2 nucleus, this detection implies that: (1) both ionizing radiation and relativistic particles are escaping through holes in the torus perpendicular to the radio jet; and/or (2) the torus is also outflowing, as proposed by recent models of tori as winds from the outer parts of an accretion flow; or (3) the torus is absent in NGC 5929.

At other locations the gas kinematics is dominated by rotation in a disk, although some evidences of interaction of the radio jet with the emitting gas are seen as a broadening of the line profiles at the locations of the radio structures.

The flux distributions for the [P II], [Fe II], H I and H$_2$ emission lines show that the line emission is more extended along the $PA = 60/240°$, extending to up to 1.5″ to both sides of the nucleus, while to the perpendicular direction ($PA = -30/150°$) the emission is extended to 0.7″ from the nucleus. The flux distributions of all emission lines show a good correlation with radio the radio structures, with the two peak of emission associated to the soutwestern and northeastern radio knots. Some differences are observed among distinct emission lines. While the [Fe II] and H$_2$ emission peak at the location of the soutwestern radio structure at 0.6″ from the nucleus, the H I recombination lines present the their highest fluxes at the location of the northeastern radio hotspot at 0.5″ from the nucleus. Another difference is that the H$_2$ emission is less collimated than that for other lines, being more extended perpendicularly to the radio jet. A detailed analysis of the line emission and kinematics will be presented in Riffel, Storchi-Bergmann & Riffel (2014b).

References

Riffel, R. A., Storchi-Bergmann, T., & Riffel, R. 2014a, *ApJ*, 780, 24.
Riffel, R. A., Storchi-Bergmann, T., & Riffel, R. 2014b, *in preparation*.

Galaxies in 3D across the Universe
Proceedings IAU Symposium No. 309, 2014
B. L. Ziegler, F. Combes, H. Dannerbauer, M. Verdugo, eds.

© International Astronomical Union 2015
doi:10.1017/S1743921314010333

Spiral Galaxy in 3D as Seen with SpIOMM

Laurie Rousseau-Nepton, Carmelle Robert and Laurent Drissen

Université Laval & Centre de Recherche en Astrophysique du Québec, Québec, Canada

Abstract. Using the imaging Fourier transform spectrograph (FTS) SpIOMM we study 7 nearby spiral galaxies. The large database of spectra obtained around Hα and Hβ is ideal to study the star forming regions and warm ionized medium (WIM) with a high spatial resolution (∼ 50-150 pc).

Keywords. galaxies: nearby spiral, galaxies: evolution - galaxies: star formation

1. SpIOMM

SpIOMM, the imaging FTS of the Mont-Mégantic Observatory, offers a large FOV(12'x12'), a good spatial resolution of 1.1" (435 500 pixels and spectra), and a spectral coverage in selected bandpasses of the visible with an adjustable resolution from 1 to 20 000. Two data cubes have been obtained (∼5 hrs/cube) for each galaxy covering the spectral domains of 4750-5150 A (R∼650) and 6500-6800 A (R∼2000).

2. HII Regions and the WIM

HII regions (HIIR) are detected over the whole galactic disks. For example, we identified 566 HIIR with a SNR >8 on NGC628. Also, emission from a WIM (Haffner & al. 1999; Blanc & al. 2009) can be detected beyond the HIIR boundaries (Fig. 1). The WIM can be related to the HIIR escaping photons, but also to SNe, PNe, and other sources including AGB stars, AGN, and shocks. These ionizing sources are not necessary physically linked together and produce very different physical conditions in the surrounding gas. For NGC 628, the metallicity gradient calculated after taking into account the effect of the WIM becomes steeper (Fig. 2; Rousseau-Nepton *et al.* in prep).

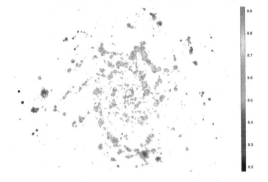

Figure 1. NGC 628 metallicity map using the O3N2 indicator for all spaxels above 4σ.

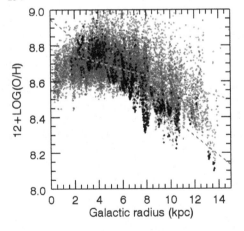

Figure 2. NGC 628 metallicity gradient. Blue dots and the orange dashed line are measurement in all spaxels above 4σ. Black dots and the red dashed line are from selected pixels in HII regions.

References

Blanc, G. A., *et al.* 2009, *ApJ*, 704, 842
Haffner, L. M., *et al.* 1999, *ApJ*, 523, 223

Galaxies in 3D across the Universe
Proceedings IAU Symposium No. 309, 2014
B. L. Ziegler, F. Combes, H. Dannerbauer, M. Verdugo, eds.

© International Astronomical Union 2015
doi:10.1017/S1743921314010345

Kennicutt-Schmidt relation in the HI dominated regime

Sambit Roychowdhury,[1] Mei-Ling Huang,[1] Guinevere Kauffmann,[1] and Jayaram N. Chengalur[2]

[1] Max-Planck-Institut für Astrophysik, Karl-Schwarzschild-Str. 1, 85748 Garching, Germany
email: sambit@mpa-garching.mpg.de
[2] NCRA-TIFR, Post Bag 3, Ganeshkhind, Pune 411 007, India

Abstract. We investigate the existence of a Kennicutt-Schmidt type relation between $\Sigma_{\rm SFR}$ and $\Sigma_{\rm gas,atomic}$ in the HI dominated regions of nearby spirals and dwarf irregulars.

1. Procedure and Results

We study the Kennicuut-Schmidt relation (Kennicutt 1998) in the HI dominated interstellar medium (ISM) of nearby star-forming galaxies, the disk-averaged version of which was studied recently for dwarfs (Roychowdhury *et al.* 2014). The samples used for this work were: (i) dwarf irregular galaxies from the FIGGS survey (Begum *et al.* 2008), (ii) spiral galaxies from the THINGS survey with CO observations from the HERACLES survey from Leroy *et al.* (2013). We determine the relation between $\Sigma_{\rm gas,atomic}$ and $\Sigma_{\rm SFR}$ at the resolution of the CO maps in regions 1 kpc in size for the THINGS galaxy sample, and at 400 pc resolution in regions 400 pc in size for FIGGS and THINGS galaxies.

In the HI dominated regime the low star formation rates (SFRs) measured are affected by the stochastically sampling of the high mass end of the IMF and bursty non-continuous SF history, when SFR calibrations assume continuous star formation during the previous Myr or so. da Silva, Fumagalli & Krumholz (2014) have simulated the variation of SFR measured using different tracers with variation in the underlying true SFR and star formation history, and find that FUV is a more trustworthy tracer. We use FUV from public *GALEX* images to trace SFR, and use public *Spitzer* 24 μm fluxes to account for internal dust extinction by following the 'composite' calibration by Hao *et al.* (2011).

For THINGS galaxies we choose HI dominated region by excluding regions with detected CO emission. In order to ensure that we are sampling enough SFR to be unaffected by the uncertainties in low SFR measurements described above, we determine the average $\Sigma_{\rm SFR}$ in each $\Sigma_{\rm gas,atomic}$ bin. We find that this results in $\Sigma_{\rm SFR}$ falling off with $\Sigma_{\rm gas,atomic}$ much less steeply than what was seen by Bigiel *et al.* (2010) for the very outskirts of spirals. We find power law slopes between 1.3 and 1.6 for all samples in HI dominated ISM, but gas consumption timescales of 10 Gyrs and more.

References

Begum, A., *et al.* 2008, *MNRAS*, 386, 1667
Bigiel, F., *et al.* 2010, *AJ*, 140, 1194
da Silva, R. L., Fumagalli, M., & Krumholz, M. R., 2014, arXiv:1403.4605
Hao, C.-N., *et al.* 2011, *ApJ*, 741, 124
Kennicutt, Jr. R. C. 1998, *ApJ*, 498, 541
Leroy, A. K., *et al.* 2013, *AJ*, 146, 19
Roychowdhury, S., *et al.* 2014, *MNRAS*, in press

Galaxies in 3D across the Universe
Proceedings IAU Symposium No. 309, 2014
B. L. Ziegler, F. Combes, H. Dannerbauer, M. Verdugo, eds.

© International Astronomical Union 2015
doi:10.1017/S1743921314010357

An Infrared Luminous Merger with Two Bipolar Molecular Outflows : ALMA View of NGC 3256

K. Sakamoto[1], S. Aalto[2], F. Combes[3], A. Evans[4,5] and A. Peck[4]

[1] ASIAA,Taiwan, [2] Univ. Chalmers, OSO, Sweden, [3] Observatoire de Paris, France,
[4] NRAO, USA, [5] Univ. Virginia, USA

Abstract. We found with ALMA 3D spectroscopy two bipolar molecular outflows in the luminous infrared galaxy NGC 3256. Each of the two merger nuclei has its own bipolar outflow.

Keywords. galaxies: active - galaxies: evolution - ISM: jets and outflows

Merging of disk galaxies tends to drive gas in the progenitor galaxies to their centers and thereby trigger intense star formation and/or nuclear activities there. The resulting high luminosity nuclei often have gas outflows, usually (but not necessarily always) perpendicular to the remnant disks. The already complex 3D structures of mergers are further complicated with this out-of-plane gas distribution and motion. It therefore needs high-quality data to recognize and characterize a luminous merger with its outflow(s).

Fig. 1 shows our recent ALMA observations of an infrared luminous merger NGC 3256, the most luminous galaxy within z = 0.01. In addition to the large-scale rotation with the major axis p.a.∼ 75° and velocity range within $V_{sys} \pm 200$ km s^{-1} (left panel), there are high-velocity emission around the two nuclei (middle panel). The one around the southern merger nucleus is clearly bipolar and distinct from the large-scale gas motion (right panels). Subtracting it (in mind) in the middle panel, one can see another bipolar high-velocity gas around the northern nucleus. Our analysis suggests that the northern high-velocity gas is a starburst-driven molecular superwind with a wide opening angle and that the southern high-V gas is a well collimated bipolar molecular jet plausibly driven by an AGN in the southern nucleus (Sakamoto *et al.* 2014).

Spatially-resolved 3D studies like this are needed to accurately characterize the feedback processes in mergers through their evolutionary stages.

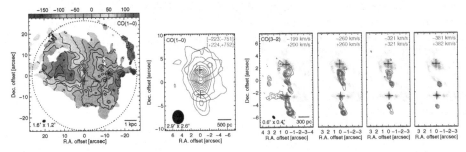

Figure 1. (Left) Overall CO velocity field. (Middle) CO(1-0) high-velocity gas. (Right 4 panels) CO(3-2) high-velocity gas. The two nuclei are marked with + signs.

References

Sakamoto, K., Aalto, S., Combes, F., Evans, A., & Peck, A. 2014, submitted to *ApJ*

Galaxies in 3D across the Universe
Proceedings IAU Symposium No. 309, 2014
B. L. Ziegler, F. Combes, H. Dannerbauer, M. Verdugo, eds.

© International Astronomical Union 2015
doi:10.1017/S1743921314010369

Spatially resolved stellar populations with SAMI

Nicholas Scott[1,2] and the SAMI team

[1]Sydney Institute for Astronomy, The University of Sydney, 44 Rosehill Street, Redfern, NSW, 2016 Australia
[2]ARC Centre of Excellent for All-Sky Astronomy (CAASTRO)
email: nscott@physics.usyd.edu.au

Abstract. Using data from the SAMI Galaxy Survey we measure azimuthally averaged stellar age and metallicity profiles for ~ 500 galaxies, using both luminosity-weighted Lick indices and mass-weighted full spectral fitting. We find a weak trend for steeper (i.e. more negative) metallicity gradients in more massive galaxies, however, below stellar masses $\sim 10^{10.5}$ M$_\odot$, the scatter in metallicity gradient increases dramatically.

Keywords. galaxies: abundances - galaxies: evolution - galaxies: formation

1. Introduction

The spatial distribution of the stellar population of a galaxy holds important clues as to its formation and assembly. Age gradients can reveal when and where new gas is accreting onto a galaxy. Metallicity gradients can be used to distinguish between stars that formed in-situ and stars that have been accreted onto a galaxy. In particular, a change in metallicity gradient, from relatively steep to shallow or flat is a strong indicator of a change from predominantly in-situ (or monolithic collapse-like) to accreted stellar populations (e.g. Font *et al.* 2011).

2. Sample and Methodologies

The Sydney-AAO Multi-object Integral-field spectrograph (SAMI) is an innovative new instrument on the 4m Anglo-Australian Telescope. It uses a fused-fibre "hexabundle" design to take simultaneous integral-field spectroscopy of up to 13 targets. SAMI provides simultaneous blue wavelength coverage from 3800 to 5700 Å at R$\sim$ 1700 and red wavelength coverage from 6300 to 7500 Å at R$\sim$ 4500. The SAMI Galaxy Survey (SGS) is a large program with SAMI targeting ~ 3400 galaxies in the redshift range $0.01 < z < 0.1$, spanning a broad range of stellar mass and local environmental density. This study uses the first ~ 20 per cent of observations from the SGS.

We extract stellar population information by radially binning our spatially resolved spectroscopy. For each radial bin we measure stellar ages, t, and metallicities, [Z/H], through two complementary techniques:

(*a*) **Luminosity-weighted age and metallicity.** We measure 5 Lick (Worthey *et al.* 1994) spectral indices, Hβ, Mgb, Fe5015, Fe5270 and Fe5335. Following Kuntschner *et al.* (2010) we use the stellar population models of Schiavon (2007) to derive [Z/H] and t.

(*b*) **Mass-weighted age and metallicity.** Following McDermid *et al.* (2014), we fit weighted combinations of Single Stellar Population (SSP) model template spectra from Vazdekis *et al.* (2010) to our observations. From the relative weights assigned to each template we then determine the mean mass-weighted [Z/H] and t.

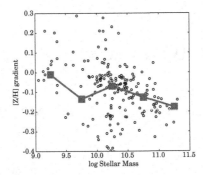

Figure 1. Mass-weighted [Z/H] gradient vs. galaxy stellar mass. The large points and solid line indicate the mean values within 5 narrow bins of stellar mass. Below $M_\star \simeq 10^{10.5}$ $M_\odot$ the range of gradients increases dramatically. Above this mass we find a weak trend of increasing steep (i.e. more negative) gradients with increasing stellar mass.

3. Results

We are able to extract age and metallicity gradients for $\gtrsim 75$ per cent of galaxies in our partial SGS sample. The success rate for the luminosity-weighted and mass-weighted techniques is comparable. For a subset of ~ 200 of our galaxies, for which we measure [Z/H] in multiple radial bins with small uncertainties, and for which we have obtained reliable effective radii, R_e, and stellar masses, $M_\star$, we derive metallicity gradients, Δ [Z/H] = Δ [Z/H] / Δ R_e (Figure 1). We find that Δ [Z/H] is typically negative, i.e. metallicity decreases with galactic radius. We find a weak trend of increasingly steep (i.e. more negative Δ [Z/H]) in more massive galaxies. Below stellar masses $\sim 10^{10.5}$ the scatter in this relation increases dramatically.

4. Outlook

With the full SGS sample of ~ 3400 galaxies we will be able to derive age and metallicity profiles for thousands of galaxies out to $1 - 2$ R_e, an order of magnitude more than any previous study. For the first time, we will be able to explore the dependence of stellar population gradients on stellar mass and environmental density in a large sample of galaxies, which will provide important constraints for the upcoming generation of cosmological galaxy formation models.

Acknowledgements

This research was conducted by the Australian Research Council Centre of Excellence for All-sky Astrophysics (CAASTRO), through project number CE110001020.

References

Font, A. S., McCarty, I. G., Crain, R. A., Theuns, T., Schaye, J., Wiersma, R. P. C., & Dalla Vecchia, C., , 2010, *MNRAS*, 416, 2802
Kuntschner, H., *et al.*, 2010, *MNRAS*, 408, 97
McDermid, R., *et al.*, 2014, *MNRAS*, submitted
Schiavon, R. P., 2007, *ApJS*, 171, 146
Vazdekis, A., *et al.*, 2010, *MNRAS*, 404, 1639
Worthey, G., Faber, S. M., Gozalez, J. J., & Burstein, D., 1994, *ApJS*, 94, 687

Galaxies in 3D across the Universe
Proceedings IAU Symposium No. 309, 2014
B. L. Ziegler, F. Combes, H. Dannerbauer, M. Verdugo, eds.

© International Astronomical Union 2015
doi:10.1017/S1743921314010370

Environmental dependence of galaxy formation explored by near-infrared spectroscopy of two protoclusters at z>2

Rhythm Shimakawa[1,2], Tadayuki Kodama[2,3], Ken-ichi Tadaki[3], Masao Hayashi[3], Yusei Koyama[4] and Ichi Tanaka[1]

email: rhythm@naoj.org

[1] Subaru telescope, National Astronomical Observatory of Japan
650 North A'ohoku Place, Hilo, HI 96720, USA
[2] Department of Astronomy, School of Science, Graduate University for Advanced Studies
Mitaka, Tokyo 181-8588, Japan
[3] Optical and Infrared Astronomy Division, National Astronomical Observatory
Mitaka, Tokyo 181-8588, Japan
[4] Institute of Space Astronomical Science, Japan Aerospace Exploration Agency
Sagamihara, Kanagawa, 252-5210, Japan

Abstract. Protoclusters at high redshifts are the ideal laboratories to study how the environmental dependences of galaxy properties seen in local Universe were initially set up when the progenitors of present-day early-type galaxies were in their early formation phases. We have conducted a deep near-infrared spectroscopy of Hα emitters (HAEs) associated with two protoclusters (PKS 1138–262 at $z = 2.16$ and USS 1558–003 at $z = 2.53$) with the Multi-Object Infrared Camera and Spectrograph (MOIRCS) on the Subaru telescope.

As a result, the cluster membership of 27 and 36 HAEs are newly confirmed in these two protoclusters, respectively. The inferred dynamical masses of the protocluster cores are consistent with being the typical progenitors of present-day most massive clusters (Shimakawa *et al.* 2014a). Also, those HAEs in the protoclusters show much higher [OIII]/Hβ ratios than local star forming galaxies. It is probably caused by the combination of their much higher specific star formation rates, lower gaseous metallicities and redshift evolution of inter-stellar medium.

We also find that the mass-metallicity relation in the protocluster galaxies is offset to higher metallicity compared to those of field galaxies at a given stellar mass at M$_\star < 10^{11}$ M$_\odot$ (Shimakawa *et al.* 2014b). This trend is compatible with the recent work (Kulas *et al.* 2013). The mass-metallicity relation is regulated not only by star formation history hence metal production history, but also by inflow and outflow processes that are known to be very active at $z > 2$ (Steidel *et al.* 2010). It suggests that the higher gaseous metallicities of protocluster galaxies may be caused by those gas transfer processes that are dependent on surrounding environments.

Keywords. galaxies: clusters: general — galaxies: evolution — galaxies: formation

Acknowledgements

This work is supported in part by a Grant-in-Aid for the Scientific Research (Nos. 21340045 and 24244015) by the Japanese Ministry of Education, Culture, Sports, Science and Technology.

References

Shimakawa, R., Kodama, T., Tadaki, K.-i., *et al.*, 2014a, *MNRAS*, 441, L1
Shimakawa, R., Kodama, T., Tadaki, K.-i., *et al.*, 2014b, arXiv:1406.5219
Kulas, K. R., McLean, I. S., Shapley, A. E., *et al.*, 2013, *ApJ*, 774, 130
Steidel, C. C., Erb, D. K., Shapley, A. E., *et al.*, 2010, *ApJ*, 717, 289

Galaxies in 3D across the Universe
Proceedings IAU Symposium No. 309, 2014
B. L. Ziegler, F. Combes, H. Dannerbauer, M. Verdugo, eds.

© International Astronomical Union 2015
doi:10.1017/S1743921314010382

A strong clustering of FIR-selected galaxies in the AKARI All-Sky Survey

Tomoko Suzuki[1,2], Tsutomu T. Takeuchi[2], Agnieszka Pollo[3,4] and Shinki Oyabu[2]

[1]Department of Astronomical Science, The Graduate University for Advanced Studies, Mitaka, Tokyo 181-8588, Japan
email: suzuki.tomoko@nao.ac.jp
[2]Department of Particle and Astrophysical Science, Nagoya University, Furo-cho, Chikusa-ku, Nagoya 464-8602, Japan
[3]National Centre for Nuclear Research, Hoża 69, 00-681 Warsaw, Poland
[4]Astronomical Observatory of the Jagiellonian University, Orla 171, 30-001 Cracow, Poland

Abstract. Various previous galaxy surveys have revealed that different types of galaxies show different spatial distributions. By comparing their clustering strengths, we can investigate the relationship between the environments where galaxies reside and the properties of galaxies. This provides us vital information on when, where, and how galaxies are formed. In this study, we focus on FIR-selected galaxies from AKARI All-Sky Survey, which are considered to be dusty star-forming galaxies. We measure their power spectrum and compare our final result with previous studies using different galaxy samples detected at other wavelengths. It is found that there is the very clear wavelength dependence of the clustering strength.

Keywords. galaxies: formation - galaxies: evolution - galaxies: star formation - infrared: galaxies

1. Results

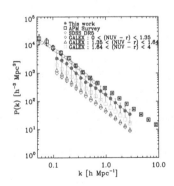

Figure 1. Comparison of our result with power spectra from other galaxy samples.

We obtained 18,077 galaxy candidates from AKARI/ FIS Bright Source Catalog v.1.0 [1]. Our result derived from FIR-selected galaxy sample is shown as red filled circles in Fig.1. It is well approximated by a single power-law and is consistent with the previous study using IRAS PSCz catalog [3]. In Fig.1, we compare our result with three galaxy samples from APM survey [4], SDSS [5], and GALEX [6]. It is found that FIR-selected galaxies have lower clustering strength than optical-selected galaxies, but on the other hand, their clustering strength is higher than UV-selected galaxies. FIR-selected galaxies with strong dust emission are expected to be more actively star-forming than UV-selected galaxies on average [7]. Thus, our result suggests that more actively star-forming galaxies distribute in denser regions. This maybe reflecting the spatial propagation of star-forming activities from dense regions to less dense regions, or this maybe due to some external environmental effects inducing dusty star formation in dense environments. As future work, it is necessary to resolve the internal structure of FIR-selected galaxies with AO imaging and/or 3D spectroscopy to investigate the physical processes occurring in each galaxy.

References

[1] Yamamura, *et al.* 2010

[2] Pollo, *et al.* 2013a, EP&S, 65, 273

[3] Hamilton, & Tegmark 2002, *MNRAS*, 330, 506

[4] Padilla, & Baugh 2003, *MNRAS*, 343, 796

[5] Percival, *et al.* 2007, *ApJ*, 657, 645

[6] Heinis, *et al.* 2009, *ApJ*, 698, 1838

[7] Buat, *et al.* 2007, *ApJS*, 173, 404

Galaxies in 3D across the Universe
Proceedings IAU Symposium No. 309, 2014
B. L. Ziegler, F. Combes, H. Dannerbauer, M. Verdugo, eds.
© International Astronomical Union 2015
doi:10.1017/S1743921314010394

NGC 5128's Globular Cluster System: Is There a Dark Side?

Matthew Taylor[1], Thomas Puzia[1], Matias Gomez[2] and Kristin Woodley[3]

[1]Institute of Astrophysics, Pontificia Universidad Católica de Chile, Avenida Vicuña Mackenna 4860, Santiago, Chile
[1]email: mtaylor@astro.puc.cl; tpuzia@astro.puc.cl
[2]Depto. de Ciencias Fisicas, Universidad Andres Bello, República 220, Santiago, Chile
email: matiasgomez@unab.cl
[3]University of California, Santa Cruz, UC Observatories, 1156 High Street, Santa Cruz, USA

Abstract. Dynamical mass estimates ($\mathcal{M}_{\mathrm{dyn}}$) and mass-to-light ratios (Υ_{dyn}) were derived for a sample of NGC 5128's globular clusters (GCs). We find two distinct sequences in the Υ_{dyn}-$\mathcal{M}_{\mathrm{dyn}}$ plane, which are well fit by power laws of the forms $\Upsilon_{\mathrm{dyn}} \propto \mathcal{M}_{\mathrm{dyn}}^{0.33\pm0.04}$ and $\Upsilon_{\mathrm{dyn}} \propto \mathcal{M}_{\mathrm{dyn}}^{0.91\pm0.04}$. The former traces bright "classical" GCs, and the latter represents objects that are dynamically dominated by large dark matter components or by abnormally massive central black holes.

Keywords. galaxies:individual:NGC 5128; galaxies:globular clusters:general

1. Introduction

Large globular cluster (GC) systems commonly found around giant galaxies are useful tracers of the formation histories of their hosts (Brodie & Strader 2006). GCs populate a narrow plane defined by their structural parameters (Djorgovski 1995) and are accepted to lack dark matter (DM), but this may not be theoretically expected (e.g. Peebles 1984).

2. Conclusions

NGC 5128 seems to have two types of high-$\mathcal{M}_{\mathrm{dyn}}$ ($\gtrsim 10^6 M_\odot$) GCs that follow distinct power-laws in the Υ_{dyn}-$\mathcal{M}_{\mathrm{dyn}}$ plane. One population is simply the bright tail of the GC luminosity function, i.e. bright but otherwise "classical" GCs. The other group is made up of objects with significant, concentrated DM components and/or they host massive central BHs that dominate the cluster dynamics out to large radii. Other artificial effects were ruled out, so we suggest that the histories of these "high-Υ_{dyn}" clusters are markedly different from "classical" GCs, with potentially important cosmological implications.

Acknowledgements

We acknowledge the Vicerrectoría de Investigación, and Institute of Astrophysics Graduate Fund at Pontificia Universidad Católica de Chile, FONDECYT Regular Project Grant 1121005, and BASAL Center for Astrophysics and Associated Technologies PFB06.

References

Brodie, J. P. & Strader, J., 2006, *ARA&A*, 44, 193
Djorgovski, S., 1995, *ApJ*, 438, L29
Peebles, P. J. E.., 1984, *ApJ*, 277, 470

Galaxies in 3D across the Universe
Proceedings IAU Symposium No. 309, 2014
B. L. Ziegler, F. Combes, H. Dannerbauer, M. Verdugo, eds.
© International Astronomical Union 2015
doi:10.1017/S1743921314010400

The Angular Momentum Dichotomy

Adelheid Teklu[1], Rhea-Silvia Remus[1], Klaus Dolag[1,2] and Andreas Burkert[1,3]

[1]University Observatory Munich, Scheinerstr. 1, 81679 Munich, Germany
[2]MPI for Astrophysics, Karl-Schwarzschild-Str. 1, 85741 Garching, Germany
[3]MPI for Extraterrestrial Physics, Giessenbachstr. 1, 85748 Garching, Germany
email: ateklu@usm.uni-muenchen.de

Abstract. In the context of the formation of spiral galaxies the evolution and distribution of the angular momentum of dark matter halos have been discussed for more than 20 years, especially the idea that the specific angular momentum of the halo can be estimated from the specific angular momentum of its disk (e.g. Fall & Efstathiou (1980), Fall (1983) and Mo *et al.* (1998)). We use a new set of hydrodynamic cosmological simulations called *Magneticum Pathfinder* which allow us to split the galaxies into spheroidal and disk galaxies via the circularity parameter ε, as commonly used (e.g. Scannapieco *et al.* (2008)). Here, we focus on the dimensionless spin parameter $\lambda = J|E|^{1/2}/(GM^{5/2})$ (Peebles 1969, 1971), which is a measure of the rotation of the total halo and can be fitted by a lognormal distribution, e.g. Mo *et al.* (1998). The spin parameter allows one to compare the relative angular momentum of halos across different masses and different times. Fig. 1 reveals a dichotomy in the distribution of λ at all redshifts when the galaxies are split into spheroids (dashed) and disk galaxies (dash-dotted). The disk galaxies preferentially live in halos with slightly larger spin parameter compared to spheroidal galaxies. Thus, we see that the λ of the whole halo reflects the morphology of its central galaxy. For more details and a larger study of the angular momentum properties of disk and spheroidal galaxies, see Teklu *et al.* (in prep.).

Keywords. dark matter – galaxies: evolution – galaxies: formation – galaxies: halos

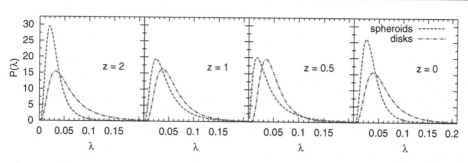

Figure 1. λ-distribution for spheroids (red dashed) and disk galaxies (blue dash-dotted) at four different redshifts. The dichotomy occurs at all redshifts. The spheroids have generally lower λ-values and thus less specific angular momentum than the disks.

References

Fall, M. & Efstathiou, G., 1980, *MNRAS*, 193, 189
Fall, M., 1983, IAUS, 100, 391
Mo, H., Mao, S., & White, S., 1998, *MNRAS*, 295, 319
Peebles, P., 1969, *ApJ*, 155, 393
Peebles, P., 1971, *A&A*, 11, 377
Scannapieco, C., Tissera, P., White, S., & Springel, V., 2008, *MNRAS*, 389, 1137

Galaxies in 3D across the Universe
Proceedings IAU Symposium No. 309, 2014
B. L. Ziegler, F. Combes, H. Dannerbauer, M. Verdugo, eds.

© International Astronomical Union 2015
doi:10.1017/S1743921314010412

Radio Recombination Line studies on M82 from LOFAR HBA observations

M. C. Toribio[1], L. K. Morabito[2], J. B. R. Oonk[1,2], F. Salgado[2], A. G. G. M. Tielens[2] and H. J. A. Röttgering[2]

[1]Netherlands Institute for Radio Astronomy (ASTRON),
Postbus 2, 7990 AA Dwingeloo, The Netherlands
email: toribio@astron.nl
[2]Leiden Observatory, Leiden University,
P. O. Box 9513, 2300 RA Leiden, The Netherlands

Abstract. We continue our search for extragalactic Carbon Radio Recombination Lines (CRRL) in M82 with the LOw Frequency ARray (LOFAR; van Haarlem *et al.* 2013). The goal of our project is to determine the physical conditions of the cold neutral gas in this object, for which low frequency radio recombination lines can provide a sensitive probe.

Keywords. galaxies: ISM

LOFAR is capable of detecting CRRLs in extragalactic sources. The first ever extragalactic detection of these lines was achieved by us using data taken with LOFAR's Low Band Antenna (LBA) in the 48-64 MHz frequency range in the nearby starbursting galaxy M82 (Morabito *et al.* 2014). By studying the frequency behaviour of these CRRLs at low frequencies (< 1 GHz) we can constrain the physical properties (temperature, density and abundances) of the cold, neutral medium (e.g., Walmsley & Watson 1982).

We are now expanding our search for CRRLs to the frequency range 110-190 MHz with the LOFAR High Band Antennae (HBA). First tests indicate that we reach a spectral rms $\sim 10^{-3}$ per 6 km/s. By stacking ~ 20 lines we should be able to detect CRRLs from M82 at these frequencies. Futhermore, the unprecedented sensitiviy and resolution of the HBAs ($\sim$5 arcsec) will allow us to identify the location of the CRRL emitting gas.

Acknowledgements

MCT acknowledges financial support from LKBF. LKM acknowledges financial support from NWO Top LOFAR project, project n. 614.001.006. JBRO acknowledges financial support from NWO Top LOFAR-CRRL project, project n. 614.001.351.

References

van Haarlem, M. P., Wise, M. W., Gunst, A. W., *et al.* 2013, *A&A*, 556, A2

Morabito, L. K., Oonk, J. B. R., Salgado, F., Toribio, M. C., Tielens, A. G. G. M, & and Röttgering, H. 2014, *Proc. of the IAU Symposium No. 309: Galaxies in 3D across the Universe*, B. L. Ziegler, F. Combes, H. Dannerbauer, M. Verdugo, eds. (Cambridge University Press)

Walmsley, C. M. & Watson, W. D. 1982, *ApJ*, 260, 317

Galaxies in 3D across the Universe
Proceedings IAU Symposium No. 309, 2014
B. L. Ziegler, F. Combes, H. Dannerbauer, M. Verdugo, eds.

© International Astronomical Union 2015
doi:10.1017/S1743921314010424

3D Study Of Magnetic Fields In NGC 6946

Anna Williams[1], George Heald[2,3], Eric Wilcots[1] and Ellen Zweibel[1]

[1]Department of Astronomy, University of Wisconsin-Madison,
475 N. Charter Street, Madison, WI, USA
email: `williams@astro.wisc.edu`, `ewilcots@astro.wisc.edu`, `zweibel@astro.wisc.edu`
[2]Astron Postbus 2, 7990 AA Dwingeloo, The Netherlands
[3]Kapteyn Astronomical Institute, Postbus 800, 9700 AV, Groningen, The Netherlands
email: `heald@astron.nl`

Abstract. Recent advancements in both radio observatories and computing have opened a new regime of 3D observations. Not only do these instruments measure emission lines and radio continuum over much larger bandpasses, but they also simultaneously observe the polarized emission over the same large bandpasses with increased sensitivity. This "polarization spectrum" can be used to recover information about the 3D structure of magnetic fields in the universe. Our combined 3-20 cm observations of NGC 6946 taken with the Westerbork Synthesis Radio Telescope provide highly sensitive diagnostics of the internal depolarization across the galaxy. We use model fitting to determine likely mechanisms for depolarization in different regions of the galaxy, and glean information about the coherent and turbulent magnetic fields in NGC 6946. We produce Faraday dispersion maps that illustrate how we can probe different depths into the galaxy at different wavelengths and display new features of the line of sight magnetic field. This work is just a sample of the new 3D studies that are possible with upgraded and new radio instruments like the VLA, ATCA, and SKA.

Keywords. galaxies: ISM, magnetic fields, structure

Magnetic fields are important ingredients in the interstellar medium of galaxies. They accelerate cosmic rays, affect star formation, and regulate the redistribution of matter and energy. Despite their ubiquitous presence, their growth and coevolution with galactic processes is not well understood. We use 3-20 cm polarization observations from the Westerbork Synthesis Radio Telescope to map the turbulent and coherent line-of-sight magnetic fields across the disk of the nearly face-on galaxy NGC 6946. We fit the observed data with the Burn slab model as described by Sokoloff *et al.* (1998). This simple model considers a slab of magnetoionized medium threaded with a coherent, line-of-sight magnetic field and cells of random magnetic fields. Our results show strong internal Faraday dispersion (Burn slab) in the south-west quadrant where there is little polarized emission at wavelengths >13 cm, reinforcing the need to consider the 3D geometry of the magnetic fields in the interstellar medium. We also find strong correspondence between Faraday dispersion and HII regions. For additional information on the model fitting procedure, results, and future work, please go to www.astro.wisc.edu/~williams for the poster PDF.

Acknowledgements

We acknowledge support from R. Beck (MPIfR), Astron, the National Science Foundation, and NASA.

References

Sokoloff, D. D., *et al.* 1998, *MNRAS*, 299, 189

Galaxies in 3D across the Universe
Proceedings IAU Symposium No. 309, 2014
B. L. Ziegler, F. Combes, H. Dannerbauer, M. Verdugo, eds.
© International Astronomical Union 2015
doi:10.1017/S1743921314010436

A 3D Search for the Interplay between AGN and Star Formation in Galaxies

Marsha Wolf[1], Eric Hooper[1], Ryan Sanders[2] and Charles Liu[3]

[1]Department of Astronomy, University of Wisconsin, Madison, WI 53706, USA
email: mwolf@astro.wisc.edu
[2]UCLA Division of Astronomy & Astrophysics, Los Angeles, CA 90095-1547, USA
[3]Department of Engineering Science and Physics, CUNY CSI, Staten Island, NY 10314, USA

Abstract. Integral field spectroscopy and radio interferometry are very powerful tools for studying the interplay between AGN and star formation (SF) in galaxies. We introduce a sample of SDSS galaxies with selection criteria designed to maximize our chances of catching both processes in action. The galaxies are post-starburst, potentially contain radio AGN, and are allowed, but not required, to have ongoing star formation. The resulting sample includes objects classified as traditional post-starbursts and ones that would have been classified as Seyferts based on their emission line properties alone. The systems span a range of merger phases from initial interaction to fully merged, providing snapshots throughout the entire sequence. We are compiling a multi-wavelength data set, including spatially resolved optical spectra from IFUs on WIYN and continuum radio maps from the VLA and GMRT. Here we present initial results on J0754+1648, an interacting system with a post-starburst region near a radio AGN surrounded by highly ionized gas. This object may be an example of SF truncated by AGN feedback.

Keywords. galaxies: interaction - galaxies: evolution - galaxies: active

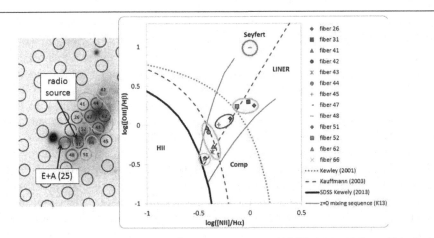

Figure 1. Layout of the SparsePak IFU (Bershady+, 2004, PASP, 116, 565) on J0754+1648 is on the left. Colored outlines mark spatial regions on the object and their corresponding locations on the BPT plot on the right (Kauffmann+, 2003, MNRAS, 346, 1055; Kewley+, 2001, ApJ, 556, 121; Kewley+, 2013, ApJ, 774, 100). The heavy black line is the mean SDSS abundance sequence and solid gray lines bound the starburst-AGN mixing sequence (Kewley+ 2013). The upper nucleus of the system, near fiber 44, is forming stars at a rate of 4 M_{sun} yr^{-1} (Sanders+, 2012, BAAS, 246.07). Fibers in this area (cyan) lie in the SF region of the BPT near the composite line. Fibers 26 & 45 that are spatially between the SF area and the AGN lie in the composite region of the BPT (blue). Fiber 31 contains the radio source and lies in the AGN/LINER region of the BPT (green). Fiber 48 has strong [OIII] emission, likely ionized by the AGN, and is fully in the Seyfert BPT region (magenta). The post-starburst (fiber 25) has a mean light-weighted age of 540 Myr and is in close proximity to the AGN and the ionized gas.

Galaxies in 3D across the Universe
Proceedings IAU Symposium No. 309, 2014
B. L. Ziegler, F. Combes, H. Dannerbauer, M. Verdugo, eds.

© International Astronomical Union 2015
doi:10.1017/S1743921314010448

Gaseous environment in LLAGN: modes of interaction with compact star nuclear population

M. Zajaček[1,2], V. Karas[1], L. Šubr[2], D. Kunneriath[1] and A. Eckart[3,4]

[1] Academy of Sciences, Boční II 1401, CZ-14100 Prague, Czech Republic
email: michal.zajacek31@gmail.com
[2] Charles University in Prague, V Holešovičkách 2, CZ-18000 Prague, Czech Republic
[3] Universität zu Köln, Zülpicher Strasse 77, D-50937 Köln, Germany
[4] MPIfR, Auf dem Hügel 69, D-53121 Bonn, Germany

Abstract. The Galactic center, which serves as a paradigm of low-luminosity active galactic nuclei (LLAGN), hosts the Nuclear star cluster (NSC) that contains both young and more evolved stars. So far the population of the end-products of stellar evolution has not been observationally confirmed and studied, although there are hints of its presence. We study the distribution of interaction modes of a hypothetical population of neutron stars within the sphere of influence of the Sgr A* supermassive black hole (SMBH). The comparison of our models with future observations could be used to constrain the 3D structure of the Galactic center.

1. Summary of the results

Interferometric data have been employed to explore the ionized medium near the centre of the Milky Way (Zhao *et al.* 2010). We use previous results (Kunneriath *et al.* 2012) about physical conditions of Sgr A West (the Minispiral) to estimate the probability of passages of the putative population of neutron stars of the NSC through the regions of enhanced density, and to assess different modes of the mutual interaction.

Streams of the Minispiral gas are ionized by ultraviolet radiation of massive OB stars present in the NSC. Based on the inferred dynamic mass in the central parsec ($\sim 10^6 M_\odot$), it is estimated that at least $\sim 10^4$ neutron stars should move in the sphere of influence of the Sgr A* SMBH ($\sim 2\,\mathrm{pc}$) (Morris 1993). According to our Monte Carlo simulations, a fraction of this unexplored population ($\lesssim 10\%$) propagates through denser ionized medium concentrated along the three arms of the Minispiral. Based on the density and the temperature of the gaseous environment inferred from observations, we analyse interaction regimes of neutron stars passing through this medium. Spectral features are expected to develop within the Minispiral due to non-thermal emission from bow-shocks and pulsar wind nebulae of strongly magnetized stars, and these could be revealed with the improved resolution in the near future (Zajaček 2014a; Zajaček *et al.* 2014b). The procedure may be applied to other galactic nuclei hosting NSC and the resulting distribution of interaction regimes is expected to vary across different galaxy types.

References

Kunneriath, D., Eckart, A., Vogel, S. N., *et al.*, 2012, *A&A*, 538, A127
Morris, M., 1993, *ApJ*, 408, 496
Zajaček, M., 2014a, MSc. Thesis, Charles University in Prague, Czech Republic
Zajaček, M., Karas, V., Kunneriath, D., 2014b, Proc. of IBWS 2014, Acta Polytechnica, submitted
Zhao, J.-H., Blundell, R., Moran, J. M., *et al.*, 2010, *ApJ*, 723, 1097

Galaxies in 3D across the Universe
Proceedings IAU Symposium No. 309, 2014
B. L. Ziegler, F. Combes, H. Dannerbauer, M. Verdugo, eds.

© International Astronomical Union 2015
doi:10.1017/S174392131401045X

Star formation enhancement characteristics in interacting galaxies

J. Zaragoza-Cardiel[1,2], J. E. Beckman[1,2,3], J. Font[1,2], A. Camps-Fariña[1,2], B. García-Lorenzo[1,2] and S. Erroz-Ferrer[1,2]

[1]Instituto de Astrofísica de Canarias, C/ Vía Láctea s/n, 38205 La Laguna, Tenerife, Spain
email: jzc@iac.es
[2]Department of Astrophysics, University of La Laguna, E-38200 La Laguna, Tenerife, Spain
[3]CSIC, 28006 Madrid, Spain

Abstract. We have observed 12 interacting galaxies using the Fabry-Perot interferometer GHαFaS (Galaxy Hα Fabry-Perot system) on the 4.2m William Herschel Telescope (La Palma). We have extracted the physical properties (sizes, Hα luminosity and velocity dispersion) of 236 HII regions for the full sample of interacting galaxies. We have derived the physical properties of 664 HII regions for a sample of 28 isolated galaxies observed with the same instrument in order to compare both populations of HII regions, finding that there are brighter and denser star forming regions in the interacting galaxies compared with the isolated galaxies sample.

Keywords. galalaxies: interactions – galaxies: kinematics and dynamics – stars: formation

---

The study of galaxy interactions kinematics, and comparison with the properties of the star forming regions reveals two populations of HII regions in Arp 270 (Zaragoza-Cardiel *et al.* 2013), and of HII regions and GMCs in the Antennae galaxies (Zaragoza-Cardiel *et al.* 2014). We compare a sample of 12 interacting galaxies with 28 isolated galaxies in order to understand the role played by the interaction in star formation enhancement. We see in Figure 1 (left) that the luminosity-radius relation for the brightest HII regions in interacting galaxies has a steeper slope than that for isolated galaxies. In Figure 1 (right) we see that the luminosity functions are different in the high luminosity range.

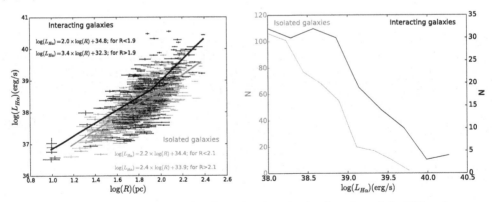

Figure 1. Left: $L_{H\alpha}$ versus R (size). Right: Luminosity function of the HII regions.

References

Zaragoza-Cardiel, J., Font-Serra, J., Beckman, J. E., *et al.* 2013, *MNRAS*, 432, 998
Zaragoza-Cardiel, J., Font, J., Beckman, J. E., *et al.* 2014, arXiv:1409.1251

Galaxies in 3D across the Universe
Proceedings IAU Symposium No. 309, 2014
B. L. Ziegler, F. Combes, H. Dannerbauer, M. Verdugo, eds.

© International Astronomical Union 2015
doi:10.1017/S1743921314010461

Galaxy population study of the 26 most massive galaxy clusters within the SPT footprint

Alfredo Zenteno

Cerro Tololo Inter-American Observatory, Casilla
603, La Serena, Chile
email: azenteno@ctio.noao.edu

Abstract. We present optical properties of the 26 most massive galaxy clusters in the South Pole Telescope 2500 sq-deg footprint. We find a general consistency between our results and results found in the literature on samples built with different selection techniques. Most interesting, we find a preference for an evolution in the slope of the Schechter function, α, with its value increasing at higher redshift.

Keywords. cluster of galaxies - galaxies: evolution - galaxies: formation

1. Introduction

The South Pole Telescope team has constructed a catalog of hundreds of galaxy clusters using the Sunyaev-Zel'dovich effect on the cosmic microwave background (Bleem *et al.* 2014). Two characteristics of the sample are that the selection function is nearly redshift independent and mass limited, making the SPT cluster sample ideal for evolutionary studies of the galaxy population. We construct the luminosity function (LF), the radial profile (RP) and the Halo Occupation number (HON), for the red and the total galaxy populations for the 26 most massive systems (Williamson *et al.* 2011), and quantified the evolution of these optical properties in a wide redshift range.

2. Results

We fit an NFW profile to the RP to find the galaxy concentration c_g. We observe that red galaxies seems to hint to a c_g evolution, with red galaxies more concentrated at lower redshift, but with a low significance. When all galaxies are used, c_g seems independent of redshift, with a value of ~ 2.3. To study the LF we fit for the Schechter function parameters ϕ^*, m^* and α. We find that ϕ^* evolves as $E^2(z)$, consistently with the self-similar expectation, while m^* evolves as predicted for a passively evolving population (Bruzual & Charlot 2003). On the other hand, α seems to evolve from $\sim -0.9/-0.7$ at redshift 1 to $\sim 1.15/ \ 0.98$ at redshift zero for the total/red population, at 2.3σ level. We find the HON versus mass consistent with previous studies (Lin *et al.* 2004), and the slope of the HON versus redshift consistent with zero, within 1σ.

References

Bleem, L. E., Stalder, B., de Haan, T., *et al.* 2014, arXiv:1409.0850
Bruzual G., Charlot S., 2003, *MNRAS*, 344, 1000
Lin, Y.-T., Mohr, J. J., & Stanford, S. A. 2004, *ApJ*, 610, 745
Williamson, R., Benson, B. A., High, F. W., *et al.* 2011, *ApJ*, 738, 139

Galaxies in 3D across the Universe
Proceedings IAU Symposium No. 309, 2014
B. L. Ziegler, F. Combes, H. Dannerbauer, M. Verdugo, eds.
© International Astronomical Union 2015
doi:10.1017/S1743921314010473

Bar effects on ionized gas properties and dust content in galaxy centers

A. Zurita[1,2], E. Florido[1,2], I. Pérez[1,2], P. Coelho[3] and D. A. Gadotti[4]

[1]Dpto. Física Teórica y del Cosmos, Universidad de Granada, 18071, Granada, Spain
[2]Instituto Carlos I de Física Teórica y Computacional, Granada, Spain
[3]Universidade Cruzeiro do Sul, Rua Galvão Bueno, 868, 01506-000, São Paulo, SP, Brazil
[4]European Southern Observatory, Casilla 19001, Santiago 19, Chile
email: azurita@ugr.es

Abstract. Observations and simulations indicate that bars are important agents to transfer material towards galaxy centers. However, observational studies devoted to investigate the effects of bars in galaxy centers are not yet conclusive. We have used a sample (Coelho & Gadotti 2011) of nearby face–on galaxies with available spectra (SDSS database) to investigate the footprints of bars in galaxy centers by analysing the central ionized gas properties of barred and unbarred galaxies separately. We find statistically significant differences in the Hβ Balmer extinction, star formation rate per unit area, in the [S II]λ6717/[S II]λ6731 line ratio, and notably in the N2 parameter (N2 $= \log([$N II$]\lambda$6583/Hα)). A deeper analysis reflects that these differences are only relevant for the less massive bulges ($\lesssim 10^{10}$M$_\odot$). These results have important consequences for studies on bulge formation and galaxy evolution.

Keywords. galaxies: spirals - galaxies: morphology - galaxies: evolution - galaxies: bulges

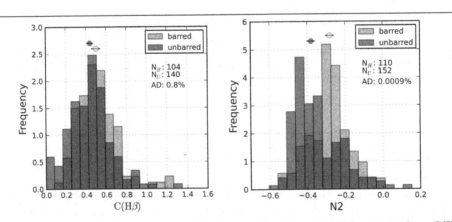

Figure 1. Comparative histograms for the Balmer extinction at the Hβ emission line, C(Hβ), and N2 $= \log([$N II$]\lambda$6584/Hα) for all non–AGN barred (yellow, hatched) and unbarred (purple) galaxies of the sample. The number of objects in each sub–sample and the P–values from the Anderson–Darling k–sample test are shown below each figure legend. The distributions are different for barred and unbarred galaxies (P–values below 5%), more importantly in the case of N2, which is on average $\sim$0.10 dex larger in barred than in unbarred galaxies.

We acknowledge support from the Spanish "Ministerio de Economía y Competitividad" and "Junta de Andalucía" via grants AYA2011-24728 and FQM-108.

References

Coelho, P. & Gadotti, G. A., 2011, *ApJ*, 74, L13

Galaxies in 3D across the Universe
Proceedings IAU Symposium No. 309, 2014
B. L. Ziegler, F. Combes, H. Dannerbauer, M. Verdugo, eds.

© International Astronomical Union 2015
doi:10.1017/S1743921314010497

WIYN's New Unique Multi-size Fiber IFUs

E. J. Hooper[1,2], M. J. Wolf[2], M. A. Bershady[2], A. D. Eigenbrot[2],
C. M. Wood[2], S. A. Buckley[2], M. P. Smith[2], C. Corson[1], G. Y. Zhu[3],
A. Vang[2], J. S. Gallagher III[2] and A. I. Sheinis[4]

[1] WIYN Observatory, 950 N. Cherry Ave., Tucson, AZ 85719, USA; email: ehooper@wiyn.org
[2] Astronomy Department, University of Wisconsin-Madison, Madison, WI 53706, USA
[3] Astronomy & Space Science, Nanjing Univ., 22 Hankou Rd, Nanjing, 210093, Jiangsu, China
[4] Australian Astronomical Observatory, PO Box 915, North Ryde NSW 1670, Australia

Abstract. Two new integral field units (IFUs) were installed recently on the WIYN Observatory's 3.5-meter telescope at Kitt Peak. These unique IFUs contain fibers of different sizes in the same head. This design allows smaller fibers to sample regions of higher surface brightness, providing higher spatial resolution while maintaining adequate signal-to-noise (S/N). Conversely, larger fibers maintain S/N at the expense of spatial resolution in the lower surface brightness regions of galaxies. The new IFUs were built with funds from NSF award ATI-0804576.

Keywords. instrumentation: spectrographs - techniques: imaging spectroscopy

1. Introduction

The first integral field unit (IFU) on the WIYN 3.5-meter telescope (www.wiyn.org) was DensePak (Barden *et al.* 1998), with 91 fibers densely packed in a rectangle, followed by a second IFU called SparsePak containing a densely packed core surrounded by a more sparse square halo of fibers. SparsePak's 82 fibers, each 5″ in diameter, gave it the largest specific grasp (the product of telescope area and fiber solid angle) of any imaging spectrograph, and among the best combinations of total grasp and spectral power, in the world at the time (Bershady *et al.* 2004). All of the WIYN IFUs feed the Bench Spectrograph located below the telescope (Bershady *et al.* 2008; Knezek *et al.* 2010).

Designers have chosen between high grasp systems with low spatial resolution (e.g., SparsePak) and lower grasp units with high spatial resolution (e.g., NIFS on Gemini). However, a customized design would have small fibers for high surface brightness regions and larger fibers for fainter areas in an astronomical source. The Washburn Astronomy Lab at the University of Wisconsin-Madison designed and built the world's first IFUs with different sized fibers (variable pitch) in the same fiber bundle (M. Bershady, PI).

2. IFU Design

Two new variable pitch IFUs were created (Figure 1; Wood *et al.* 2012). Each has its own light collecting head that can be swapped quickly at the telescope. The first IFU, HexPak, has a central core of 0.9″ diameter fibers (18 total), surrounded by a hexagonal halo of 2.8″ fibers (84 total; 40″ extent on sky). HexPak is designed for spheroidal or face-on disk galaxies, plus AGN host galaxies. The second IFU, GradPak, is an approximately rectangular array ($\sim 35'' \times 55''$), consisting of 11 rows of fibers ranging in size from $1.9'' - 5.6''$ (90 fibers total). It is designed for observing edge-on galaxies, with smaller fibers near the mid-plane and larger fibers sampling fainter emission perpendicular to the disk. Each fiber head contains separate sky fibers of every size.

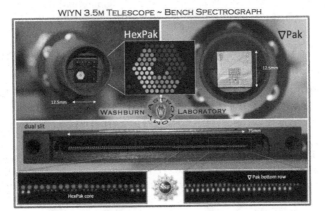

Figure 1. Top: photographs of HexPak (left) and GradPak (right). Sky fibers are above and to the right in HexPak; they occur in two clumps in GradPak. **Bottom:** fibers from both IFUs are arranged into parallel psuedo-slits on a common foot at the spectrograph entrance.

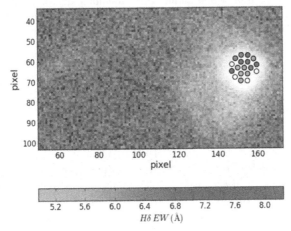

Figure 2. Hδ equivalent width derived from HexPak data of the galaxy G515 ($z = 0.0875$). Only the inner core of small fibers is shown; this entire core is roughly the same size as one SparsePak fiber. Each fiber corresponds to a physical size of 1.5 kpc.

3. Early Results

Preliminary results from early science data already demonstrate the advantages of the new IFUs. Figure 2 shows Hδ strength in the inner HexPak fibers overlaid on an image of the post-merger, post-starburst galaxy G515, which likely hosts a low power radio AGN (Liu *et al.* 2007). The strength and large extent of the $H\delta$ signature is consistent with a strong starburst resulting from a near equal mass merger (e.g., Snyder *et al.* 2011).

References

Barden, S., *et al.* 1998, *SPIE*, 3355, 892
Bershady, M., *et al.* 2004, *ApJS*, 156, 311
Bershady, M., *et al.* 2005, *PASP*, 116, 820
Bershady, M., *et al.* 2008, *SPIE*, 7014, 70140H
Knezek, P., *et al.* 2010, *SPIE*, 7735, 77357D
Liu, C., *et al.* 2007, *ApJ*, 658, 249
Snyder, G., *et al.* 2011, *ApJ*, 741, 77
Wood, C., *et al.* 2012, *SPIE*, 8446, 84462W

Galaxies in 3D across the Universe
Proceedings IAU Symposium No. 309, 2014
B. L. Ziegler, F. Combes, H. Dannerbauer, M. Verdugo, eds.
© International Astronomical Union 2015
doi:10.1017/S1743921314010503

Using 3D Spectroscopy to Probe the Orbital Structure of Composite Bulges

Peter Erwin[1,2], Roberto Saglia, Jens Thomas[1,2],
Maximilian Fabricius[1,2], Ralf Bender[1,2], Stephanie Rusli[1,2],
Nina Nowak[3], John E. Beckman[4] and Juan Carlos Vega Beltrán[4]

[1] Max-Planck-Insitut für extraterrestrische Physik, Giessenbachstr., 85748 Garching, Germany
[2] Universitäts-Sternwarte München, Scheinerstrasse 1, 81679 München, Germany
[3] Stockholm University, Department of Astronomy, Oskar Klein Centre, SE-10691 Stockholm, Sweden
[4] Instituto de Astrofísica de Canarias, C/ Via Láctea s/n, 38200 La Laguna, Tenerife, Spain

Abstract. Detailed imaging and spectroscopic analysis of the centers of nearby S0 and spiral galaxies shows the existence of "composite bulges", where both classical bulges and disky pseudobulges coexist in the same galaxy. As part of a search for supermassive black holes in nearby galaxy nuclei, we obtained VLT-SINFONI observations in adaptive-optics mode of several of these galaxies. Schwarzschild dynamical modeling enables us to disentangle the stellar orbital structure of the different central components, and to distinguish the differing contributions of kinematically hot (classical bulge) and kinematically cool (pseudobulge) components in the same galaxy.

Keywords. galaxies: elliptical and lenticular, cD - galaxies: evolution - galaxies: formation

1. Introduction

Although the standard picture of the stellar structure of disk galaxies combines a disk and a central bulge, recent studies have suggested a dichotomy between galaxies which host *classical* bulges – round, kinematically hot, and presumed to originate from violent mergers at high redshift – and those with *pseudobulges*, where the central excess stellar light is from a flattened, kinematically cool structure, presumed to originate from some long-term, internal ("secular") processes.

We have recently found evidence that some disk galaxies can harbor both a classical bulge *and* a disky pseuduobulge (we use the term "disky pseudobulges" to distinguish them from bar-derived box/peanut structures, which are sometimes also called pseudobulges). Evidence for this includes a combination of highly flattened isophotes, disky substructures (spirals, nuclear rings, nuclear bars), and stellar kinematics dominated by rotation in the disky pseudobulge, and rounder isophotes and stellar kinematics dominated by velocity dispersion in the classical-bulge region; see Nowak *et al.* (2010) and Erwin *et al.* (2014) for details.

As part of our SINFONI Search for Supermassive Black Holes (S³BH), we observed approximately 30 disk and elliptical galaxies with the SINFONI IFU on the VLT, using natural- or laser-guide-star adaptive optics to obtain 3D *K*-band spectroscopy of the galaxy centers; our sample includes three well-defined examples of composite-bulge galaxies. We combine the high-resolution stellar kinematics derived from this data with larger-scale, ground-based spectroscopy and *HST* and ground-based imaging to measure SMBH masses via Schwarzschild dynamical modeling (e.g., Nowak *et al.* 2007, 2008, 2010; Rusli *et al.* 2011, 2013).

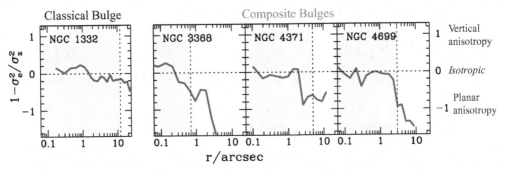

Figure 1. Results of Schwarzschild modeling for classical-bulge S0 galaxy NGC 1332 (Rusli *et al.* 2011, left) and three composite-bulge galaxies, based on VLT-SINFONI AO data. The plots, based on mass-weighted averages of stellar orbits within $\pm 23°$ of the equatorial plane for each galaxy, show equatorial-vs-vertical stellar anisotropy as a function of radius; the equatorial term combines radial and tangential dispersions: $\sigma_e^2 = (\sigma_R^2 + \sigma_\phi^2)/2$. Vertical dashed lines mark the approximate photometric transition between the classical bulge and the disk (for NGC 1332) or between the classical bulge and the disky pseudobulge (other three galaxies), with the shading indicating the classical-bulge region. For all four galaxies, the classical bulge is dominated by isotropic velocity dispersion, while the disk or disky pseudobulge regions show planar-dominant anisotropy.

As part of the Schwarzschild modeling process, we obtain weighted libraries of stellar orbits for the central galaxy regions; these can be used to explore the relative contributions of ordered (rotational) and random stellar motions within classical and disky pseudobulge regions. Fig. 1 shows part of this analysis, plotting the planar/vertical anisotropy of 3D stellar orbits as a function of radius for an S0 with a purely classical bulge (NGC 1332) and for three composite-bulge galaxies. The anisotropy term measures the relative amounts of "equatorial" dispersion (that is, radial and azimuthal dispersions added in quadrature) versus vertical dispersion (with respect to the equatorial plane). In all four galaxies, the dispersion is approximately isotropic within the classical-bulge region, and shifts to an equatorial-dominant state in the disk outside (for NGC 1332) or in the disky pseudobulge region (for the composite-bulge galaxies). This is additional evidence supporting the argument that what we identify as "classical bulges" in the composite-bulge systems are isotropic, pressure-supported components similar to low-luminosity ellipticals, while the disky pseudobulges have stellar kinematics similar to those of large-scale disks.

Full details of this study are presented in Erwin *et al.* (2014).

References

Erwin, P., Saglia, R. P., Fabricius, M., Thomas, J., Nowak, N., Rusli, S., Bender, R., Vega Beltrán, J. C., & Beckman, J. E., 2014, *MNRAS*, submitted

Nowak, N., Saglia, R. P., Thomas, J., Bender, R., Pannella, M., Gebhardt, K., & Davies, R. I., 2007, *MNRAS*, 379, 909

Nowak, N., Saglia, R. P., Thomas, J., Bender, R., Davies, R. I., & Gebhardt, K., R. I., 2008, *MNRAS*, 391, 1629

Nowak, N., Thomas, J., Erwin, P., Saglia, R. P., Bender, R., & Davies, R. I., 2010, *MNRAS*, 403, 646

Rusli, S. P., Thomas, J., Erwin, P., Saglia, R. P., Nowak, N., & Bender, R., 2011, *MNRAS*, 410, 1223

Rusli, S. P., Thomas, J., Saglia, R. P., Fabricius, M., Erwin, P., Bender, R., Nowak, N., Lee, C. H., Riffeser, A., & Sharp, R., 2013, *AJ*, 146, 45

Galaxies in 3D across the Universe
Proceedings IAU Symposium No. 309, 2014
B. L. Ziegler, F. Combes, H. Dannerbauer, M. Verdugo, eds.

© International Astronomical Union 2015
doi:10.1017/S1743921314010515

Three-Dimensional Spectroscopy and Star Formation Histories of Field E+A Galaxies

Charles T. Liu[1], Marsha Wolf[2], Eric J. Hooper[2] and Joshua Bather[3]

[1]Astrophysical Observatory, Dept. of Eng. Science and Physics, City Univ. of New York,
College of Staten Island, 2800 Victory Blvd, Staten Island, NY 10314 USA
[2]Dept. of Astronomy, Univ. of Wisconsin, 475 N. Charter Street, Madison, WI 53706-1582 USA
[3]School of Physics and Astronomy, Univ. of St. Andrews, North Haugh, KY16 9SS Scotland
email: `charles.liu@csi.cuny.edu, cliu@amnh.org`

Abstract. We present the initial results of an integral field spectroscopic survey of E+A galaxies in the field, which combined with radio continuum measurements and multi-wavelength photometry and imaging provides significant insight into the dynamical and star formation histories of these transitioning post-starburst systems. We focus on the E+A galaxy known as G515 ($z = 0.088$), a massive merger remnant that began its star formation quenching process $\sim$1.0 Gyr ago. Its relatively young stellar population contrasts with its light profile and kinematics, which are more consistent with a slowly-rotating, early-type galaxy.

Keywords. galaxies: starburst and post-starburst - galaxies: evolution - galaxies: formation

E+A (i.e., post-starburst or 'k+a') galaxies represent some of the best examples of a system whose star formation is on the verge of being quenched. Thus, they can be valuable signposts of an important but poorly understood process of galaxy evolution. We present the first results of integral field spectroscopy and radio continuum measurements of a sample of E+A galaxies in the field with redshift $0 < z < 0.2$, with the goal of examining the properties of this significant evolutionary phase. We focus particularly on J152426.55+080907 at $z = 0.088$, also known as 'G515' or 'Flagellan,' for which more than 20 years of spectroscopic data are available (Liu *et al.* 2007). On several-kpc scales, its optical light is dominated by stellar populations varying in age from 0.5 to 1 Gyr. Optical line emission, particularly in [OII], also appears to be highly variable in both intensity and spatial location. We also see substantial evidence of outflows; and multi-epoch radio observations show weak and significantly variable radio continuum emission, possibly indicative of an AGN nearing the end of its duty cycle. On the other hand, 3-D spectroscopy of G515 obtained with the Sparsepak and Hexpak IFUs at the WIYN telescope at Kitt Peak show that its light profile and kinematics are more typical of a "slow rotator" early-type galaxy. This is perhaps surprising, given its recent merger activity and observed levels of subsequent evolution.

Our results indicate that E+A galaxies apparently do not always fade gradually from "star-forming" to "quenched" states. Rather, they may cease their activity unevenly, in fits and starts, evolving rapidly over very short timescales.

We gratefully acknowledge support of this work from NSF grants AST-1004583 to the City University of New York and AST-10004591 to the American Museum of Natural History, and from the Office of Academic Affairs at the College of Staten Island.

Reference

Liu, C. T., Hooper, E., O'Neil, K., Thompson, D., Wolf, M., & Lisker, T., 2007, *ApJ*, 658, 249

Galaxies in 3D across the Universe
Proceedings IAU Symposium No. 309, 2014
B. L. Ziegler, F. Combes, H. Dannerbauer, M. Verdugo, eds.

© International Astronomical Union 2015
doi:10.1017/S1743921314010527

Chemo-Kinematic Survey of z ~ 1 Star Forming Galaxies using Keck OSIRIS LGS-AO

Etsuko Mieda[1], Shelley A. Wright[1,2], James E. Larkin[3], Lee Armus[4], and Stephanie Juneau[5]

[1]Department of Astronomy and Astrophysics, University of Toronto, ON, Canada, M5S3H4
email: mieda@astro.utoronto

[2]Dunlap Institute for Astronomy & Astrophysics, Toronto, ON, Canada, M5S3H4

[3]Department of Physics and Astronomy, University of California, Los Angeles, CA, USA, 90095

[4]Spitzer Science Center, California Institute of Technology, Pasadena, CA, 91125

[5]CEA Saclay, DSM/IRFU/Sap. F-91191 Gif-sur-Yvette, France

Abstract. We present first results from the Intermediate Redshift OSIRIS Chemo-Kinematic Survey (IROCKS) of z ~ 1 star forming galaxies (Mieda *et al.* in prep). We have targeted Hα and [NII] emission lines in J-band and have spatially resolved the galaxies at sub-kilo parsec scale. We have combined our sample with deep HST continuum images, and are able to reveal the dynamics, morphologies, metallicity distribution, emission-line diagnostics, and star formation rates of galaxies spanning this crucial z ~ 1 epoch.

1. Introduction

The majority of high redshift galaxy studies using IFS + AO have focused on z > 1.3 and under-sampled z ~ 1 due to instrumental and AO performance. The z ~ 1 regime is a critical epoch to understand galaxy evolution; it is when the star formation rate density starts to rapidly drop (Hopkins & Beacom 2006), and the stellar mass density becomes more established to present-day values (e.g. Ilbert *et al.* 2013).

2. Conclusions

We have observed z ~ 1 star forming galaxies using Keck OSIRIS LGS-AO. We selected galaxies in many fields: GOODS-S, GOODS-S, DEEP2, and UDS. We detected over ten sources, and it is now the biggest IFS + AO z ~ 1 samples. Our samples span $0.83 \lesssim z \lesssim 0.98$. Hα and [NII] emission lines show $50 \lesssim \sigma \lesssim 80$ km/s and $-0.7 \lesssim log_{10}([NII]/H\alpha) \lesssim -0.3$. Using Planck cosmology, Kennicutt (1998) + Chabrier (2003), and E(B-V) = 0.15, SFR of our samples span an order of magnitude, $2 \lesssim SFR_{H\alpha} \lesssim 20$ M$_\odot$/yr.

Acknowledgments

The Dunlap Institute is funded the through an endowment established by the David Dunlap family and the University of Toronto.

References

Chabrier, G., 2003, *PASP*, 115, 763
Hopkins, A. M. & Beacom, J. F., 2006, *ApJ*, 651, 142
Ilbert, O., *et al.* 2013, *A&A*, 556, A55
Kennicutt, Jr. R. C., 1998, *ApJ*, 498, 541
Planck Collaboration, 2013, arXiv

Galaxies in 3D across the Universe
Proceedings IAU Symposium No. 309, 2014
B. L. Ziegler, F. Combes, H. Dannerbauer, M. Verdugo, eds.

© International Astronomical Union 2015
doi:10.1017/S1743921314010539

A 3D analysis of the metal distribution in the compact group of galaxies HCG 31

Sergio Torres-Flores[1], Claudia Mendes de Oliveira[2],
Mayte Alfaro-Cuello[1], Eleazar Rodrigo Carrasco[3], Duilia de Mello[4]
and Philippe Amram[5]

[1] Departamento de Física, Universidad de La Serena,
Av. Cisternas 1200 Norte, La Serena, Chile
email: storres@dfuls.cl
[2] Instituto de Astronomia, Geofísica e Ciências Atmosféricas da Universidade de São Paulo,
Cidade Universitária, CEP:05508-900, São Paulo, SP, Brazil
[3] Gemini Observatory/AURA, Southern Operations Center, Casilla 603, La Serena, Chile
[4] Catholic University of America, Washington, DC 20064, USA
[5] Aix Marseille Université, CNRS, LAM (Laboratoire d'Astrophysique de Marseille)
UMR 7326, 13388, Marseille, France

Abstract. We present new Gemini/GMOS integral field unit observations of the central region of the merging compact group of galaxies HCG 31. Using this data set, we derive the oxygen abundances for the merging galaxies HCG 31A and HCG 31C. We found a smooth metallicity gradient between the nuclei of these galaxies, suggesting a mixing of metals between these objects. These results are confirmed by high-resolution Fabry-Perot data, from which we infer that gas is flowing between HCG 31A and HCG 31C.

1. Main results

We obtained GMOS/IFU spectra covering the central region of HCG 31A and HCG 31C. We then used the N2 metallicity calibrator defined by Marino *et al.* (2013) to derive the oxygen abundance between the merging galaxies HCG 31A and HCG 31C. We found that HCG 31A and HCG 31C display values of $12+\log(O/H)=8.44\pm0.03$ and $12+\log(O/H)=8.22\pm0.03$, respectively, with a smooth metallicity gradient between these galaxies. Using high-resolution archival Fabry-Perot data (Amram *et al.* 2007), we infer that there is gas flowing between these galaxies, detected by a shift in the radial velocities of the Hα emission lines along the line that joins HCG 31A and HCG 31C. This fact suggests a gas mixing between HCG 31AC, which can be responsible in producing the observed metal distribution.

Acknowledgements

ST-F acknowledges the financial support of FONDECYT through a project "Iniciación en la Investigación", under contract 11121505 and the support of the project CONICYT PAI/ACADEMIA 7912010004. MAC acknowledges the financial support of DIULS, through a "Concurso de Apoyo a Tesis 2013", number PT13146. CMdO acknowledges support from FAPESP and CNPq.

References

Amram, P., Mendes de Oliveira, C., Plana, H., *et al.* 2007, *A&A*, 471, 753
Marino, R. A., Rosales-Ortega, F. F., Sánchez, S. F., *et al.* 2013, *A&A*, 559A, 114

Galaxies in 3D across the Universe
Proceedings IAU Symposium No. 309, 2014
B. L. Ziegler, F. Combes, H. Dannerbauer, M. Verdugo, eds.

© International Astronomical Union 2015
doi:1017/S1743921314010540

The Distribution of Mass in (Disk) Galaxies: Maximal or Not?

Stéphane Courteau[1]

Department for Physics, Engineering Physics and Astrophysics, Queen's University,
Kingston, ON K7L 3N6, Canada
email: courteau@astro.queensu.ca

Abstract. The relative distribution of matter in galaxies ought to be one of the most definitive predictions of galaxy formation models yet its validation is challenged by numerous observational, theoretical, and operational challenges. All galaxies are believed to be dominated by an invisible matter component in their outskirts. A debate has however been blazing for the last two decades regarding the relative fraction of baryons and dark matter in the inner parts of galaxies: whether galaxies are centrally dominated by baryons ("maximal disk") is of issue. Some of those debates have been misconstrued on account of operational confusion, such as dark matter fractions being measured and compared at different radii. All galaxies are typically baryon-dominated (maximal) at the center and dark-matter dominated (sub-maximal) in their outskirts; for low-mass galaxies ($V_{\rm tot} \lesssim 200$ km s^{-1}), the mass of the dark halo equals the stellar mass at least within 2 disk scale lengths, the transition occurs at larger effective radii for more massive galaxies. An ultimate goal for galaxy structure studies is to achieve *accurate* data-model comparisons for the relative fractions of baryonic to total matter at any radius.

Keywords. galaxies: spiral - galaxies: dark matter - galaxies: formation

1. Introduction

Ever since the first speculations about the presence of "dark matter" in galaxies (Kapteyn 1922; Oort 1932), astronomers have endeavored to characterize the fraction of visible to total mass at all radii in galaxies†. Because mass is an integrated quantity which accounts for all the material within a given radius, the question about local gravitating mass is perhaps best posed by considering the ratios of local densities or velocities. In this brief review, we examine the various reports of $\mathcal{F}(R) \equiv V_{\rm baryon}/V_{\rm tot}$ in disk galaxies, where V_{bar} is the inferred circular velocity ascribed to the baryonic material and $V_{\rm tot}$ is the measured circular velocity at a given radius. V_{bar} is typically obtained by multiplying a galaxy light profile, $L(R)$, by a suitable mass-to-light ratio, $(M_*/L)(R)$. Any difference between $V_{\rm baryon}$ and $V_{\rm tot}$ is viewed as a signature for the presence of dark matter.

How galaxies ultimately arrange their baryonic and non-baryonic matter is the result of numerous complex mechanisms involving their mass accretion history, the depth of the potential well, the initital mass function (which ultimately affects feedback and quenching processes), dynamical friction, dynamical instabilities, and more. Analytical models of galaxy formation strive to predict $\mathcal{F}(\mathcal{R})$ (Dalcanton *et al.* 1997; Dutton *et al.* 2007; Mo *et al.* 2010) with mitigated success in light of the above challenges; numerical simulations suffer additional limitations too.

---

† Courteau *et al.* (2014) report in their review that Kapteyn (1922) is the first reference to "dark matter" in the astrophysical literature.

Can observations provide unique determinations of $\mathcal{F}(\mathcal{R})$ as ideal constraints to galaxy formation models? Not yet, but, after decades of muddled debates and incomplete data, a clear picture seems to be emerging.

2. Mass Models

Mass models of disk galaxies, where the observed rotation curve is decomposed into its principal gas, stars, and dark matter components, have historically been embraced as the ideal method to separate baryons and dark matter as a function of radius (Bosma 1978; Carignan & Freeman 1985; van Albada *et al.* 1985). The pros and cons of this approach are addressed in the review on "Galaxy Masses" by (Courteau *et al.* 2014). As that review stresses (see also Dutton *et al.* 2005), fundamental degeneracies between the disk and dark halo models prevent a unique baryon/dark matter decomposition based on rotation curve data alone. Other information or methods must be considered to break these degeneracies.

Among others, the quantity V_{bar} is obtained indirectly via a stellar or baryonic mass-to-light ratio (M_*/L) which is itself inferred from stellar population studies or dynamical stability analysis. Current stellar population mass-to-light ratios carry an uncertainty of a factor 2 at best (Conroy 2013; Courteau *et al.* 2014). The latter uncertainty is unfortunately large enough to encompass a full range of baryon fractions at a given radius. A maximal disk obeys

$$\mathcal{F} \equiv V_{\mathrm{disk}}(R_{\mathrm{max}})/V_{\mathrm{tot}}(R_{\mathrm{max}}) > 0.85, \tag{2.1}$$

where V_{disk} is the inferred velocity of the disk (stars and gas), V_{tot} is the total observed velocity, and R_{max} is the radius at which V_{disk} reaches its peak value. Therefore, for most disk galaxy mass models, stellar population M/L ratio uncertainties are such that both maximal and sub-maximal disk solutions are allowed (see Fig. 12 of Courteau *et al.* 2014).

3. Other Methods, Same Results?

To circumvent the loopholes of mass modeling, Courteau & Rix (1999) observed that the residuals of the velocity-luminosity relation, $\Delta \log(L)$, contrasted against those of the size-luminosity, $\Delta \log R(L)$, relation could provide a tighter constraints on F measured at the peak of the rotation curve†. It can be easily shown that $\Delta \log V(L)/\Delta \log R(L) = -/+$ 0.5 for baryon (-0.5) and dark matter (+0.5) dominated galaxies (Courteau & Rix 1999; Reyes *et al.* 2011). Fig. 1 gives an example for such residual estimates measured at R=2.2 disk scale lengths from a typical sample of spiral galaxies where $\Delta \log V(L)/\Delta \log R(L) \simeq$ 0; indeed, all spiral galaxies obey this relation (Courteau & Rix 1999; Courteau *et al.* 2007; Dutton *et al.* 2007; Reyes *et al.* 2011). The null slope essentially indicates that both baryons and dark matter contribute to the galaxy dynamics at $2.2R_d$. Various galaxy structure models whereby the dark matter is compressed adiabatically by the cooling baryons as the galaxy stabilizes dynamically yield a mapping between $\Delta \log V(L)/$ $\Delta \log R(L)$ and the baryon fraction $\mathcal{F}$. These models yield, on average for bright spiral

---

† If a disk is assumed to be exponential, its rotation curve will peak at 2.2 disk scale lengths, R_d. However, only $\sim 20\%$ of all disk galaxies have near exponential disks (Courteau *et al.* 2007). Thus the so-called $2.2R_d$ peak in the baryonic rotation curve is an ill-defined metric since most disk galaxies have Freeman Type-II profiles. A more practical fiducial radius that enables the comparison of both late and early-type galaxies in a non-parametric fashion, is the half-light, or effective, radius, R_e. For an exponential disk, $2.2R_d = 1.3R_e$. To avoid additional confusion, $\mathcal{F}(\mathcal{R})$ should use either metric radii (however spoiled by distance errors) or radii relative to R_e.

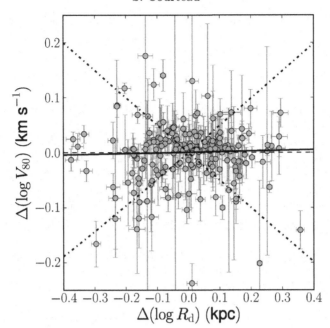

Figure 1. Correlation between velocity residuals $\Delta(\log V_{80})$ and disk size offsets $\Delta(\log R_d)$ from a velocity-luminosity and size-luminosity relations for a sample of typical spiral galaxies covering a broad range of surface brightnesses. The best-fit linear relation for $\Delta(\log V(L))/\Delta(\log R(L))$ has a slope consistent with zero (solid line). Predicted trends for a pure self-gravitating disk model (slope $= -0.5$) and a pure NFW DM halo model (slope $= +0.5$) are also shown (dot–dashed lines). Figure taken from Reyes *et al.* (2011); in agreement with Courteau & Rix (1999) and Dutton *et al.*(2007).

galaxies, $\mathcal{F}(2.2R_d) = 0.6 - 0.72$. The lower value applies if only adiabatic contraction is at play; a higher value is found if other mechanisms (e.g. feedback) oppose the adiabatic contraction (Courteau & Rix 1999; Dutton *et al.* 2007). By virtue of Eq. (2.1) above, galaxy disks are thus, *on average*, sub-maximal at $2.2R_d$. Since mass scale as V_{tot}^2, $M_{DM}/M_{tot} = 50\text{-}70\%$ at $2.2R_d$ ($= 1.3R_e$ for a pure exponential disks). Baryon/DM equality in disk galaxies is thus roughly achieved at $1R_e$. Reyes *et al.* (2011) showed that $\mathcal{F}$ is controlled by the surface density of the baryonic material; i.e. if the latter is high, so is $\mathcal{F}$ (see also Fig. 23 in Bovy & Rix 2013).

A similar result is obtained via the "velocity dispersion" method of (van der Kruit 1988), as applied by many for a full suite of disk galaxies (Bottema 1993), edge-on systems (Kregel *et al.* 2005), or face-on spiral galaxies Bershady *et al.* (2011); Martinsson *et al.* (2013). For a self-gravitating, radially exponential disk with vertical profile of the form $\rho(R, z) = \rho(R, 0)\text{sech}^2(z/z_0)$, van der Kruit (1988) and Bottema (1993) showed that the peak circular velocity of the stellar disk, V_{disk}, measured at $R = 2.2R_d$ can be related to the vertical velocity dispersion, V_z, and the intrinsic thickness (or scale height) of the disk, z_0, via:

$$V_{disk}(R_{peak}) = c\left\langle V_z^2\right\rangle_{R=0}^{1/2}\sqrt{\frac{R_d}{z_0}}. \tag{3.1}$$

where $c \simeq 0.88(1 - 0.28z_0/R_d)$ (Bershady *et al.* 2011). A more detailed discussion of Eq. (3.1) is presented in van der Kruit & Freeman (2011) (see their Section 3.2.4).

$V_{disk}(R_{peak})$, the contribution of the baryonic disk to the total velocity where the rotation curve reaches its peak (at R_{peak}), can thus be measured directly **if** the disk scale length, R_d, the scale height, z_0, and the vertical component of the velocity dispersion, V_z, are known. For face-on systems, V_z and R_d can be measured but z_0 must be inferred; and vice versa for edge-ons. Van der Kruit's method is thus statistical in nature and relies on a number of (potentially noisy) scaling transformations. Bovy & Rix (2013) also point out that the velocity dispersions measured from integrated light, as in most applications of Eq. (3.1) for external galaxies so far, do not trace the older, dynamically-relaxed stellar populations very well. The younger, brighter, stellar populations traced by the light-weighted spectra tend to have a lower velocity dispersion therefore biasing $V_{\rm disk}(R_{\rm peak})$ low and thus favoring sub-maximal disks. Sample selection also plays a substantial role (see Fig. 2 below).

The DiskMass project has made use of van der Kruit's formalism to extract $V_{\rm disk}(R_{\rm peak})$ for a sample of 46 nearly face-on (inclinations $\simeq$ 30 degrees) galaxies with rotation velocities between 100 km s^{-1} and 250 km s^{-1}(Bershady *et al.* 2011). This survey uses integral-field spectroscopy to measure stellar and gas kinematics using the custom-built SparsePak and PPAK instruments. For 30 DiskMass galaxies covering a range of structural properties, Martinsson *et al.* (2013) report that the fraction $\mathcal{F}$ ranges from 0.45 to 0.85 (the maximum disk threshold) and increases with luminosity, surface brightness, rotation speed, and redder color. The average value of their sample is $\mathcal{F} = 0.57 \pm 0.07$, in agreement with Courteau & Rix (1999).

Similar and broader results have been achieved through very different techniques for disk dominated systems: such as fluid dynamical modeling (Weiner *et al.* 2001; Kranz *et al.* 2003; Athanassoula 2014, see the latter reference for a more critical outlook) and strong gravitational lensing (Barnabè *et al.* 2012; Dutton *et al.* 2013). For instance, because gravitational lensing is sensitive to projected mass while dynamics trace the mass within ellipsoids, their combination provides additional geometrical dimensions to disentangle the baryons and dark matter in galaxies; this approach is especially efficient for highly-inclined disk galaxies (Dutton 2014). These methods, reviewed in Courteau *et al.* (2014), are consistent *on average* with $\mathcal{F}(2.2R_d) = 0.6 - 0.7$, but also show a broad range of $\mathcal{F}$ as a function of maximum circular velocity or disk size. In their hydrodynamical modeling of five grand design non-barred galaxies, Kranz *et al.* (2003) presciently suggested that galaxy disks appear to be maximal if $V_{max} > 200$ km s^{-1}, sub-maximal otherwise. The later summarizes rather well the more complete picture about the mass distribution in galaxies that is now clearly emerging.

Figure 2 is a compilation of dark matter fractions, $f_{DM} = 1 - \mathcal{F}^2 \approx (V_{DM}/V_{tot})^2$, at $\sim 2.2R_d$ for various galaxy samples discussed above (others are discussed below), as a function of total circular velocity.

The DiskMass survey (blue points) predominently samples small-to-average size galaxies thus avoiding the regime of disk maximality (at $2.2R_d$). Issues about additional sensitivity to young versus old stellar populations addressed above may also explain the DiskMass dark matter fractions above the mean model (Dutton *et al.* 2011). The cyan pentagons represent the parameter space for the Tully-Fisher residual analysis of the Courteau *et al.* (2007) sample of spiral galaxies with (open pentagon) and without (filled pentagon) adiabatic contraction (Dutton *et al.* 2007). The Courteau and DiskMass samples target essentially the same sub-maximal galaxy types. Unlike those samples, the SWELLS gravitational lensing survey is intrinsically biased towards high mass, predominently maximal, systems (Barnabè *et al.* 2012; Dutton *et al.* 2013). The Milky Way study of Bovy & Rix (2013) uses a method similar to that of (van der Kruit 1988) but with the added benefit that stellar populations can be disentangled, thus providing less biased measurements of the true underlying structure (scale length, rotation and dispersion)

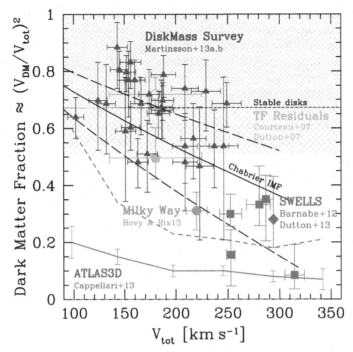

Figure 2. Dark matter fraction at $\sim 2.2R_d$ as a function of total circular velocity for samples discussed in the text. A clear trend is observed such that the dark matter fraction increases with circular speed; indeed below 250 km s^{-1}, dark matter dominates ($> 50\%$) the total mass budget of the galaxy at $2.2\,R_d$. The yellow shading shows the stability criteria (Efstathiou *et al.* 1982) for stellar disks ($f_{DM} \simeq 0.67$; short-dashed black line and above) and gaseous disks ($f_{DM} \simeq 0.51$). The black line represents a model for galaxy scaling relations assuming a Chabrier initial mass function (Dutton *et al.* 2011); the long-dashed black lines are the 1σ dispersion in galaxy sizes. Figure prepared with Aaron Dutton.

of the dynamical tracer. Their analysis results in a low mass-weighted disk scale length, $R_d = 2.15\pm0.14$ kpc, and a correspondingly low estimate of $f_{DM} = 0.31\pm0.07$ at $2.2R_d$†.

The short-dash boundary delineates the region where baryon-heavy disk galaxies become disk unstable. The criterion for stability is $V_{circ}(R_d/GM_{disk})^{1/2} \gtrsim 1.1$ (Efstathiou *et al.* 1982)‡. The yellow shading shows the stability criteria for stellar ($f_{DM} > 0.67$) and gaseous ($f_{DM} > 0.51$) disks.

Figure 2 (left) also shows the distribution of 260 ATLAS3D early-type galaxies (Cappellari *et al.* 2013), using $V_{circ} = 1.38\sigma_8$ to transform ATLAS3D velocity dispersion estimates, σ_8, to circular velocity V_{circ} (Dutton *et al.* 2011). This confirms that ATLAS3D probes ETG dynamics in a regime proper to maximal "baryons".

The baryon-DM equality by mass in disk galaxies is achieved roughly at $1R_e$. While ATLAS3D spectral maps do not extend beyond $1.5R_e$, deeper dynamical studies reaching

† Various investigations of the Milky Way mass density profiles are reported in Table 1 of Courteau *et al.* (2014). The early studies of Dehnen & Binney (1998) and Englmaier & Gerhard (1999) already pointed to a maximal MW disk. Measurements of $\mathcal{F}$ for the Milky Way hinge significantly on estimates of its disk scale length. A high estimate of R_d biases $\mathcal{F}$ high.

‡ See Foyle *et al.* (2008) for an alternate, numerical derivation, of this stability criterion as controlled by the ratio of the disc mass fraction, m_d, to the halo spin parameter, λ. If $\lambda < m_d$, the galaxy will be bar unstable. All galaxy mass models ought to be tested for the stability of their disk against the formation of a central bar (Widrow *et al.* 2003).

out to $\sim$ 4-5 R_e suggest that the baryon-DM mass equality in ETGs occurs at $\sim 2R_e$ or beyond (Thomas *et al.* 2007; Deason *et al.* 2012; Morganti *et al.* 2013; Oguri *et al.* 2014), albeit with significant scatter.

4. Future Prospects

The trends observed in Figure 2 and the characterization of baryon-DM transitions in galaxies are significant constraints that galaxy formation models have yet to reproduce. This is partly due to additional complications such as the degeneracy between the baryon or dark matter fractions, the stellar initial mass function, adiabatic contraction, and various counter-acting non-gravitational effects (e.g. feedback, dynamical friction) (Dutton *et al.* 2011; Trujillo-Gomez *et al.* 2011; Oguri *et al.* 2014). The advent of new extensive dynamical models of galaxies based on wide-field integral field surveys, such as CALIFA, MaNGA, SAMI (out to $\sim 2R_e$) and SLUGGS (reaching $\sim 6-10R_e$) heralds a promising future for mapping the mass distributions at all radii in galaxies. Model depenednces and biases could at least be partially offset with reduced sampling errors.

Let us no longer ask if galaxies are maximal or not (they all are!); let us simply decompose accurately the mass profiles of galaxies into their major components (e.g. gas, stars, and dark matter, or bar, bulge, disks, and halos) from the center out to the largest galactocentric radii, and using as many dynamical tracers as possible, in order to derive a reliable continuum of dynamical properties across the Hubble sequence and over time.

Acknowledgements

I acknowledge support from the Natural Science and Engineering Research Council of Canada for funding via a generous Research Discovery Grant. Figure 2 was prepared with Aaron Dutton. Aaron Romanowsky suggested valuable references too.

References

Cappellari, M., McDermid, R. M., Alatalo, K., *et al.* 2013, *MNRAS*, 432, 1862
Athanassoula, E. 2014, *MNRAS*, 438, L81
Barnabè, M., Dutton, A. A., Marshall, P. J., *et al.* 2012, *MNRAS*, 423, 1073
Bershady, M. A., Martinsson, T. P. K., Verheijen, M. A. W., *et al.* 2011, *ApJ*, 739, LL47
Bosma, A. 1978, "The distribution and kinematics of neutral hydrogen in spiral galaxies of various morphological types," Ph.D. thesis (University of Groningen).
Bottema, R. 1993, *A&A*, 275, 16
Bovy, J. & Rix, H.-W. 2013, *ApJ*, 779, 115
Carignan, C. & Freeman, K. C. 1985, *ApJ*, 294, 494
Conroy, C. 2013, *ARA&A*, 51, 393
Courteau, S., Cappellari, M., de Jong, R. S., *et al.* 2014, *Reviews of Modern Physics*, 86, 47
Courteau, S., Dutton, A. A., van den Bosch, F. C., *et al.* 2007, *ApJ*, 671, 203
Courteau, S. & Rix, H.-W. 1999, *ApJ*, 513, 561
Dalcanton, J. J., Spergel, D. N., & Summers, F. J. 1997, *ApJ*, 482, 659
Deason, A. J., Belokurov, V., Evans, N. W., & McCarthy, I. G. 2012, *ApJ*, 748, 2
Dehnen, W. & Binney, J. 1998, *MNRAS*, 294, 429
Dutton, A. A., in "Galaxy Masses as Constraints of Formation Models", Proceedings IAU Symposium No. 311, 2014 M. Cappellari & S. Courteau, eds.
Dutton, A. A., Treu, T., Brewer, B. J., *et al.* 2013, *MNRAS*, 428, 3183
Dutton, A. A., Conroy, C., van den Bosch, F. C., *et al.* 2011, *MNRAS*, 416, 322
Dutton, A. A., van den Bosch, F. C., Dekel, A., & Courteau, S. 2007, *ApJ*, 654, 27

Dutton, A. A., Courteau, S., de Jong, R., & Carignan, C. 2005, *ApJ*, 619, 218

Efstathiou, G., Lake, G., & Negroponte, J. 1982, *MNRAS*, 199, 1069

Englmaier, P. & Gerhard, O. 1999, *MNRAS*, 304, 512

Foyle, K., Courteau, S., & Thacker, R. J. 2008, *MNRAS*, 386, 1821

Kapteyn, J. C. 1922, *ApJ*, 55, 302

Kranz, T., Slyz, A., & Rix, H.-W. 2003, *ApJ*, 586, 143

Kregel, M., van der Kruit, P. C., & Freeman, K. C. 2005, *MNRAS*, 358, 503

Martinsson, T. P. K., Verheijen, M. A. W., Westfall, K. B., *et al.* 2013, *A&A*, 557, AA131

Mo, H., van den Bosch, F. C., & White, S. 2010, *Galaxy Formation and Evolution*, by Houjun Mo , Frank van den Bosch , Simon White, Cambridge, UK: Cambridge University Press, 2010,

Morganti, L., Gerhard, O., Coccato, L., Martinez-Valpuesta, I., & Arnaboldi, M. 2013, *MNRAS*, 431, 3570

Oguri, M., Rusu, C. E., & Falco, E. E. 2014, *MNRAS*, 439, 2494

Oort, J. H. 1932, *Bull. Astron. Inst. Neth.*, 6, 249

Reyes, R., Mandelbaum, R., Gunn, J. E., Pizagno, J., & Lackner, C. N. 2011, *MNRAS*, 417, 2347

Thomas, J., Saglia, R. P., Bender, R., *et al.* 2007, *MNRAS*, 382, 657

Trujillo-Gomez, S., Klypin, A., Primack, J., & Romanowsky, A. J. 2011, *ApJ*, 742, 16

van Albada, T. S., Bahcall, J. N., Begeman, K., & Sancisi, R. 1985, *ApJ*, 295, 305

van der Kruit, P. C. & Freeman, K. C. 2011, *ARA&A*, 49, 301

van der Kruit, P. C. 1988, *A&A*, 192, 117

Weiner, B. J., Sellwood, J. A., & Williams, T. B. 2001, *ApJ*, 546, 931

Widrow, L. M., Perrett, K. M., & Suyu, S. H. 2003, *ApJ*, 588, 311

Author Index